T0344476

There are many interactions between noncommutative algebra and representation theory on the one hand and classical algebraic geometry on the other, with important applications in both directions. The aim of this book is to provide a comprehensive introduction to some of the most significant topics in this area, including noncommutative projective algebraic geometry, deformation theory, symplectic reflection algebras, and noncommutative resolutions of singularities.

The book is based on lecture courses in Noncommutative Algebraic Geometry given by the authors at a Summer Graduate School at MSRI in 2012 and, as such, is suitable for advanced graduate students and early postdocs. In keeping with the lectures on which the book is based, a large number of exercises are provided, for which partial solutions are included.

Mathematical Sciences Research Institute
Publications

64

Noncommutative Algebraic Geometry

Mathematical Sciences Research Institute Publications

Noncommutative
Algebraic Geometry

Gwyn Bellamy

University of Glasgow

Daniel Rogalski

University of California, San Diego

Travis Schedler

Imperial College London

J. Toby Stafford

University of Manchester

Michael Wemyss

Michael Wemyss

CAMBRIDGE
UNIVERSITY PRESS

Gwyn Bellamy   Daniel Rogalski   Travis Schedler
gwyn.bellamy@glasgow.ac.uk   drogalsk@math.edu   schedler@math.utexas.edu

J. Toby Stafford   Michael Wemyss
toby.stafford@manchester.ac.uk   m.wemyss@ed.ac.uk

Silvio Levy (*Series Editor*)
Mathematical Sciences Research Institute
levy@msri.org

The Mathematical Sciences Research Institute wishes to acknowledge support by the National Science Foundation and the *Pacific Journal of Mathematics* for the publication of this series.

CAMBRIDGE
UNIVERSITY PRESS

University Printing House, Cambridge CB2 8BS, United Kingdom

One Liberty Plaza, 20th Floor, New York, NY 10006, USA

477 Williamstown Road, Port Melbourne, VIC 3207, Australia

4843/24, 2nd Floor, Ansari Road, Daryaganj, Delhi - 110002, India

79 Anson Road, #06-04/06, Singapore 079906

Cambridge University Press is part of the University of Cambridge.

It furthers the University's mission by disseminating knowledge in the pursuit of education, learning and research at the highest international levels of excellence.

www.cambridge.org
Information on this title: www.cambridge.org/9781107129542

© Mathematical Sciences Research Institute 2016

This publication is in copyright. Subject to statutory exception and to the provisions of relevant collective licensing agreements, no reproduction of any part may take place without the written permission of Cambridge University Press.

First published 2016

A catalogue record for this publication is available from the British Library

Library of Congress Cataloging in Publication data
Names: Bellamy, Gwyn.
Title: Noncommutative algebraic geometry / Gwyn Bellamy,
University of Glasgow [and four others].
Description: New York, NY : Cambridge University Press, [2016] | Series: Mathematical Sciences Research Institute publications ; 64 | Includes bibliographical references and index.
Identifiers: LCCN 2016018480 | ISBN 9781107129542 (hardback : alk. paper)
Subjects: LCSH: Geometry, Algebraic. | Noncommutative algebras.
Classification: LCC QA564 .N6645 2016 | CDD 516.3/5–dc23
LC record available at https://lccn.loc.gov/2016018480

ISBN 978-1-107-12954-2 Hardback   978-1-107-57003-0

Cambridge University Press has no responsibility for the persistence or accuracy of URLs for external or third-party internet websites referred to in this publication, and does not guarantee that any content on such websites is, or will remain, accurate or appropriate.

Preface

These notes are based on the lecture courses in Noncommuative Algebraic Geometry given by Bellamy, Rogalski, Schedler and Wemyss at the Summer Graduate School at the Mathematical Sciences Research Institute (MSRI) in Berkeley, California, in June 2012. This school served in part as an introductory conference to the MSRI program on Interactions between Noncommutative Noncommutative Algebra, Representation Theory and Algebraic Geometry that was held at MSRI in January–May 2013.

We would like to thank Jackie Blue, Riz Mayodong, Megan Nguyen and Stephanie Yurus who made our stay at MSRI so enjoyable, but most especially we extend our thanks to Chris Marshall and Hélène Barcelo who did so much to make the conference run smoothly. We also thank MSRI and the NSF for their financial support; in particular, part of this material is based on work supported by the National Science Foundation under Grant No. 0932078 000, while the authors were in residence at MSRI during the summer of 2012.

Thanks also to all the students who so enthusiastically attended the course and worked so hard. We hope that it was as productive and enjoyable for them as it was for the lecturers.

Contents

Introduction

J. Toby Stafford

There are multiple interactions between noncommutative algebra and representation theory on the one hand and classical algebraic geometry on the other, and the aim of this book is to expand upon this interplay. One of the most obvious areas of interaction is in noncommutative algebraic geometry, where the ideas and techniques of algebraic geometry are used to study noncommutative algebra. An introduction to this material is given in Chapter I. Many of the algebras that appear naturally in that, and other, areas of mathematics are deformations of commutative algebras, and so in Chapter II we provide a comprehensive introduction to that theory. One of the most interesting classes of algebras to have appeared recently in representation theory, and discussed in Chapter III, is that of symplectic reflection algebras. Finally, one of the strengths of these topics is that they have applications back in the commutative universe. Illustrations of this appear throughout the book, but one particularly important instance is that of noncommutative (crepant) resolutions of singularities. This forms the subject of Chapter IV.

These notes have been written up as an introduction to these topics, suitable for advanced graduate students or early postdocs. In keeping with the lectures upon which the book is based, we have included a large number of exercises, for which we have given partial solutions at the end of book. Some of these exercises involve computer computations, and for these we have either included the code or indicated web sources for that code.

We now turn to the individual topics in this book. Throughout the introduction k will denote an algebraically closed base field and all algebras will be k-algebras.

I. Noncommutative projective geometry. This subject seeks to use the results and intuition from algebraic geometry to understand noncommutative algebras. There are many different versions of noncommutative algebraic geometry, but the one that concerns us is noncommutative projective algebraic geometry, as introduced by Artin, Tate, and Van den Bergh [9, 10].

As is true of classical projective algebraic geometry, we will be concerned with *connected graded (cg) k-algebras* A. This means that (1) $A = \bigoplus_{n \geq 0} A_n$ with $A_n A_m \subseteq A_{n+m}$ for all $n, m \geq 0$ and (2) $A_0 = k$. For the rest of the introduction we will also assume, for simplicity, that A is generated as a k-algebra by the finite-dimensional vector space A_1. Write $A_+ = \bigoplus_{n>0} A_n$ for the irrelevant ideal.

The starting point to this theory appears in work of Artin and Schelter [5], who were interested in classifying noncommutative analogues A of the polynomial ring $k[x, y, z]$ or, as we will describe later, noncommutative analogues of $\mathbb{P}^2$. So, what should the definition be? The first basic condition is that A should have finite global dimension gldim $A = m$, in the sense that every finitely generated A-module M should have a finite projective resolution. This hypothesis is insufficient by itself; for example the free algebra $k\{x, y\}$ has global dimension one. So we also demand that A have *polynomially bounded growth* in the sense that the function $p(n) = \dim_k A_n$ is bounded above by some polynomial function of n. This is still not enough to eliminate rings like $k\{x, y\}/(xy)$ that have rather unpleasant properties. The insight of Artin and Schelter was to add a *Gorenstein* condition: $\mathrm{Ext}^i_A(k, A) = \delta_{i,m} k$, where k is the trivial (right) A-module A/A_+. In the commutative case this condition is equivalent to the ring having finite injective dimension, hence weaker than having finite global dimension, yet in many ways in the noncommutative setting it is a more stringent condition. Algebras with these three properties — global dimension m, polynomially bounded growth and the Gorenstein condition — are now called *Artin–Schelter regular* or *AS-regular rings of dimension m*. These algebras appear throughout noncommutative algebraic geometry and form the underlying theme for Chapter I. All references in this subsection are to that chapter.

Artin–Schelter regular algebras of dimension 2 are easily classified; this is the content of Theorem 2.2.1. In fact there are just two examples: the *quantum plane* $k_q[x, y] := k\{x, y\}/(xy - qyx)$ for $q \in k \smallsetminus \{0\}$ and the *Jordan plane* $k_J[x, y] := k\{x, y\}/(xy - yx - y^2)$. (Since we are concerned with projective rather than affine geometry, we probably ought to call them the quantum and Jordan projective lines, but we will stick to these more familiar names.) It is straightforward to analyse the properties of these rings using elementary methods.

So it was the case of dimension 3 that interested Artin and Schelter, and here things are not so simple. The Gorenstein condition enables one to obtain detailed information about the projective resolution of the trivial module $k = A/A_+$. In many cases this is enough to describe the algebra in considerable detail, and in particular to give a basis for the algebra. However there was one algebra, now called the *Sklyanin algebra*, that Artin and Schelter could not completely understand (this algebra is described in terms of generators and relations in Example 1.3.4 but its precise description is not so important here). It was the elucidation of this and closely related algebras that required the introduction of geometric techniques through the work of Artin, Tate, and Van den Bergh [9, 10].

The idea is as follows. Given a commutative cg domain A the (closed) points of the corresponding projective variety $\mathrm{Proj}(A)$ can be identified with the maximal nonirrelevant graded prime ideals; under our hypotheses these are the graded ideals P such that $A/P \cong k[x]$ is a polynomial ring in one variable. In the noncommutative case, this is too restrictive — for example if q is not a root of unity, then $k_q[x, y]$ has just two such ideals; (x) and (y). Instead one works module-theoretically and defines a *point module* to be a right A-module $M = \bigoplus_{n \geq 0} M_n$ such that $M = M_0 A$ and $\dim_k M_n = 1$ for all $n \geq 0$. Of course, when A is commutative, these are the factor rings we just mentioned, but they are more subtle in the noncommutative setting and are discussed in detail in Sections 3 and 4. The gist is as follows. Let A be an AS regular algebra of dimension 3. Then the point modules for A are in one-to-one correspondence with (indeed, parametrised by) a scheme E, known naturally enough as the *point scheme* of A. This scheme further comes equipped with the extra data of an automorphism σ and a line bundle $\mathcal{L}$. From these data one can construct an algebra, known as the *twisted homogeneous coordinate ring* $B = B(E, \mathcal{L}, \sigma)$ of E. If $A = k[x_0, x_1, x_2]$ were a commutative polynomial ring in three variables, then $E = \mathbb{P}^2$ and B would simply be A. In the noncommutative case E will either be a surface (indeed either $\mathbb{P}^2$ or $\mathbb{P}^1 \times \mathbb{P}^1$) in which case $A = B$ or, more interestingly, E could be a curve inside one of those surfaces. The interesting case is when E is an elliptic curve, as is the case for the Sklyanin algebra we mentioned before. This also helps explain why the Sklyanin algebra caused such a problem in the original work of Artin and Schelter: elliptic curves are not so easily approached by the sorts of essentially linear calculations that were integral to their work.

The beauty of this theory is that the geometry of the point scheme E can be used to describe the twisted homogeneous coordinate ring $B = B(E, \mathcal{L}, \sigma)$ and its modules in great detail. Moreover, for an AS regular algebra A of dimension 3, the ring B is a factor $B = A/gA$ of A, and the pleasant properties of B lift to give a detailed description of A and ultimately to classify the AS-regular algebras of dimension 3. This process is outlined in Section 3.2. An important and surprising consequence is that these algebras A are all noetherian domains; thus every right (or left) ideal of A is finitely generated.

We study twisted homogeneous coordinate rings in some detail since they are one of the basic notions in the subject, with numerous applications. A number of these applications are given in Section 5. For example, if A is a domain for which $\dim_k A_n$ grows linearly, then, up to a finite-dimensional vector space, A is a twisted homogeneous coordinate ring (see Theorem 5.1.1 for the details). One consequence of this is that the module structure of the algebra A is essentially that of a commutative ring. To explain the module theory we need some more notation.

If A is a commutative cg algebra then one ignores the irrelevant ideal A_+ in constructing the projective variety $\mathrm{Proj}(A)$. This means we should ignore finite-dimensional modules when relating that geometry to the module structure of A. This holds in the noncommutative case as well. Assume that A is noetherian, which is the case that interests us, and let $\mathrm{gr}\, A$ denote the category of finitely generated graded A-modules $M = \bigoplus_{n \in \mathbb{Z}} M_i$ (thus $M_i A_j \subseteq M_{i+j}$ for all i and j). The category $\mathrm{qgr}(A)$ is defined to be the quotient category of $\mathrm{gr}\, A$ by the finite-dimensional modules; see Definition 4.0.7 for more details. A surprisingly powerful intuition is to regard $\mathrm{qgr}(A)$ as the category of coherent sheaves on the (nonexistent) space $\mathrm{Proj}(A)$. Similarly, there are strong arguments for saying that the AS regular algebras A of dimension 3 (or at least those for which $\dim_k A_1 = 3$) are the coordinate rings of the noncommutative $\mathbb{P}^2$'s.

The fundamental result relating algebraic geometry to twisted homogeneous coordinate rings is the following theorem of Artin and Van den Bergh [11], which is itself a generalisation of a result of Serre [198]: under a condition called σ-ampleness of the sheaf $\mathcal{L}$, the category qgr $B(E, \mathcal{L}, \sigma)$ is equivalent to the category of coherent sheaves on E. In particular, if A is a domain for which $\dim_k A_n$ grows linearly with n, as was the case two paragraphs ago, then qgr A will be equivalent to the category of coherent sheaves on a projective curve: sometimes this is phrased as saying that noncommutative curves are commutative!

Structure of Chapter I. The main aim of this chapter is to give the reader a firm understanding of the mathematics behind the above outline, and we have kept the geometric prerequisites to a minimum. Thus, in Section 1 we emphasise techniques for calculating the basis (or more generally the Hilbert series) of a graded algebra given by generators and relations. Section 2 introduces the Artin–Schelter regular algebras and, again, we emphasise how to use the Gorenstein condition to understand some of the basic examples. Of course this does not work everywhere, so Section 3 introduces point modules, the corresponding point scheme and shows how to compute this in explicit examples. Section 4 then describes the corresponding twisted homogeneous coordinate rings, while Section 5 outlines the applications of these techniques to the classifications of noncommutative curves and particular classes of noncommutative surfaces.

II. Deformations of algebras in noncommutative geometry. For simplicity, in discussing Chapter II we will assume that the base field k has characteristic zero. A great many algebras appearing in noncommutative algebra, and certainly most of the ones described in this book, are deformations of commutative algebras. For example, if $k_q[x, y] = k\{x, y\}/(xy - qyx)$ is the quantum plane mentioned above then it is easy to see that this algebra has basis $\{x^i y^j\}$. Thus, as q passes from 1 to a general element of k, it is natural to regard this algebra as *deforming* the multiplication of the algebra $k[x, y]$. In fact there are many different ways of deforming algebras and some very deep results about when this is possible. This is the topic of Chapter II. Once again, all references in this subsection are to that chapter.

Here are a couple of illustrative examples. Given a finite-dimensional Lie algebra $\mathfrak{g}$ over the field k, with Lie bracket $\{-, -\}$, its enveloping algebra $U\mathfrak{g}$ is defined to be the factor of the tensor algebra $T\mathfrak{g}$ on $\mathfrak{g}$ modulo the relations $xy - yx - \{x, y\}$ for $x, y \in \mathfrak{g}$. One can also form the symmetric algebra Sym $\mathfrak{g}$ on $\mathfrak{g}$, which is nothing more than the polynomial ring in $\dim_k \mathfrak{g}$ variables. Perhaps the most basic theorem on enveloping algebras is the PBW or Poincaré–Birkhoff–Witt Theorem: if one filters $U\mathfrak{g} = \bigcup_{n \geq 0} \Lambda_{\leq n}$ by assigning $\mathfrak{g} + k$ to $\Lambda_{\leq 1}$, then Sym $\mathfrak{g}$ is isomorphic to the associated graded ring gr $U\mathfrak{g} = \bigoplus \Lambda_{\leq n}/\Lambda_{\leq(n-1)}$. We interpret this as saying that $U\mathfrak{g}$ is a *filtered deformation of* Sym $\mathfrak{g}$. A similar phenomenon occurs with the *Weyl algebra*, or ring of linear differential operators on $\mathbb{C}^n$. This is the ring with generators $\{x_i, \partial_i : 1 \leq i \leq n\}$ with relations $\partial_i x_i - x_i \partial_i = 1$ and all other generators commuting. Again one can filter this algebra by putting the x_i and ∂_j into degree one and its associated graded ring is then the polynomial ring $\mathbb{C}[x_1, \ldots, y_n]$ in $2n$ variables. As is indicated in Section 1, there are numerous other examples of filtered deformations of commutative algebras, including more general rings of differential operators and even some of the algebras from Chapter III.

The commutative rings B that arise as the associated graded rings $B = \operatorname{gr} A = \bigoplus \Lambda_{\leq n}/\Lambda_{\leq(n-1)}$ of filtered rings $A = \bigcup \Lambda_{\leq n}$ automatically have the extra structure of a Poisson algebra. Indeed, given non-zero elements $\bar{a} \in \Lambda_{\leq n}/\Lambda_{\leq(n-1)}$ and $\bar{b} \in \Lambda_{\leq m}/\Lambda_{\leq(m-1)}$, with preimages $a, b \in A$ then we define a new bracket $\{\bar{a}, \bar{b}\} = ab - ba$ mod $\Lambda_{\leq(m+n-1)}$. It is routine to see that this is actually a *Poisson bracket* in the sense that it is a Lie bracket satisfying the Leibniz identity $\{ab, c\} = a\{b, c\} + b\{a, c\}$. The algebra $\operatorname{gr} A$ is then called a *Poisson algebra*.

One can ask if the reverse procedure holds: Given a commutative Poisson algebra B, can one deform it to a noncommutative algebra A in such a way that the Poisson structure on B is induced from the multiplication in A? This is better phrased in terms of infinitesimal and formal deformations, but see Corollary 2.6.6 for the connection. To describe these deformations, pick an augmented base commutative ring R with augmentation ideal R_+, which for us means either $R = k[\![h]\!]$ or $R = k[h]/(h^n)$ with $R_+ = hR$. Then a *(flat) deformation* of B over R is (up to some technicalities) an R-algebra A, isomorphic to $B \otimes_k R$ as an R-module, such that $A \otimes_R R/R_+ = B$ as k-algebras. In other words, a deformation of B over R is an algebra $(B \otimes_k R, \cdot)$ such that $a \cdot b = ab$ mod R_+. An *infinitesimal deformation* of B is a flat deformation over $R = k[h]/(h^2)$, while a *formal deformation* is the case when $R = k[\![h]\!]$. In both cases the multiplication on A induces a Poisson structure on B by $\{\bar{a}, \bar{b}\} = h^{-1}(ab - ba)$ (which does make sense in the infinitesimal case) and we require that this is the given Poisson structure on B. Remarkably, these concepts are indeed equivalent: *Poisson structures on the coordinate ring B of a smooth affine variety X correspond bijectively to formal deformations of B.* However it takes much more work to make this precise (in particular one needs to work with appropriate equivalence classes on the two sides) and much of Sections 3 and 4 is concerned with setting this up. The original result here is Kontsevich's famous formality theorem, which was first proved at the level of $\mathbb{R}^n$ or more generally C^∞ manifolds. Kontsevich also outlined how to extend this to smooth affine (and some nonaffine) algebraic varieties, while a thorough study in the global algebraic setting was accomplished by Yekutieli and others; see Section 4.6 for the details.

The starting point to deformation theory is that deformations are encoded in Hochschild cohomology. To be a little more precise, let B be a k-algebra with opposite ring B^{op} and set $B^e = B \otimes_k B^{\mathrm{op}}$. The infinitesimal deformations of B are encoded by the second Hochschild cohomology group $HH^2(B) = \operatorname{Ext}^2_{B^e}(B, B)$, while the obstructions to extending these deformations to higher-order ones (i.e., those where $R = k[h]/(h^2)$ is replaced by some $R = k[h]/(h^n)$) are contained within the third Hochschild cohomology group HH^3. This is made precise in Section 3 and put into a more general context in Section 4.

One disadvantage of formal deformation theorems like Kontsevich's is that the multiplication on the deformation is very complicated to describe, yet in many concrete examples (like enveloping algebras) there is actually quite a simple formula for that multiplication. The possible deformations also have a range of different properties, and can often be "smoother" than the original commutative algebra. A illustrative example is given by the fixed ring $\mathbb{C}[x, y]^G$ where the generator σ of $G = \mathbb{Z}/(2)$ acts by -1 on x and y. This has many interesting deformations, including the factor $\overline{U} = U(\mathfrak{sl}_2)/(\Omega)$ of the enveloping algebra of $\mathfrak{sl}_2$ by its Casimir element. (Here $\overline{U}$ is smooth in the sense that, for instance, it has finite global dimension, whereas the global dimension of $\mathbb{C}[x, y]^G$ is infinite.) The ring $\overline{U}$ in turn has many different interpretations; for example, as the ring of global differential operators on the projective line (see Theorems 1.8.2 and (1.J)) or as a spherical subalgebra of a Cherednik algebra in Chapter III.

This example can be further generalised to the notion of a Calabi–Yau algebra. These algebras are ubiquitous in this book. The formal definition is given in Definition 3.7.9 but here we simply note that connected graded Calabi–Yau algebras are a special case of AS regular algebras (see Section 5.5.3 of Chapter I). In particular the polynomial ring $\mathbb{C}[x_1, \ldots, x_n]$ is Calabi–Yau, as are many of its deformations, including Weyl algebras and many enveloping algebras. Further examples are provided by $U(\mathfrak{sl}_2)/(\Omega)$ and the symplectic reflection algebras of Chapter III, as well as various noncommutative resolutions of Chapter IV. As these examples suggest, Calabi–Yau algebras can frequently be written as deformations of commutative rings or at least of rings that are "close" to commutative. This is discussed in Section 5 and has important applications to both commutative and noncommutative algebras, as is explained in the next two subsections.

Structure of Chapter II. The aim of the chapter is to give an introduction to deformation theory. Numerous motivating examples appear in Section 1, including enveloping algebras, rings of differential operators and Poisson algebras. The basic concepts of formal deformation theory and Hochschild (co)homology appear in Section 2, while the relationship between these concepts is examined in greater depth in Section 3. These ideas are considerably generalised in Section 4, in order to give the appropriate context for Kontsevich's formality theorem. The ramifications of this result and a hint to its proof are also given there. Finally, Section 5 discusses Calabi–Yau algebras and their applications to deformation theory, such as to quantizations of isolated hypersurface singularities.

III. Symplectic reflection algebras. A fascinating class of algebras that have only recently been discovered (the first serious treatment appears in the seminal paper of Etingof and Ginzburg [99] from 2002) are the *symplectic reflection algebras*, also known in a special case as *rational Cherednik algebras*. They have many interactions with, and applications to, other parts of mathematics and are also related to deformation theory, noncommutative algebraic geometry and noncommutative resolutions. As such, they form a natural class of algebras to study in depth in this book, and we do so in Chapter III. Once again, all references in this subsection are to that chapter.

We first describe these algebras as deformations. Let G be a finite subgroup of $\mathrm{GL}(V)$ for a finite-dimensional vector space V, say over $\mathbb{C}$ for simplicity. Then G acts naturally on the coordinate ring $\mathbb{C}[V]$ and a classic theorem of Chevalley–Shephard–Todd says that the quotient variety $V/G = \operatorname{Spec} \mathbb{C}[V]^G$ is smooth if and only if G is a complex reflection group (see Section 1 for the definitions). There is a symplectic analogue of reflection groups, where V is now symplectic and $G \subset \operatorname{Sp}(V)$. (The simplest case, of type A_{n-1}, is when $G = S_n$ is the symmetric group acting naturally on $\mathbb{C}^n \oplus (\mathbb{C}^n)^*$ by simultaneous permutations of the coordinates.) The variety V/G will not now be smooth; for example in the A_1 case V/G is the surface $xy = z^2$. However there are some very natural noncommutative deformations of $\mathbb{C}[V/G] := \mathbb{C}[V]^G$; notably $\overline{U}_\lambda = U(\mathfrak{sl}_2)/(\Omega - \lambda)$, where Ω is again the Casimir element and $\lambda \in \mathbb{C}$. For all but one choice of λ, the ring $\overline{U}_\lambda$ has finite global dimension and can be regarded as a smooth noncommutative deformation of $\mathbb{C}[V/G]$.

This generalises to any symplectic reflection group. Given such a group $G \subset \operatorname{Sp}(V)$, one can form the invariant ring $\mathbb{C}[V]^G$ and the *skew group ring* $\mathbb{C}[V] \rtimes G$; this is the same abelian group as the ordinary group ring $\mathbb{C}[V]G$, except that the multiplication is twisted: $gf = f^g g$ for $f \in \mathbb{C}[V]$ and $g \in G$. Then Etingof and Ginzburg [99] showed that one can deform $\mathbb{C}[V] \rtimes G$ into a noncommutative algebra, called *the symplectic reflection algebra* $H_{t,c}(G)$, depending on two parameters t and c. The trivial idempotent $e = \sum_{g \in G} g|G|^{-1}$ still lives in this ring and the *spherical subalgebra* $e H_{t,c}(G)e$ is then a deformation of $\mathbb{C}[V]^G$. Crucially, these algebras are filtered deformations in the sense of Chapter II and so, under a natural filtration, one has an analogue of the PBW Theorem: $\operatorname{gr} H_{t,c}(G) = \mathbb{C}[V] \rtimes G$ and $\operatorname{gr} e H_{t,c}(G)e = \mathbb{C}[V]^G$.

The parameter t can always be scaled and so can be chosen to be either 0 or 1. These cases are very different. For most of the chapter we will work in the case $t = 1$ and write $H_c(G) = H_{1,c}(G)$.

The rings $H_c(G)$ are typically defined in terms of generators and relations, which are not easy to unravel (see Definition 1.2.1 and Equation 1.C). However, in the A_1 case, $e H_c(G)e = \overline{U}_\lambda$ for some $\lambda \in \mathbb{C}$, and all such λ occur. In general the properties of the spherical subalgebras $e H_{t,c}(G)e$ are reminiscent of those of a factor ring of an enveloping algebra of a semisimple Lie algebra, and this analogy will guide much of the exposition.

This similarity is most apparent in the special case of Cherednik algebras. Here one takes a complex reflection group $W \subseteq \operatorname{GL}(\mathfrak{h})$ for a complex vector space $\mathfrak{h}$. Then W acts naturally on $V = \mathfrak{h} \times \mathfrak{h}^*$ and defines a symplectic reflection group. The *rational Cherednik algebra* is then the corresponding symplectic reflection algebra $H_{t,c}(W)$. Inside $H_{t,c}(W)$, one has copies of $\mathbb{C}[\mathfrak{h}]$ and $\mathbb{C}[\mathfrak{h}^*]$ as well as the group ring $\mathbb{C}W$ and the PBW Theorem can be refined to give a triangular decomposition $H_{t,c}(G) \cong \mathbb{C}[\mathfrak{h}] \otimes_{\mathbb{C}} \mathbb{C}W \otimes_{\mathbb{C}} \mathbb{C}[\mathfrak{h}^*]$ as vector spaces.

The Cherednik algebra $H_{t,c}(W)$ can also regarded as a deformation of the skew group ring $A_n \rtimes W$ of the Weyl algebra; in this case the spherical subalgebra $e H_{t,c}(W)e$ becomes a deformation of the fixed ring A_n^W. This is most readily seen through the Dunkl embedding of $H_c(W)$ into a localisation $\mathcal{D}(\mathfrak{h}_{\operatorname{reg}}) \rtimes W$ of $A_n \rtimes W$ (see Subsection 1.8 for the details). However, in many ways the intuition from Lie theory is more fruitful;

for example, $H_c(W)$ can have finite-dimensional representations, whereas $A_n \rtimes W$ is always simple and so cannot have any such representations.

The triangular decomposition is particularly useful for representation theory, since one has natural analogues of the Verma modules and Category $\mathcal{O}$, which are so powerful in the representation theory of semisimple Lie algebras (see, for example [83]). For Cherednik algebras, Category $\mathcal{O}$ consists of the full subcategory of finitely generated $H_c(W)$-modules on which $\mathbb{C}[\mathfrak{h}^*]$ acts locally nilpotently. The most obvious such modules are the *standard modules* $\Delta(\lambda) = H_c(W) \otimes_{\mathbb{C}[\mathfrak{h}^*]} \lambda$, where λ is an irreducible representation of W on which $\mathbb{C}[\mathfrak{h}^*]$ is given a trivial action. The structure of these modules is very similar to that of Verma modules; for example, each Δ_λ has a unique simple factor module and these define all the simple objects in Category $\mathcal{O}$. The general theory of Category $\mathcal{O}$-modules is given in Section 2. In Type A_{n-1}, when W is the symmetric group S_n, one can get a much more complete description of these modules, as is explained in Section 3. For example, it is known exactly when $H_c(S_n)$ has a finite-dimensional simple module (curiously, $H_c(S_n)$ can never have more than one such module). Moreover, the composition factors of the $\Delta(\lambda)$ and character formulae for the simple modules in Category $\mathcal{O}$ are known. The answers are given in terms of some beautiful combinatorics relating two fundamental bases of representations of certain quantum groups (more precisely, the level-one Fock spaces for quantum affine Lie algebras of Type A).

Section 4 deals with the Knizhnik–Zamolodchikov (KZ) functor. This remarkable functor allows one to relate Category $\mathcal{O}$ to modules over yet another important algebra, in this case the *cyclotomic Hecke algebra* $H_q(W)$ related to W. At its heart the KZ functor is quite easy to describe. Recall that the Dunkl embedding identifies $H_c(W)$ with a subalgebra of $\mathcal{D}(\mathfrak{h}_{\text{reg}}) \rtimes W$, and in fact $\mathcal{D}(\mathfrak{h}_{\text{reg}}) \rtimes W$ is then a localisation of $H_c(W)$. The key idea behind the KZ functor is that one can also localise the given module to obtain a $(\mathcal{D}(\mathfrak{h}_{\text{reg}}) \rtimes W)$-module. At this point powerful results from the theory of $\mathcal{D}$-modules can be applied and these results ultimately lead to modules over the Hecke algebra.

When we first defined the symplectic reflection algebras $H_{t,c}(G)$ there was the second parameter t and the representation-theoretic results we have described so far have all been concerned with the case $t \neq 0$. The case $t = 0$, which is the topic of the final Section 5, has a rather different flavour. The reason is that $H_{0,t}(G)$ is now a finite module over its centre $Z_c(G) = Z(H_{0,c}(G))$.

We again give a thorough description of the representation theory of $H_{0,t}(G)$ although this has a much more geometric flavour with a strong connection to Poisson and even symplectic geometry. The Poisson structure on $\operatorname{Spec} Z_c(G)$ comes from the fact that the parameter t gives a quantization of $Z_c(G)$! A key observation here is that the simple $H_{0,t}(G)$-modules are finite-dimensional, of dimension bounded by $|G|$ (see Theorem 5.1.4). Moreover they have maximal dimension precisely when their central annihilator is a smooth point of $\operatorname{Spec} Z_c(G)$. So, the geometry of that space and the representations of $H_{0,c}(G)$ are intimately connected.

The final topic in this chapter relates to Chapter IV and concerns applications of $H_{0,c}(G)$ back to algebraic geometry. An important geometric question is to understand *resolutions of singularities* $\pi : \widetilde{X} \to X$ of a singular space X (thus $\widetilde{X}$ should be nonsingular and the birational map π should be an isomorphism outside the singular subset of X). Here we are interested in the case $X = Y = V/G$. In this setting, the smooth locus of Y is even symplectic, and so one would further like the resolution $\pi : \widetilde{Y} \to Y$ to be *a symplectic resolution* in the sense that $\widetilde{Y}$ is symplectic and π is an isomorphism of symplectic spaces away from the singular set. Remarkably the question of when this happens has been determined using $H_{0,c}(G)$ — it happens if and only if Spec $Z_c(G)$ is smooth for some value of c. This can be made more precise. Recall that the spherical subalgebra $eH_{0,c}(G)e$ is a deformation of $\mathbb{C}[Y] = \mathbb{C}[V]^G$. Indeed, $eH_{0,c}(G)e \cong Z_c(G)$ is even commutative and so is a commutative deformation of $\mathbb{C}[Y]$. Thus Y has a symplectic resolution if and only if $eH_{0,c}(G)e$ is a smooth deformation of $\mathbb{C}[Y]$ (see Theorem 5.8.3). Completing this circle of ideas we note that symplectic reflection algebas have even been used to determine the groups G for which $Y = V/G$ has a symplectic resolution of singularities.

Structure of Chapter III. The aim of the chapter is to give an introduction to the construction and representation theory of symplectic reflection algebras $H_{t,c}(G)$. The basic definitions and structure theorems, including the PBW Theorem and deformation theory, are given in Section 1. In Section 2 the representation theory of the Cherednik algebra $H_c(W) = H_{1,c}(W)$ is discussed, with emphasis on Category $\mathcal{O}$-modules. In particular, Category $\mathcal{O}$ is shown to be highest weight category. These results can be considerably refined when $W = S_n$ is a symmetric group, and this case is studied in detail in Section 3. Here one can completely describe the characters of simple $\mathcal{O}$-modules and the composition factors of the standard modules. This is achieved by relating $H_c(S_n)$ to certain Schur and quantum algebras. The KZ functor is described in Section 4 and again relates the representation theory of $H_c(W)$ to other subjects: in this case the theory of $\mathcal{D}$-modules and, ultimately, to cyclotomic Hecke algebras $H_q(W)$. This allows one to prove subtle and nontrivial results about both the Cherednik and Hecke algebras. The final Section 5 studies the representation theory of the symplectic reflection algebras $H_{0,c}(G)$, with particular reference to their Poisson geometry and symplectic leaves. The application of these algebras to the theory of symplectic resolutions of quotient singularities is discussed briefly.

IV. Noncommutative resolutions. As we have just remarked, a fundamental problem in algebraic geometry is to understand the resolution of singularities $\pi : \widetilde{X} \to X$ of a singular space $X = \operatorname{Spec} R$. Even for nonsymplectic singularities, one can sometimes resolve the singularity by means of a noncommutative space and this can provide more information about the commutative resolutions. This theory is described in Chapter IV.

The resolution π is obtained by blowing up an ideal I of R related in some way to the singular subspace of Y. Unfortunately, the ideal I is not unique and even among rings R of (Krull) dimension three there are standard examples where different ideals I', I'' give rise to nonisomorphic resolutions of singularities. However, through work of

Bridgeland [42] and Bridgeland–King–Reid [44], these resolutions are closely related (more precisely, derived equivalent). The motivation for this chapter comes from work of Van den Bergh who, in [230, 231], abstracted Bridgeland's work by showing that one can find a noncommutative ring A that is actually derived equivalent to *both* these resolutions. Thus it is reasonable to think of A (or perhaps its category of modules) as a noncommutative resolution of singularities of Y. The purpose of this chapter is to explain how to construct such a ring A, to outline some of the methods that are used to extract the geometry, and to discuss the geometric applications.

For the rest of this introduction fix a commutative Gorenstein algebra R and set $X = \operatorname{Spec} R$ (in fact much of the theory works for Cohen–Macaulay rather than Gorenstein rings, but the theory is more easily explained in the Gorenstein case, and this also fits naturally with the other parts of the book). For simplicity, we assume throughout this introduction that R is also a normal, local domain. Then a *noncommutative crepant resolution* or *NCCR* for X (or R) is a ring A satisfying

(1) $A = \operatorname{End}_R(M)$ for some reflexive R-module M,
(2) A is a Cohen–Macaulay (CM) R-module, and
(3) $\operatorname{gldim} A = \dim X$.

Before discussing this definition, here is a simple but still very important example: let $G \subset \operatorname{SL}(2, \mathbb{C})$ be a finite subgroup acting naturally on $S = \mathbb{C}[x, y]$ and set $R = S^G$. Then it is easy to show that the skew group ring $A = S \rtimes G$, in the sense of Chapter III, is isomorphic to $\operatorname{End}_R(S)$, and it follows that A is a NCCR for the quotient singularity $X = \operatorname{Spec}(R) = \mathbb{C}^2/G$. This construction can be considerably generalised (to a polynomial ring in $n \geq 2$ variables, in particular), but the present case provides a rich supply of examples that we use throughout the chapter, not least because the finite subgroups of $\operatorname{SL}(2, \mathbb{C})$ are classified and easy to manipulate. See Sections 1 and 5 in particular.

Let us now explain some aspects of the definition of a NCCR; further details can be found in Section 2. First, the hypothesis that A be CM corresponds to the geometric property of crepancy (which is one reason these are called NC Crepant Resolutions). The definition of crepancy is harder to motivate, and is discussed in Section 4, but for symplectic singularities like the variety $Y = V/G$ from the last subsection, crepancy is equivalent to the resolution being symplectic. Since we want to obtain a smooth resolution of the given singular space it is very natural to require that $\operatorname{gldim} A < \infty$. Unfortunately, as also occurred in Chapter I, this is too weak an assumption in a noncommutative setting and so we require the stronger hypothesis (3). This is actually the same as demanding that all the simple A-modules have the same homological dimension and is in turn equivalent to demanding that A satisfy a nongraded version of the Artin–Schelter condition from Chapter I (see Corollary 4.6.3 for the details). So, once again, the definition is quite natural given the general philosophy of the book. Although NCCRs are not unique, they are at least Morita equivalent in dimension 2 (meaning that the categories of modules are equivalent) and derived equivalent in dimension 3 (meaning that the derived categories of modules are equivalent). See Theorems 2.3.4 and 2.4.2, respectively.

So, what does this have to do with *commutative* resolutions of singularities? The main motivation both for the definition of NCCRs and for this chapter is to prove that, in dimensions 2 and 3, an NCCR A of R leads to a commutative crepant resolution of singularities $\pi : \widetilde{X} \to X = \text{Spec } R$ (see Theorem 5.4.1 for a more precise version). Combined with the derived equivalence mentioned above this easily proves the result of Bridgeland with which we began this discussion. In order to explain this transition from noncommutative algebra to geometry, we need two further concepts: Geometric Invariant Theory (GIT) to construct appropriate smooth spaces and then a further discussion on derived categories and tilting to utilise those spaces.

The basic ideas of quiver GIT are described in Section 3. To set the stage, fix an NCCR $A = \text{End}_R(M)$ of R. From this we are hoping to construct a smooth variety Y that will provide our crepant resolution $\widetilde{X} \to X$. This is easiest to explain if we also assume that A is a *quiver with relations* $A = kQ/J$, although this is not strictly necessary. A brief review of quivers and their representations is given in the chapter's Appendix (page 297), so here we just mention that Q is a quiver, or finite directed graph, and kQ denotes the *path algebra* or set of all possible paths around the quiver. The relations J will be ignored for the moment. A *representation* $\mathbb{V} = (V_i, f_a)$ of Q (or, equivalently, of kQ) consists of a finite-dimensional vector space V_i at each vertex $i \in Q_0$ and a linear map f_a at each arrow a of Q. The *dimension vector* of $\mathbb{V}$ is simply the vector $\alpha = (\dim_k V_i : i \in Q_0)$. It is easy to obtain a smooth space from this — one simply takes the vector space

$$\mathcal{R} = \mathcal{R}_\alpha = \text{Rep}(kQ, \alpha) = \{\text{representations of } kQ \text{ of dimension vector } \alpha\}.$$

Unfortunately, one needs to take the isomorphism classes of such representations, and this corresponds to the set of G-orbits in $\mathcal{R}$ for $G = \prod_{i \in Q_0} \text{GL}(\dim V_i)$. This space will typically be unpleasant to work with and will certainly not be smooth and so the idea of GIT is to find a better quotient with nicer properties. This is achieved by first introducing a little extra data in the form of a character χ of G and open subsets $\mathcal{R}^s \subseteq \mathcal{R}^{ss}$ of $\mathcal{R}$ called the χ-(semi)stable representations. One then constructs a quotient $q : \mathcal{R}^{ss} \to \mathcal{R} /\!/_\chi G$ of $\mathcal{R}^{ss}$ that actually parametrises the isomorphism classes of χ-stable representations. It is called the *moduli space of χ-stable representations of kQ of dimension vector α*. If $A = kQ/J$ is a quiver with relations then the same constructions work except that one has to be mindful of the relations. These moduli spaces are surprisingly easy to compute for specific quivers and a number of examples are given in Section 3. Crucially, they are often smooth and so they are candidates for our desired resolutions.

At this stage the categories we are given are the category $\text{mod}(A)$ of finitely generated A-modules and the category of coherent sheaves $\text{coh}(Y)$ for one of these moduli spaces Y. Unfortunately, it is almost impossible for them to be equivalent; for example, $\text{mod}(A)$ always has enough projective modules but $\text{coh}(Y)$ almost never does, and this is why we can only hope for an equivalence of derived categories. This is achieved in Section 4, where the basic theory of derived categories is described, as well as the basic technique, called tilting, for producing equivalences between such categories.

Finally, in Section 5 we pull everything together by looking in detail at two important cases. First we relate NCCRs to minimal resolutions of ADE surface singularities: this is the case when the singular variety equals $\mathrm{Spec}(\mathbb{C}[x, y]^G)$, for a finite group $G \subseteq \mathrm{SL}(2, \mathbb{C})$. Secondly, we show how NCCRs induce crepant resolutions in dimension three—this result was really Van den Bergh's original motivation for defining NCCRs, since it implies the theorem of Bridgeland with which we began the discussion.

Structure of Chapter IV. The aim of the chapter is to give an introduction to NCCRs. In Section 1 we introduce the basic ideas by explicitly constructing potential NCCRs of the form $A = \mathrm{End}_R(M)$ for one particular ring R. The ring we pick is $R = \mathbb{C}[\![a, b, c]\!]/(ab - c^3)$, although the computations work more generally. This leads to the general noncommutative concepts, including the definition of NCCRs in Section 2. In that section NCCRs are also shown to be Morita equivalent in dimension 2 and derived equivalent in dimension 3. Section 3 gives a brief overview of quiver Geometric Invariant Theory, which then allows us to extract geometric objects from NCCRs. Section 4 introduces derived categories and the homological techniques needed to relate the geometry with the NCCRs. In particular, we show that NCCRs are Calabi–Yau. Finally, Section 5 applies this to describe minimal resolutions of ADE surface singularities and crepant resolutions of 3-folds.

CHAPTER I

Noncommutative projective geometry

Daniel Rogalski

Introduction

These notes are a significantly expanded version of the author's lectures at the graduate workshop "Noncommutative algebraic geometry" held at the Mathematical Sciences Research Institute in June 2012. The main point of entry to the subject we chose was the idea of an Artin–Schelter regular algebra. The introduction of such algebras by Artin and Schelter motivated many of the later developments in the subject. Regular algebras are sufficiently rigid to admit classification in dimension at most 3, yet this classification is nontrivial and uses many interesting techniques. There are also many open questions about regular algebras, including the classification in dimension 4.

Intuitively, regular algebras with quadratic relations can be thought of as the coordinate rings of noncommutative projective spaces; thus, they provide examples of the simplest, most fundamental noncommutative projective varieties. In addition, regular algebras provide some down-to-earth examples of Calabi–Yau algebras. This is a class of algebras defined by Ginzburg more recently, which is related to several of the other lecture courses given at the workshop.

Section 1 reviews some important background and introduces noncommutative Gröbner bases. We also include as part of Exercise Set 1 a few exercises using the computer algebra system GAP. Section 2 presents some of the main ideas of the theory of Artin–Schelter regular algebras. Then, using regular algebras as examples and motivation, in Sections 3 and 4 we discuss two important aspects of the geometry of noncommutative graded rings: the parameter space of point modules for a graded algebra, and the noncommutative projective scheme associated to a noetherian graded ring. Finally, in Section 5 we discuss some aspects of the classification of noncommutative curves and surfaces, including a review of some more recent results.

We have tried to keep these notes as accessible as possible to readers of varying backgrounds. In particular, Sections 1 and 2 assume only some basic familiarity with noncommutative rings and homological algebra, as found for example in [117] and [193]. Only knowledge of the concept of a projective space is needed to understand the main ideas about point modules in the first half of Section 3. In the final two sections, however, we will of necessity assume that the reader has a more thorough background in algebraic geometry, including the theory of schemes and sheaves as in Hartshorne's textbook [128].

We thank MSRI for holding the graduate workshop, and we thank the student participants and our co-organizers for the feedback and inspiration they provided. We are indebted to Toby Stafford, from whom we first learned this subject in a graduate course at the University of Michigan. Other sources that have influenced these notes include some lecture notes of Darrin Stephenson [215], and the survey article of Stafford and Van den Bergh [214]. We thank all of these authors. We also thank Jonathan Conder, Susan Elle, Jason Gaddis, Matthew Grimm, Brendan Nolan, Stephan Weispfenning, and Robert Won for reading earlier versions of these notes and giving helpful comments.

1. Review of basic background and the Diamond Lemma

1.1. Graded algebras. In this section, we review several topics in the theory of rings and homological algebra which are needed before we can discuss Artin–Schelter regular algebras in Section 2. We also include an introduction to noncommutative Gröbner bases and the Diamond Lemma.

Throughout these notes we work for simplicity over an algebraically closed base field k. Recall that a k-algebra is a (not necessarily commutative) ring A with identity which has a copy of k as a subring of its center; then A is also a k-vector space such that scalar multiplication $\cdot$ satisfies $(\lambda \cdot a)b = \lambda \cdot (ab) = a(\lambda \cdot b)$ for all $\lambda \in k$, $a, b \in A$. (The word algebra is sometimes used for objects with nonassociative multiplication, in particular Lie algebras, but for us all algebras are associative.)

DEFINITION 1.1.1. A k-algebra A is $\mathbb{N}$-*graded* if it has a k-vector space decomposition $A = \bigoplus_{n \geq 0} A_n$ such that $A_i A_j \subseteq A_{i+j}$ for all $i, j \geq 0$. We say that A is *connected* if $A_0 = k$. An element x in A is *homogeneous* if $x \in A_n$ for some n. A right or left ideal I of A is called *homogeneous* or *graded* if it is generated by homogeneous elements, or equivalently if $I = \bigoplus_{n \geq 0} (I \cap A_n)$.

EXAMPLE 1.1.2. Recall that the *free algebra* in n generators $x_1, \ldots, x_n$ is the ring $k\langle x_1, \ldots, x_n \rangle$, whose underlying k-vector space has as basis the set of all words in the variables x_i, that is, expressions $x_{i_1} x_{i_2} \ldots x_{i_m}$ for some $m \geq 1$, where $1 \leq i_j \leq n$ for all j. The *length* of a word $x_{i_1} x_{i_2} \ldots x_{i_m}$ is m. We include among the words a symbol 1, which we think of as the empty word, and which has length 0. The product of two words is concatenation, and this operation is extended linearly to define an associative product on all elements.

The free algebra $A = k\langle x_1, \ldots, x_n \rangle$ is connected $\mathbb{N}$-graded, where A_i is the k-span of all words of length i. For a more general grading, one can put *weights* $d_i \geq 1$ on the variables x_i and define A_i to be the k-span of all words $x_{i_1} \ldots x_{i_m}$ such that $\sum_{j=1}^{m} d_{i_j} = i$.

DEFINITION 1.1.3. A k-algebra A is *finitely generated* (as an algebra) if there is a finite set of elements $a_1, \ldots, a_n \in A$ such that the set

$$\{a_{i_1} a_{i_2} \ldots a_{i_m} \mid 1 \leq i_j \leq n, m \geq 1\} \cup \{1\}$$

spans A as a k-space. It is clear that if A is finitely generated and $\mathbb{N}$-graded, then it has a finite set of homogeneous elements that generate it. Then it is easy to see that a connected $\mathbb{N}$-graded k-algebra A is finitely generated if and only if there is a degree preserving surjective ring homomorphism $k\langle x_1, \ldots, x_n \rangle \to A$ for some free algebra $k\langle x_1, \ldots, x_n \rangle$ with some weighting of the variables, and thus $A \cong k\langle x_1, \ldots x_n \rangle / I$ for some homogeneous ideal I. If I is generated by finitely many homogeneous elements (as a 2-sided ideal), say $I = (f_1, \ldots, f_m)$, then we say that A is *finitely presented*, and we call $k\langle x_1, \ldots, x_n \rangle / (f_1, \ldots, f_m)$ a *presentation* of A with generators $x_1, \ldots, x_n$ and relations $f_1, \ldots, f_m$.

DEFINITION 1.1.4. For the sake of brevity, in these notes we say that an algebra A is *finitely graded* if it is connected $\mathbb{N}$-graded and finitely generated as a k-algebra. Note that if A is finitely graded, then $\dim_k A_n < \infty$ for all n, since this is true already for the free algebra.

In Section 1.3 below, we will give a number of important examples of algebras defined by presentations.

1.2. Graded modules, GK-dimension, and Hilbert series.

DEFINITION 1.2.1. Let A be an $\mathbb{N}$-graded k-algebra. A right A-module M is *graded* if M has a k-space decomposition $M = \bigoplus_{n \in \mathbb{Z}} M_n$ such that $M_i A_j \subseteq M_{i+j}$ for all $i \in \mathbb{Z}$, $j \in \mathbb{N}$.

Given a graded A-module M, we define $M(i)$ to be the graded module which is isomorphic to M as an abstract A-module, but which has degrees shifted so that $M(i)_n = M_{i+n}$. Any such module is called a *shift* of M. (Note that if we visualize the pieces of M laid out along the integer points of the usual number line, then to obtain $M(i)$ one shifts all pieces of M to the left i units if i is positive, and to the right $|i|$ units if i is negative.)

A homomorphism of A-modules $\phi : M \to N$ is a *graded homomorphism* if $\phi(M_n) \subseteq N_n$ for all n.

We will mostly be concerned with graded A-modules M which are finitely generated. In this case, we can find a finite set of homogeneous generators of M, say $m_1, \ldots, m_r$ with $m_i \in M_{d_i}$, and thus define a surjective graded right A-module homomorphism $\bigoplus_{i=1}^r A(-d_i) \to M$, where the 1 of the ith summand maps to the generator m_i. This shows that any finitely generated graded A-module M over a finitely graded algebra A has $\dim_k M_n < \infty$ for all n and $\dim_k M_n = 0$ for $n \ll 0$, and so the following definition makes sense.

DEFINITION 1.2.2. Let A be finitely graded. If M is a finitely generated graded A-module, then the *Hilbert series* of M is the formal Laurent series

$$h_M(t) = \sum_{n \in \mathbb{Z}} (\dim_k M_n) t^n.$$

We consider the Hilbert series of a finitely generated graded module as a generating function (in the sense of combinatorics) for the integer sequence $\dim_k M_n$, and it is useful to manipulate it in the ring of Laurent series $\mathbb{Q}((t))$.

EXAMPLE 1.2.3. The Hilbert series of the commutative polynomial ring $k[x]$ is $1 + t + t^2 + \ldots$, which in the Laurent series ring has the nicer compact form $1/(1 - t)$. More generally, if $A = k[x_1, \ldots, x_m]$ then $h_A(t) = 1/(1 - t)^m$. On the other hand, the free associative algebra $A = k\langle x_1, \ldots, x_m \rangle$ has Hilbert series

$$h_A(t) = 1 + mt + m^2 t^2 + \cdots = 1/(1 - mt).$$

In particular, $\dim_k A_n$ grows exponentially as a function of n if $m \geq 2$. In Exercise 1.6.1, the reader is asked to prove more general versions of these formulas for weighted polynomial rings and free algebras.

DEFINITION 1.2.4. If A is a finitely generated (not necessarily graded) k-algebra, the Gelfand–Kirillov (GK) dimension of A is defined to be

$$\text{GKdim}(A) = \limsup_{n \to \infty} \log_n (\dim_k V^n),$$

where V is any finite dimensional k-subspace of A which generates A as an algebra and has $1 \in V$. The algebra A has *exponential growth* if $\limsup_{n \to \infty} (\dim_k V^n)^{1/n} > 1$; otherwise, clearly $\limsup_{n \to \infty} (\dim_k V^n)^{1/n} = 1$ and we say that A has *subexponential growth*. The book [157] is the main reference for the basic facts about the GK-dimension. In particular, the definitions above do not depend on the choice of V [157, Lemma 1.1, Lemma 2.1]. Also, if A is a commutative finitely generated algebra, then $\text{GKdim } A$ is the same as the Krull dimension of A [157, Theorem 4.5(a)].

If A is finitely graded, then one may take V to be $A_0 \oplus \cdots \oplus A_m$ for some m, and using this one may prove that $\text{GKdim } A = \limsup_{n \to \infty} \log_n (\sum_{i=0}^{n} \dim_k A_i)$ [157, Lemma 6.1]. This value is easy to calculate if we have a formula for the dimension of the ith graded piece of A. In fact, in most of the examples in which we are interested below, there is a polynomial $p(t) \in \mathbb{Q}(t)$ such that $p(n) = \dim_k A_n$ for all $n \gg 0$, in which case p is called the *Hilbert polynomial* of A. When p exists then it easy to see that $\text{GKdim}(A) = \deg(p) + 1$. For example, for the commutative polynomial ring $A = k[x_1, \ldots, x_m]$, one has $\dim_k A_n = \binom{n+m-1}{m-1}$, which agrees with a polynomial $p(n)$ of degree $m - 1$ for all $n \geq 0$, so that $\text{GKdim } A = m$.

We briefly recall the definitions of noetherian rings and modules.

DEFINITION 1.2.5. A right module M is *noetherian* if it has the ascending chain condition (ACC) on submodules, or equivalently if every submodule of M is finitely generated. A ring is right noetherian if it is noetherian as a right module over itself, or equivalently if it has ACC on right ideals. The left noetherian property is defined analogously, and a ring is called noetherian if it is both left and right noetherian.

The reader can consult [117, Chapter 1] for more information on the noetherian property. By the Hilbert basis theorem, the polynomial ring $k[x_1, \ldots, x_m]$ is noetherian, and thus all finitely generated commutative k-algebras are noetherian. On the other hand, a free algebra $k\langle x_1, \ldots, x_m \rangle$ in $m \geq 2$ variables is not noetherian, and consequently noncommutative finitely generated algebras need not be noetherian. The noetherian property still holds for many important noncommutative examples of interest, but one often has to work harder to prove it.

1.3. Some examples, and the use of normal elements.

EXAMPLES 1.3.1.

(1) For any constants $0 \neq q_{ij} \in k$, the algebra

$$A = k\langle x_1, x_2, \ldots, x_n \rangle / (x_j x_i - q_{ij} x_i x_j \mid 1 \leq i < j \leq n)$$

is called a *quantum polynomial ring*. The set $\{x_1^{i_1} x_2^{i_2} \ldots x_n^{i_n} \mid i_1, i_2, \ldots, i_n \geq 0\}$ is a k-basis for A, as we will see in Example 1.4.2. Then A has the same Hilbert series as a commutative polynomial ring in n variables, and thus $h_A(t) = 1/(1-t)^n$.

(2) The special case $n = 2$ of (1), so that

$$A = k\langle x, y \rangle / (yx - qxy)$$

for some $0 \neq q$, is called the *quantum plane*.

(3) The algebra

$$A = k\langle x, y \rangle / (yx - xy - x^2)$$

is called the *Jordan plane*. We will also see in Example 1.4.2 that if A is the Jordan plane, then $\{x^i y^j \mid i, j \geq 0\}$ is a k-basis for A, and so $h_A(t) = 1/(1-t)^2$.

(Note that we often use a variable name such as x to indicate both an element in the free algebra and the corresponding coset in the factor ring; this is often convenient in a context where there is no chance of confusion.)

All of the examples above have many properties in common with a commutative polynomial ring in the same number of generators. For example, they all have the Hilbert series of a polynomial ring and they are all noetherian domains. The standard way to verify these facts is to express these examples as iterated Ore extensions. We omit a discussion of Ore extensions here, since the reader can find a thorough introduction to these elsewhere, for example in [117, Chapters 1-2]. Instead, we mention a different method, which will also apply to some important examples which are not iterated Ore extensions. Given a ring A, an element $x \in A$ is *normal* if $xA = Ax$ (and hence the right or left ideal generated by x is an ideal). Certain properties can be lifted from a factor ring of a graded ring to the whole ring, when one factors by an ideal generated by a homogeneous normal element.

LEMMA 1.3.2. *Let A be a finitely graded k-algebra, and let $x \in A_d$ be a homogeneous normal element for some $d \geq 1$.*

(1) *If x is a nonzerodivisor in A, then if A/xA is a domain then A is a domain.*

(2) *If A/xA is noetherian, then A is noetherian.*

PROOF. We ask the reader to prove part (1) as Exercise 1.6.2.

We sketch the proof of part (2), which is [9, Theorem 8.1]. First, by symmetry we need only show that A is right noetherian. An easy argument, which we leave to the reader, shows that it suffices to show that every *graded* right ideal of A is finitely generated. Suppose that A has an infinitely generated graded right ideal. By Zorn's lemma, we may choose a maximal element of the set of such right ideals, say I. Then A/I is a noetherian right A-module. Consider the short exact sequence

$$0 \to (Ax \cap I)/Ix \to I/Ix \to I/(Ax \cap I) \to 0. \tag{1.A}$$

We have $I/(Ax \cap I) \cong (I + Ax)/Ax$, which is a right ideal of A/Ax and hence is noetherian by hypothesis. Now $(Ax \cap I) = Jx$ for some subset J of A which is easily checked to be a graded right ideal of A. Then $(Ax \cap I)/Ix = Jx/Ix \cong Mx$, where $M = J/I$ is a noetherian A-module since it is a submodule of A/I. Given an A-submodule P of Mx, it is easy to check that $P = Nx$ where $N = \{m \in M \mid mx \in P\}$. Thus since M is noetherian, so is Mx. Then (1.A) shows that I/Ix is a noetherian A-module, in particular finitely generated. Thus we can choose a finitely generated graded right ideal $N \subseteq I$ such that $N + Ix = I$. An easy induction proof shows that $N + Ix^n = I$ for all $n \geq 0$, and then one gets $N_m = I_m$ for all $m \geq 0$ since x has positive degree. In particular, I is finitely generated, a contradiction. (Alternatively, once one knows I/Ix is finitely generated, it follows that I is finitely generated by the graded Nakayama lemma described in Lemma 1.5.1 below.) $\qquad\square$

COROLLARY 1.3.3. *The algebras in Examples 1.3.1 are noetherian domains.*

PROOF. Consider for example the quantum plane $A = k\langle x, y \rangle / (yx - qxy)$. Then it is easy to check that x is a normal element and that $A/xA \cong k[y]$. We know $h_A(t) = 1/(1-t)^2$ and $h_{A/xA}(t) = 1/(1-t)$. Obviously $h_{xA}(t)$ is at most as large as $t/(1-t)^2$, with equality if and only if x is a nonzerodivisor in A. Since $h_A(t) = h_{xA}(t) + h_{A/xA}(t)$, equality is forced, so x is a nonzerodivisor in A. Since $k[y]$ is a noetherian domain, so is A by Lemma 1.3.2. The argument for the general skew polynomial ring is a simple inductive version of the above, and the argument for the Jordan plane is similar since x is again normal. $\qquad\square$

EXAMPLE 1.3.4. The algebra

$$S = k\langle x, y, z \rangle / (ayx + bxy + cz^2, axz + bzx + cy^2, azy + byz + cx^2),$$

for any $a, b, c \in k$, is called a Sklyanin algebra, after E. J. Sklyanin. As long as a, b, c are sufficiently general, S also has many properties in common with a polynomial ring in 3 variables, for example S is a noetherian domain with $h_S(t) = 1/(1-t)^3$. These facts are much more difficult to prove than for the other examples above, since S does not have such a simply described k-basis of words in the generators. In fact S does have a normal element g of degree 3 (which is not easy to find) and in the end Lemma 1.3.2 can be applied, but it is hard to show that g is a nonzerodivisor and the factor ring S/gS takes effort to analyze. Some of the techniques of noncommutative geometry that are the subject of this course, in particular the study of point modules, were developed precisely to better understand the properties of Sklyanin algebras. See the end of Section 3 for more details.

1.4. The Diamond Lemma. In general, it can be awkward to prove that a set of words that looks like a k-basis for a presented algebra really is linearly independent. The *Diamond Lemma* gives an algorithmic method for this. It is part of the theory of noncommutative Gröbner bases, which is a direct analog of the theory of Gröbner bases which was first studied for commutative rings. George Bergman gave an important exposition of the method in [34], which we follow closely below. The basic idea behind the method goes back farther, however; in particular, the theory is also known by the name Gröbner–Shirshov bases, since A. I. Shirshov was one of its earlier proponents.

Consider the free algebra $F = k\langle x_1, \ldots, x_n \rangle$. While we stick to free algebras on finitely many indeterminates for notational convenience, everything below goes through easily for arbitrarily many indeterminates. Fix an ordering on the variables, say $x_1 < x_2 < \cdots < x_n$. Also, we choose a total order on the set of words in the x_i which extends the ordering on the variables and has the following properties: (i) for all words u, v, and w, if $w < v$, then $uw < uv$ and $wu < vu$, and (ii) for each word w the set of words $\{v \mid v < w\}$ is finite. We call such an order *admissible*.

If we assign weights to the variables, and thus assign a degree to each word, then one important choice of such an ordering is the degree lex order, where $w < v$ if w has smaller degree than v or if w and v have the same degree but w comes earlier than v in the dictionary ordering with respect to the fixed ordering on the variables. For example, in $k\langle x, y \rangle$ with $x < y$ and variables of weight 1, the beginning of the degree lex order is

$$1 < x < y < x^2 < xy < yx < y^2 < x^3 < x^2y < xyx < xy^2 < \cdots$$

Given an element f of the free algebra, its *leading word* is the largest word under the ordering which appears in f with nonzero coefficient. For example, the leading word of $xy + 3x^2y + 5xyx$ in $k\langle x, y \rangle$ with the degree lex ordering is xyx.

Now suppose that the set $\{g_\sigma\}_{\sigma \in S} \subseteq F$ generates an ideal I of F (as a 2-sided ideal). By adjusting each g_σ by a scalar we can assume that each g_σ has a leading term with coefficient 1, and so we can write $g_\sigma = w_\sigma - f_\sigma$, where w_σ is the leading word of g_σ and f_σ is a linear combination of words v with $v < w_\sigma$. A word is *reduced* (with respect to the fixed set of relations $\{g_\sigma\}_{\sigma \in S}$) if it does not contain any of the w_σ as a subword. If a word w is not reduced, but say $w = uw_\sigma v$, then w is equal modulo I to $uf_\sigma v$, which is a linear combination of strictly smaller words. Given an index $\sigma \in S$ and words u, v, the corresponding *reduction* $r = r_{u\sigma v}$ is the k-linear endomorphism of the free algebra which takes $w = uw_\sigma v$ to $uf_\sigma v$ and sends every other word to itself. Since every word has finitely many words less than it, it is not hard to see that given any element h of the free algebra, some finite composition of reductions will send h to a k-linear combination of reduced words. Since a reduction does not change the class of an element modulo I, we see that the images of the reduced words in $k\langle x_1, x_2, \ldots, x_n \rangle / I$ are a k-spanning set. The idea of noncommutative Gröbner bases is to find good sets of generators g_σ of the ideal I such that the images of the corresponding reduced words in $k\langle x_1, \ldots, x_n \rangle / I$ are k-independent, and thus a k-basis. The Diamond Lemma gives a convenient way to verify if a set of generators g_σ has this property. Moreover, if a set of generators does not, the Diamond Lemma leads to a (possibly non-terminating) algorithm for enlarging the set of generators to one that does.

The element h of the free algebra is called *reduction-unique* if given any two finite compositions of reductions s_1 and s_2 such that $s_1(h)$ and $s_2(h)$ consist of linear combinations of reduced words, we have $s_1(h) = s_2(h)$. In this case we write $\mathrm{red}(h)$ for this uniquely defined linear combination of reduced words. Suppose that w is a word which can be written as $w = tvu$ for some nonempty words t, u, v, where $w_\sigma = tv$ and $w_\tau = vu$ for some $\sigma, \tau \in S$. We call this situation an *overlap* ambiguity. Then there are (at least) two different possible reductions one can apply to reduce w, namely $r_1 = r_{1\sigma u}$ and $r_2 = r_{t\tau 1}$. If there exist compositions of reductions s_1, s_2 with the property that $s_1 \circ r_1(w) = s_2 \circ r_2(w)$, then we say that this ambiguity is *resolvable*. Similarly, we have an *inclusion* ambiguity when we have $w_\sigma = tw_\tau u$ for some words t, u and some $\sigma, \tau \in S$. Again, the ambiguity is called resolvable if there are compositions of reductions s_1 and s_2 such that $s_1 \circ r_{1\sigma 1}(w) = s_2 \circ r_{t\tau u}(w)$.

We now sketch the proof of the main result underlying the method of noncommutative Gröbner bases.

THEOREM 1.4.1. *(Diamond Lemma) Suppose that* $\{g_\sigma\}_{\sigma \in S} \subseteq F = k\langle x_1, \ldots, x_n\rangle$ *generates an ideal* I, *where* $g_\sigma = w_\sigma - f_\sigma$ *with* w_σ *the leading word of* g_σ *under some fixed admissible ordering on the words of* F. *Consider reductions with respect to this fixed set of generators of* I. *Then the following are equivalent:*

(1) *All overlap and inclusion ambiguities among the* g_σ *are resolvable.*
(2) *All elements of* $k\langle x_1, \ldots, x_n\rangle$ *are reduction-unique.*
(3) *The images of the reduced words in* $k\langle x_1, \ldots, x_n\rangle / I$ *form a* k-*basis.*

When any of these conditions holds, we say that $\{g_\sigma\}_{\sigma \in S}$ *is a Gröbner basis for the ideal* I *of* F.

PROOF. (1) $\implies$ (2) First, it is easy to prove that the set of reduction-unique elements of $k\langle x_1, \ldots, x_n\rangle$ is a k-subspace, and that $\mathrm{red}(-)$ is a linear function on this subspace (Exercise 1.6.3). Thus it is enough to prove that all words are reduction-unique. This is proved by induction on the ordered list of words, so assume that w is a word such that all words v with $v < w$ are reduction-unique (and so any linear combination of such words is). Suppose that $r = r_k \circ \cdots \circ r_2 \circ r_1$ and $r' = r'_j \circ \cdots \circ r'_2 \circ r'_1$ are two compositions of reductions such that $r(w)$ and $r'(w)$ are linear combinations of reduced words. If $r_1 = r'_1$, then since $r_1(w) = r'_1(w)$ is reduction-unique by the induction hypothesis, clearly $r(w) = r'(w)$. Suppose instead that $r_1 = r_{y\sigma uz}$ and $r'_1 = r_{yt\tau z}$ where $w = yw_\sigma uz = ytw_\tau z$ and the subwords w_σ and w_τ overlap. By the hypothesis that all overlap ambiguities resolve, there are compositions of reductions s_1 and s_2 such that $s_1 \circ r_{1\sigma u}(w_\sigma u) = s_2 \circ r_{t\tau 1}(tw_\tau)$. Then replacing each reduction $r_{a\rho b}$ among those occurring in the compositions s_1 and s_2 by $r_{ya\rho bz}$, we obtain compositions of reductions s'_1 and s'_2 such that $v = s'_1 \circ r_{y\sigma uz}(w) = s'_2 \circ r_{yt\tau z}(w)$. Since $r_{y\sigma uz}(w), r_{yt\tau z}(w)$, and v are reduction-unique, we get $r(w) = \mathrm{red}(r_{y\sigma uz}(w)) = \mathrm{red}(v) = \mathrm{red}(r_{yt\tau z}(w)) = r'(w)$. Similarly, if $w = ytw_\sigma uz = yw_\tau z$ and $r_1 = r_{yt\sigma uz}$ and $r'_1 = r_{y\tau z}$, then using the hypothesis that all inclusion ambiguities resolve we get $r(w) = r'(w)$. Finally, if $w = yw_\sigma uw_\tau z$ and $r_1 = r_{yw_\sigma uw_\tau z}$, $r_1' = r_{yw_\sigma u\tau z}$, then $r_1(w) = yf_\sigma uw_\tau z$ and $r_1'(w) = yw_\sigma uf_\tau z$ are linear combinations of reduction-unique words. Then since $\text{red}(-)$ is linear, we get $r(w) = \text{red}(yf_\sigma uw_\tau z) = \text{red}(yf_\sigma uf_\tau z) = \text{red}(yw_\sigma uf_\tau z) = r'(w)$. Thus $r(w) = r'(w)$ in all cases, and w is reduction-unique, completing the induction step.

(2) $\implies$ (3) Let $F = k\langle x_1, \ldots, x_n \rangle$. By hypothesis there is a well-defined k-linear map red : $F \to F$. Let V be its image $\text{red}(F)$, in order words the k-span of all reduced words. Obviously $\text{red}|_V$ is the identity map $V \to V$ and so red is a projection; thus $F \cong K \oplus V$ as k-spaces, where $K = \ker(\text{red})$. We claim that $K = I$. First, every element of I is a linear combination of expressions $ug_\sigma v$ for words u and v. Obviously the reduction $r_{ug_\sigma v}$ sends $ug_\sigma v$ to 0 and thus $ug_\sigma v \in K$; since K is a subspace we get $I \subseteq K$. Conversely, since every reduction changes an element to another one congruent modulo I, we must have $K \subseteq I$. Thus $K = I$ as claimed. Finally, since $F \cong I \oplus V$, clearly the basis of V consisting of all reduced words has image in F/I which is a k-basis of F/I.

The reverse implications, (3) $\implies$ (2) and (2) $\implies$ (1), are left to the reader; see Exercise 1.6.3. □

EXAMPLE 1.4.2. Let $A = k\langle x, y, z \rangle/(f_1, f_2, f_3)$ be a quantum polynomial ring in three variables, with $f_1 = yx - pxy$, $f_2 = zx - qxz$, and $f_3 = zy - ryz$. Taking $x < y < z$ and degree lex order, the leading terms of these relations are yx, zx, zy. There is one ambiguity among these three leading words, the overlap ambiguity zyx. Reducing the zy first and then continuing to do more reductions, we get

$$zyx = ryzx = rqyxz = rqpxyz,$$

while reducing the yx first we get

$$zyx = pzxy = pqxzy = pqrxyz.$$

Thus the ambiguity is resolvable, and by Theorem 1.4.1 the set $\{f_1, f_2, f_3\}$ is a Gröbner basis for the ideal it generates. The same argument applies to a general quantum polynomial ring in n variables: choosing degree lex order with $x_1 < \cdots < x_n$, there is one overlap ambiguity $x_k x_j x_i$ for each triple of variables with $x_i < x_j < x_k$, which resolves by the same argument as above. Thus the corresponding set of reduced words, $\{x_1^{i_1} x_2^{i_2} \ldots x_n^{i_n} \mid i_j \geq 0\}$, is a k-basis for the quantum polynomial ring A. In particular, A has the same Hilbert series as a polynomial ring in n variables, $h_A(t) = 1/(1-t)^n$, as claimed in Example 1.3.1(1).

An even easier argument shows that the Jordan plane in Examples 1.3.1 has the claimed k-basis $\{x^i y^j \mid i, j \geq 0\}$: taking $x < y$, its single relation $yx - xy - x^2$ has leading term yx and there are no ambiguities.

EXAMPLE 1.4.3. Let $A = k\langle x, y \rangle/(yx^2 - x^2y, y^2x - xy^2)$. Taking degree lex order with $x < y$, there is one overlap ambiguity y^2x^2 in the Diamond Lemma, which resolves, as is easily checked. Thus the set of reduced words $\{x^i (yx)^j y^k \mid i, j, k \geq 0\}$ is a k-basis for A. Then A has the same Hilbert series as a commutative polynomial ring in variables of weights 1, 1, 2, namely $h_A(t) = 1/(1 - t)^2(1 - t^2)$ (Exercise 1.6.1).

If a generating set $\{g_\sigma\}_{\sigma \in S}$ for an ideal I has non-resolving ambiguities and so is not a Gröbner basis, the proof of the Diamond Lemma also leads to an algorithm for expanding the generating set to get a Gröbner basis. Namely, suppose the overlap ambiguity $w = w_\sigma u = t w_\tau$ does not resolve. Then for some compositions of reductions s_1 and s_2, $h_1 = s_1 \circ r_{1\sigma u}(w)$ and $h_2 = s_2 \circ r_{t\tau 1}(w)$ are distinct linear combinations of reduced words. Thus $0 \neq h_1 - h_2 \in I$ is a new relation, whose leading word is necessarily different from any leading word w_ρ with $\rho \in S$. Replace this relation with a scalar multiple so that its leading term has coefficient 1, and add this new relation to the generating set of I. The previously problematic overlap now obviously resolves, but there may be new ambiguities involving the leading word of the new relation, and one begins the process of checking ambiguities again. Similarly, a nonresolving inclusion ambiguity will lead to a new relation in I with new leading word, and we add this new relation to the generating set. This process may terminate after finitely many steps, and thus produce a set of relations with no unresolving ambiguities, and hence a Gröbner basis. Alternatively, the process may not terminate. It is still true in this case that the infinite set of relations produced by repeating the process infinitely is a Gröbner basis, but this is not so helpful unless there is a predictable pattern to the new relations produced at each step, so that one can understand what this infinite set of relations actually is.

EXAMPLE 1.4.4. Let $A = k\langle x, y, z \rangle / (z^2 - xy - yx, zx - xz, zy - yz)$. The reader may easily check that $g_1 = z^2 - xy - yx$, $g_2 = zx - xz$, $g_3 = zy - yz$ is not a Gröbner basis under degree lex order with $x < y < z$, but attempting to resolve the overlap ambiguities $z^2 x$ and $z^2 y$ lead by the process described above to new relations $g_4 = yx^2 - x^2 y$ and $g_5 = y^2 x - xy^2$ such that $\{g_1, \ldots, g_5\}$ is a Gröbner basis (Exercise 1.6.4).

EXAMPLE 1.4.5. Let $A = k\langle x, y \rangle / (yx - xy - x^2)$ be the Jordan plane, but take degree lex order with $y < x$ instead so that x^2 is now the leading term. It overlaps itself, and one may check that the overlap ambiguity does not resolve. In this case the algorithm of checking overlaps and adding new relations never terminates, but there is a pattern to the new relations added, so that one can write down the infinite Gröbner basis given by the infinite process. The reader may attempt this calculation by hand, or by computer (Exercise 1.6.7(2)). This example shows that whether or not the algorithm terminates is sensitive even to the choice of ordering.

1.5. Graded Ext and minimal free resolutions. The main purpose of this section is to describe the special features of projective resolutions and Ext for graded modules over a graded ring. Although we remind the reader of the basic definitions, the reader encountering Ext for the first time might want to first study the basic concept in a book such as [193]. Some facts we need below are left as exercises for a reader with some experience in homological algebra, or they may be taken on faith.

Recall that a right module P over a ring A is *projective* if, whenever A-module homomorphisms $f : M \to N$ and $g : P \to N$ are given, with f surjective, then there exists a homomorphism $h : P \to M$ such that $f \circ h = g$. It is a basic fact that a module P is projective if and only if there is a module Q such that $P \oplus Q$ is a free module; in particular, free modules are projective. A *projective resolution* of an A-module M is a complex of A-modules and A-module homomorphisms,

$$\cdots \to P_n \overset{d_{n-1}}{\to} P_{n-1} \to \cdots \overset{d_1}{\to} P_1 \overset{d_0}{\to} P_0 \to 0, \tag{1.B}$$

together with a surjective *augmentation map* $\epsilon : P_0 \to M$, such that each P_i is projective, and the sequence

$$\cdots \to P_n \overset{d_{n-1}}{\to} P_{n-1} \to \cdots \overset{d_1}{\to} P_1 \overset{d_0}{\to} P_0 \overset{\epsilon}{\to} M \to 0 \tag{1.C}$$

is exact. Another way of saying that (1.B) is a projective resolution of M is to say that it is a complex $P_\bullet$ of projective A-modules, with homology groups $H_i(P_\bullet) = 0$ for $i \neq 0$ and $H_0(P_\bullet) \cong M$. Since every module is a homomorphic image of a free module, every module has a projective resolution.

Given right A-modules M and N, there are abelian groups $\text{Ext}_A^i(M, N)$ for each $i \geq 0$. To define them, take any projective resolution of M, say

$$\cdots \to P_n \overset{d_{n-1}}{\to} P_{n-1} \to \cdots \overset{d_1}{\to} P_1 \overset{d_0}{\to} P_0 \to 0,$$

and apply the functor $\text{Hom}_A(-, N)$ to the complex (which reverses the direction of the maps), obtaining a complex of abelian groups

$$\cdots \leftarrow \text{Hom}_A(P_n, N) \overset{d_{n-1}^*}{\leftarrow} \text{Hom}_A(P_{n-1}, N) \leftarrow \cdots \overset{d_0^*}{\leftarrow} \text{Hom}_A(P_0, N) \leftarrow 0.$$

Then $\text{Ext}^i(M, N)$ is defined to be the ith homology group $\ker d_i^* / \text{im}\, d_{i-1}^*$ of this complex. These groups do not depend up to isomorphism on the choice of projective resolution of M, and moreover $\text{Ext}_A^0(M, N) \cong \text{Hom}_A(M, N)$ [193, Corollary 6.57, Theorem 6.61].

For the rest of the section we consider the special case of the definitions above where A is finitely graded. A *graded free module* over a finitely graded algebra A is a direct sum of shifted copies of A, that is $\bigoplus_{\alpha \in I} A(i_\alpha)$. A graded module M is *left bounded* if $M_n = 0$ for $n \ll 0$. For any $m \geq 0$, we write $A_{\geq m}$ as shorthand for $\bigoplus_{n \geq m} A_n$. Because A has a unique homogeneous maximal ideal $A_{\geq 1}$, in many ways the theory for finitely graded algebras mimics the theory for local rings. For example, we have the following graded version of Nakayama's lemma.

LEMMA 1.5.1. *Let A be a finitely graded k-algebra. Let M be a left bounded graded A-module. If $M A_{\geq 1} = M$, then $M = 0$. Also, a set of homogeneous elements $\{m_i\} \subseteq M$ generates M as an A-module if and only if the images of the m_i span $M/M A_{\geq 1}$ as a $A/A_{\geq 1} = k$-vector space.*

PROOF. The first statement is easier to prove than the ungraded version of Nakayama's lemma: if M is nonzero, and d is the minimum degree such that $M_d \neq 0$, then $M A_{\geq 1}$

is contained in degrees greater than or equal to $d + 1$, so $MA_{\geq 1} = M$ is impossible. The second statement follows by applying the first statement to $N = M/\left(\sum m_i A\right)$. $\quad\square$

Given a left bounded graded module M over a finitely graded algebra A, a set $\{m_i\} \subseteq M$ of homogeneous generators is said to *minimally* generate M if the images of the m_i in $M/MA_{\geq 1}$ are a k-basis. In this case we can construct a surjective graded A-module homomorphism $\phi : \bigoplus_i A(-d_i) \to M$, where $d_i = \deg(m_i)$ and the 1 of the ith copy of A maps to $m_i \in M$. The k-vector space map $\bigoplus_i A/A_{\geq 1}(-d_i) \to M/MA_{\geq 1}$ induced by ϕ is an isomorphism. We call ϕ a *minimal* surjection of a graded free module onto M. A graded free resolution of M of the form

$$\cdots \to \bigoplus_i A(-a_{n,i}) \xrightarrow{d_{n-1}} \cdots \xrightarrow{d_1} \bigoplus_i A(-a_{1,i}) \xrightarrow{d_0} \bigoplus_i A(-a_{0,i}) \to 0 \qquad (1.\text{D})$$

is called *minimal* if d_i is a minimal surjection onto $\operatorname{im} d_i$ for all $i \geq 0$, and the augmentation map $\epsilon : \bigoplus_i A(-a_{0,i}) \to M$ is a minimal surjection onto M. For any left bounded graded module M over A, the kernel of a minimal surjection of a graded free module onto M is again left bounded. Thus, by induction a minimal graded free resolution of M always exists.

In general, projective resolutions of a module M are unique only up to homotopy [193, Theorem 6.16], but for minimal graded free resolutions we have the following stronger form of uniqueness.

LEMMA 1.5.2. *Let M be a left bounded graded right module over a finitely graded algebra A.*

(1) *A graded free resolution $P_\bullet$ of M is minimal if and only if $d_i : P_{i+1} \to P_i$ has image inside $P_i A_{\geq 1}$ for each $i \geq 0$.*

(2) *Any two minimal graded free resolutions $P_\bullet$ and $Q_\bullet$ of M are isomorphic as complexes; that is, there are graded module isomorphisms $f_i : P_i \to Q_i$ for each i giving a chain map. In particular, the ranks and shifts of the free modules occurring in a free resolution of M are invariants of M.*

PROOF. We provide only a sketch, leaving some of the details to the reader. The proof is similar to the proof of the analogous fact for commutative local rings, for example as in [92, Theorem 20.2].

(1) Consider a minimal surjection ϕ of a graded free module P onto a left bounded module N, and the resulting exact sequence $0 \to K \xrightarrow{f} P \xrightarrow{\phi} N \to 0$, where $K = \ker \phi$. Applying $- \otimes_A A/A_{\geq 1}$ gives an exact sequence

$$K/KA_{\geq 1} \xrightarrow{\bar{f}} P/PA_{\geq 1} \xrightarrow{\bar{\phi}} N/NA_{\geq 1} \to 0.$$

By definition $\bar{\phi}$ is an isomorphism, forcing $\bar{f} = 0$ and $K \subseteq PA_{\geq 1}$. If $P_\bullet$ is a minimal projective resolution, then each $\operatorname{im} d_i$ is the kernel of some minimal surjection and so $\operatorname{im} d_i \subseteq P_i A_{\geq 1}$ for each $i \geq 0$ by the argument above. The converse is proved by essentially the reverse argument.

(2) As in part (1), consider the exact sequence $0 \to K \xrightarrow{f} P \xrightarrow{\phi} N \to 0$ with ϕ a minimal surjection, and suppose that there is another minimal surjection $\psi : Q \to N$ leading to an exact sequence $0 \to K' \xrightarrow{g} Q \xrightarrow{\psi} N \to 0$. If $\rho : N \to N$ is any graded isomorphism, then there is an induced isomorphism $\overline{\psi}^{-1} \circ \overline{\rho} \circ \overline{\phi} : P/PA_{\geq 1} \to Q/QA_{\geq 1}$, which lifts by projectivity to an isomorphism of graded free modules $h : P \to Q$. Then h restricts to an isomorphism $K \to K'$. Part (2) follows by applying this argument inductively to construct the required isomorphism of complexes, beginning with the identity map $M \to M$. $\qquad\square$

In the graded setting that concerns us in these notes, it is most appropriate to use a graded version of Hom. For graded modules M, N over a finitely graded algebra A, let $\mathrm{Hom}_{\mathrm{gr}\text{-}A}(M, N)$ be the vector space of graded (that is, degree preserving) module homomorphisms from M to N. Then we define $\underline{\mathrm{Hom}}_A(M, N) = \bigoplus_{d \in \mathbb{Z}} \mathrm{Hom}_{\mathrm{gr}\text{-}A}(M, N(d))$, as a graded vector space. It is not hard to see that there is a natural inclusion $\underline{\mathrm{Hom}}_A(M, N) \subseteq \mathrm{Hom}_A(M, N)$, and that this is an equality when M is finitely generated (or more generally, generated by elements in some finite set of degrees). We are primarily concerned with finitely generated modules in these notes, and so in most cases there is no difference between $\underline{\mathrm{Hom}}$ and Hom, but for consistency we will use the graded Hom functors $\underline{\mathrm{Hom}}$ throughout. Similarly as in the ungraded case, for graded modules M and N we define $\underline{\mathrm{Ext}}^i_A(M, N)$ by taking a graded free resolution of M, applying the functor $\underline{\mathrm{Hom}}_A(-, N)$, and taking the ith homology. Then $\underline{\mathrm{Ext}}^i_A(M, N)$ is a graded k-vector space.

As we will see, the graded free resolution of the *trivial module* $k = A/A_{\geq 1}$ of a finitely graded algebra A will be of primary importance. In the next example, we construct such a resolution explicitly. In general, given a right A-module M we sometimes write M_A to emphasize the side on which A is acting. Similarly, $_AN$ indicates a left A-module. This is especially important for bimodules such as the trivial module k, which has both a natural left and right A-action.

EXAMPLE 1.5.3. Let $A = k\langle x, y \rangle / (yx - xy - x^2)$ be the Jordan plane. We claim that the minimal graded free resolution of k_A has the form

$$0 \to A(-2) \xrightarrow{d_1 = \begin{pmatrix} -y-x \\ x \end{pmatrix}} A(-1)^{\oplus 2} \xrightarrow{d_0 = (x\ y)} A \to 0. \tag{1.E}$$

Here, we think of the free right modules as column vectors of elements in A, and the maps as left multiplications by the indicated matrices. Thus d_1 is $c \mapsto \left(\begin{smallmatrix} (-y-x)c \\ xc \end{smallmatrix} \right)$ and d_0 is $\left(\begin{smallmatrix} a \\ b \end{smallmatrix} \right) \mapsto xa + yb$. Note that left multiplication by a matrix is a right module homomorphism.

If we prove that (1.E) is a graded free resolution of k_A, it is automatically minimal by Lemma 1.5.2. Recall that A is a domain with k-basis $\{x^i y^j \mid i, j \geq 0\}$ (Examples 1.3.1 and Corollary 1.3.3). It is easy to see that the sequence above is a complex; this amounts to the calculation

$$\begin{pmatrix} x & y \end{pmatrix} \begin{pmatrix} -y-x \\ x \end{pmatrix} = (x(-y-x) + yx) = (0),$$

using the relation. The injectivity of d_1 is clear because A is a domain, and the cokernel of d_0 is obviously isomorphic to $k = A/A_{\geq 1}$. Exactness in the middle may be checked in the following way. Suppose we have an element $\binom{a}{b} \in \ker d_0$. Since

$$\operatorname{im} d_1 = \left\{ \begin{pmatrix} (-y-x)h \\ xh \end{pmatrix} \middle| h \in A \right\},$$

using that $\{x^i y^j\}$ is a k-basis for A, after subtracting an element of $\operatorname{im}(d_1)$ from $\binom{a}{b}$ we can assume that $b = f(y)$ is a polynomial in y only. Then $xa + yb = xa + yf(y) = 0$, which using the form of the basis again implies that $a = b = 0$. In Section 2 we will prove a general result, Lemma 2.1.3, from which exactness in the middle spot follows automatically.

One may always represent maps between graded free modules as matrices, as in the previous result, but one needs to be careful about conventions. Given a finitely graded k-algebra A, it is easy to verify that any graded right A-module homomorphism $\phi : \bigoplus_{i=1}^{m} A(-s_i) \to \bigoplus_{j=1}^{n} A(-t_j)$ between two graded free right A-modules of finite rank can be represented as left multiplication by an $n \times m$ matrix of elements of A, where the elements of the free modules are thought of as column vectors. On the other hand, any graded left module homomorphism $\psi : \bigoplus_{i=1}^{m} A(-s_i) \to \bigoplus_{j=1}^{n} A(-t_j)$ can be represented as *right* multiplication by an $m \times n$ matrix of elements of A, where the elements of the free modules are thought of as *row* vectors. A little thought shows that the side we multiply the matrix on, and whether we consider row or column vectors, is forced by the situation. The following is an easy reinterpretation of Lemma 1.5.2(1):

LEMMA 1.5.4. *Let A be a finitely graded k-algebra, and let M be a left bounded graded right A-module. If M has a graded free resolution $P_\bullet$ with all P_i of finite rank, then the resolution is minimal if and only if all entries of the matrices representing the maps d_i lie in $A_{\geq 1}$.* □

It will be important for us to understand the action of the functor $\underline{\operatorname{Hom}}_A(-, A)$ on maps between graded free modules. If N is a graded right module over the finitely graded algebra A, then $\underline{\operatorname{Hom}}_A(N, A)$ is a left A-module via $[a \cdot \psi](x) = a\psi(x)$. Given a homomorphism $\phi : N_1 \to N_2$ of graded right A-modules, then the induced map $\underline{\operatorname{Hom}}_A(N_2, A) \to \underline{\operatorname{Hom}}_A(N_1, A)$ given by $f \mapsto f \circ \phi$ is a left A-module homomorphism. Then from the definition of $\underline{\operatorname{Ext}}$, it is clear that $\underline{\operatorname{Ext}}^i_A(N, A)$ is a graded left A-module, for each graded right module N. A similar argument shows that in general, if M and N are right A-modules and B is another ring, then $\underline{\operatorname{Ext}}^i_A(M, N)$ obtains a left B-module structure if N is a (B, A)-bimodule, or a right B-module structure if M is a (B, A)-bimodule (use [193, Proposition 2.54] and a similar argument), but we will need primarily the special case we have described explicitly.

LEMMA 1.5.5. *Let A be a finitely graded k-algebra.*

(1) *For any graded free right module $\bigoplus_{i=1}^{m} A(-s_i)$, there is a canonical graded left A-module isomorphism*

$$\underline{\operatorname{Hom}}_A\Big(\bigoplus_{i=1}^{m} A(-s_i), A\Big) \cong \bigoplus_{i=1}^{m} A(s_i).$$

(2) *Given a graded right module homomorphism*

$$\phi : P = \bigoplus_{i=1}^{m} A(-s_i) \to Q = \bigoplus_{j=1}^{n} A(-t_j),$$

represented by left multiplication by the matrix M, then applying $\underline{\operatorname{Hom}}(-, A)$ gives a left module homomorphism $\phi^* : \underline{\operatorname{Hom}}_A(Q, A) \to \underline{\operatorname{Hom}}_A(P, A)$, which we can canonically identify with a graded left-module map $\phi^* : \bigoplus_{j=1}^{n} A(t_j) \to \bigoplus_{i=1}^{m} A(s_i)$, using part (1). Then ϕ^* is given by right multiplication by the same matrix M.

PROOF. The reader is asked to verify this as Exercise 1.6.5.   □

We now revisit Example 1.5.3 and calculate $\underline{\operatorname{Ext}}_A^i(k, A)$.

EXAMPLE 1.5.6. Let $A = k\langle x, y\rangle/(yx - xy - x^2)$ be the Jordan plane, and consider the graded free resolution (1.E) of k_A. Applying $\underline{\operatorname{Hom}}_A(-, A)$ and using Lemma 1.5.5, we get the complex of left modules

$$0 \leftarrow A(2) \overset{d_1^*=\binom{-y-x}{x}}{\longleftarrow} A(1)^{\oplus 2} \overset{d_0^*=(x\ y)}{\longleftarrow} A \leftarrow 0,$$

where as described above, the free modules are row vectors and the maps are right multiplication by the matrices. By entirely analogous arguments as in Example 1.5.3, we get that this complex is exact except at the $A(2)$ spot, where d_1^* has image $A_{\geq 1}(2)$. Thus $\underline{\operatorname{Ext}}^i(k_A, A) = 0$ for $i \neq 2$, and $\underline{\operatorname{Ext}}^2(k_A, A) \cong {}_A k(2)$.

Notice that this actually shows that the complex above is a graded free resolution of the left A-module ${}_A k(2)$. We will see in the next section that this is the key property of an Artin–Schelter regular algebra: the minimal free resolutions of ${}_A k$ and k_A are interchanged (up to shift of grading) by applying $\underline{\operatorname{Hom}}_A(-, A)$.

Let A be a finitely graded algebra. We will be primarily concerned with the category of graded modules over A, and so we will work exclusively throughout with the following graded versions of projective and global dimension. Given a $\mathbb{Z}$-graded right A-module M, its *projective dimension* proj. dim(M) is the minimal n such that there is a graded projective resolution of M of length n, that is

$$0 \to P_n \overset{d_{n-1}}{\to} P_{n-1} \to \cdots \overset{d_1}{\to} P_1 \overset{d_0}{\to} P_0 \to 0,$$

in which all projectives P_i are graded and all of the homomorphisms of the complex, as well as the augmentation map, are graded homomorphisms. If no such n exists, then proj. dim$(M) = \infty$. The *right global dimension* of A, r. gl. dim(A), is defined to be the supremum of the projective dimensions of all $\mathbb{Z}$-graded right A-modules, and the *left global dimension* l. gl. dim(A) is defined analogously in terms of graded left modules.

PROPOSITION 1.5.7. *Let A be a finitely graded k-algebra. Then*

$$\text{r. gl. dim}(A) = \text{proj. dim}(k_A) = \text{proj. dim}(_A k) = \text{l. gl. dim}(A),$$

and this number is equal to the length of the minimal graded free resolution of k_A.

PROOF. The reader familiar with the basic properties of Tor can attempt this as Exercise 1.6.6, or see [9, Section 2] for the basic idea.   □

Because of the result above, for a finitely graded algebra we will write the common value of r. gl. dim(A) and l. gl. dim(A) as gl. dim(A) and call this the *global dimension* of A. For example, the result above, together with Example 1.5.3, shows that the Jordan plane has global dimension 2.

1.6. Exercise Set 1.

EXERCISE 1.6.1. Let $A = k[x_1, \ldots, x_n]$ be a commutative polynomial algebra with weights deg $x_i = d_i$, and let $F = k\langle x_1, \ldots, x_n \rangle$ be a free associative algebra with weights deg $x_i = d_i$.
(1) Prove that $h_A(t) = 1/p(t)$ where $p(t) = \prod_{i=1}^{n}(1 - t^{d_i})$. (Hint: induct on n).
(2) Prove that $h_F(t) = 1/q(t)$ where $q(t) = 1 - \sum_{i=1}^{n} t^{d_i}$. (Hint: write $h_F(t) = \sum a_i t^i$ and prove by counting that $a_j = \sum_{i=1}^{n} a_{j-d_i}$).

EXERCISE 1.6.2. Prove Lemma 1.3.2(1). Also, use Lemma 1.3.2 to prove that Example 1.4.3 is a noetherian domain (consider the element $xy - yx$).

EXERCISE 1.6.3. Complete the proof of Theorem 1.4.1. Namely, prove that the set of reduction-unique elements is a k-subspace and that the operator red($-$) is linear on this subspace, and prove the implications (3) $\implies$ (2) and (2) $\implies$ (1).

EXERCISE 1.6.4. Verify the calculations in Example 1.4.4, find explicitly the set of reduced words with respect to the Grobner basis $g_1, \ldots, g_5$, and show that the Hilbert series of the algebra is $1/(1 - t)^3$.

EXERCISE 1.6.5. Prove Lemma 1.5.5.

EXERCISE 1.6.6. (This exercise assumes knowledge of the basic properties of Tor). Prove Proposition 1.5.7, using the following outline.
Assume that the global dimension of A is equal to the supremum of the projective dimensions of all finitely generated graded A-modules. (The general argument that it suffices to consider finitely generated modules in the computation of global dimension is due to Auslander, see [15, Theorem 1].) In particular, we only need to look at left bounded graded A-modules.
(1) Show that if M_A is left bounded, then its minimal graded free resolution $P_\bullet$ is isomorphic to a direct summand of any graded projective resolution. Thus the length of the minimal graded free resolution is equal to proj. dim(M_A).
(2) Show that if M_A is left bounded, then

$$\text{proj. dim}(M_A) = \max\{i \mid \text{Tor}_A^i(M_A, {}_A k) \neq 0\}.$$

(3) Show that if M is left bounded then $\mathrm{proj.\,dim}(M_A) \leq \mathrm{proj.\,dim}(_Ak)$. A left-sided version of the same argument shows that $\mathrm{proj.\,dim}(_AN) \leq \mathrm{proj.\,dim}(k_A)$ for any left bounded graded left module N.

(4) Conclude that $\mathrm{proj.\,dim}(_Ak) = \mathrm{proj.\,dim}(k_A)$ and that this is the value of both $\mathrm{l.\,gl.\,dim}(A)$ and $\mathrm{r.\,gl.\,dim}(A)$.

EXERCISE 1.6.7. In this exercise, the reader will use the computer algebra system GAP to get a feel for how one can calculate noncommutative Gröbner bases using software. To do these exercises you need a basic GAP installation, together with the noncommutative Gröbner bases package (GBNP). After opening a GAP window, the following code should be run once:

```
LoadPackage("GBNP");
SetInfoLevel(InfoGBNP,0);
SetInfoLevel(InfoGBNPTime,0);
```

(1) Type in and run the following code, which shows that the three defining relations of a quantum polynomial ring are already a Gröbner basis. It uses degree lex order with $x < y < z$, and stops after checking all overlaps (and possibly adding new relations if necessary) up to degree 12. The vector $[1, 1, 1]$ is the list of weights of the variables.

```
A:=FreeAssociativeAlgebraWithOne(Rationals, "x", "y", "z");
x:=A.x;; y:=A.y;; z:=A.z;; o:=One(A);;
uerels:=[y*x-2*x*y, z*y-3*y*z, z*x-5*x*z];
uerelsNP:=GP2NPList(uerels);;
PrintNPList(uerelsNP);
GBNP.ConfigPrint(A);
GB:=SGrobnerTrunc(uerelsNP, 12, [1,1,1]);;
PrintNPList(GB);
```

(2) Change the first three lines of the code in (1) to

```
A:=FreeAssociativeAlgebraWithOne(Rationals, "y", "x");
x:=A.x;; y:=A.y;; o:=One(A);;
uerels:=[y*x-x*y-x*x];
```

and run it again. This attempts to calculate a Gröbner basis for the Jordan plane with the ordering $y < x$ on the variables. Although it stops in degree 12, the calculation suggests that this calculation would continue infinitely, and the pattern to the new relations produced suggest what the infinite Gröbner basis produced by the process is. (One can easily prove for certain that this infinite set of relations is inded a Gröbner basis by induction.) Calculate the corresponding k-basis of reduced words and verify that it gives the same Hilbert series $1/(1-t)^2$ we already know.

(3) Change the third line in code for (1) to

```
uerels:=[2*y*x+3*x*y+5*z*z, 2*x*z+3*z*x+5*y*y, 2*z*y+3*y*z+5*x*x];
```

and change the number 12 in the seventh line to 8. Run the code again and scroll through the output. Notice how quickly the new relations added in the Diamond

Lemma become unwieldy and complicated. Certainly the calculation suggests that the algorithm is unlikely to terminate, and it is hard to discern any pattern in the new relations added by the algorithm. This shows the resistance of the Sklyanin algebra to Gröbner basis methods.

(4) Replace the third line in (1) with

```
uerels:=[z*x+x*z, y*z+z*y, z*z-x*x-y*y];
```

and run the algorithm again. This is an example where the original relations are not a Gröbner basis, but the algorithm terminates with a finite Gröbner basis. This basis can be used to prove that the algebra has Hilbert series $1/(1-t)^3$, similarly as in Exercise 1.6.4.

2. Artin–Schelter regular algebras

With the background of Section 1 in hand, we are now ready to discuss regular algebras.

DEFINITION 2.0.8. Let A be a finitely graded k-algebra, and let $k = A/A_{\geq 1}$ be the trivial module. We say that A is *Artin–Schelter regular* if

(1) $\mathrm{gl.\,dim}(A) = d < \infty$;

(2) $\mathrm{GKdim}(A) < \infty$; and

(3) as left A-modules,

$$\underline{\mathrm{Ext}}^i_A(k_A, A_A) \cong \begin{cases} 0 & \text{if } i \neq d, \\ {}_A k(\ell) & \text{if } i = d, \end{cases}$$

for some shift of grading $\ell \in \mathbb{Z}$.

We call an algebra satisfying (1) and (3) (but not necessarily (2)) *weakly Artin–Schelter regular*.

This definition requires some further discussion. From now on, we will abbreviate "Artin–Schelter" to "AS", or sometimes even omit the "AS", but the reader should be aware that there are other notions of regularity for rings. We want to give a more intuitive interpretation of condition (3), which is also known as the *(Artin–Schelter) Gorenstein* condition. By Proposition 1.5.7, the global dimension of A is the same as the length of the minimal graded free resolution of k_A. Thus condition (1) in Definition 2.0.8 is equivalent to the requirement that the minimal graded free resolution of the trivial module k_A has the form

$$0 \to P_d \to P_{d-1} \to \cdots \to P_1 \to P_0 \to 0, \tag{2.A}$$

for some graded free modules P_i. It is not immediately obvious that the P_i must have finite rank, unless we also happen to know that A is noetherian. It is a fact, however, that for a weakly AS-regular algebra the P_i must have finite rank ([218, Proposition 3.1]). This is not a particularly difficult argument, but we will simply assume this result here, since in the main examples we consider it will obviously hold by direct computation. The minimality of the resolution means that the matrices representing the maps in the complex $P_\bullet$ have all of their entries in $A_{\geq 1}$ (Lemma 1.5.4).

To calculate $\underline{\mathrm{Ext}}_A^i(k_A, A_A)$, we apply $\underline{\mathrm{Hom}}_A(-, A)$ to (2.A) and take homology. Thus putting $P_i^* = \underline{\mathrm{Hom}}_A(P_i, A)$, we get a complex of graded free left modules

$$0 \leftarrow P_d^* \leftarrow P_{d-1}^* \leftarrow \cdots \leftarrow P_1^* \leftarrow P_0^* \leftarrow 0. \tag{2.B}$$

By Lemma 1.5.5, the matrices representing the maps of free left modules in this complex are the same as in (2.A) and so also have all of their entries in $A_{\geq 1}$. The AS-Gorenstein condition (3) demands that the homology of this complex should be 0 in all places except at P_d^*, where the homology should be isomorphic to $_A k(\ell)$ for some ℓ. This is equivalent to (2.B) being a graded free resolution of $_A k(\ell)$ as a left module. Moreover, this is necessarily a minimal graded free resolution because of the left-sided version of Lemma 1.5.4. Equivalently, applying the shift operation $(-\ell)$ to the entire complex,

$$0 \leftarrow P_d^*(-\ell) \leftarrow P_{d-1}^*(-\ell) \leftarrow \cdots \leftarrow P_1^*(-\ell) \leftarrow P_0^*(-\ell) \leftarrow 0$$

is a minimal graded free resolution of $_A k$.

Thus parts (1) and (3) of the definition of AS-regular together assert that the minimal free resolutions of k_A and $_A k$ are finite in length, consist of finite rank free modules ([218, Proposition 3.1]), and that the operation $\underline{\mathrm{Hom}}_A(-, A_A)$ together with a shift of grading sends the first to the second. Furthermore, the situation is automatically symmetric: one may also show that $\underline{\mathrm{Hom}}_A(-, {}_A A)$ sends a minimal free resolution of $_A k$ to a shift of a minimal free resolution of k_A, or equivalently that $\underline{\mathrm{Ext}}_A^i({}_A k, {}_A A)$ is trivial for $i \neq d$ and is isomorphic to $k_A(\ell)$ as a right module if $i = d$ (Exercise 2.4.1).

When we talk about the dimension of an AS-regular algebra, we mean its global dimension. However, all known examples satisfying Definition 2.0.8 have $\mathrm{gl.\,dim}(A) = \mathrm{GKdim}(A)$, so there is no chance for confusion. The term weakly AS-regular is not standard. Rather, some papers simply omit condition (2) in the definition of AS-regular. It is useful for us to have distinct names for the two concepts here. There are many more examples of weakly AS-regular algebras than there are of AS-regular algebras. For example, see Exercise 2.4.4 below.

Condition (2) of Definition 2.0.8 can also be related to the minimal free resolution of k: if the ranks and shifts of the P_i occurring in (2.A) are known, then one may determine the Hilbert series of A, as follows. Recall that when we have a finite length complex of finite-dimensional vector spaces over k, say

$$C: \quad 0 \to V_n \to V_{n-1} \to \cdots \to V_1 \to V_0 \to 0,$$

then the alternating sum of the dimensions of the V_i is the alternating sum of the dimensions of the homology groups, in other words

$$\sum (-1)^i \dim_k V_i = \sum (-1)^i \dim_k H_i(C). \tag{2.C}$$

Now apply (2.C) to the restriction to degree n of the minimal free resolution (2.A) of k_A, obtaining

$$\sum_i (-1)^i \dim_k (P_i)_n = \begin{cases} 0 & \text{if } n \neq 0, \\ 1 & \text{if } n = 0. \end{cases}$$

These equations can be joined together into the single power series equation

$$1 = h_k(t) = h_{P_0}(t) - h_{P_1}(t) + \cdots + (-1)^d h_{P_d}(t). \tag{2.D}$$

Moreover, writing $P_i = \bigoplus_{j=1}^{m_i} A(-s_{i,j})$ for some integers m_i and $s_{i,j}$, then because $h_{A(-s)}(t) = h_A(t) t^s$ for any shift s, we have $h_{P_i}(t) = \sum_j h_A(t) t^{s_{i,j}}$. Thus (2.D) can be solved for $h_A(t)$, yielding

$$h_A(t) = 1/p(t), \quad \text{where } p(t) = \sum_{i,j} (-1)^i t^{s_{i,j}}. \tag{2.E}$$

Note also that $P_0 = A$, while all other P_i are sums of negative shifts of A. Thus the constant term of $p(t)$ is equal to 1.

The GK-dimension of A is easily determined from knowledge of $p(t)$, as follows.

LEMMA 2.0.9. *Suppose that A is a finitely graded algebra with Hilbert series $h_A(t) = 1/p(t)$ for some polynomial $p(t) \in \mathbb{Z}[t]$ with constant term 1. Then precisely one of the following occurs:*

(1) *All roots of $p(t)$ in $\mathbb{C}$ are roots of unity, $\mathrm{GKdim}(A) < \infty$, and $\mathrm{GKdim}(A)$ is equal to the multiplicity of 1 as a root of $p(t)$; or else*

(2) *$p(t)$ has a root r with $|r| < 1$, and A has exponential growth; in particular, $\mathrm{GKdim}(A) = \infty$.*

PROOF. This is a special case of [218, Lemma 2.1, Corollary 2.2]. Since $p(t)$ has constant term 1, we may write $p(t) = \prod_{i=1}^{m} (1 - r_i t)$, where the $r_i \in \mathbb{C}$ are the reciprocals of the roots of $p(t)$.

Suppose first that $|r_i| \leq 1$ for all i. Since the product of the r_i is the leading coefficient of $p(t)$ (up to sign), it is an integer. This forces $|r_i| = 1$ for all i. This also implies that the leading coefficient of $p(t)$ must be ± 1, so each root $1/r_i$ of $p(t)$ is an algebraic integer. Since the geometric series $1/(1 - r_i t) = 1 + r_i t + r_i^2 t^2 + \ldots$ has coefficients of norm 1, and $1/p(t)$ is the product of m such series, it easily follows that $1/p(t) = \sum a_n t^n$ has $|a_n| \leq \binom{n+m-1}{m-1}$ for all $n \geq 0$, so $\mathrm{GKdim}(A) \leq m$. It follows from a classical theorem of Kronecker that algebraic integers on the unit circle are roots of unity. References for this fact and for the argument that the precise value of $\mathrm{GKdim}(A)$ is determined by the multiplicity of 1 as a root of $p(t)$ may be found in [218, Corollary 2.2].

If instead $|r_i| > 1$ for some i, then we claim that A has exponential growth. By elementary complex analysis the series $\sum a_n t^n$ has radius of convergence $r = \left(\limsup_{n \to \infty} |a_n|^{1/n} \right)^{-1}$. Since A is finitely generated as an algebra, A is generated by $V = A_0 \oplus \cdots \oplus A_d$ for some $d \geq 1$. Then it is easy to prove that $A_n \subseteq V^n$ for all $n \geq 1$. If A has subexponential growth, then by definition we have $\limsup_{n \to \infty} (\dim_k V^n)^{1/n} = 1$, and this certainly forces $\limsup_{n \to \infty} a_n^{1/n} \leq 1$. In particular, the radius of convergence of $h_A(t)$ is at least as large as 1, and so it converges at $1/r_i$. Plugging in $1/r_i$ into $p(t) h_A(t) = 1$ now gives $0 = 1$, a contradiction. $\square$

EXAMPLE 2.0.10. Consider the Hilbert series calculated in Exercise 1.6.1. Applying Lemma 2.0.9 shows that a weighted polynomial ring in n variables has GK-dimension

n, and that a weighted free algebra in more than one variable has exponential growth (it is easy to see from the intermediate value theorem that a polynomial $1 - \sum_{i=1}^{n} t^{d_i}$ has a real root in the interval $(0, 1)$ when $n \geq 2$.)

2.1. Examples. Our next goal is to present some explicit examples of AS-regular algebras. One interpretation of regular algebras is that they are the noncommutative graded algebras most analogous to commutative (weighted) polynomial rings. This intuition is reinforced by the following observation.

EXAMPLE 2.1.1. Let $A = k[x_1, \ldots, x_n]$ be a commutative polynomial ring in n variables, with any weights $\deg x_i = d_i$. Then the minimal free resolution of k is

$$0 \to A^{r_n} \to A^{r_{n-1}} \to \cdots \to A^{r_1} \to A^{r_0} \to 0,$$

where the basis of the free A-module A^{r_j} is naturally indexed by the jth wedge in the exterior algebra in n symbols $e_1, e_2, \ldots, e_n$, that is,

$$\Lambda^j = k\{e_{i_1} \wedge e_{i_2} \wedge \cdots \wedge e_{i_j} \mid i_1 < i_2 < \cdots < i_j\},$$

and $A^{r_j} = \Lambda^j \otimes_k A$. In particular, $r_j = \binom{n}{j}$. The differentials are defined on basis elements of the free modules by

$$d(e_{i_1} \wedge \cdots \wedge e_{i_j}) = \sum_k (-1)^{k+1} (e_{i_1} \wedge \cdots \wedge \widehat{e_{i_k}} \wedge \cdots \wedge e_{i_j}) x_{i_k},$$

where as usual $\widehat{e_{i_k}}$ means that element is removed. The various shifts on the summands of each A^{r_j} which are needed to make this a graded complex are not indicated above, but are clearly uniquely determined by the weights of the x_i.

The complex $P_\bullet$ above is well-known and is called the *Koszul complex*: it is standard that it is a free resolution of k_A [54, Corollary 1.6.14]. It is also graded once the appropriate shifts are introduced, and then it is minimal by Lemma 1.5.4. The complex $\underline{\mathrm{Hom}}_A(P_\bullet, A_A)$ is isomorphic to a shift of the Koszul complex again [54, Proposition 1.6.10] (of course, left and right modules over A are canonically identified), and thus the weighted polynomial ring $k[x_1, \ldots, x_n]$ is AS-regular.

In fact, it is known that a commutative AS-regular algebra must be isomorphic to a weighted polynomial ring; see [54, Exercise 2.2.25].

EXAMPLE 2.1.2. The Jordan and quantum planes from Examples 1.3.1 are AS-regular of dimension 2. Indeed, the calculations in Examples 1.5.3 and 1.5.6 immediately imply that the Jordan plane is regular, and the argument for the quantum plane is very similar. More generally, the quantum polynomial ring in n variables from Examples 1.3.1 is regular; we won't give a formal proof here, but the explicit resolution in case $n = 3$ appears in Example 2.3.2 below.

Before we give some examples of AS-regular algebras of dimension 3, it is useful to prove a general result about the structure of the first few terms in a minimal free resolution of k_A. We call a set of generators for an k-algebra *minimal* if no proper subset generates the algebra. Similarly, a minimal set of generators of an ideal of the free algebra is one such that no proper subset generates the same ideal (as a 2-sided ideal).

LEMMA 2.1.3. *Let $F = k\langle x_1, \ldots, x_n \rangle$ be a weighted free algebra, where $\deg x_i = e_i$ for all i, and let $A = F/I$ for some homogeneous ideal I. Let $\{g_j \mid 1 \leq j \leq r\}$ be a set of elements in I, with $\deg g_j = s_j$, and write $g_j = \sum_i x_i g_{ij}$ in F. Let $\overline{x}$ be the image in A of an element x of F. Then the complex*

$$\bigoplus_{j=1}^{r} A(-s_j) \xrightarrow{d_1 = \left(\overline{g_{ij}}\right)} \bigoplus_{i=1}^{n} A(-e_i) \xrightarrow{d_0 = \left(\overline{x_1} \quad \cdots \quad \overline{x_n}\right)} A \to 0 \qquad (2.F)$$

is the beginning of a minimal free resolution of k_A if and only if the $\{\overline{x_i}\}$ minimally generate A as a k-algebra and the $\{g_j\}$ minimally generate the ideal I of relations.

PROOF. It is straightforward to check that a set of homogeneous elements of $A_{\geq 1}$ generates $A_{\geq 1}$ as a right ideal if and only if it is a generating set for A as a k-algebra. Thus d_0 has image $A_{\geq 1}$ and the homology at the A spot is k_A. It also follows that d_0 is a minimal surjection onto its image $A_{\geq 1}$ if and only if the $\{\overline{x_i}\}$ are a minimal generating set for A as a k-algebra.

Now consider the right A-submodule $X = \ker d_0 = \{(h_1, \ldots, h_n) \mid \sum_i \overline{x_i} h_i = 0\}$ of $\bigoplus_{i=1}^{n} A(-e_i)$. We also define $Y = \{(h_1, \ldots, h_n) \mid \sum_i x_i h_i \in I\} \subseteq \bigoplus_{i=1}^{n} F(-e_i)$. Suppose that we are given a set of elements $\{f_{ij} \mid 1 \leq i \leq n, 1 \leq j \leq s\} \subseteq F$ such that $f_j = \sum_i x_i f_{ij} \in I$ for all j. We leave it to the reader to prove that the following are all equivalent:

(1) $\{f_j\}$ generates I as a 2-sided ideal of F.
(2) $I = x_1 I + \cdots + x_n I + \sum_j f_j F$.
(3) $Y = I^{\oplus n} + \sum_j (f_{1j}, \ldots, f_{nj})F$.
(4) $X = \sum_j (\overline{f_{1j}}, \ldots, \overline{f_{nj}})A$.

It follows that the set $\{g_j\}$ generates I as a 2-sided ideal if and only if $\operatorname{im} d_1 = X = \ker d_0$. Then $\{g_j\}$ minimally generates I if and only if d_1 is a minimal surjection onto $\ker d_0$. $\square$

REMARK 2.1.4. As remarked earlier, the graded free modules in the minimal graded free resolution of k_A for a weakly AS-regular algebra A must be of finite rank [218, Proposition 3.1]. Thus Lemma 2.1.3, together with the uniqueness of minimal graded free resolutions up to isomorphism (Lemma 1.5.2), shows that any weakly AS-regular algebra is finitely presented.

Proving that an algebra of dimension 3 is AS-regular is often straightforward if it has a nice Gröbner basis.

EXAMPLE 2.1.5. Fix nonzero $a, b \in k$, and let

$$A = k\langle x, y, z \rangle / (zy + (1/b)x^2 - (a/b)yz, zx - (b/a)xz - (1/a)y^2, yx - (a/b)xy).$$

We leave it to the reader to show that under degree lex order with $x < y < z$, the single overlap ambiguity zyx resolves, and so the algebra A has a k-basis $\{x^i y^j z^k \mid i, j, k \geq 0\}$ and Hilbert series $1/(1-t)^3$.

We claim that a minimal graded free resolution of k_A has the form

$$0 \to A(-3) \xrightarrow{u} A(-2)^{\oplus 3} \xrightarrow{M} A(-1)^{\oplus 3} \xrightarrow{v} A \to 0,$$

where

$$u = \begin{pmatrix} x \\ (-a/b)y \\ z \end{pmatrix}; \quad M = \begin{pmatrix} (1/b)x & -(b/a)z & -(a/b)y \\ -(a/b)z & -(1/a)y & x \\ y & x & 0 \end{pmatrix}; \quad \text{and } v = \begin{pmatrix} x & y & z \end{pmatrix}.$$

Here, as usual the maps are left multiplications by the matrices and the free modules are represented by column vectors. It is straightforward, as always, to check this is a complex. It is easy to see that $\{x, y, z\}$ is a minimal generating set for A, and that the three given relations are a minimal generating set of the ideal of relations. Thus exactness at the A and $A(-1)^{\oplus 3}$ spots follows from Lemma 2.1.3. Exactness at the $A(-3)$ spot requires the map $A(-3) \to A(-2)^{\oplus 3}$ to be injective. If this fails, then there is $0 \neq w \in A$ such that $zw = yw = xw = 0$. However, the form of the k-basis above shows that $xw = 0$ implies $w = 0$.

We lack exactness only possibly at the $A(-2)^{\oplus 3}$ spot. We can apply a useful general argument in this case: if an algebra is known in advance to have the correct Hilbert series predicted by a potential free resolution of k_A, and the potential resolution is known to be exact except in one spot, then exactness is automatic at that spot. In more detail, if our complex has a homology group at $A(-2)^{\oplus 3}$ with Hilbert series $q(t)$, then a similar computation as the one preceding Lemma 2.0.9 will show that $h_A(t) = (1 - q(t))/(1-t)^3$. But we know that $h_A(t) = 1/(1-t)^3$ by the Gröbner basis computation, and this forces $q(t) = 0$. Thus the complex is a free resolution of k_A, as claimed.

Now applying $\underline{\mathrm{Hom}}_A(-, A_A)$, using Lemma 1.5.5 and shifting by (-3), we get the complex of left modules

$$0 \leftarrow A \xleftarrow{u} A(-1)^{\oplus 3} \xleftarrow{M} A(-2)^{\oplus 3} \xleftarrow{v} A(-3) \leftarrow 0,$$

where the free modules are now row vectors and the maps are given by right multiplication by the matrices. The proof that this is a free resolution of $_Ak$ is very similar, using a left-sided version of Lemma 2.1.3, and noting that the form of the k-basis shows that $wz = 0$ implies $w = 0$, to get exactness at $A(-3)$.

Thus A is a regular algebra of global dimension 3 and GK-dimension 3.

Not every regular algebra of dimension 3 which is generated by degree 1 elements has Hilbert series $1/(1-t)^3$. Here is another example.

EXAMPLE 2.1.6. Let $A = k\langle x, y\rangle/(yx^2 - x^2y, y^2x - xy^2)$, as in Example 1.4.3. As shown there, A has k-basis $\{x^i(yx)^jy^k \mid i, j, k \geq 0\}$ and thus the Hilbert series of A is $1/(1-t)^2(1-t^2)$. We claim that the minimal graded free resolution of k_A has the following form:

$$0 \to A(-4) \xrightarrow{\binom{y}{-x}} A(-3)^{\oplus 2} \xrightarrow{\begin{pmatrix} -xy & -y^2 \\ x^2 & yx \end{pmatrix}} A(-1)^{\oplus 2} \xrightarrow{(x\ y)} A \to 0.$$

This is proved in an entirely analogous way as in Example 2.1.5. The proof of the AS-Gorenstein condition also follows in the same way as in that example, and so A is AS-regular of global and GK-dimension 3. Notice from this example that the integer ℓ in Definition 2.0.8 can be bigger than the dimension of the regular algebra in general.

Next, we give an example which satisfies conditions (1) and (2) of Definition 2.0.8, but not the AS-Gorenstein condition (3).

EXAMPLE 2.1.7. Let $A = k\langle x, y\rangle/(yx)$. The Diamond Lemma gives that this algebra has Hilbert series $1/(1-t)^2$ and a k-basis $\{x^i y^j \mid i, j \geq 0\}$, so $\mathrm{GKdim}(A) = 2$. The basis shows that left multiplication by x is injective; together with Lemma 2.1.3 we conclude that

$$0 \to A(-2) \xrightarrow{\binom{0}{x}} A(-1)^{\oplus 2} \xrightarrow{(x\ y)} A \to 0$$

is a minimal free resolution of k_A. In particular, Proposition 1.5.7 yields gl. dim $A = 2$.

Applying $\underline{\mathrm{Hom}}(-, A_A)$ to this free resolution and applying the shift (-2) gives the complex of left A-modules

$$0 \leftarrow A \xleftarrow{\binom{0}{x}} A(-1)^{\oplus 2} \xleftarrow{(x\ y)} A(-2) \leftarrow 0,$$

using Lemma 1.5.5. It is apparent that this is not a minimal resolution of $_A k$: the fact that the entries of $\binom{0}{x}$ do not span $kx + ky$ prevents this complex from being isomorphic to the minimal graded free resolution constructed by the left-sided version of Lemma 2.1.3. More explicitly, the homology at the A spot is A/Ax, and $Ax \neq A_{\geq 1}$ (in fact $Ax = k[x]$). So A fails the AS-Gorenstein condition and hence is not AS-regular. It is obvious that A is not a domain, and one may also check that A is not noetherian (Exercise 2.4.3).

The example above demonstrates the importance of the Gorenstein condition in the noncommutative case. There are many finitely graded algebras of finite global dimension with bad properties, and for some reason adding the AS-Gorenstein condition seems to eliminate these bad examples. It is not well-understood why this should be the case. In fact, the answers to the following questions are unknown:

QUESTION 2.1.8. Is an AS-regular algebra automatically noetherian? Is an AS-regular algebra automatically a domain? Must an AS-regular algebra have other good homological properties, such as Auslander-regularity and the Cohen–Macaualay property?

We omit the definitions of the final two properties; see, for example, [164]. The questions above have a positive answer for all AS-regular algebras of dimension at most 3 (which are classified), and for all known AS-regular algebras of higher dimension. It is known that if a regular algebra of dimension at most 4 is noetherian, then it is a domain [10, Theorem 3.9], but there are few other results of this kind.

In the opposite direction, one could hope that given a finitely graded algebra of finite global dimension, the AS-Gorenstein condition might follow from some other more basic assumption, such as the noetherian property. There are some results along these lines: for example, Stephenson and Zhang proved that a finitely graded noetherian algebra which is quadratic (that is, with relations of degree 2) and of global dimension 3 is automatically AS-regular [219, Corollary 0.2]. On the other hand, the author and Sierra have recently given examples of finitely graded quadratic algebras of global dimension 4 with Hilbert series $1/(1-t)^4$ which are noetherian domains, but which fail to be AS-regular [187].

2.2. Classification of regular algebras of dimension 2. We show next that AS-regular algebras A of dimension 2 are very special. For simplicity, we only study the case that A is generated as an k-algebra by A_1, in which case we say that A is *generated in degree 1*. The classification in the general case is similar (Exercise 2.4.5).

THEOREM 2.2.1. *Let A be a finitely graded algebra which is generated in degree 1, such that* gl. $\dim(A) = 2$.

(1) *If A is weakly AS-regular, then $A \cong F/(g)$, where $F = k\langle x_1, \ldots, x_n \rangle$ and $g = \sum_{i=1}^{n} x_i \tau(x_i)$ for some bijective linear transformation $\tau \in \mathrm{End}_k(F_1)$ and some $n \geq 2$.*
(2) *If A is AS-regular, then $n = 2$ above and A is isomorphic to either the Jordan or quantum plane of Examples 1.3.1.*

PROOF. (1) Since A is weakly regular, it finitely presented (Remark 2.1.4). Thus $A \cong F/(g_1, \ldots, g_r)$, where $F = k\langle x_1, \ldots, x_n \rangle$ with $\deg x_i = 1$, and $\deg g_i = s_i$. We may assume that the x_i are a minimal set of generators and that the g_i are a minimal set of homogeneous relations. The minimal resolution of k_A has length 2, and so by Lemma 2.1.3, it has the form

$$0 \to \bigoplus_{j=1}^{r} A(-s_j) \xrightarrow{(g_{ij})} \bigoplus_{i=1}^{n} A(-1) \xrightarrow{\begin{pmatrix} x_1 & x_2 & \cdots & x_n \end{pmatrix}} A \to 0,$$

where $g_j = \sum_i x_i g_{ij}$. Applying $\underline{\mathrm{Hom}}_A(-, A_A)$ gives

$$0 \leftarrow \bigoplus_{j=1}^{r} A(s_j) \xleftarrow{(g_{ij})} \bigoplus_{i=1}^{n} A(1) \xleftarrow{\begin{pmatrix} x_1 & x_2 & \cdots & x_n \end{pmatrix}} A \leftarrow 0, \qquad (2.\mathrm{G})$$

which should be the minimal graded free resolution of $_A k(\ell)$ for some ℓ, by the AS-Gorenstein condition. On the other hand, by the left-sided version of Lemma 2.1.3, the minimal free resolution of $_A k$ has the form

$$0 \leftarrow A \xleftarrow{\begin{pmatrix} x_1 \\ x_2 \\ \cdots \\ x_n \end{pmatrix}} \bigoplus_{i=1}^{n} A(-1) \xleftarrow{(h_{ji})} \bigoplus_{j=1}^{r} A(-s_j) \leftarrow 0, \qquad (2.\mathrm{H})$$

where $g_j = \sum_i h_{ji} x_i$. Since minimal graded free resolutions are unique up to isomorphism of complexes (Lemma 1.5.2(2)), comparing the ranks and shifts of the free modules in (2.G) and (2.H) immediately implies that $r = 1$, $s_1 = 2$, and $\ell = 2$. Thus the minimal free resolution of k_A has the form

$$0 \to A(-2) \xrightarrow{\begin{pmatrix} y_1 \\ \cdots \\ y_n \end{pmatrix}} \bigoplus_{i=1}^{n} A(-1) \xrightarrow{\begin{pmatrix} x_1 & \cdots & x_n \end{pmatrix}} A \to 0 \qquad (2.I)$$

for some $y_i \in A_1$, and there is a single relation $g = \sum_{i=1}^{n} x_i y_i$ such that $A \cong F/(g)$. Once again using that (2.G) is a free resolution of $_A k(2)$, the $\{y_i\}$ must span A_1; otherwise $\sum A y_i \neq A_{\geq 1}$. Thus the $\{y_i\}$ are a basis of A_1 also and $g = \sum x_i \tau(x_i)$ for some linear bijection τ of F_1. If $n = 1$, then we have $A \cong k[x]/(x^2)$, which is well-known to have infinite global dimension (or notice that the complex (2.I) is not exact at $A(-2)$ in this case), a contradiction. Thus $n \geq 2$.

(2) The shape of the free resolution in part (1) implies by (2.E) that $h_A(t) = 1/(1 - nt + t^2)$. By Lemma 2.0.9, A has finite GK-dimension if and only if all of the roots of $(1 - nt + t^2)$ are roots of unity. It is easy to calculate that this happens only if $n = 1$ or 2, and $n = 1$ is already excluded by part (1). Thus $n = 2$, and $A \cong \langle x, y \rangle/(g)$, where g has the form $x\tau(x) + y\tau(y)$ for a bijection τ. By Exercise 2.4.3 below, any such algebra is isomorphic to the quantum plane or the Jordan plane. Conversely, the regularity of the quantum and Jordan planes was noted in Example 2.1.2. $\qquad \square$

In fact, part (1) of the theorem above also has a converse: any algebra of the form $A = k\langle x_1, \ldots, x_n \rangle/(\sum_i x_i \tau(x_i))$ with $n \geq 2$ is weakly AS-regular (Exercise 2.4.4). When $n \geq 3$, these are non-noetherian algebras of exponential growth which have other interesting properties [250].

2.3. First steps in the classification of regular algebras of dimension 3. The classification of AS-regular algebras of dimension 3 was a major achievement. The basic framework of the classification was laid out by Artin and Schelter in [5], but to complete the classification required the development of the geometric techniques in the work of Artin, Tate, and Van den Bergh [9, 10]. We certainly cannot give the full details of the classification result in these notes, but we present some of the easier first steps now, and give a glimpse of the main idea of the rest of the proof at the end of Section 3.

First, a similar argument as in the global dimension 2 case shows that the possible shapes of the free resolutions of k_A for regular algebras A of dimension 3 are limited. We again focus on algebras generated in degree 1 for simplicity. For the classification of regular algebras of dimension 3 with generators in arbitrary degrees, see the work of Stephenson [216, 217].

LEMMA 2.3.1. *Let A be an AS-regular algebra of global dimension 3 which is generated in degree 1. Then exactly one of the following holds:*

(1) $A \cong k\langle x_1, x_2, x_3\rangle/(f_1, f_2, f_3)$, where the f_i have degree 2, and $h_A(t) = 1/(1-t)^3$;

   or

(2) $A \cong k\langle x_1, x_2\rangle/(f_1, f_2)$, where the f_i have degree 3, and $h_A(t) = 1/(1-t)^2(1-t^2)$.

PROOF. As in the proof of Theorem 2.2.1, A is finitely presented. Let $A \cong k\langle x_1, \ldots, x_n\rangle/(f_1, \ldots, f_r)$ be a minimal presentation. By Lemma 2.1.3, the minimal free resolution of k_A has the form

$$0 \to P_3 \to \bigoplus_{i=1}^{r} A(-s_i) \to \bigoplus_{i=1}^{n} A(-1) \to A \to 0,$$

where s_i is the degree of the relation f_i. A similar argument as in the proof of Theorem 2.2.1 using the AS-Gorenstein condition implies that the ranks and shifts of the graded free modules appearing must be symmetric: namely, necessarily $r = n$, $P_3 = A(-\ell)$ for some ℓ, and $s_i = \ell - 1$ for all i, so that the resolution of k_A now has the form

$$0 \to A(-s-1) \to \bigoplus_{i=1}^{n} A(-s) \to \bigoplus_{i=1}^{n} A(-1) \to A \to 0.$$

In particular, from (2.E) we immediately read off the Hilbert series

$$h_A(t) = 1/(-t^{s+1} + nt^s - nt + 1).$$

By Lemma 2.0.9, since A has finite GK-dimension, $p(t) = -t^{s+1} + nt^s - nt + 1$ has only roots of unity for its zeros. Note that $p(1) = 0$, and that $p'(1) = -(s+1) + sn - n = sn - n - s - 1$. If $n + s > 5$, then $sn - n - s - 1 = (s-1)(n-1) - 2 > 0$, so that $p(t)$ is increasing at $t = 1$. Since $\lim_{t\to\infty} p(t) = -\infty$, $p(t)$ has a real root greater than 1 in this case, a contradiction. Thus $n + s \leq 5$. The case $n = 1$ gives $A \cong k[x]/(x^s)$, which is easily ruled out for having infinite global dimension, similarly as in the proof of Theorem 2.2.1. If $s = 1$, then the relations have degree 1, and the chosen generating set is not minimal, a contradiction. If $s = n = 2$ then an easy calculation shows that the power series $h_A(t) = 1/(1 - 2t + 2t^2 - t^3)$ has first few terms $1 + 2t + 2t^2 + t^3 + 0t^4 + 0t^5 + t^6 + \ldots$; in particular, $A_4 = 0$ and $A_6 \neq 0$. Since A is generated in degree 1, $A_i A_j = A_{i+j}$ for all $i, j \geq 0$, and so $A_4 = 0$ implies $A_{\geq 4} = 0$, a contradiction. This leaves the cases $n = 3, s = 2$ and $n = 2, s = 3$.   □

Based on the degree in which the relations occur, we say that A is a *quadratic* regular algebra in the first case of the lemma above, and that A is *cubic* in the second case. We have seen regular algebras of both types already (Examples 2.1.5 and 2.1.6). Note that the cubic case has the same Hilbert series as a commutative polynomial ring in three variables with weights 1, 1, 2; such a commutative weighted polynomial ring is of course AS-regular, but not generated in degree 1. In general, no AS-regular algebra is known which does not have the Hilbert series of a weighted polynomial ring.

Suppose that A is an AS-regular algebra of global dimension 3, generated in degree 1. By Lemma 2.3.1, we may write the minimal graded free resolution of k_A in the form

$$0 \to A(-s-1) \xrightarrow{u} A(-s)^{\oplus n} \xrightarrow{M} A(-1)^{\oplus n} \xrightarrow{v} A \to 0, \qquad (2.J)$$

where $v = (x_1, \ldots, x_n)$ $(n = 2$ or $n = 3)$ and $u = (y_1, \ldots, y_n)^t$, and M is an $n \times n$ matrix of elements of degree $s - 1$, where $s = 5 - n$. The y_i are also a basis of $kx_1 + \cdots + kx_n$, by a similar argument as in the proof of Theorem 2.2.1. Recall that the minimal resolution of k_A is unique only up to isomorphism of complexes. In particular, it is easy to see we can make a linear change to the free basis of $A(-s)^{\oplus n}$, which will change u to $u' = Pu$ and M to $M' = MP^{-1}$, for some $P \in \mathrm{GL}_n(k)$. In this way we can make $u' = (x_1, \ldots, x_n)^t = v^t$. Our resolution now has the form

$$0 \to A(-s-1) \xrightarrow{v^t} A(-s)^{\oplus n} \xrightarrow{M'} A(-1)^{\oplus n} \xrightarrow{v} A \to 0. \qquad (2.\mathrm{K})$$

In terms of the canonical way to construct the beginning of a graded free resolution given by Lemma 2.1.3, all we have done above is replaced the relations with some linear combinations. Now change notation back and write M for M'. Since the entries of M consist of elements of degree $s - 1$ and the relations of A have degree s, the entries of M and v lift uniquely to homogeneous elements of the free algebra $F = k\langle x_1, \ldots, x_n \rangle$, where $A \cong F/I$ is a minimal presentation. The n entries of the product vM, taking this product in the ring F, are a minimal generating set of the ideal I of relations, as we saw in the proof of Lemma 2.1.3. Since applying $\underline{\mathrm{Hom}}_A(-, A_A)$ to the complex $(2.\mathrm{K})$ gives a shift of a minimal free resolution of $_A k$, this also forces the n entries of the column vector Mv^t, with the product taken in F, to be a minimal generating set of the ideal I. Thus there is a matrix $Q \in \mathrm{GL}_n(k)$ such that $QMv^t = (vM)^t$, as elements of F.

EXAMPLE 2.3.2. Consider a quantum polynomial ring

$$A = k\langle x, y, z \rangle / (yx - pxy, xz - qzx, zy - ryz).$$

A minimal graded free resolution of k_A has the form

$$0 \to A(-3) \xrightarrow{\begin{pmatrix} rz \\ py \\ qx \end{pmatrix}} A(-2)^{\oplus 3} \xrightarrow{\begin{pmatrix} -py & z & 0 \\ x & 0 & -rz \\ 0 & -qx & y \end{pmatrix}} A(-1)^{\oplus 3} \xrightarrow{(x\ y\ z)} A \to 0,$$

by a similar argument as in Example 2.1.5. By a linear change of basis of $A(-2)^{\oplus 3}$, this can be adjusted to have the form

$$0 \to A(-3) \xrightarrow{\begin{pmatrix} x \\ y \\ z \end{pmatrix}} A(-2)^{\oplus 3} \xrightarrow{M = \begin{pmatrix} 0 & pz & -rpy \\ -rqz & 0 & rx \\ qy & -pqx & 0 \end{pmatrix}} A(-1)^{\oplus 3} \xrightarrow{(x\ y\ z)} A \to 0,$$

as in $(2.\mathrm{K})$. Then in the free algebra, one easily calculates that $QMv^t = (vM)^t$, where $v = (x\ y\ z)$, and $Q = \mathrm{diag}(q/p, p/r, r/q)$.

In Artin and Schelter's original work on the classification of regular algebras [5], the first step was to classify possible solutions of the equation $QMv^t = (vM)^t$ developed above, where M has entries of degree $s - 1$ and Q is an invertible matrix. Since one is happy to study the algebras A up to isomorphism, by a change of variables one can change the matrices v^t, M, v in the resolution $(2.\mathrm{K})$ to $P^t v^t$, $P^{-1}M(P^t)^{-1}$, and vP for some $P \in \mathrm{GL}_n(k)$, changing Q to $P^{-1}QP$. Thus one may assume that Q is in Jordan canonical form. Artin and Schelter show in this way that there are a finite number of parametrized families of algebras containing all regular algebras up to isomorphism, where each family consists of a set of relations with unknown parameters, such as the Sklyanin family $S(a, b, c)$ given in Example 1.3.4 (for which Q is the identity matrix). It is a much more difficult problem to decide for which values of the parameters in a family one actually gets an AS-regular algebra, and the Sklyanin family is one of the hardest in this respect: no member of the family was proved to be regular in [5]. In fact, these algebras are all regular except for a few special values of a, b, c. At the end of Section 3, we will outline the method that was eventually used to prove this in [9].

2.4. Exercise Set 2.

EXERCISE 2.4.1. Show that if P is a graded free right module of finite rank over a finitely graded algebra A, then there is a canonical isomorphism

$$\mathrm{Hom}_A(\mathrm{Hom}_A(P_A, A_A), {}_A A) \cong P_A.$$

Using this, prove that if A satisfies (1) and (3) of Definition 2.0.8, then A automatically also satisfies the left-sided version of (3), namely that

$$\underline{\mathrm{Ext}}^i_A({}_A k, {}_A A) \cong \begin{cases} 0 & \text{if } i \neq d, \\ k_A(\ell) & \text{if } i = d. \end{cases}$$

EXERCISE 2.4.2. Let A be the universal enveloping algebra of the Heisenberg Lie algebra $L = kx + ky + kz$, which has bracket defined by $[x, z] = [y, z] = 0, [x, y] = z$. Explicitly, A has the presentation $k\langle x, y, z\rangle/(zx - xz, zy - yz, z - xy - yx)$, and thus A is graded if we assign weights $\deg x = 1$, $\deg y = 1$, $\deg z = 2$. Note that z can be eliminated from the set of generators, yielding the minimal presentation $A \cong k\langle x, y\rangle/(yx^2 - 2xyx + x^2y, y^2x - 2yxy + xy^2)$. Now prove that A is a cubic AS-regular algebra of dimension 3.

EXERCISE 2.4.3. Consider rings of the form $A = k\langle x, y\rangle/(f)$ where $0 \neq f$ is homogeneous of degree 2. Then f can be written uniquely in the form $f = x\tau(x) + y\tau(y)$ for some nonzero linear transformation $\tau : kx + ky \to kx + ky$. Write $A = A(\tau)$.

(1) The choice of τ can be identified with a matrix $B = (b_{ij})$ by setting $\tau(x) = b_{11}x + b_{12}y$ and $\tau(y) = b_{21}x + b_{22}y$. Show that there is a graded isomorphism $A(\tau) \cong A(\tau')$ if and only if the corresponding matrices B and B' are *congruent*, that is $B' = C^t BC$ for some invertible matrix C.

(2) Show that A is isomorphic to $k\langle x, y\rangle/(f)$ for one of the four following f's: (i) $f = yx - qxy$ for some $0 \neq q \in k$ (the quantum plane); (ii) $f = yx - xy - x^2$ (the Jordan plane); (iii) $f = yx$; or (iv) $f = x^2$. (Hint: show that the corresponding matrices are a complete set of representatives for congruence classes of nonzero 2×2 matrices.)

(3) Examples (i), (ii), (iii) above all have Hilbert series $1/(1-t)^2$, GK-dimension 2, and global dimension 2, as we have already seen (Examples 1.4.2, 2.1.2, 2.1.7). In case (iv), find the Hilbert series, GK-dimension, and global dimension of the algebra. Show in addition that both algebras (iii) and (iv) are not noetherian.

EXERCISE 2.4.4. Prove a converse to Theorem 2.2.1(1). Namely, show that if $n \geq 2$ and $\tau : F_1 \to F_1$ is any linear bijection of the space of degree 1 elements in $F = k\langle x_1, \ldots, x_n \rangle$, then $A = F/(f)$ is weakly AS-regular, where $f = \sum_i x_i \tau(x_i)$. (See [250] for a more extensive study of this class of rings.)

(Hint: if the term x_n^2 does not appear in f, then under the degree lex order with $x_1 < \cdots < x_n$, f has leading term of the form $x_n x_i$ for some $i < n$. Find a k-basis of words for A and hence $h_A(t)$. Find a potential resolution of k_A and use the Hilbert series to show that it is exact. If instead x_n^2 appears in f, do a linear change of variables to reduce to the first case.)

EXERCISE 2.4.5. Classify AS-regular algebras of global dimension 2 which are not necessarily generated in degree 1. More specifically, show that any such A is isomorphic to $k\langle x, y \rangle/(f)$ for some generators with weights $\deg x = d_1$, $\deg y = d_2$, where either (i) $f = xy - qyx$ for some $q \in k^\times$, or else (ii) $d_1 i = d_2$ for some i and $f = yx - xy - x^{i+1}$. Conversely, show that each of the algebras in (i), (ii) is AS-regular. (Hint: mimic the proof of Theorem 2.2.1.)

EXERCISE 2.4.6. Suppose that A is AS-regular of global dimension 3. Consider a free resolution of k_A of the form (2.K), and the corresponding equation $QMv^t = (vM)^t$ (as elements of the free algebra $F = k\langle x_1, \ldots, x_n \rangle$). Let s be the degree of the relations of A, and let $d = s + 1$.

Consider the element $\pi = vMv^t$, again taking the product in the free algebra. Write $\pi = \sum_{(i_1, \ldots i_d) \in \{1, 2, \ldots, n\}^d} \alpha_{i_1, \ldots, i_d} x_{i_1} \ldots x_{i_d}$, where $\alpha_{i_1, \ldots, i_d} \in k$. Let τ be the automorphism of the free algebra F determined by the matrix equation $(x_1, \ldots, x_n)Q^{-1} = (\tau(x_1), \ldots, \tau(x_n))$.

(1) Prove that $\pi = \sum_{(i_1, \ldots i_d)} \alpha_{i_1, \ldots, i_d} \tau(x_{i_d}) x_{i_1} \ldots x_{i_{d-1}}$. (In recent terminology, this says that π is a *twisted superpotential*. For instance, see [39].) Conclude that $\tau(\pi) = \pi$.

(2) Show that τ preserves the ideal of relations I (where $A = F/I$) and thus induces an automorphism of A. (This is an important automorphism of A called the *Nakayama automorphism*.)

3. Point modules

In this section, we discuss one of the ways that geometry can be found naturally, but perhaps unexpectedly, in the underlying structure of noncommutative graded rings. Point modules and the spaces parameterizing them were first studied by Artin, Tate, and Van den Bergh [9] in order to complete the classification of AS-regular algebras of dimension 3. They have turned out to be important tools much more generally, with many interesting applications.

DEFINITION 3.0.7. Let A be a finitely graded k-algebra that is generated in degree 1. A *point module* for A is a graded right module M such that M is cyclic, generated in degree 0, and has Hilbert series $h_M(t) = 1/(1-t)$, in other words $\dim_k M_n = 1$ for all $n \geq 0$.

In this section, we are interested only in graded modules over a finitely graded algebra A, and so all homomorphisms of modules will be graded (degree preserving) unless noted otherwise. In particular, when we speak of isomorphism classes of point modules, we will mean equivalence classes under the relation of graded isomorphism.

The motivation behind the definition of point module comes from commutative projective geometry. Recall that the projective space $\mathbb{P}^n$ over k consists of equivalence classes of $n + 1$-tuples $(a_0, a_1, \ldots, a_n) \in k^{n+1}$ such that not all a_i are 0, where two $n + 1$-tuples are equivalent if they are nonzero scalar multiples of each other. The equivalence class of the point $(a_0, a_1, \ldots, a_n)$ is written as $(a_0 : a_1 : \cdots : a_n)$. Each point $p = (a_0 : a_1 : \cdots : a_n) \in \mathbb{P}^n$ corresponds to a homogeneous ideal $I = I(p)$ of $B = k[x_0, \ldots, x_n]$, where $I_d = \{f \in B_d \mid f(a_0, a_1, \ldots, a_n) = 0\}$. It is easy to check that B/I is a point module of B, since vanishing at a point is a linear condition on the elements of B_d. Conversely, if M is a point module of B, then since M is cyclic and generated in degree 0, we have $M \cong B/J$ for some homogeneous ideal J of B. Necessarily $J = \mathrm{Ann}(M)$, so J is uniquely determined by M and is the same for any two isomorphic point modules M and M'. Since J_1 is an n-dimensional subspace of $kx_0 + \cdots + kx_n$, it has a 1-dimensional orthogonal complement, in other words there is a unique up to scalar nonzero vector $p = (a_0 : \cdots : a_n)$ such that $f(a_0, \ldots, a_n) = 0$ for all $f \in J_1$. It is straightforward to prove that I is generated as an ideal by I_1 (after a change of variables, one can prove this for the special case $I = (x_0, \ldots, x_{n-1})$, which is easy.) Since $J_1 = I_1$ we conclude that $J \supseteq I$ and hence $J = I$ since they have the same Hilbert series.

To summarize the argument of the previous paragraph, the isomorphism classes of point modules for the polynomial ring B are in bijective correspondence with the points of the associated projective space. This correspondence generalizes immediately to any finitely graded commutative algebra which is generated in degree 1. Namely, given any homogeneous ideal J of $B = k[x_0, \ldots, x_n]$, one can consider $A = k[x_0, \ldots, x_n]/J$ and the corresponding closed subset of projective space

$$\mathrm{Proj}\, A = \{(a_0 : a_1 : \cdots : a_n) \in \mathbb{P}^n \mid f(a_0, \ldots, a_n) = 0 \text{ for all homogeneous } f \in J\}.$$

Then the point modules of A are precisely the point modules of B killed by J, that is those point modules B/I such that $J \subseteq I$. So there is a bijective correspondence between isomorphism classes of point modules of A and points in $X = \mathrm{Proj}\, A$. We may also say that X *parametrizes* the point modules for A, in a sense we leave informal for now but make precise later in this section.

Many noncommutative graded rings also have nice parameter spaces of point modules.

EXAMPLE 3.0.8. Consider the quantum plane $A = k\langle x, y\rangle/(yx - qxy)$ for some $q \neq 0$. We claim that its point modules are parametrized by $\mathbb{P}^1$, just as is true for a commutative polynomial ring in two variables. To see this, note that if M is a point module for A, then $M \cong A/I$ for some homogeneous *right* ideal I of A, with $\dim_k I_n = \dim_k A_n - 1$ for all $n \geq 0$. Also, I is uniquely determined by knowledge of the (graded) isomorphism class of the module M, since $I = \mathrm{Ann}\, M_0$. Thus it is enough to parametrize such homogeneous right ideals I. But choosing any $0 \neq f \in I_1$, the right ideal fA has Hilbert series $t/(1-t)^2$ (since A is a domain) and so A/fA has Hilbert series $1/(1-t)^2 - t/(1-t)^2 = 1/(1-t)$, that is, it is already a point module. Thus $fA = I$. Then the point modules up to isomorphism are in bijective correspondence with the 1-dimensional subspaces of the 2-dimensional space A_1, that is, with a copy of $\mathbb{P}^1$. The same argument shows the same result for the Jordan plane.

It is natural to consider next the point modules for a quantum polynomial ring in more than two variables, as in Examples 1.3.1. The answer is more complicated than one might first guess—we discuss the three variable case in Example 3.0.12 below. First, we will develop a general method which in theory can be used to calculate the parameter space of point modules for any finitely presented algebra generated in degree 1. In the commutative case, we saw that the point modules for $k[x_0, \ldots, x_n]/J$ were easily determined as a subset of the point modules for $k[x_0, \ldots, x_n]$. In the noncommutative case it is natural to begin similarly by examining the point modules for a free associative algebra.

EXAMPLE 3.0.9. Let $A = k\langle x_0, \ldots, x_n \rangle$ be a free associative algebra with $\deg x_i = 1$ for all i. Fix a graded k-vector space of Hilbert series $1/(1-t)$, say $M = km_0 \oplus km_1 \oplus \ldots$, where m_i is a basis vector for the degree i piece. We think about the possible graded A-module structures on this vector space. If M is an A-module, then

$$m_i x_j = \lambda_{i,j} m_{i+1} \tag{3.A}$$

for some $\lambda_{i,j} \in k$. It is clear that these constants $\lambda_{i,j}$ determine the entire module structure, since the x_j generate the algebra. Conversely, since A is free on the generators x_i, it is easy to see that any choice of arbitrary constants $\lambda_{i,j} \in k$ does determine an A-module structure on M via the formulas (3.A). Since a point module is by definition cyclic, if we want constants $\lambda_{i,j}$ to define a point module, we need to make sure that for each i, some x_j actually takes m_i to a nonzero multiple of m_{i+1}. In other words, M is cyclic if and only if for each i, $\lambda_{i,j} \neq 0$ for some j. Also, we are interested in classifying point modules up to isomorphism. It is easy to check that point modules determined by sequences $\{\lambda_{i,j}\}$, $\{\lambda'_{i,j}\}$ as above are isomorphic precisely when for each i, the nonzero vectors $(\lambda_{i,0}, \ldots, \lambda_{i,n})$ and $(\lambda'_{i,0}, \ldots, \lambda'_{i,n})$ are scalar multiples (scale the basis vectors m_i to compensate). Thus we can account for this by considering each $(\lambda_{i,0} : \cdots : \lambda_{i,n})$ as a point in projective n-space.

In conclusion, the isomorphism classes of point modules over the free algebra A are in bijective correspondence with $\mathbb{N}$-indexed sequences of points in $\mathbb{P}^n$,

$$\{(\lambda_{i,0} : \cdots : \lambda_{i,n}) \in \mathbb{P}^n \mid i \geq 0\},$$

or in other words points of the infinite product $\mathbb{P}^n \times \mathbb{P}^n \times \cdots = \prod_{i=0}^{\infty} \mathbb{P}^n$.

Next, we show that we can parametrize the point modules for a finitely presented algebra by a subset of the infinite product in the example above. Suppose that $f \in k\langle x_0, \ldots, x_n \rangle$ is a homogeneous element of degree m, say $f = \sum_w a_w w$, where the sum is over words of degree m. Consider a set of $m(n+1)$ commuting indeterminates $\{y_{ij} \mid 1 \leq i \leq m, 0 \leq j \leq n\}$ and the polynomial ring $B = k[y_{ij}]$. The *multilinearization* of f is the element of B given by replacing each word $w = x_{i_1} x_{i_2} \dots x_{i_m}$ occurring in f by $y_{1,i_1} y_{2,i_2} \dots y_{m,i_m}$. Given any such multilinearization g and a sequence of points $\{p_i = (a_{i,0} : \dots : a_{i,n}) \mid 1 \le i \le m\} \subseteq \prod_{i=1}^m \mathbb{P}^n$, the condition $g(p_1, \dots, p_m) = 0$, where $g(p_1, \dots, p_m)$ means the evaluation of g by substituting $a_{i,j}$ for $y_{i,j}$, is easily seen to be well-defined.

PROPOSITION 3.0.10. *Let* $A \cong k\langle x_0, \dots, x_n \rangle / (f_1, \dots, f_r)$ *be a finitely presented connected graded k-algebra, where* $\deg x_i = 1$ *and the* f_j *are homogeneous of degree* $d_j \ge 2$. *For each* f_i, *let* g_i *be the multilinearization of* f_i.

(1) *The isomorphism classes of point modules for* A *are in bijection with the closed subset* X *of* $\prod_{i=0}^\infty \mathbb{P}^n$ *given by*

$$\{(p_0, p_1, \dots) \mid g_j(p_i, p_{i+1}, \dots, p_{i+d_j-1}) = 0 \text{ for all } 1 \le j \le r, i \ge 0\}.$$

(2) *Consider for each* $m \ge 1$ *the closed subset* X_m *of* $\prod_{i=0}^{m-1} \mathbb{P}^n$ *given by*

$$\{(p_0, p_1, \dots, p_{m-1}) \mid g_j(p_i, p_{i+1}, \dots, p_{i+d_j-1}) = 0 \text{ for all } 1 \le j \le r, 0 \le i \le m - d_j\}.$$

The natural projection onto the first m *coordinates defines a map* $\phi_m : X_{m+1} \twoheadrightarrow X_m$ *for each* m. *Then* X *is equal to the inverse limit* $\varprojlim X_m$ *of the* X_m *with respect to the maps* ϕ_m. *In particular, if* ϕ_m *is a bijection for all* $m \ge m_0$, *then the isomorphism classes of point modules of* A *are in bijective correspondence with the points of* X_{m_0}.

PROOF. (1) Let $F = k\langle x_0, \dots, x_n \rangle$ and $J = (f_1 \dots, f_r)$, so that $A \cong F/J$. Clearly the isomorphism classes of point modules for A correspond to those point modules of F which are annihilated by J. Write a point module for F as $M = km_0 \oplus km_1 \oplus \dots$, as in Example 3.0.9, where $m_i x_j = \lambda_{i,j} m_{i+1}$. Thus M corresponds to the infinite sequence of points $(p_0, p_1, \dots)$, where $p_j = (\lambda_{j,0} : \dots : \lambda_{j,n})$. The module M is a point module for A if and only if $m_i f_j = 0$ for all i, j. If $w = x_{i_1} x_{i_2} \dots x_{i_d}$ is a word, then $m_i w = \lambda_{i,i_1} \lambda_{i+1,i_2} \dots \lambda_{i+d-1,i_d} m_{i+d}$, and so $m_i f_j = 0$ if and only if the multilinearization g_j of f_j satisfies $g_j(p_i, \dots, p_{i+d_j-1}) = 0$.

(2) This is a straightforward consequence of part (1). $\square$

In many nice examples the inverse limit in part (2) above does stabilize (that is, ϕ_m is a bijection for all $m \ge m_0$), and so some closed subset X_{m_0} of a finite product of projective spaces parametrizes the isomorphism classes of point modules.

EXAMPLE 3.0.11. Let $k = \mathbb{C}$ for simplicity in this example. Let

$$S = k\langle x, y, z \rangle / (azy + byz + cx^2, axz + bzx + cy^2, ayx + bxy + cz^2)$$

be the Sklyanin algebra for some $a, b, c \in k$, as in Example 1.3.4. Consider the closed subset $X_2 \subseteq \mathbb{P}^2 \times \mathbb{P}^2 = \{(x_0 : y_0 : z_0), (x_1 : y_1 : z_1)\}$ given by the vanishing of the multilinearized relations

$$az_0 y_1 + by_0 z_1 + cx_0 x_1, \quad ax_0 z_1 + bz_0 x_1 + cy_0 y_1, \quad ay_0 x_1 + bx_0 y_1 + cz_0 z_1.$$

Let E be the projection of X_2 onto the first copy of $\mathbb{P}^2$. To calculate E, note that the 3 equations can be written in the matrix form

$$\begin{pmatrix} cx_0 & az_0 & by_0 \\ bz_0 & cy_0 & ax_0 \\ ay_0 & bx_0 & cz_0 \end{pmatrix} \begin{pmatrix} x_1 \\ y_1 \\ z_1 \end{pmatrix} = 0. \tag{3.B}$$

Now given $(x_0 : y_0 : z_0) \in \mathbb{P}^2$, there is at least one solution $(x_1 : y_1 : z_1) \in \mathbb{P}^2$ to this matrix equation if and only if the matrix on the left has rank at most 2, in other words is not invertible. Moreover, if the matrix has rank exactly 2 then there is exactly one solution $(x_1 : y_1 : z_1) \in \mathbb{P}^2$. Taking the determinant of the matrix, we see that the locus of $(x_0 : y_0 : z_0) \in \mathbb{P}^2$ such that the matrix is singular is the solution set E of the equation

$$(a^3 + b^3 + c^3)x_0 y_0 z_0 - abc(x_0^3 + y_0^3 + z_0^3) = 0.$$

One may check that E is a nonsingular curve (as in [128, Section I.5]), as long as

$$abc \neq 0 \text{ and } ((a^3 + b^3 + c^3)/3abc)^3 \neq 1,$$

and then E is an elliptic curve since it is the vanishing of a degree 3 polynomial [128, Example V.1.5.1]. We will assume that a, b, c satisfy these constraints.

A similar calculation can be used to find the second projection of X_2. The three multilinearized relations also can be written in the matrix form

$$\begin{pmatrix} x_0 & y_0 & z_0 \end{pmatrix} \begin{pmatrix} cx_1 & az_1 & by_1 \\ bz_1 & cy_1 & ax_1 \\ ay_1 & bx_1 & cz_1 \end{pmatrix} = 0.$$

Because the 3×3 matrix here is simply the same as the one in (3.B) with the subscript 0 replaced by 1, an analogous argument shows that the second projection is the same curve E.

One may show directly, given our constraints on a, b, c, that for each point $(x_0 : x_1 : x_2) \in E$ the corresponding matrix in (3.B) has rank exactly 2. In particular, for each $p \in E$ there is a unique $q \in E$ such that $(p, q) \in X_2$. Thus $X_2 = \{(p, \sigma(p)) \mid p \in E\}$ for some bijective function σ. It is easy to see that σ is a regular map by finding an explicit formula: the cross product of the first two rows of the matrix in (3.B) will be orthogonal to both rows and hence to all rows of the matrix when it has rank 2. This produces the formula

$$\sigma(x_0 : y_0 : z_0) = (a^2 z_0 x_0 - bcy_0^2 : b^2 y_0 z_0 - acx_0^2 : c^2 x_0 y_0 - abz_0^2),$$

which holds on the open subset of E for which the first two rows of the matrix are linearly independent. One gets similar formulas by taking the other possible pairs of rows, and since at each point of E some pair of rows is linearly independent, the map is regular.

It now follows from Proposition 3.0.10 that $X = \{(p, \sigma(p), \sigma^2(p), \dots) \mid p \in E\}$ is the subset of the infinite product $\prod_{i=0}^{\infty} \mathbb{P}^2$ parametrizing the point modules. Thus the maps ϕ_m of Proposition 3.0.10 are isomorphisms for $m \geq 2$ and $X_2 \cong E$ is already in bijective correspondence with the isomorphism classes of point modules for the Sklyanin algebra S (with parameters a, b, c satisfying the constraints above).

EXAMPLE 3.0.12. Let

$$A = k\langle x, y, z \rangle / (zy - ryz, xz - qzx, yx - pxy)$$

be a quantum polynomial ring, where $p, q, r \neq 0$.

The calculation of the point modules for this example is similar as in Example 3.0.11, but a bit easier, and so we leave the details to the reader as Exercise 3.3.1, and only state the answer here. Consider $X_2 \subseteq \mathbb{P}^2 \times \mathbb{P}^2$, the closed set cut out by the multilinearized relations $z_i y_{i+1} - r y_i z_{i+1} = 0$, $x_i z_{i+1} - q z_i x_{i+1} = 0$, and $y_i x_{i+1} - p x_i y_{i+1} = 0$. It turns out that if $pqr = 1$, then $X_2 = \{(s, \sigma(s)) \mid s \in \mathbb{P}^2\}$, where $\sigma : \mathbb{P}^2 \to \mathbb{P}^2$ is an automorphism. If instead $pqr \neq 1$, then defining $E = \{(a : b : c) \mid abc = 0\} \subseteq \mathbb{P}^2$, we have $X_2 = \{(s, \sigma(s)) \mid s \in E\}$ for some automorphism σ of E. Note that E is a union of three lines in this case.

In either case, we see that X_2 is the graph of an automorphism of a subset E of $\mathbb{P}^2$, and as in Example 3.0.11 it follows that the maps ϕ_m of Proposition 3.0.10 are isomorphisms for $m \geq 2$. Thus E is in bijection with the isomorphism classes of point modules.

The examples above demonstrate a major difference between the commutative and noncommutative cases. If A is a commutative finitely graded k-algebra, which is generated in degree 1 and a domain of GK-dimension $d + 1$, then the projective variety Proj A parametrizing the point modules has dimension d. When A is noncommutative, then even if there is a nice space parameterizing the point modules, it may have dimension smaller than d. For instance, the quantum polynomial ring A in Example 3.0.12 has GK-dimension 3, but a 1-dimensional parameter space of point modules when $pqr \neq 1$; similarly, the Sklyanin algebra for generic a, b, c in Example 3.0.11 has point modules corresponding to an elliptic curve. In fact, the examples above are fairly representative of the possibilities for AS-regular algebras A of dimension 3, as we will see later in this section.

3.1. The formal parametrization of the point modules. The details of this section require some scheme theory. The reader less experienced with schemes can skim this subsection and then move to the next, which describes how the theory of point modules is used to help classify AS-regular algebras of dimension 3.

Up until now, we have just studied the isomorphism classes of point modules of an algebra as a set, and shown that in many cases these correspond bijectively to the points of some closed subset of a projective variety. Of course, it should mean something stronger to say that a variety or scheme parametrizes the point modules—there should be a natural geometry intrinsically attached to the set of point modules, which is given by that scheme. We now sketch how this may be made formal.

It is not hard to see why the set of point modules for a finitely graded algebra A has a natural topology, following the basic idea of the Zariski topology. As we have noted several times, every point module is isomorphic to A/I for a uniquely determined graded right ideal I. Then given a graded right ideal J of A, the set of those point modules A/I such that I contains J can be declared to be a closed subset in the set X of isomorphism classes of point modules, and the family of such closed subsets made the basis of a topology. But we really want a scheme structure, not just a topological space, so a more formal construction is required. The idea is to use the "functor of points", which is fundamental to the study of moduli spaces in algebraic geometry. An introduction to this concept can be found in [93, Section I.4 and Chapter VI].

To use this idea, one needs to formulate the objects one is trying to parametrize over arbitrary commutative base rings. Let A be a finitely graded k-algebra. Given a commutative k-algebra R, an R-point module for A is a graded $R \otimes_k A$-module M (where $R \otimes_k A$ is graded with R in degree 0) which is cyclic, generated in degree 0, has $M_0 = R$, and such that M_n is a locally free R-module of rank 1 for all $n \geq 0$. Clearly a k-point module is just a point module in the sense we have already defined. Now for each commutative k-algebra R we let $P(R)$ be the set of isomorphism classes of R-point modules for A. Then P is a functor from the category of commutative k-algebras to the category of sets, where given a homomorphism of k-algebras $\phi : R \to S$, the function $P(\phi) : P(R) \to P(S)$ is defined by tensoring up, that is $M \mapsto S \otimes_R M$. We call P the *point functor* for A.

Given any k-scheme X, there is also a corresponding functor h_X from k-algebras to sets, defined on objects by $R \mapsto \mathrm{Hom}_{k-\mathrm{schemes}}(\mathrm{Spec}\, R, X)$. The functor h_X acts on a homomorphism of k-algebras $\phi : R \to S$ (which corresponds to a morphism of schemes $\widetilde{\phi} : \mathrm{Spec}\, S \to \mathrm{Spec}\, R$) by sending it to $h_X(\phi) : \mathrm{Hom}_{k-\mathrm{schemes}}(\mathrm{Spec}\, R, X) \to \mathrm{Hom}_{k-\mathrm{schemes}}(\mathrm{Spec}\, S, X)$, where $h_X(\phi)(f) = f \circ \widetilde{\phi}$. We say that a functor from k-algebras to sets is *representable* if it is naturally isomorphic to the functor h_X for some scheme X. It is a basic fact that the functor h_X uniquely determines the scheme X; this is a version of Yoneda's lemma [93, Proposition VI-2]. If the point functor P above associated to the finitely graded algebra A is naturally isomorphic to h_X for some scheme X, then we say that the scheme X *parametrizes* the point modules for the algebra A, or that X is a *fine moduli space* for the point modules of A. Note that morphisms of schemes from $\mathrm{Spec}\, k$ to X are in bijective correspondence with the closed points of X, and so in particular the closed points of X then correspond bijectively to the isomorphism classes of k-point modules.

The previous paragraph shows how to formalize the notion of the point modules for A being parametrized by a scheme, but in practice one still needs to understand whether there exists a k-scheme X which represents the point functor P. In fact, Proposition 3.0.10, which showed how to find the point modules as a set in terms of the relations for the algebra, also gives the idea for how to find the representing scheme. For each $m \geq 0$ we define X_m to be the *subscheme* of $\prod_{i=0}^{m-1} \mathbb{P}^n$ defined by the vanishing of the multilinearized relations, as in Proposition 3.0.10(2) (previously, we only considered X_m as a subset). A *truncated R-point module of length $m+1$ for A* is an $R \otimes_k A$ module $M = \bigoplus_{i=0}^{m} M_i$ with $M_0 = R$, which is generated in degree 0, and such that M_i is locally free of rank 1 over R for $0 \leq i \leq m$. Then one can define the truncated point functor P_m which sends a commutative k-algebra R to the set $P_m(R)$

of isomorphism classes of truncated R-point modules of length $m + 1$. A rather formal argument shows that the elements of $P_m(R)$ are in natural bijective correspondence with elements of $\mathrm{Hom}_{k-\mathrm{schemes}}(\mathrm{Spec}\ R, X_m)$, and thus X_m represents the functor P_m (see [9, Proposition 3.9]). Just as in Proposition 3.0.10, one has morphisms of schemes $\phi_m : X_{m+1} \to X_m$ for each m, induced by projecting onto the first m coordinates. In nice cases, ϕ_m is an isomorphism for all $m \geq m_0$, and in such cases the projective scheme X_{m_0} represents the point functor P.

When it is not true that ϕ_m is an isomorphism for all large m, one must work with an inverse limit of schemes $\varprojlim X_m$ as the object representing the point functor P. Such objects are rather unwieldy, and so it is useful to understand when the inverse limit does stabilize (that is, when ϕ_m is an isomorphism for $m \gg 0$). One cannot expect it to stabilize in complete generality, as this already fails for the free algebra as in Example 3.0.9. Artin and Zhang gave an important sufficient condition for stabilization of the inverse limit in [14], which we describe now.

DEFINITION 3.1.1. A noetherian k-algebra A is *strongly noetherian* if for all commutative noetherian k-algebras C, the base extension $A \otimes_k C$ is also noetherian.

THEOREM 3.1.2. [14, Corollary E4.12] Let A be a finitely graded algebra which is strongly noetherian and generated in degree 1. Then there is m_0 such that the maps of schemes $\phi_m : X_{m+1} \to X_m$ described above are isomorphisms for all $m \geq m_0$, and the point modules for A are parametrized by the projective scheme X_{m_0}.

In fact, Artin and Zhang studied in [14] the more general setting of Hilbert schemes, where one wishes to parametrize factors M of some fixed finitely generated graded A-module Q with a given fixed Hilbert function $f : n \mapsto \dim_k M_n$. Theorem 3.1.2 is just a special case of [14, Theorem E4.3], which shows that under the same hypotheses, there is a projective scheme parametrizing such factors for any Q and f. The study of these more general moduli spaces is also useful: for example, the *line modules*—cyclic modules generated in degree 0 with Hilbert series $1/(1 - t)^2$—have an interesting geometry for AS-regular algebras of dimension 4 [200].

The strong noetherian property is studied in detail in [6]. Many nice algebras are strongly noetherian. For example, no non-strongly noetherian AS-regular algebras are known (although as we noted in Question 2.1.8 above, it has not been proved that an AS-regular algebra must even be noetherian in general). On the other hand, there are families of finitely graded algebras which are noetherian, but not strongly noetherian, and for which the inverse limit of truncated point schemes does not stabilize. The simplest such examples are known as *naïve blowups*, which will be described in Section 5.

3.2. Applications of point modules to regular algebras. We now give an overview of how point modules were used in the classification of AS-regular algebras of dimension 3 by Artin, Tate and Van den Bergh. The details can be found in [9].

Let A be an AS-regular algebra of global dimension 3 which is generated in degree 1. By Lemma 2.3.1, we know that A is either quadratic or cubic. The method we are about to describe works quite uniformly in the two cases, but for simplicity it is easiest to consider only the quadratic case in the following discussion. Thus we assume that A has three generators and 3 quadratic relations, and we let $X_2 \subseteq \mathbb{P}^2 \times \mathbb{P}^2$ be the subscheme defined by the multilinearizations of the three relations, as in Proposition 3.0.10.

We have seen several examples already where X_2 is the graph of an automorphism of a closed subscheme of $\mathbb{P}^2$, and the first step is to show that this is always the case. First, using the matrix method of Example 3.0.11, it is straightforward to show that the first and second projections of X_2 are equal to a common closed subscheme $E \subseteq \mathbb{P}^2$, and that either $E = \mathbb{P}^2$ or else E is a degree 3 divisor in $\mathbb{P}^2$, in other words the vanishing of some cubic polynomial (Exercise 3.3.4). It is important to work with subschemes here; for example, for the enveloping algebra of the Heisenberg Lie algebra of Exercise 2.4.2, E turns out to be a triple line. Showing that X_2 is the graph of an automorphism of E is more subtle, and is done using some case-by-case analysis of the form of the relations [9, Section 5].

Once one has $X_2 = \{(p, \sigma(p)) \mid p \in E\}$ for some automorphism $\sigma : E \to E$, one knows that E parametrizes the point modules for A. Since E comes along with an embedding $i : E \subseteq \mathbb{P}^2$, there is also an invertible sheaf on E defined by the pullback $\mathcal{L} = i^*(\mathcal{O}(1))$, where $\mathcal{O}(1)$ is the twisting sheaf of Serre [128, Section II.5]. From the data $(E, \mathcal{L}, \sigma)$ one may construct a *twisted homogeneous coordinate ring* $B = B(E, \mathcal{L}, \sigma)$. We will study this construction in more detail in the next section; for the purposes of this outline, it is important only to know that this is a certain graded ring built out of the geometry of E, whose properties can be analyzed using geometric techniques. In particular, using algebraic geometry one can prove the following facts: (i) B is generated in degree 1; (ii) B is noetherian; and (iii) B has Hilbert series $1/(1-t)^3$ in case $E = \mathbb{P}^2$, while B has Hilbert series $(1 - t^3)/(1-t)^3$ in case E is a cubic curve in $\mathbb{P}^2$.

The idea of the remainder of the proof is to study a canonical ring homomorphism $\phi : A \to B(E, \mathcal{L}, \sigma)$ built out of the geometric data coming from the point modules. The construction of ϕ is quite formal, and it is automatically an isomorphism in degree 1. Since B is generated in degree 1, ϕ is surjective. When $E = \mathbb{P}^2$, it follows from the Hilbert series that $A \cong B$, and so $A \cong B(\mathbb{P}^2, \mathcal{O}(1), \sigma)$. These particular twisted homogeneous coordinate rings are known to have a fairly simple structure; for example, they can also be described as Zhang twists of commutative polynomial rings in three variables (see Definition 4.1.1 below). In the other case, where E is a cubic curve, one wants to show that there is a normal nonzerodivisor $g \in A_3$ such that $gA = \ker \phi$ and thus $A/gA \cong B$. If one happened to know in advance that A was a domain, for example via a Gröbner basis method, then this would easily follow from the known Hilbert series of A and B. However, one does not know this in general, particularly for the Sklyanin algebra. Instead, this requires a detailed analysis of the presentation $k\langle x, y, z\rangle/J$ of B, again using geometry [9, Sections 6, 7]. In particular, one proves that J is generated as a two-sided ideal by three quadratic relations and one cubic relation. The cubic relation provides the needed element g. This also shows that A is uniquely determined by B, since the three quadratic relations for B are all of the relations for A. SL: short page

The final classification of the quadratic AS-regular algebras A which have a cubic curve E as a point scheme is as follows: there is a simple characterization of the possible geometric triples $(E, \mathcal{L}, \sigma)$ which can occur, and the corresponding rings A are found by forgetting the cubic relation of each such $B(E, \mathcal{L}, \sigma)$ [9, Theorem 3].

One immediate corollary of this classification is the fact that any regular algebra of dimension 3 is noetherian, since B is noetherian and either $A \cong B$ or $A/gA \cong B$ (use Lemma 1.3.2). It is also true that all AS-regular algebras of dimension 3 are domains [10, Theorem 3.9], though Lemma 1.3.2 does not immediately prove this in general, since the ring B is a domain only if E is irreducible. For the Sklyanin algebra with generic parameters as in Example 3.0.11, however, where E is an elliptic curve, the ring B is a domain, and hence A is a domain by Lemma 1.3.2 in this case. While this is not the only possible method for proving these basic properties of the Sklyanin algebra (see [224] for another approach), there is no method known which does not use the geometry of the elliptic curve E in some way.

3.3. Exercise Set 3.

EXERCISE 3.3.1. Fill in the details of the calculation of X_2 in Example 3.0.12, for instance using the matrix method of Example 3.0.11. In particular, show that X_2 is the graph of an automorphism σ of $E \subseteq \mathbb{P}^2$, and find a formula for σ.

EXERCISE 3.3.2. Let A be the cubic regular algebra $k\langle x, y \rangle / (y^2 x - xy^2, yx^2 - x^2 y)$. Calculate the scheme parametrizing the point modules for A. (Hint: study the image of $X_3 \subseteq \mathbb{P}^1 \times \mathbb{P}^1 \times \mathbb{P}^1$ under the projections $p_{12}, p_{23} : (\mathbb{P}^1)^{\times 3} \to (\mathbb{P}^1)^{\times 2}$.)

EXERCISE 3.3.3. Find the point modules for the algebra $A = k\langle x, y \rangle / (yx)$, by calculating explicitly what sequences $(p_0, p_1, p_2, \ldots)$ of points in $\mathbb{P}^1$ as in Proposition 3.0.10 are possible. Show that the map $\phi_n : X_{n+1} \to X_n$ from that proposition is not an isomorphism for all $n \geq 1$.

EXERCISE 3.3.4. Let A be AS-regular of global dimension 3, generated in degree 1. Choose the free resolution of k_A in the special form (2.J), namely

$$0 \to A(-s-1) \xrightarrow{v^t} A(-s)^{\oplus n} \xrightarrow{M} A(-1)^{\oplus n} \xrightarrow{v} A \to 0,$$

Let A be presented by the particular relations which are the entries of the vector vM (taking this product in the free algebra). We also know that $(vM)^t = QMv^t$ in the free algebra, for some $Q \in \mathrm{GL}_3(k)$.

(1) Generalizing Examples 3.0.12 and 3.0.11, show that if A is quadratic then $X_2 \subseteq \mathbb{P}^2 \times \mathbb{P}^2$ has equal first and second projections $E = p_1(X_2) = p_2(X_2) \subseteq \mathbb{P}^2$, and either $E = \mathbb{P}^2$ or E is the vanishing of a cubic polynomial in $\mathbb{P}^2$.

(2) Formulate and prove an analogous result for a cubic regular algebra A.

4. Noncommutative projective schemes

In Section 3, we saw that the parameter space of point modules is one important way that geometry appears naturally in the theory of noncommutative graded rings. In this section, we consider the more fundamental question of how to assign a geometric object to a noncommutative ring, generalizing the way that Proj A is assigned to A when A is commutative graded. One possible answer, which has led to a very fruitful theory, is the idea of a noncommutative projective scheme defined by Artin and Zhang [13]. In a nutshell, the basic idea is to give up on the actual geometric space, and instead generalize only the category of coherent sheaves to the noncommutative case. The lack of an actual geometric space is less of a problem than one might at first think. In fact, the study of the category of coherent sheaves on a commutative projective scheme (or its derived category) is of increasing importance in commutative algebraic geometry as an object of interest in its own right.

To begin, we will quickly review some relevant notions from the theory of schemes and sheaves, but this and the next section are primarily aimed at an audience already familiar with the basics in Hartshorne's book [128]. We also assume that the reader has familiarity with the concept of an abelian category.

Recall that a scheme X is a locally ringed space with an open cover by affine schemes $U_\alpha = \operatorname{Spec} R_\alpha$ [128, Section II.2]. We are primarily interested here in schemes of finite type over the base field k, so that X has such a cover by finitely many open sets, where each R_α is a finitely generated commutative k-algebra. The most important way of producing projective schemes is by taking Proj of a graded ring. Let A be a finitely graded commutative k-algebra, generated by $A_1 = kx_0 + \cdots + kx_n$. For each i, we can localize the ring A at the multiplicative system of powers of x_i, obtaining a ring A_{x_i} which is now $\mathbb{Z}$-graded (since the inverse of x_i will have degree -1). Then the degree 0 piece of A_{x_i} is a ring notated as $A_{(x_i)}$. The projective scheme Proj A has an open cover by the open affine schemes $U_i = \operatorname{Spec} A_{(x_i)}$ [128, Proposition II.2.5]. Recall that a sheaf $\mathcal{F}$ on a scheme X is called *quasi-coherent* if there is an open cover of X by open affine sets $U_\alpha = \operatorname{Spec} R_\alpha$, such that $\mathcal{F}(U_\alpha)$ is the sheaf associated to an R_α-module M_α for each α [128, Section II.5]. The sheaf $\mathcal{F}$ is *coherent* if each M_α is a finitely generated module. If $X = \operatorname{Proj} A$ as above, then we can get quasi-coherent sheaves on X as follows. If M is a $\mathbb{Z}$-graded A-module, then there is a sheaf $\widetilde{M}$ where $\widetilde{M}(U_i) = M_{(x_i)}$ is the degree 0 piece of the localization M_{x_i} of M at the powers of x_i [128, Proposition II.5.11]. The sheaf $\widetilde{M}$ is coherent if M is finitely generated.

The constructions above demonstrate how crucial localization is in commutative algebraic geometry. The theory of localization for noncommutative rings is more limited: there is a well-behaved localization only at certain sets of elements called Ore sets (see [117, Chapter 9]). In particular, the set of powers of an element in a noncommutative ring is typically not an Ore set, unless the element is a normal element. Thus it is problematic to try to develop a general theory of noncommutative schemes based around the notion of open affine cover. There has been work in this direction for rings with "enough" Ore sets, however [236].

The actual space with a topology underlying a scheme is built out of prime ideals. In particular, given a finitely graded commutative k-algebra A, as a set Proj A is the set of all homogeneous prime ideals of A, excluding the *irrelevant ideal* $A_{\geq 1}$. As a topological space, it has the Zariski topology, so the closed sets are those of the form

$V(I) = \{P \in \text{Proj } A \mid P \supseteq I\}$, as I varies over all homogeneous ideals of A. Recall that for a not necessarily commutative ring R, an ideal P is called *prime* if $IJ \subseteq P$ for ideals I, J implies that $I \subseteq P$ or $J \subseteq P$. Thus one can define the space of homogeneous non-irrelevant prime ideals, with the Zariski topology, for any finitely graded k-algebra A. For noncommutative rings, this is often a space which is too small to give a good geometric intuition. The reader may verify the details of the following example in Exercise 4.5.1.

EXAMPLE 4.0.5. Let $A = k\langle x, y\rangle/(yx - qxy)$ be the quantum plane, where $q \in k$ is not a root of unity. Then the homogeneous prime ideals of A are $(0), (x), (y), (x, y) = A_{\geq 1}$. If $A = k\langle x, y\rangle/(yx - xy - x^2)$ is the Jordan plane, then the homogeneous prime ideals of A are $(0), (x), (x, y) = A_{\geq 1}$.

On the other hand, both the Jordan and quantum planes have many properties in common with a commutative polynomial ring $k[x, y]$, and one would expect them to have an associated projective geometry which more closely resembles $\mathbb{P}^1 = \text{Proj } k[x, y]$.

We now begin to discuss the theory of noncommutative projective schemes, which finds a way around some of the difficulties mentioned above. The key idea is that there are ways of defining and studying coherent sheaves on commutative projective schemes without explicit reference to an open affine cover. Let A be a finitely graded commutative k-algebra, and let $X = \text{Proj } A$. For a $\mathbb{Z}$-graded module M, for any $n \in \mathbb{Z}$ we define $M_{\geq n} = \bigoplus_{i \geq n} M_i$, and call this a *tail* of M. It is easy to see that any two tails of a finitely generated graded A-module M lead to the same coherent sheaf $\widetilde{M}$ on X. Namely, given $n \in \mathbb{Z}$ and a finitely generated graded module M, we have the short exact sequence

$$0 \to M_{\geq n} \to M \to M/M_{\geq n} \to 0,$$

where the last term is finite dimensional over k. Since localization is exact, and localization at the powers of an element $x_i \in A_1$ kills any finite-dimensional graded module, we see that the $A_{(x_i)}$-modules $(M_{\geq n})_{(x_i)}$ and $M_{(x_i)}$ are equal. Thus $\widetilde{M_{\geq n}} = \widetilde{M}$. In fact, we have the following stronger statement.

LEMMA 4.0.6. *Let $X = \text{Proj } A$ for a commutative finitely graded k-algebra A which is generated in degree 1.*

(1) *Every coherent sheaf on X is isomorphic to $\widetilde{M}$ for some finitely generated graded A-module M [128, Proposition II.5.15].*

(2) *Two finitely generated $\mathbb{Z}$-graded modules M, N satisfy $\widetilde{M} \cong \widetilde{N}$ as sheaves if and only if there is an isomorphism of graded modules $M_{\geq n} \cong N_{\geq n}$ for some $n \in \mathbb{Z}$ [128, Exercise II.5.9].*

We can interpret the lemma above in the following way: the coherent sheaves on a projective scheme $\text{Proj } A$ can be defined purely in terms of the $\mathbb{Z}$-graded modules over A, by identifying those finitely generated $\mathbb{Z}$-graded modules with isomorphic tails. The scheme $\text{Proj } A$ and its open cover play no role in this description. One can use this idea to define a noncommutative analog of the category of coherent sheaves, as follows.

DEFINITION 4.0.7. Let A be a noetherian finitely graded k-algebra. Let gr-A be the abelian category of finitely generated $\mathbb{Z}$-graded right A-modules. Let tors-A be its full subcategory of graded modules M with $\dim_k M < \infty$; we call such modules *torsion* in this context. We define a new abelian category qgr-A. The objects in this category are the same as the objects in gr-A, and we let $\pi : $ gr-$A \to$ qgr-A be the identity map on objects. The morphisms are defined for $M, N \in$ gr-A by

$$\mathrm{Hom}_{\mathrm{qgr}\text{-}A}(\pi(M), \pi(N)) = \lim_{n \to \infty} \mathrm{Hom}_{\mathrm{gr}\text{-}A}(M_{\geq n}, N), \tag{4.A}$$

where the direct limit on the right is taken over the maps of abelian groups

$$\mathrm{Hom}_{\mathrm{gr}\text{-}A}(M_{\geq n}, N) \to \mathrm{Hom}_{\mathrm{gr}\text{-}A}(M_{\geq n+1}, N)$$

induced by the inclusion homomorphisms $M_{\geq n+1} \to M_{\geq n}$.

The pair (qgr-A, $\pi(A)$) is called the *noncommutative projective scheme* associated to the graded ring A. The object $\pi(A)$ is called the *distinguished object* and plays the role of the structure sheaf. The map $\pi : $ gr-$A \to$ qgr-A is a functor, the *quotient functor*, which sends the morphism $f : M \to N$ to $f|_{M_{\geq 0}} \in \mathrm{Hom}(M_{\geq 0}, N)$ in the direct limit.

The passage from gr-A to qgr-A in the definition above is a special case of a more general abstract construction called a *quotient category*. See [13, Section 2] for more discussion. It may seem puzzling at first that gr-A and qgr-A are defined to have the same set of objects. However, some objects that are not isomorphic in gr-A become isomorphic in qgr-A, so there is indeed a kind of quotienting of the set of isomorphism classes of objects. For example, it is easy to see from the definition that $\pi(M) \cong \pi(M_{\geq n})$ in the category qgr-A, for any graded module M and $n \in \mathbb{Z}$ (Exercise 4.5.2).

EXAMPLE 4.0.8. If A is a commutative finitely graded k-algebra, generated by A_1, then there is an equivalence of categories $\Phi : $ qgr-$A \to$ coh X, where coh X is the category of coherent sheaves on $X = \mathrm{Proj}\, A$, and where $\Phi(\pi(A)) = \mathcal{O}_X$ [128, Exercise II.5.9]. Thus commutative projective schemes (or more properly, their categories of coherent sheaves) are special cases of noncommutative projective schemes.

The definition of a noncommutative projective scheme is indicative of a general theme in noncommutative geometry. Often there are many equivalent ways of thinking about a concept in the commutative case. For example, the idea of a point module of a commutative finitely graded algebra B and the idea of a (closed) point of Proj B (that is, a maximal element among nonirrelevant homogeneous prime ideals) are just two different ways of getting at the same thing. But their analogues in the noncommutative case are very different, and point modules are more interesting for the quantum plane, say, than prime ideals are. Often, in trying to generalize commutative concepts to the noncommutative case, one has to find a way to formulate the concept in the commutative case, often not the most obvious one, whose noncommutative generalization gives the best intuition.

We have now defined a noncommutative projective scheme associated to any connected finitely graded k-algebra. This raises many questions, for instance, what are these categories like? What can one do with this construction? We will first give some more examples of these noncommutative schemes, and then we will discuss some of the interesting applications.

4.1. Examples of noncommutative projective schemes. It is natural to study the category qgr-A in the special case that A is an Artin–Schelter regular algebra. We will see that for such A of global dimension 2, the category qgr-A is something familiar. For A of global dimension 3 one obtains new kinds of categories, in general.

First, we study a useful general construction.

DEFINITION 4.1.1. Let A be an $\mathbb{N}$-graded k-algebra. Given a graded algebra automorphism $\tau : A \to A$, the *Zhang twist* of A by τ is a new algebra $B = A^\tau$. It has the same underlying graded k-vector space as A, but a new product $\star$ defined on homogeneous elements by $a \star b = a\tau^m(b)$ for $a \in B_m, b \in B_n$.

The reader may easily verify that the product on A^τ is associative. Given an $\mathbb{N}$-graded k-algebra A, let Gr-A be its abelian category of $\mathbb{Z}$-graded A-modules. One of the most important features of the twisting construction is that it preserves the category of graded modules.

THEOREM 4.1.2. [249, Theorem 1.1] For any $\mathbb{N}$-graded k-algebra A with graded automorphism $\tau : A \to A$, there is an equivalence of categories $F : \text{Gr-}A \to \text{Gr-}A^\tau$. The functor F is given on objects by $M \mapsto M^\tau$, where M^τ is the same as M as a $\mathbb{Z}$-graded k-space, but with new A^τ-action $\star$ defined by $x \star b = x\tau^m(b)$, for $x \in M_m, b \in A_n$. The functor F acts as the identity map on morphisms.

PROOF. Given $M \in \text{Gr-}A$, the proof that M^τ is indeed an A^τ-module is formally the same as the proof that A^τ is associative, and it is obvious that F is a functor.

Now $\tau : A^\tau \to A^\tau$, given by the same underlying map as $\tau : A \to A$, is an automorphism of A^τ as well. Thus we can define the twist $(A^\tau)^{\tau^{-1}}$, which is isomorphic to A again via the identity map of the underlying k-space. Thus we also get a functor $G : \text{Gr-}A^\tau \to \text{Gr-}A$ given by $N \mapsto N^{\tau^{-1}}$. The reader may easily check that F and G are inverse equivalences of categories. $\square$

The reader may find more details about the twisting construction defined above in [249]. In fact, Zhang defined a slightly more general kind of twist depending on a *twisting system* instead of just a choice of graded automorphism, and proved that these more general twists of a connected finitely graded algebra A with $A_1 \neq 0$ produce all $\mathbb{N}$-graded algebras B such that Gr-A and Gr-B are equivalent categories [249, Theorem 1.2].

Suppose now that A is finitely graded and noetherian. Given the explicit description of the equivalence Gr-$A \to \text{Gr-}A^\tau$ in Theorem 4.1.2, it is clear that it restricts to the subcategories of finitely generated modules to give an equivalence gr-$A \to \text{gr-}A^\tau$. Moreover, the subcategories of modules which are torsion (that is, finite-dimensional over k) also correspond, and so one gets an equivalence qgr-$A \to \text{qgr-}A^\tau$, under which the distinguished objects $\pi(A)$ and $\pi(A^\tau)$ correspond. Since a commutative graded ring typically has numerous noncommutative Zhang twists, this can be used to give many examples of noncommutative graded rings whose noncommutative projective schemes are simply isomorphic to commutative projective schemes.

EXAMPLE 4.1.3. Let $A = k[x, y]$ be a polynomial ring, with graded automorphism $\sigma : A \to A$ given by $\sigma(x) = x, \sigma(y) = y - x$. In $B = A^{\sigma}$ we calculate $x \star y = x(y - x)$, $y \star x = xy$, $x \star x = x^2$. Thus we have the relation

$$x \star y - y \star x + x \star x = 0.$$

Then there is a surjection $C = k\langle x, y \rangle / (yx - xy - x^2) \to B$, where C is the Jordan plane from Examples 1.3.1. Since $h_C(t) = 1/(1 - t)^2$ and $h_B(t) = h_A(t) = 1/(1 - t)^2$, $B \cong C$. Now there is an equivalence of categories qgr-$B \simeq$ qgr-A by the remarks above. Since qgr-$A \simeq \operatorname{coh} \mathbb{P}^1$ by Example 4.0.8, we see that the noncommutative projective scheme qgr-C associated to the Jordan plane is just the commutative scheme $\mathbb{P}^1$.

A similar argument shows that the quantum plane C from Examples 1.3.1 is also isomorphic to a twist of $k[x, y]$, and so qgr-$C \simeq \operatorname{coh} \mathbb{P}^1$ for the quantum plane also.

EXAMPLE 4.1.4. Consider the quantum polynomial ring

$$A = k\langle x, y, z \rangle / (yx - pxy, zy - qyz, xz - rzx)$$

from Example 3.0.12. It is easy to check that A is isomorphic to a Zhang twist of $k[x, y, z]$ if and only if $pqr = 1$ (Exercise 4.5.3). Thus when $pqr = 1$, one has qgr-$A \simeq$ qgr-$k[x, y, z] \simeq \operatorname{coh} \mathbb{P}^2$. We also saw in Example 3.0.12 that A has point modules parametrized by $\mathbb{P}^2$ in this case. This can also also be proved using Zhang twists: it is clear that the isomorphism classes of point modules of B and B^{τ} are in bijection under the equivalence of categories gr-$B \simeq$ gr-B^{τ}, for any finitely graded algebra B.

On the other hand, when $pqr \neq 1$, we saw in Example 3.0.12 that the point modules of A are parametrized by a union of three lines. In this case it is known that qgr-A is not equivalent to the category of coherent sheaves on any commutative projective scheme, although we do not prove this assertion here. It is still appropriate to think of the noncommutative projective scheme qgr-A as a kind of noncommutative $\mathbb{P}^2$, but this noncommutative projective plane has only a one-dimensional subscheme of closed points on it.

4.2. Coordinate rings. Above, we associated a category qgr-A to a noetherian finitely graded algebra A. It is important that one can also work in the other direction and construct graded rings from categories.

DEFINITION 4.2.1. Let $\mathcal{C}$ be an abelian category such that each $\operatorname{Hom}_{\mathcal{C}}(\mathcal{F}, \mathcal{G})$ is a k-vector space, and such that composition of morphisms is k-bilinear; we say that $\mathcal{C}$ is a k-*linear* abelian category. Let $\mathcal{O}$ be an object in $\mathcal{C}$ and let $s : \mathcal{C} \to \mathcal{C}$ be an *autoequivalence*, that is, an equivalence of categories from $\mathcal{C}$ to itself. Then we define an $\mathbb{N}$-graded k-algebra $B = B(\mathcal{C}, \mathcal{O}, s) = \bigoplus_{n \geq 0} \operatorname{Hom}_{\mathcal{C}}(\mathcal{O}, s^n \mathcal{O})$, where the product is given on homogeneous elements $f \in B_m, g \in B_n$ by $f \star g = s^n(f) \circ g$.

In some nice cases, the construction above allows one to recover a finitely graded algebra A from the data of its noncommutative projective scheme (qgr-A, $\pi(A)$) and a natural shift functor.

PROPOSITION 4.2.2. *Let A be a noetherian finitely graded k-algebra with $A \neq k$. Let $\mathcal{C} = $ qgr-A and let $s = (1)$ be the autoequivalence of $\mathcal{C}$ induced from the shift functor $M \mapsto M(1)$ on modules $M \in$ gr-A. Let $\mathcal{O} = \pi(A)$.*

(1) *There is a canonical homomorphism of rings $\phi : A \to B(\mathcal{C}, \mathcal{O}, s)$. The kernel $\ker \phi$ is finite-dimensional over k and is 0 if A is a domain.*

(2) *If $\underline{\operatorname{Ext}}^1_A(k_A, A_A)$ is finite-dimensional over k, then $\operatorname{coker} \phi$ is also finite-dimensional; this is always the case if A is commutative. Moreover, if $\underline{\operatorname{Ext}}^1_A(k_A, A_A) = 0$, then ϕ is surjective.*

PROOF. Let $B = B(\mathcal{C}, \mathcal{O}, s)$. Then

$$B = \bigoplus_{n \geq 0} \operatorname{Hom}_{\mathcal{C}}(\mathcal{O}, s^n(\mathcal{O})) = \bigoplus_{n \geq 0} \varinjlim_{i \to \infty} \operatorname{Hom}_{\text{gr-}A}(A_{\geq i}, A(n)) = \varinjlim_{i \to \infty} \underline{\operatorname{Hom}}_A(A_{\geq i}, A)_{\geq 0}.$$

Thus there is a map $\phi : A \to B$ which sends a homogeneous element $x \in A_n$ to the corresponding element $l_x \in \underline{\operatorname{Hom}}_A(A, A)$ in the $i = 0$ part of this direct limit, where $l_x(y) = xy$. It is easy to see from the definition of the multiplication in B that ϕ is a ring homomorphism. For each exact sequence $0 \to A_{\geq i} \to A \to A/A_{\geq i} \to 0$ we can apply $\underline{\operatorname{Hom}}_A(-, A)_{\geq 0}$, and write the corresponding long exact sequence in $\underline{\operatorname{Ext}}$. It is easy to see that the direct limit of exact sequences of abelian groups is exact, so we obtain an exact sequence

$$0 \to \varinjlim_{i \to \infty} \underline{\operatorname{Hom}}_A(A/A_{\geq i}, A)_{\geq 0} \to A \xrightarrow{\phi} B \to \varinjlim_{i \to \infty} \underline{\operatorname{Ext}}^1_A(A/A_{\geq i}, A)_{\geq 0} \to 0.$$

The proof is finished by showing that $\varinjlim_{i \to \infty} \underline{\operatorname{Ext}}^j_A(A/A_{\geq i}, A)_{\geq 0}$ is finite dimensional (or 0) as long as $\underline{\operatorname{Ext}}^j_A(k, A)$ is finite dimensional (or 0, respectively). We ask the reader to complete the proof in Exercise 4.5.4. Note that $\underline{\operatorname{Hom}}_A(k, A)$ is certainly finite-dimensional, since A is noetherian, and $\underline{\operatorname{Hom}}_A(k, A) = 0$ if A is a domain (since $A \neq k$). The claim that $\underline{\operatorname{Ext}}^1_A(k, A)$ is always finite-dimensional in the commutative case is also part of Exercise 4.5.4. $\square$

The proposition above shows that for a noetherian finitely graded domain A, one recovers A in large degree from its category qgr-A, as long as $\underline{\operatorname{Ext}}^1_A(k, A)$ is finite-dimensional. This kind of interplay between categories and graded rings is very useful.

Interestingly, it is not always true for a noncommutative finitely graded algebra that $\underline{\operatorname{Ext}}^1_A(k, A)$ is finite-dimensional. We ask the reader to work through an example, which was first given by Stafford and Zhang in [213], in Exercise 4.5.5. More generally, Artin and Zhang defined the χ-*conditions* in [13], as follows. A finitely graded noetherian algebra A satisfies χ_i if $\underline{\operatorname{Ext}}^j_A(k, M)$ is finite-dimensional for all finitely generated $\mathbb{Z}$-graded A-modules M and for all $j \leq i$; the algebra A satisfies χ if it satisfies χ_i for all $i \geq 0$. If A satisfies χ_1, this is enough for the map $\phi : A \to B(\mathcal{C}, \mathcal{O}, s)$ of

Proposition 4.2.2 to have finite-dimensional cokernel, but other parts of the theory (such as the cohomology we discuss below) really work well only for graded rings satisfying the full χ condition. Fortunately, χ holds for many important classes of rings, for example noetherian AS-regular algebras [13, Theorem 8.1].

4.3. Twisted homogeneous coordinate rings.
Twisted homogeneous coordinate rings already made a brief appearance at the end of Section 3, in the outline of the classification of AS-regular algebras of dimension 3. Because such rings are defined purely geometrically, the analysis of their properties often reduces to questions of commutative algebraic geometry. These rings also occur naturally (for example, in the study of regular algebras, as we have already seen), and in some cases would be very difficult to study and understand without the geometric viewpoint.

We now give the precise definition of twisted homogeneous coordinate rings, relate them to the general coordinate ring construction of the previous section, and work through an example.

DEFINITION 4.3.1. Let X be a projective scheme defined over the base field k. Let $\mathcal{L}$ be an invertible sheaf on X, and let $\sigma : X \to X$ be an automorphism of X. We use the notation $\mathcal{F}^\sigma$ for the pullback sheaf $\sigma^*(\mathcal{F})$. Let $\mathcal{L}_n = \mathcal{L} \otimes \mathcal{L}^\sigma \otimes \cdots \otimes (\mathcal{L})^{\sigma^{n-1}}$ for each $n \geq 1$, and let $\mathcal{L}_0 = \mathcal{O}_X$. Let $H^0(X, \mathcal{F})$ be the global sections $\mathcal{F}(X)$ of a sheaf $\mathcal{F}$. For any sheaf, there is a natural pullback of global sections map $\sigma^* : H^0(X, \mathcal{F}) \to H^0(X, \mathcal{F}^\sigma)$.

We now define a graded ring $B = B(X, \mathcal{L}, \sigma) = \bigoplus_{n \geq 0} B_n$, called the *twisted homogeneous coordinate ring* associated to this data. Set $B_n = H^0(X, \mathcal{L}_n)$ for $n \geq 0$, and define the multiplication on $B_m \otimes_k B_n$ via the chain of maps

$$H^0(X, \mathcal{L}_m) \otimes H^0(X, \mathcal{L}_n) \xrightarrow{1 \otimes (\sigma^m)^*} H^0(X, \mathcal{L}_m) \otimes H^0(X, \mathcal{L}_n^{\sigma^m})$$

$$\xrightarrow{\mu} H^0(X, \mathcal{L}_m \otimes \mathcal{L}_n^{\sigma^m}) = H^0(X, \mathcal{L}_{m+n}),$$

where μ is the natural multiplication of global sections map.

We can also get these rings as a special case of the construction in the previous section. Consider $\mathcal{C} = \text{coh } X$ for some projective k-scheme X. If $\mathcal{L}$ is an invertible sheaf on X, then $\mathcal{L} \otimes_{\mathcal{O}_X} -$ is an autoequivalence of $\mathcal{C}$. For any automorphism σ of X, the pullback map $\sigma^*(-)$ is also an autoequivalence of the category $\mathcal{C}$. It is known that in fact an arbitrary autoequivalence of $\mathcal{C}$ must be a composition of these two types, in other words it must have the form $s = (\mathcal{L} \otimes \sigma^*(-))$ for some $\mathcal{L}$ and σ [11, Proposition 2.15], [13, Corollary 6.9]. Now we may define the ring $B = B(\mathcal{C}, \mathcal{O}_X, s)$, and an exercise in tracing through the definitions shows that this ring is isomorphic to the twisted homogeneous coordinate ring $B(X, \mathcal{L}, \sigma)$ defined above.

It is also known when qgr-B recovers the category coh X. Recalling that $\mathcal{L}_n = \mathcal{L} \otimes \mathcal{L}^\sigma \otimes \cdots \otimes \mathcal{L}^{\sigma^{n-1}}$, then $\mathcal{L}$ is called σ-*ample* if for any coherent sheaf $\mathcal{F}$, one has $H^i(X, \mathcal{F} \otimes \mathcal{L}_n) = 0$ for all $n \gg 0$ and all $i \geq 1$. When $\sigma = 1$, this is just one way to define the ampleness of $\mathcal{L}$ in the usual sense [128, Proposition III.5.3]. In fact, Keeler completely characterized σ-ampleness [150, Theorem 1.2]. In particular, when X has at least one σ-ample sheaf, then $\mathcal{L}$ is σ-ample if and only if $\mathcal{L}_n$ is ample for some $n \geq 1$, so it is easy to find σ-ample sheaves in practice. When $\mathcal{L}$ is σ-ample, then $B = B(X, \mathcal{L}, \sigma)$ is noetherian and there is an equvialence of categories qgr-$B \simeq \text{coh } X$ [11, Theorems 1.3, 1.4], and so this construction produces many different "noncommutative coordinate rings" of X.

It can be difficult to get a intuitive feel for the twisted homogeneous coordinate ring construction, so we work out the explicit details of a simple example.

EXAMPLE 4.3.2. We calculate a presentation for $B(\mathbb{P}^1, \mathcal{O}(1), \sigma)$, where $\sigma : \mathbb{P}^1 \to \mathbb{P}^1$ is given by the explicit formula $(a : b) \mapsto (a : a + b)$, and $\mathcal{O}(1)$ is the twisting sheaf of Serre as in [128, Sec. II.5].

Let $R = k[x, y]$ and $\mathbb{P}^1 = \text{Proj } R$, with its explicit open affine cover $U_1 = \text{Spec } R_{(x)} = \text{Spec } k[u]$, and $U_2 = \text{Spec } R_{(y)} = \text{Spec } k[u^{-1}]$, where $u = yx^{-1}$. Then the field of rational functions of $\mathbb{P}^1$ is explicitly identified with the field $k(u)$, the fraction field of both $R_{(x)}$ and $R_{(y)}$. Let $\mathcal{K}$ be the constant sheaf on $\mathbb{P}^1$ whose value is $k(u)$ on every nonempty open set. In doing calculations with a twisted homogeneous coordinate ring on an integral scheme, it is useful to embed all invertible sheaves explicitly in the constant sheaf of rational functions $\mathcal{K}$, which is always possible by [128, Prop. II.6.13]. The sheaf $\mathcal{O}(1)$ is defined abstractly as the coherent sheaf $\widetilde{R(1)}$ associated to the graded module $R(1)$ as in [128, Section II.5], but it is not hard to see that $\mathcal{O}(1)$ is isomorphic to the subsheaf $\mathcal{L}$ of $\mathcal{K}$ generated by the global sections $1, u$, whose sections on the two open sets of the cover are

$$\mathcal{L}(U_1) = 1k[u] + uk[u] = k[u] \quad \text{and} \quad \mathcal{L}(U_2) = 1k[u^{-1}] + uk[u^{-1}] = uk[u^{-1}].$$

Then the space of global sections of $\mathcal{L}$ is just the intersection of the sections on the two open sets, namely $H^0(\mathbb{P}^1, \mathcal{L}) = k + ku$.

The automorphism of $\mathbb{P}^1$ induces an automorphism of the field $k(u)$, defined on a rational function $f : \mathbb{P}^1 \dashrightarrow k$ by $f \mapsto f \circ \sigma$. We call this automorphism σ also, and it is straightforward to calculate the formula $\sigma(u) = u + 1$. Then given an invertible subsheaf $\mathcal{M} \subseteq \mathcal{K}$, such that $\mathcal{M}$ is generated by its global sections $V = H^0(\mathbb{P}^1, \mathcal{M})$, the subsheaf of $\mathcal{K}$ generated by the global sections $\sigma^n(V)$ is isomorphic to the pullback $\mathcal{M}^{\sigma^n}$. In our example, letting $V = k + ku$, then $\mathcal{L}^{\sigma^i}$ is the subsheaf of $\mathcal{K}$ generated by $\sigma^i(V) = V$ (that is, $\mathcal{L}^{\sigma^i} = \mathcal{L}$) and $\mathcal{L}_n = \mathcal{L} \otimes \mathcal{L}^{\sigma} \otimes \cdots \otimes \mathcal{L}^{\sigma^{n-1}}$ is the sheaf generated by $V_n = V\sigma(V) \ldots \sigma^{n-1}(V) = V^n = k + ku + \cdots + ku^n$. Also, $H^0(\mathbb{P}^1, \mathcal{L}_n) = V_n$, by intersecting the sections on each of the two open sets, as for $n = 1$ above. (We caution that in more general examples, $\mathcal{L}_n$ is not isomorphic to $\mathcal{L}^{\otimes n}$.)

Finally, with all of our invertible sheaves $\mathcal{M}, \mathcal{N}$ explicitly embedded in $\mathcal{K}$ as above, and thus their global sections embedded in $k(u)$, the pullback of global sections map $\sigma^* : H^0(\mathbb{P}^1, \mathcal{M}) \to H^0(\mathbb{P}^1, \mathcal{M}^{\sigma})$ is simply given by applying the automorphism σ of $k(u)$, and the multiplication of sections map $H^0(\mathbb{P}^1, \mathcal{M}) \otimes H^0(\mathbb{P}^1, \mathcal{N}) \to H^0(\mathbb{P}^1, \mathcal{M} \otimes \mathcal{N})$ is simply multiplication in $k(u)$. Thus $B = B(\mathbb{P}^1, \mathcal{L}, \sigma) = \bigoplus_{n \geq 0} V_n$, with multiplication on homogeneous elements $f \in V_m, g \in V_n$ given by $f \star g = f\sigma^m(g)$. It easily follows that B is generated in degree 1, and putting $v = 1, w = u \in V_1$, we immediately calculate $v \star w = (u + 1), w \star v = u, v \star v = 1$, giving the relation $v \star w = w \star v + v \star v$. It follows that there is a surjective graded homomorphism $A = k\langle x, y \rangle / (xy - yx - x^2) \to B$. But A is easily seen to be isomorphic to the Jordan plane from Examples 1.3.1 (apply the change of variable $x \mapsto -x, y \mapsto y$). So A and B have the same Hilbert series, and we conclude that $A \cong B$.

The quantum plane also arises as a twisted homogeneous coordinate ring of $\mathbb{P}^1$, by a similar calculation as in the previous example. Of course, this is not the simplest way to describe the Jordan and quantum planes. The point is that many less trivial examples, such as the important case $B(E, \mathcal{L}, \sigma)$ where E is an elliptic curve, do not arise in a more naïve way. The twisted homogeneous coordinate ring formalism is the simplest way of defining these rings, and their properties are most easily analyzed using geometric techniques. For more details about twisted homogeneous coordinate rings, see the original paper of Artin and Van den Bergh [11] and the work of Keeler [150].

4.4. Further applications. Once one has a category qgr-A associated to a finitely graded noetherian k-algebra A, one can try to formulate and study all sorts of geometric concepts, such as points, lines, closed and open subsets, and so on, using this category. As long as a geometric concept for a projective scheme can be phrased in terms of the category of coherent sheaves, then one can attempt to transport it to noncommutative projective schemes. For example, see [207], [208] for some explorations of the notions of open and closed subsets and morphisms for noncommutative schemes.

If X is a commutative projective k-scheme, then for each point $x \in X$ there is a corresponding skyscraper sheaf $k(x) \in \text{coh } X$, with stalks $\mathcal{O}_x \cong k$ and $\mathcal{O}_y = 0$ for all closed points $y \neq x$. This is obviously a simple object in the abelian category coh X (that is, it has no subobjects other than 0 and itself) and it is not hard to see that such skyscraper sheaves are the only simple objects in this category. Since simple objects of coh X correspond to points of X, one may think of the simple objects of qgr-A in general as the "points" of a noncommutative projective scheme. This connects nicely with Section 3, as follows.

EXAMPLE 4.4.1. Let M be a point module for a finitely graded noetherian k-algebra A which is generated in degree 1. We claim that $\pi(M)$ is a simple object in the category qgr-A. First, the only graded submodules of M are 0 and the tails $M_{\geq n}$ for $n \geq 0$. Also, all tails of M have $\pi(M_{\geq n}) = \pi(M)$ in qgr-A (Exercise 4.5.2). Given a nonzero subobject of $\pi(M)$, it has the form $\pi(N)$ for some graded A-module N, and the monomorphism $\pi(N) \to \pi(M)$ corresponds to some nonzero element of $\text{Hom}_{\text{gr-}A}(N_{\geq n}, M)$, whose image must therefore be a tail of M. But then the map $\pi(N) \to \pi(M)$ is an epimorphism and hence an isomorphism. Thus $\pi(M)$ is simple, as claimed.

A similar proof shows that given an arbitrary finitely generated $\mathbb{Z}$-graded module M of A, $\pi(M)$ is a simple object in qgr-A if and only if every graded submodule N of M has $\dim_k M/N < \infty$ (or equivalently, $M_{\geq n} \subseteq N$ for some n). Such modules are called *1-critical* (with respect to Krull dimension; see [117, Chapter 15]). It is possible for a finitely graded algebra A to have such modules M which are bigger than point modules; for example, some AS-regular algebras of dimension 3 have 1-critical modules M with

$\dim_k M_n = d$ for all $n \gg 0$, some $d > 1$ [10, Note 8.43]. The corresponding simple object $\pi(M)$ in qgr-A is sometimes called a *fat point*.

One of the most important tools in algebraic geometry is the cohomology of sheaves. One possible approach is via Čech cohomology [128, Section III.4], which is defined using an open affine cover of the scheme, and thus doesn't generalize in an obvious way to noncommutative projective schemes. However, the modern formulation of sheaf cohomology due to Grothendieck, which uses injective resolutions, generalizes easily. Recall that if X is a (commutative) projective scheme and $\mathcal{F}$ is a quasi-coherent sheaf on X, then one defines its cohomology groups by $H^i(X, \mathcal{F}) = \text{Ext}^i_{\text{Qcoh } X}(\mathcal{O}_X, \mathcal{F})$ [128, Section III.6]. Although the category of quasi-coherent sheaves does not have enough projectives, it has enough injectives, so such Ext groups can be defined using an injective resolution of $\mathcal{F}$. It is not sufficient to work in the category of coherent sheaves here, since injective sheaves are usually non-coherent.

Given a finitely graded noetherian k-algebra A, we defined the category qgr-A, which is an analog of coherent sheaves. We did not define an analog of quasi-coherent sheaves above, for reasons of simplicity only. In general, one may define a category Qgr-A, by starting with the category Gr-A of all $\mathbb{Z}$-graded A-modules and defining an appropriate quotient category by the subcategory of torsion modules, which in this case are the direct limits of finite-dimensional modules. The category Qgr-A is the required noncommutative analog of the category of quasi-coherent sheaves. We omit the precise definition of Qgr-A, which requires a slightly more complicated definition of the Hom sets than (4.A); see [13, Section 2]. Injective objects and injective resolutions exist in Qgr-A, so Ext is defined. Thus the natural generalization of Grothendieck's definition of cohomology is $H^i(\text{Qgr-}A, \mathcal{F}) = \text{Ext}^i_{\text{Qgr-}A}(\pi(A), \mathcal{F})$, for any object $\mathcal{F} \in$ Qgr-A. See [13, Section 7] for more details.

Once the theory of cohomology is in place, many other concepts related to cohomology can be studied for noncommutative projective schemes. To give just one example, there is a good analog of Serre duality, which holds in a number of important cases [246]. In another direction, one can study the bounded derived category $D^b(\text{qgr-}A)$, and ask (for example) when two such derived categories are equivalent, as has been studied for categories of commutative coherent sheaves.

We close this section by connecting it more explicitly with Section 2. Since AS-regular algebras A are intuitively noncommutative analogs of (weighted) polynomial rings, their noncommutative projective schemes qgr-A should be thought of as analogs of (weighted) projective spaces. Thus these should be among the most fundamental noncommutative projective schemes, and this gives another motivation for the importance of regular algebras. One difficult aspect of the noncommutative theory, however, is a lack of a general way to find projective embeddings. Many important examples of finitely graded algebras A are isomorphic to factor algebras of AS-regular algebras B, and thus qgr-A can be thought of as a closed subscheme of the noncommutative projective space qgr-B. However, there is as yet no theory showing that some reasonably general class of graded algebras must arise as factor algebras of AS-regular algebras.

4.5. Exercise Set 4.

EXERCISE 4.5.1. Show that the graded prime ideals of the quantum plane

$$k\langle x, y\rangle/(yx - qxy)$$

(for q not a root of 1) and the Jordan plane $k\langle x, y\rangle/(yx - xy - x^2)$ are as claimed in Example 4.0.5.

EXERCISE 4.5.2. Let A be a connected finitely graded noetherian k-algebra. Given two finitely generated $\mathbb{Z}$-graded A-modules M and N, prove that $\pi(M) \cong \pi(N)$ in qgr-A if and only if there is $n \geq 0$ such that $M_{\geq n} \cong N_{\geq n}$ in gr-A. In other words, show that two modules are isomorphic in the quotient category if and only if they have isomorphic tails.

EXERCISE 4.5.3. Prove that the quantum polynomial ring of Example 4.1.4 is isomorphic to a Zhang twist of $R = k[x, y, z]$ if and only if $pqr = 1$. (Hint: what are the degree 1 normal elements of R^σ for a given graded automorphism σ?)

EXERCISE 4.5.4. Complete the proof of Proposition 4.2.2, in the following steps.

(1) Let $D = \lim_{i \to \infty} \underline{\mathrm{Ext}}^j(A/A_{\geq i}, A)_{\geq 0}$. Show that if $\underline{\mathrm{Ext}}_A^j(k, A)$ is finite dimensional, then D is also finite-dimensional. Show moreover that if $\underline{\mathrm{Ext}}^j(k, A) = 0$, then $D = 0$. (Hint: consider the long exact sequence in Ext associated to $0 \to K \to A/A_{\geq i+1} \to A/A_{\geq i} \to 0$, where K is isomorphic to a direct sum of copies of $k_A(-i)$.)
(2) If A is commutative, show that $\underline{\mathrm{Ext}}_A^j(k, A)$ is finite-dimensional for all $j \geq 0$. (Hint: calculate Ext in two ways: with a projective resolution of k_A, and with an injective resolution of A_A.)

EXERCISE 4.5.5. Assume that char $k = 0$. Let $B = k\langle x, y\rangle/(yx - xy - x^2)$ be the Jordan plane. Let $A = k + By$, which is a graded subring of B. It is known that the ring A is noetherian (see [213, Theorem 2.3] for a proof).

(1) Show that A is the *idealizer* of the left ideal By of B, that is, $A = \{z \in B \mid Byz \subseteq By\}$.
(2) Show that as a graded right A-module, $(B/A)_A$ is isomorphic to the infinite direct sum $\bigoplus_{n \geq 1} k_A(-n)$.
(3) Show for each $n \geq 1$ that the natural map $\underline{\mathrm{Hom}}_A(A, A) \to \underline{\mathrm{Hom}}_A(A_{\geq 1}, A)$ is not surjective in degree n, by finding an element in the latter group corresponding to $x \in B_n \setminus A_n$. Conclude that the cokernel of the homomorphism $A \to B(\text{qgr-}A, \pi(A), (1))$ constructed in Proposition 4.2.2 has infinite dimension over k. Similarly, conclude that $\dim_k \underline{\mathrm{Ext}}_A^1(k, A) = \infty$.

5. Classification of noncommutative curves and surfaces

Many of the classical results in algebraic geometry focus on the study of curves and surfaces, for example as described in [128, Chapters IV, V]. Naturally, one would like to develop a comparatively rich theory of noncommutative curves and surfaces, especially their classification. The canonical reference on noncommutative curves and surfaces is the survey article by Stafford and Van den Bergh [214], which describes the state of the subject as of 2001. While there are strong classification results for noncommutative curves, the classification of noncommutative surfaces is very much a work in progress. In this section we describe some of the theory of noncommutative curves and surfaces, including some more recent work not described in [214], especially the special case of *birationally commutative* surfaces. We then close with a brief overview of some other recent themes in noncommutative projective geometry. By its nature this section is more of a survey, so we will be able to give fewer details, and do not include exercises.

5.1. Classification of noncommutative projective curves. While there is no single definition of what a noncommutative curve or surface should be, one obvious approach to the projective case is to take finitely graded noetherian algebras A with $GKdim(A) = d + 1$, and consider the corresponding noncommutative projective schemes qgr-A as the d-dimensional noncommutative projective schemes. In this way curves correspond to algebras of GK-dimension 2 and surfaces to algebras of GK-dimension 3. In much of the preceding we have concentrated on domains A only, in which case one can think of qgr-A as being an analog of an integral projective scheme, or a variety. We continue to focus on domains here.

To study noncommutative projective (integral) curves, we consider finitely graded domains A with $GKdim(A) = 2$. Artin and Stafford proved very strong results about the structure of these, as we will see in the next theorem. First, we need to review a few more definitions. Given a k-algebra R with automorphism $\sigma : R \to R$, the skew-Laurent ring $R[t, t^{-1}; \sigma]$ is a k-algebra whose elements are Laurent polynomials $\sum_{i=a}^{b} r_i t^i$ with $a \leq b$ and $r_i \in R$, and with the unique associative multiplication rule determined by $ta = \sigma(a)t$ for all $a \in R$ (see [117, Chapter 1]). Assume now that A is a finitely graded domain with $A \neq k$ and $GKdim(A) < \infty$. In this case, one can localize A at the set of nonzero homogeneous elements in A, obtaining its *graded quotient ring* $Q = Q_{gr}(A)$. Since every homogeneous element of Q is a unit, $Q_0 = D$ is a division ring. Moreover, if $d \geq 1$ is minimal such that $Q_d \neq 0$, then choosing any $0 \neq t \in Q_d$ the elements in Q are Laurent polynomials of the form $\sum_{i=m}^{n} a_i t^i$ with $a_i \in D$. If $\sigma : D \to D$ is the automorphism given by conjugation by t, that is $a \mapsto tat^{-1}$, then one easily sees that $Q \cong D[t, t^{-1}; \sigma]$. For example, if A is commutative and generated in degree 1, then $Q_{gr}(A) = k(X)[t, t^{-1}]$, where $k(X)$ must be the field of rational functions of $X = \operatorname{Proj} A$, since it is the field of fractions of each $A_{(x)}$ with $x \in A_1$. Thus, for a general finitely graded A the division ring D may be thought of as a noncommutative analogue of a rational function field.

The following theorem of Artin and Stafford shows that noncommutative projective integral curves are just commutative curves. Recall that our standing convention is that k is algebraically closed.

THEOREM 5.1.1. [7, Theorems 0.1, 0.2] Let A be a finitely graded domain with $GKdim A = 2$.

(1) The graded quotient ring $Q_{gr}(A)$ of A is isomorphic to $K[t, t^{-1}; \sigma]$ for some field K with $\operatorname{tr. deg}(K/k) = 1$ and some automorphism $\sigma : K \to K$.

(2) If A is generated in degree 1, then there is an injective map $\phi : A \to B(X, \mathcal{L}, \sigma)$ for some integral projective curve X with function field $k(X) = K$, some ample invertible sheaf $\mathcal{L}$ on X, and the automorphism $\sigma : X \to X$ corresponding to $\sigma : K \to K$. Moreover, ϕ is an isomorphism in all large degrees; in particular, qgr-$A \sim \operatorname{coh} X$.

Artin and Stafford also gave a detailed description of those algebras A satisfying the hypotheses of the theorem except generation in degree 1 [5, Theorem 0.4, 0.5]. A typical example of this type with σ of infinite order is the idealizer ring studied in Exercise 4.5.5. In a follow-up paper [8], Artin and Stafford classified semiprime graded algebras of GK-dimension 2; these rings are described in terms of a generalization of a twisted homogeneous coordinate ring involving a sheaf of orders on a projective curve.

Another approach to the theory of noncommutative curves is to classify categories which have all of the properties a category of coherent sheaves on a nice curve has. Reiten and Van den Bergh proved in [183] that any connected noetherian Ext-finite hereditary abelian category satisfying Serre duality over k is either the category of coherent sheaves on a sheaf of hereditary $\mathcal{O}_X$-orders, where X is a smooth curve, or else one of a short list of exceptional examples. We refer to [214, Section 7] for the detailed statement, and the definitions of some of the properties involved. Intuitively, the categories this theorem classifies are somewhat different than those arising from graded rings, since the hypotheses demand properties which are analogs of properness and smoothness of the noncommutative curve, rather than projectivity.

There are many other categories studied in the literature which should arguably be thought of as examples of noncommutative quasi-projective curves. For example, the category of $\mathbb{Z}$-graded modules over the Weyl algebra $A = k\langle x, y \rangle / (yx - xy - 1)$, where A is $\mathbb{Z}$-graded with $\deg x = 1$, $\deg y = -1$, can also be described as the quasi-coherent sheaves on a certain stack of dimension 1. See [209] for this geometric description, and [202] for more details about the structure of this category. Some other important examples are the *weighted projective lines* studied by Lenzing and others (see [162] for a survey). As of yet, there is not an overarching theory of noncommutative curves which encompasses all of the different kinds of examples mentioned above.

5.2. The minimal model program for surfaces and Artin's conjecture.

Before discussing noncommutative surfaces, we first recall the main idea of the classification of commutative surfaces. Recall that two integral surfaces are *birational* if they have isomorphic fields of rational functions, or equivalently, if they have isomorphic open subsets [128, Section I.4]. The coarse classification of projective surfaces divides them into birational equivalence classes. There are various numerical invariants for projective surfaces which are constant among all surfaces in a birational class. The most important of these is the Kodaira dimension; some others include the arithmetic and geometric genus [128, Section V.6]. There is a good understanding of the possible birational equivalence classes in terms of such numerical invariants.

For the finer classification of projective surfaces, one seeks to understand the smooth projective surfaces within a particular birational class. A fundamental theorem states that any surface in the class can be obtained from any other by a sequence of monoidal transformations, that is, blowups at points or the reverse process, blowdowns of exceptional curves [128, Theorem V.5.5]. This gives a specific way to relate surfaces within a class, and some important properties of a surface, such as the Picard group, change in a simple way under a monoidal transformation [128, Proposition V.3.2]. Every birational class has at least one *minimal model*, a smooth surface which has no exceptional curves, and every surface in the class is obtained from some sequence of blowups starting with some minimal model. In fact, most birational classes have a unique minimal model, with the exception of the classes of rational and ruled surfaces [128, Remark V.5.8.4]. For example, the birational class of rational surfaces—those with function field $k(x, y)$—has as minimal models $\mathbb{P}^2$, $\mathbb{P}^1 \times \mathbb{P}^1$, and the other Hirzebruch surfaces [128, Example V.5.8.2].

An important goal in noncommutative projective geometry is to find a classification of noncommutative surfaces, modeled after the classification of commutative surfaces described above. It is easy to find an analog of birationality: as we have already mentioned, for a finitely graded domain A its graded quotient ring has the form $Q_{\mathrm{gr}}(A) \cong D[t, t^{-1}; \sigma]$, where the division ring D plays the role of a field of rational functions. Thus the birational classification of noncommutative surfaces requires the analysis of which division rings D occur as $Q_{\mathrm{gr}}(A)_0$ for some finitely graded domain A of GK-dimension 3. In [4], Artin gave a list of known families of such division rings D and conjectured that these are all the possible ones, with a deformation-theoretic heuristic argument as supporting evidence. Artin's conjecture is still open and remains one of the important but elusive goals of noncommutative projective geometry. See [214, Section 10.1] for more details. In the remainder of this section, we focus on the other part of the classification problem, namely, understanding how surfaces within a birational class are related.

5.3. Birationally commutative surfaces. Let A be a finitely graded domain of finite GK-dimension, with graded ring of fractions $Q_{\mathrm{gr}}(A) \cong D[t, t^{-1}; \sigma]$. When $D = K$ is a field, we say that A is *birationally commutative*. By the Artin–Stafford theorem (Theorem 5.1.1), this holds automatically when GKdim$(A) = 2$, that is, for noncommutative projective curves. Of course, it is not automatic for surfaces. For example, the reader may check that the quantum polynomial ring $A = k\langle x, y, z \rangle / (yx - pxy, zy - ryz, xz - qzx)$ from Example 3.0.12 is birationally commutative if and only if $pqr = 1$. As we saw in that example, this is the same condition that implies that the scheme parametrizing the point modules is $\mathbb{P}^2$, rather than three lines.

Let A be some finitely graded domain of GK-dimension 3 with $Q_{\mathrm{gr}}(A) \cong D[t, t^{-1}; \sigma]$. The problem of classification within this birational class is to understand and relate the possible finitely graded algebras A which have a graded quotient ring Q with $Q_0 \cong D$, and their associated noncommutative projective schemes qgr-A. For simplicity, one may focus on one slice of this problem at a time and consider only those A with $Q_{gr}(A) \cong D[t, t^{-1}; \sigma]$ for the fixed automorphism σ.

We have now seen several examples where a finitely graded algebra A has a homomorphism to a twisted homogeneous coordinate ring $B(X, \mathcal{L}, \sigma)$: in the classification of noncommutative projective curves (Theorem 5.1.1), and in the sketch of the proof of the classification of AS-regular algebras of dimension 3 at the end of Section 3. In fact, this is a quite general phenomenon.

THEOREM 5.3.1. [191, Theorem 1.1] Let A be a finitely graded algebra which is strongly noetherian and generated in degree 1. By Theorem 3.1.2, the maps ϕ_m from Proposition 3.0.10 are isomorphisms for $m \geq m_0$, so that the point modules of A are parametrized by the projective scheme $X = X_{m_0}$. Then we can canonically associate to A an invertible sheaf $\mathcal{L}$ on X, an automorphism $\sigma : X \to X$, and a ring homomorphism $\phi : A \to B(X, \mathcal{L}, \sigma)$ which is surjective in all large degrees. The kernel of ϕ is the ideal of elements that kill all R-point modules of A, for all commutative k-algebras R.

The σ and $\mathcal{L}$ in the theorem above arise naturally from the data of the point modules; for example, σ is induced by the truncation shift map on point modules which sends a point module P to $P(1)_{\geq 0}$. Theorem 5.3.1 can also be interpreted in the following way: a strongly noetherian finitely graded algebra, generated in degree 1, has a unique largest factor ring determined by the point modules, and this factor ring is essentially (up to a finite dimensional vector space) a twisted homogeneous coordinate ring. Roughly, one may also think of X as the largest commutative subscheme of the noncommutative projective scheme qgr-A.

While the canonical map to a twisted homogeneous coordinate ring was the main tool in the classification of AS-regular algebras of dimension 3, it is less powerful as a technique for understanding AS-regular algebras of global dimension 4 and higher. Such algebras can have small point schemes, in which case the kernel of the canonical map is too large for the map to give much interesting information. For example, there are many examples of regular algebras of global dimension 4 whose point scheme is 0-dimensional, and presumably the point scheme of an AS-regular algebra of dimension 5 or higher might be empty. There is a special class of algebras, however, which are guaranteed to have a rich supply of point modules, and for which the canonical map leads to a strong structure result.

COROLLARY 5.3.2. [191, Theorem 1.2] Let A satisfy the hypotheses of Theorem 5.3.1, and assume in addition that A is a domain which is birationally commutative with $Q_{gr}(A) \cong K[t, t^{-1}; \sigma]$. Then the canonical map $\phi : A \to B(X, \mathcal{L}, \sigma)$ described by Theorem 5.3.1 is injective, and thus is an isomorphism in all large degrees.

The main idea behind this corollary is that the positive part of the quotient ring itself, $K[t; \sigma]$, is a K-point module for A. It obviously has annihilator 0 in A, so no nonzero elements annihilate all point modules (over all base rings).

Corollary 5.3.2 completely classifies birationally commutative algebras, generated in degree 1, which happen to be known in advance to be strongly noetherian. However, there are birationally commutative algebras of GK-dimension 3 which are noetherian but not strongly noetherian, and so a general theory of birationally commutative surfaces needs to account for these.

EXAMPLE 5.3.3. Let X be a projective surface with automorphism $\sigma : X \to X$, and let $\mathcal{L}$ be an ample invertible sheaf on X. Choose an ideal sheaf $\mathcal{I}$ defining a 0-dimensional subscheme Z of X. For each $n \geq 0$, set $\mathcal{I}_n = \mathcal{I}\mathcal{I}^\sigma \ldots \mathcal{I}^{\sigma^{n-1}}$, and let $\mathcal{L}_n = \mathcal{L} \otimes \mathcal{L}^\sigma \otimes \cdots \otimes \mathcal{L}^{\sigma^{n-1}}$ as in Definition 4.3.1. Now we define the *naïve blowup algebra*

$$R = R(X, \mathcal{L}, \sigma, Z) = \bigoplus_{n \geq 0} H^0(X, \mathcal{I}_n \otimes \mathcal{L}_n) \subseteq B = B(X, \mathcal{L}, \sigma) = \bigoplus_{n \geq 0} H^0(X, \mathcal{L}_n),$$

so that R is a subring of B. The ring R is known to be noetherian but not strongly noetherian when every point p in the support of Z lies on a *critically dense* orbit, that is, when every infinite subset of $\{\sigma^i(p) \mid i \in \mathbb{Z}\}$ has closure in the Zariski topology equal to all of X [190, Theorem 1.1].

A very explicit example of a naïve blowup algebra is the following.

EXAMPLE 5.3.4. Let $R = R(\mathbb{P}^2, \mathcal{O}(1), \sigma, Z)$, where Z is the single reduced point $(1 : 1 : 1)$, and $\sigma(a : b : c) = (qa : rb : c)$ for some $q, r \in k$, where char $k = 0$. As long as q and r are algebraically independent over $\mathbb{Q}$, the σ-orbit of $(1 : 1 : 1)$ is critically dense and R is a noetherian ring [186, Theorem 12.3]. In this case one has

$$B(\mathbb{P}^2, \mathcal{O}(1), \sigma) \cong k\langle x, y, z\rangle/(yx - qr^{-1}xy, zy - ryz, xz - q^{-1}zx),$$

and R is equal to the subalgebra of B generated by $x - z$ and $y - z$.

Historically, Example 5.3.4 was first studied by D. Jordan as a ring generated by Eulerian derivatives [145]. Later, these specific examples were shown to be noetherian in most cases [186], and last the more general notion of naïve blowup put such examples in a more general geometric context [151]. The name naïve blowup reflects the fact that the definition of such rings is a kind of twisted version of the Rees ring construction which is used to define a blowup in the commutative case [128, Section II.7]. The relationship between the noncommutative projective schemes qgr-$R(X, \mathcal{L}, \sigma, Z)$ and qgr-$B(X, \mathcal{L}, \sigma) \simeq X$ is more obscure, however, and does not have the usual geometric properties one expects of a blowup [151, Section 5].

The examples we have already seen are typical of birationally commutative surfaces, as the following result shows.

THEOREM 5.3.5. [189] Let A be a noetherian finitely graded k-algebra, generated in degree 1, with $Q_{gr}(A) \cong K[t, t^{-1}; \sigma]$, where K is a field of transcendence degree 2 over k. Assume in addition that there exists a projective surface Y with function field $K = k(Y)$, and an automorphism $\sigma : Y \to Y$ which induces the automorphism $\sigma : K \to K$. Then A is isomorphic to a naïve blowup algebra $A \cong R(X, \mathcal{L}, \sigma, Z)$, where X is a surface with $k(X) = K$, $\sigma : X \to X$ is an automorphism corresponding to $\sigma : K \to K$, the sheaf $\mathcal{L}$ is σ-ample, and every point of Z lies on a critically dense σ-orbit.

Sierra has extended this theorem to the case of algebras not necessarily generated in degree 1 [203]. In this setting, one gets a more general class of possible examples, which are a bit more technical to describe, but all of the examples are still defined in terms of sheaves on a surface X with automorphism σ. The condition in Theorem 5.3.5 that there exists an automorphism of a projective surface Y inducing σ, which is also a hypothesis in Sierra's generalization, may seem a bit mysterious. It turns out that there are automorphisms of fields of transcendence degree 2 which do not correspond to an automorphism of any projective surface with that field as its fraction field [81]. A noetherian ring A with a graded quotient ring $Q_{\mathrm{gr}}(A) = K[t, t^{-1}; \sigma]$, where σ is a field automorphism of this strange type, was constructed in [187].

The rings classified in Theorem 5.3.5 are either twisted homogeneous coordinate rings $B(X, \mathcal{L}, \sigma)$ (these are the cases where $Z = \varnothing$ and are the only strongly noetherian ones), or naïve blowup algebras inside of these. The possible smooth surfaces X occurring are all birational and so are related to each other via monoidal transformations, as described above. Each $B = B(X, \mathcal{L}, \sigma)$ has qgr-$B \simeq \mathrm{coh}\, X$, while for each $R = R(X, \mathcal{L}, \sigma, Z)$, the category qgr-$R$ can be thought of as a naïve blowup of $\mathrm{coh}\, X$. Thus within this birational class, it is true that the possible examples are all related by some kind of generalized blowup or blowdown procedures, in accordance with the intuition coming from the classification of commutative projective surfaces.

5.4. A recent application of point modules to universal enveloping algebras. Since these lectures were originally delivered, Sierra and Walton discovered a stunning new application of point modules which settled a long standing open question in the theory of enveloping algebras [204], and we wish to briefly discuss this here.

Let k have characteristic 0 and consider the infinite-dimensional Lie algebra L with k-basis $\{x_i \mid i \geq 1\}$ and bracket $[x_i, x_j] = (j - i)x_{i+j}$, which is known as the positive part of the Witt algebra. Since L is a graded Lie algebra, its universal enveloping algebra $A = U(L)$ is connected $\mathbb{N}$-graded, and has the Hilbert series of a polynomial ring in variables of weights $1, 2, 3, \ldots$, by the PBW theorem. Thus the function $f(n) = \dim_k A_n$ is actually the partition function, and from this one may see that A has infinite GK-dimension but subexponential growth. The question of whether A is noetherian arose in the work of Dean and Small [79]. Stephenson and Zhang proved that finitely graded algebras of exponential growth cannot be noetherian [218], so it is an obvious question whether a finitely graded algebra of subexponential but greater than polynomial growth could possibly be noetherian. The ring A was an obvious test case for this question. Note that a canonical map as in Theorem 5.3.1 does not necessarily exist, since one does not know if A is strongly noetherian. Nonetheless, using point modules Sierra and Walton found a factor ring A/I which is birationally commutative and can be described in terms of sheaves on a certain surface, though it is slightly more complicated than a twisted homogeneous coordinate ring. Using geometry, the authors proved that A/I is non-noetherian, so that A is non-noetherian also. The ideal I does not have obvious generators, and it is unlikely that I would have been discovered without using point modules, or that the factor ring A/I could have been successfully analyzed without using geometric techniques.

5.5. Brief overview of other topics. In this section we have described one particular thread of recent research in noncommutative projective geometry, one with which we are intimately familiar. To close, we give some very brief summaries of a few other themes of current research. This list is not meant to be comprehensive.

5.5.1. *Noncommutative projective surfaces.* The reader can find a survey of some important work on noncommutative projective surfaces in the second half of [214]. For example, there is a rich theory of noncommutative quadric surfaces, that is, noncommutative analogs of subschemes of $\mathbb{P}^3$ defined by a degree 2 polynomial; see for example [232] and [211]. We should also mention Van den Bergh's theory of noncommutative blowing up, which allows one to blow up a point lying on a commutative curve contained in a noncommutative surface, under certain circumstances [229]. These blowups do have properties analogous to commutative blowups. (Van den Bergh's blowups and naïve blowups generally apply in completely different settings.) The author, Sierra, and Stafford have studied the classification of surfaces within the birational class of the generic Sklyanin algebra [188], and shown that some of these surfaces are related via blowups of Van den Bergh's kind.

In a different direction, there is a deep theory of maximal orders on commutative surfaces, which are certain sheaves of algebras on the surface which are locally finite over their centers. Chan and Ingalls [64] laid the foundations of a minimal model program for the classification of such orders, which has been studied for many special types of orders. The reader may find an introduction to the theory in [63].

5.5.2. *Regular algebras of dimension 4.* Since the classification of AS-regular algebras of dimension 3 was acheived, much attention has been focused on regular algebras of dimension 4. The Sklyanin algebra of dimension 4–which also has point modules paramaterized by an elliptic curve, like its analog in dimension 3–was one of the first regular algebras of dimension 4 to be intensively studied, see [210] for example. There has been much interest in the problem of classification of AS-regular algebras of dimension 4, which have three possible Hilbert series [167, Proposition 1.4]. Many interesting examples of 4-dimensional regular algebras have been given with point schemes of various kinds. In addition, a number of important new constructions which produce regular algebras have been invented, for example the double Ore extensions due to Zhang and Zhang [251], and the skew graded Clifford algebras due to Cassidy and Vancliff [62]. Another interesting technique developed in [167] is the study of the A_∞-algebra structure on the Ext algebra of a AS-regular algebra. Some combination of these techniques has been used to successfully classify AS-regular algebras of global dimension 4, generated in degree 1, with a nontrivial grading by $\mathbb{N} \times \mathbb{N}$ [167], [252], [192]. The general classification problem is an active topic of research.

5.5.3. *Calabi–Yau algebras.* The notion of Calabi–Yau algebra, which was originally defined by Ginzburg [112], appeared in several of the lecture courses at the MSRI workshop. There is also a slightly more general notion called a twisted or skew Calabi–Yau algebra. In the particular setting of finitely graded algebras, the definition of twisted Calabi–Yau algebra is actually equivalent to the definition of AS-regular algebra [184, Lemma 1.2]. There has been much interesting work on (twisted) Calabi–Yau algebras which arise as factor algebras of path algebras of quivers, especially the study of when the relations of such algebras come from superpotentials. One may think of this theory as a generalization of the theory of AS-regular algebras to the non-connected graded case. The literature in this subject has grown quickly in recent years. We mention [39] as one paper close in spirit to these lectures, since it includes a study of the 4-dimensional Sklyanin algebras from the Calabi–Yau algebra point of view.

5.5.4. *Noncommutative invariant theory.* The study of the rings of invariants of finite groups acting on commutative polynomial rings is now classical. It is natural to generalize this to study the invariant rings of group actions on AS-regular algebras, the noncommutative analogs of polynomial rings. A further generalization allows a finite-dimensional Hopf algebra to act on the ring instead of a group. A number of classical theorems concerning when the ring of invariants is Gorenstein, regular, and so on, have been generalized to this context. Two recent papers from which the reader can get an idea of the theory are [154] and [65].

CHAPTER II

Deformations of algebras in noncommutative geometry

Travis Schedler

These are significantly expanded lecture notes for the author's minicourse at MSRI in June 2012. In these notes, following, e.g., [98, 156, 101], we give an example-motivated review of the deformation theory of associative algebras in terms of the Hochschild cochain complex as well as quantization of Poisson structures, and Kontsevich's formality theorem in the smooth setting. We then discuss quantization and deformation via Calabi–Yau algebras and potentials. Examples discussed include Weyl algebras, enveloping algebras of Lie algebras, symplectic reflection algebras, quasihomogeneous isolated hypersurface singularities (including du Val singularities), and Calabi–Yau algebras.

The exercises are a great place to learn the material more detail. There are detailed solutions provided, which the reader is encouraged to consult if stuck. There are some *starred* (parts of) exercises which are quite difficult, so the reader can feel free to skip these (or just glance at them).

There are a lot of remarks, not all of which are essential; so many of them can be skipped on a first reading.

We will work throughout over a field $\Bbbk$. A lot of the time we will need it to have characteristic zero; feel free to assume this always.

Acknowledgments. These notes are based on my lectures for MSRI's 2012 summer graduate workshop on noncommutative algebraic geometry. I am grateful to MSRI and the organizers of the Spring 2013 MSRI program on noncommutative algebraic geometry and representation theory for the opportunity to give these lectures; to my fellow instructors and scientific organizers Gwyn Bellamy, Dan Rogalski, and Michael Wemyss for their help and support; to the excellent graduate students who attended the workshop for their interest, excellent questions, and corrections; and to Chris Marshall and the MSRI staff for organizing the workshop. I am grateful to Daniel Kaplan and Michael Wong for carefully studying these notes and providing many corrections.

Introduction

Deformation theory is ubiquitous in mathematics: given any sort of structure it is a natural (and often deep and interesting) question to determine its deformations. In geometry there are several types of deformations one can consider. The most obvious is actual deformations, such as a family of varieties (or manifolds), X_t, parametrized by $t \in \mathbb{R}$ or $\mathbb{C}$. Many times this is either too difficult to study or there are not enough actual deformations (which are not isomorphic to the original variety X), so it makes sense to consider *infinitesimal* deformations: this is a family of structures X_t where $t \in \mathbb{C}[\varepsilon]/(\varepsilon^2)$, i.e., the type of family which can be obtained from an actual family by taking the tangent space to the deformation. Sometimes, but not always, infinitesimal deformations can be extended to higher-order deformations, i.e., one can extend the family to a family where $t \in \mathbb{C}[\varepsilon]/(\varepsilon^k)$ for some $k \geq 2$. Sometimes these extensions exist to all orders. A *formal deformation* is the same thing as a compatible family of such extensions for all $k \geq 2$, i.e., such that restriction from order k to order $j < k$ recovers the deformation at order j.

In commutative algebraic geometry, affine varieties correspond to commutative algebras (which are finitely generated and have no nilpotent elements): they are of the form $\operatorname{Spec} A$ for A commutative. In "noncommutative affine algebraic geometry," therefore, it makes sense to study deformations of associative algebras. This is the main subject of these notes. As we will see, such deformations arise from and have applications to a wide variety of subjects in representation theory (of algebras, Lie algebras, and Lie groups), differential operators and D-modules, quantization, rational homotopy theory, Calabi–Yau algebras (which are a noncommutative generalization of affine Calabi–Yau varieties), and many other subjects. Moreover, many of the important examples of noncommutative algebras studied in representation theory, such as symplectic reflection algebras (the subject of Chapter III) and many noncommutative projective spaces (the subject of Chapter I) arise in this way. The noncommutative resolutions studied in Chapter IV can also be deformed, and the resulting deformation theory should be closely related to that of commutative resolutions.

Of particular interest is the study of noncommutative deformations of commutative algebras. These are called quantizations. Whenever one has such a deformation, the first-order part of the deformation (i.e., the derivative of the deformation) recovers a Lie bracket on the commutative algebra, which is a derivation in each component. A commutative algebra together with such a bracket is called a Poisson algebra. Its spectrum is called an (affine) Poisson variety. By convention, a quantization is an associative deformation of a commutative algebra equipped with a fixed Poisson bracket, i.e., a noncommutative deformation which, to first order, recovers the Poisson bracket. Poisson brackets are very old and appeared already in classical physics (particularly Hamiltonian mechanics) and this notion of quantization is already used in the original formulation of quantum mechanics. In spite of the old history, quantization has attracted a lot of recent attention in both mathematics and physics.

One of the most important questions about quantization, which is a central topic of these notes, is of their existence: given a Poisson algebra, does there exist a quantization, and can one construct it explicitly? In the most nondegenerate case, the Poisson variety is a smooth symplectic variety; in this case, the analogous problem for C^∞ manifolds was answered in the affirmative in [78], and an important explicit construction was given in [105]. In the general case of affine algebraic Poisson varieties, the answer is negative; see Mathieu's example in Remark 2.3.14 below. A major breakthrough occurred in 1997 with Kontsevich's proof that, for arbitrary smooth C^∞ manifolds, and for real algebraic affine space $\mathbb{R}^n$, all Poisson structures can be quantized. In fact, Kontsevich constructed a natural (i.e., functorial) quantization, and indicated how to extend it to general smooth affine (or suitable nonaffine) varieties; the details of this extension and a study of the obstructions for nonaffine varieties were first completed by Yekutieli [245], see also [233], but there have been a large body of refinements to the result, e.g., in [86] for the affine setting, and in [59, 58] for a sheaf version of the global setting.

More recently, the study of Calabi–Yau algebras, mathematically pioneered by Ginzburg [112], has become extremely interesting. This is a subject of overlap of all the chapters of this book, since many of the algebras studied in all chapters are Calabi–Yau, including many of the regular algebras studied in Chapter I, all of the symplectic reflection algebras studied in Chapter III, and all of the noncommutative crepant resolutions of Gorenstein singularities studied in Chapter IV. In the commutative case, Calabi–Yau algebras are merely rings of functions on affine Calabi–Yau varieties; the noncommutative generalization is much more interesting, but shares some of the same properties. Beginning with a Calabi–Yau variety, one can consider Calabi–Yau deformations (see, e.g., [235] for a study of their moduli). In [101], this was applied to the quantization of del Pezzo surfaces: Etingof and Ginzburg first reduced the problem to the affine surface obtained by deleting an elliptic curve; this affine surface embeds into $\mathbb{C}^3$. They then deformed the ambient smooth Calabi–Yau variety $\mathbb{C}^3$, together with the hypersurface. One advantage of this is the fact that many Calabi–Yau algebras can be defined only by a single noncommutative polynomial, called the (super)potential, such that the relations for the algebra are obtained by differentiating the potential (see, e.g., [112, 39]). (It was in fact conjectured that all Calabi–Yau algebras are obtained in this way; this was proved for graded algebras [38] and more generally for completed algebras [234], but is false in general [77].) Thus the deformations of $\mathbb{C}^3$ studied in [101] are very explicit and given by deforming the potential function. To quantize the original affine surface, one then takes a quotient of such a deformed algebra by a central element. We end this chapter by explaining this beautiful construction.

Deformations of algebras are closely related to a lot of other subjects we are not able to discuss here. Notably, this includes the mathematical theory of quantum groups, pioneered in the 1980s by Drinfeld, Jimbo, and others: this is the analogue for groups of the latter, where one quantizes Poisson–Lie groups rather than Poisson varieties. Mathematically, this means one deforms Hopf algebras rather than associative algebras, and one begins with the commutative Hopf algebras of functions on a group. One can moreover consider actions of such quantum groups on noncommutative spaces, which has attracted recent attention in, e.g., [104, 66]. Quantum groups also have close relationships to the formality theorem: Etingof and Kazhdan proved in [102] that all Lie bialgebras can be quantized to a quantum group (a group analogue of Kontsevich's existence theorem), and Tamarkin gave a new proof of Kontsevich's formality theorem for $\mathbb{R}^n$ which used this result; in fact, this result was stronger, as it takes into account the cup product structure on polyvector fields $\Lambda^\bullet_{\mathcal{O}(\mathbb{R}^n)} \mathrm{Vect}(\mathbb{R}^n)$, i.e., its full differential graded Gerstenhaber algebra structure, rather than merely considering its differential graded Lie algebra structure. Moreover, using [223], Tamarkin's proof works over any field of characteristic zero, and requires only a rational Drinfeld associator rather than the Etingof–Kazhdan theorem; in this form the latter theorem also follows as a consequence [222].

A rough outline of these notes is as follows. In Section 1, we will survey some of the most basic and interesting examples of deformations of associative algebras. This serves not merely as a motivation, but also begins the study of the theory and important concepts and constructions. The reader should have in mind these examples while reading the remainder of the text. In particular, we will consider Weyl algebras and algebras of differential operators, universal enveloping algebras of Lie algebras, quantizations of the nilpotent cone, and we will conclude by explaining the Beilinson–Bernstein localization theorem, which relates all of these examples and is one of the cornerstones of geometric representation theory.

In Section 2, we will define the notions of formal deformations, Poisson structures, and deformation quantization. We culminate with the statement of Kontsevich's theorem (and its refinements) on deformation quantization of smooth Poisson manifolds and smooth affine Poisson varieties. We will come back to this in Section 4.

In Section 3, we begin a systematic study of deformation theory of algebras, focusing on their Hochschild cohomology. This allows us to classify infinitesimal deformations and the obstructions to second-order deformations, as well as some theory of deforming their modules.

In Section 4 we pass from the Hochschild cohomology to the Hochschild cochain complex, which is the structure of a differential graded Lie algebra. This allows us to classify formal deformations. We then return to the subject of Kontsevich's theorem, and explain how it follows from his more refined statement on formality of the Hochschild cochain complex as a differential graded Lie algebra.

Finally, in Section 5 we discuss Calabi–Yau algebras, which is a subject that connects all of the chapters of the book. This came up already in previous sections because our main examples up to this point are all (twisted) Calabi–Yau, including Weyl algebras, universal enveloping algebras (as well as many of their central reductions), and symplectic reflection algebras. In this section, we explain how to define and deform Calabi–Yau algebras using potentials. We then apply this to quantization of hypersurfaces in $\mathbb{C}^3$, following [101].

We stress that these notes only scratch the surface of the theory of deformations of associative algebras. Many subjects are not discussed, such as Gerstenhaber and Schack's detailed study via Hochschild cohomology and their cocycles (via lifting one by one from k-th to $(k+1)$-st order deformations); other important subjects are mentioned only in the exercises, such as the Koszul deformation principle (Theorem 2.8.5).

1. Motivating examples

In this section, we begin with the definitions of graded associative algebras and filtered deformations. We proceed with the fundamental examples of Weyl algebras and universal enveloping algebras of Lie algebras, which we define and discuss, along with the invariant subalgebras of Weyl algebras. We then introduce the concept of Poisson algebras, which one obtains from a filtered deformation of a commutative algebra, such as in the previous cases, and define a filtered quantization, which is a filtered deformation whose associated Poisson algebra is a fixed one. We consider the algebra of functions on the nilpotent cone of a semisimple Lie algebra, and explain how to construct its filtered quantization by central reductions of the universal enveloping Lie algebra. We explain how the geometry of the nilpotent cone encapsulates representations of Lie algebras, via the Beilinson–Bernstein theorem. Finally, we conclude by discussing an important example of deformations of a noncommutative algebra, namely the preprojective algebra of a quiver; this also allows us to refer to quivers in later examples in the text.

1.1. Preliminaries. A central object of study for us is a graded associative algebra. First we define a graded vector space:

DEFINITION 1.1.1. A $\mathbb{Z}$-graded vector space is a vector space $V = \bigoplus_{m \in \mathbb{Z}} V_m$. A homogeneous element is an element of V_m for some $m \in \mathbb{Z}$. For $v \in V_m$ we write $|v| = m$.

REMARK 1.1.2. There are two conventions for indicating the grading: either using a subscript, or a superscript. We will switch to the latter in later sections when dealing with dg algebras. See also Remark 2.1.1.

DEFINITION 1.1.3. A $\mathbb{Z}$-graded associative algebra is an algebra $A = \bigoplus_{m \in \mathbb{Z}} A_m$ with $A_m A_n \subseteq A_{m+n}$ for all $m, n \in \mathbb{Z}$.

A graded associative algebra is in particular a graded vector space, so we still write $|a| = n$ when $a \in A_n$.

The simplest (but very important) example of a graded associative algebra is a tensor algebra TV for V a vector space:

DEFINITION 1.1.4. The tensor algebra TV is defined as $TV := \bigoplus_{m \geq 0} T^m V$, with $T^m V := V^{\otimes m}$. The multiplication is the tensor product, and the grading is tensor degree: $(TV)_m := T^m V$.

Note that $(TV)_0 = \Bbbk$.

The next example, which is commutative, is a symmetric algebra $\operatorname{Sym} V$ for V a vector space:

DEFINITION 1.1.5. The symmetric algebra $\operatorname{Sym} V$ is defined by

$$\operatorname{Sym} V := TV/(xy - yx \mid x, y \in V).$$

It is graded again by tensor degree: $(\operatorname{Sym} V)_m$ is defined to be the image of $(TV)_m = T^m V$, also denoted $\operatorname{Sym}^m V$. The exterior algebra $\bigwedge V := TV/(x^2 \mid x \in V)$ is defined similarly, with $(\bigwedge V)_m = \bigwedge^m V$.

In other words, $(\mathrm{Sym}\,V)_m$ is spanned by monomials of length m: $v_1 \cdots v_m$ for $v_1, \ldots, v_m \in V$. Note that $(\mathrm{Sym}\,V)_m = \mathrm{Sym}^m V := T^m V / S_m$, the quotient of the vector space $T^m V$ by the action of the symmetric group S_m, which is called the m-th symmetric power of V.

Given a commutative algebra B (which is finitely generated over $\Bbbk$ and has no nilpotent elements), we can consider equivalently the affine variety $\mathrm{Spec}\,B$. We use the notation $\mathcal{O}(\mathrm{Spec}\,B) := B$ for affine varieties.

There is a geometric interpretation of the above, which the reader unfamiliar with algebraic groups can safely skip: a grading on a commutative algebra B is the same as an action of the algebraic group $\mathbb{G}_m = \Bbbk^\times$ (the group whose $\Bbbk$-points are nonzero elements of $\Bbbk$ under multiplication) on $\mathrm{Spec}\,B$, with B_m the weight space of the character $\chi(z) = z^m$ of $\mathbb{G}_m$.

REMARK 1.1.6. In this section we will deal with many such algebras satisfying the commutativity constraint $xy = yx$. There is a very important alternative commutativity constraint, called supercommutativity or graded commutativity, given by $xy = (-1)^{|x||y|} yx$; this constraint will come up in later sections. The choice of commutativity constraint is linked to how one views the grading: gradings where the former (usual) constraint is applied are often "weight" gradings, coming from an action of the multiplicative group $\mathbb{G}_m$ (otherwise known as $\Bbbk^\times$) as explained above, whereas gradings where the latter type of constraint are applied are often "homological" or "cohomological" gradings, coming up, e.g., in Hochschild cohomology of algebras.

1.2. Weyl algebras. Let $\Bbbk$ have characteristic zero. Letting $\Bbbk\langle x_1, \ldots, x_n \rangle$ denote the noncommutative polynomial algebra in variables $x_1, \ldots, x_n$, one can define the n-th Weyl algebra as

$$\mathrm{Weyl}_n := \Bbbk\langle x_1, \ldots, x_n, y_1, \ldots, y_n \rangle / ([x_i, y_j] - \delta_{ij}, [x_i, x_j], [y_i, y_j]),$$

denoting here $[a, b] := ab - ba$.

EXERCISE 1.2.1. (a) Show that, setting $y_i := -\partial_i$, one obtains a surjective homomorphism from Weyl_n to the algebra, $\mathcal{D}(\mathbb{A}^n)$, of differential operators on $\Bbbk[x_1, \ldots, x_n] = \mathcal{O}(\mathbb{A}^n)$ with polynomial coefficients (this works for any characteristic).

(b)(*) Show that, assuming $\Bbbk$ has characteristic zero, this homomorphism is an isomorphism.

More invariantly, recall that a *symplectic vector space* is a vector space equipped with a skew-symmetric, nondegenerate bilinear form (called the symplectic form).

DEFINITION 1.2.2. Let V be a symplectic vector space with symplectic form $(-, -)$. Then, the Weyl algebra $\mathrm{Weyl}(V)$ is defined by

$$\mathrm{Weyl}(V) = TV / (xy - yx - (x, y)).$$

(This definition makes sense even if $(-, -)$ is degenerate, but typically one imposes nondegeneracy.)

EXERCISE 1.2.3. For every n, we can consider the symplectic vector space V of dimension $2n$ with basis $(x_1, \ldots, x_n, y_1, \ldots, y_n)$, and form

$$(x_i, y_j) = \delta_{ij} = -(y_j, x_i), (x_i, x_j) = 0 = (y_i, y_j),$$

for all $1 \le i, j \le n$. Show that $\mathsf{Weyl}(V) = \mathsf{Weyl}_n$ (i.e., check that $\mathsf{Weyl}(V)$ is generated by the x_i and y_i, with the same relations as Weyl_n.)

The Weyl algebra deforms the algebra $\mathsf{Sym}\, V$ in the following sense:

DEFINITION 1.2.4. An increasing filtration, or just filtration, on a vector space V is a sequence of subspaces $V_{\le m} \subseteq V$ such that $V_{\le m} \subseteq V_{\le n}$ for all $m \le n$. It is called nonnegative if $V_{\le m} = 0$ whenever $m < 0$. It is called exhaustive if $V = \bigcup_m V_{\le m}$. It is called Hausdorff if $\bigcap_m V_{\le m} = 0$. A vector space with a filtration is called a filtered vector space.

We will only consider exhaustive, Hausdorff filtrations, so we omit those terms. Most of the time, we will only consider nonnegative filtrations (which immediately implies Hausdorff). We will also usually only use increasing filtrations, so we often omit that term.

DEFINITION 1.2.5. An (increasing) filtration on an associative algebra A is an increasing filtration $A_{\le m}$ such that

$$A_{\le m} \cdot A_{\le n} \subseteq A_{\le(m+n)}, \quad \forall m, n \in \mathbb{Z}.$$

An algebra equipped with such a filtration is called a filtered algebra.

DEFINITION 1.2.6. For a filtered algebra $A = \bigcup_{m \ge 0} A_{\le m}$, the associated graded algebra is $\mathrm{gr}\, A := \bigoplus_m A_{\le m}/A_{\le(m-1)}$. Let $\mathrm{gr}_m A = (\mathrm{gr}\, A)_m = A_{\le m}/A_{\le(m-1)}$.

We now return to the Weyl algebra. It is equipped with two different nonnegative filtrations, the *additive* or *Bernstein* filtration, and the *geometric* one. We first consider the additive filtration. This is the filtration by degree of noncommutative monomials in the x_i and y_i, i.e.,

$$\mathsf{Weyl}(V)_{\le n} = \mathrm{Span}\{v_1 \cdots v_m \mid v_i \in V, m \le n\}.$$

The *geometric* filtration, on the other hand, assigns the x_i degree 0 and only the y_i degree one. This filtration has the advantage that it generalizes from $\mathsf{Weyl}(V)$ to the setting of differential operators on arbitrary varieties, since there one obtains the filtration by order of differential operators, but has the disadvantage that the full symmetry group $\mathrm{Sp}(V)$ does not preserve the filtration, but only the subgroup $\mathrm{GL}(U)$, where $U = \mathrm{Span}\{x_i\}$ (which acts on $U' = \mathrm{Span}\{y_i\}$ by the inverse transpose matrix of the one on U, in the bases of the y_i and x_i, respectively).

EXERCISE 1.2.7. Still assuming $\Bbbk$ has characteristic zero, show that, with either the additive or geometric filtration, $\mathrm{gr}\,\mathsf{Weyl}(V)$ is isomorphic to the symmetric algebra $\mathsf{Sym}\, V$, but the induced gradings on $\mathsf{Sym}\, V$ are different! With the additive filtration, $\mathsf{Sym}\, V \cong \mathrm{gr}\,\mathsf{Weyl}(V)$ with the grading placing V in degree one (i.e., the usual grading on the symmetric algebra placing $\mathsf{Sym}^k V$ in degree k). With the geometric filtration, show that $V = V_0 \oplus V_1$, where V_0 is spanned by the x_i and V_1 is spanned by the y_i.

The easiest way to do this is to use Exercise 1.2.1 and to show that a basis for $\mathcal{D}(\mathbb{A}^n)$ is given by monomials of the form $f(x_1, \ldots, x_n)g(y_1, \ldots, y_n)$, where f and g are considered as commutative monomials in the commutative subalgebras $\mathbb{k}[x_1, \ldots, x_n]$ and $\mathbb{k}[y_1, \ldots, y_n]$, respectively.

EXERCISE 1.2.8. More difficult: the algebra $\mathsf{Weyl}(V)$ can actually be defined over an arbitrary field (or even commutative ring) $\mathbb{k}$. In general, show that, for V the vector space with basis the x_i and y_i as above (or the free $\mathbb{k}$-module in the case $\mathbb{k}$ is a commutative ring), a $\mathbb{k}$-linear basis of $\mathsf{Weyl}(V)$ can be obtained via monomials $f(x_1, \ldots, x_n)g(y_1, \ldots, y_n)$ as in the previous exercise, and conclude that the canonical homomorphism $\mathsf{Sym}(V) \to \mathrm{gr}\,\mathsf{Weyl}(V)$ is still an isomorphism.

To do this, first note that $\mathsf{Sym}(V) \to \mathrm{gr}\,\mathsf{Weyl}(V)$ is obviously surjective; we just have to show injectivity. Next note that, if $R \subseteq S$ are commutative rings, and V an R-module with a skew-symmetric pairing, then $\mathsf{Weyl}(V)$ can be defined, and that $\mathsf{Weyl}(V \otimes_R S) = \mathsf{Weyl}(V) \otimes_R S$. So the desired statement reduces to the case $\mathbb{k} = \mathbb{Z}$, with V the free module generated by the x_i and y_i. There one can see that the desired monomials are linearly independent by using their action by differential operators on $\mathbb{Z}[x_1, \ldots, x_n]$, i.e., sending $f(x_1, \ldots, x_n)g(y_1, \ldots, y_n)$ to $f(x_1, \ldots, x_n)g(-\partial_1, \ldots, -\partial_n)$. This implies injectivity of $\mathsf{Sym}(V) \to \mathrm{gr}\,\mathsf{Weyl}(V)$.

This motivates the following:

DEFINITION 1.2.9. A filtered deformation of a graded algebra B is a filtered algebra A such that $\mathrm{gr}(A) \cong B$ (as graded algebras).

In the case that B is commutative and A is noncommutative, we will call such a deformation a *filtered quantization*. We will give a formal definition in §1.4, once we discuss Poisson algebras.

1.2.10. *Invariant subalgebras of Weyl algebras.* Recall that the symplectic group $\mathsf{Sp}(V)$ is the group of linear transformations $V \to V$ preserving the symplectic form $(-, -)$, i.e., $\mathsf{Sp}(V) = \{T : V \to V \mid (T(v), T(w)) = (v, w), \forall v, w \in V\}$.

Observe that $\mathsf{Sp}(V)$ acts by algebra automorphisms of $\mathsf{Weyl}(V)$, and that it preserves the additive (but not the geometric) filtration. Let $\Gamma < \mathsf{Sp}(V)$ be a finite subgroup of order relatively prime to the characteristic of $\mathbb{k}$. Then, one can consider the invariant subalgebras $\mathsf{Weyl}(V)^\Gamma$ and $\mathsf{Sym}(V)^\Gamma$.

The latter can also be viewed as the algebra of polynomial functions on the singular quotient V^*/Γ. Recall that an affine variety X is defined as the spectrum $\mathsf{Spec}\,\mathcal{O}(X)$ of a commutative $\mathbb{k}$-algebra $\mathcal{O}(X)$ (to be called a variety, $\mathcal{O}(X)$ should be finitely generated over $\mathbb{k}$ and have no nilpotents). Then, for V a vector space, the spectrum $\mathsf{Spec}\,\mathsf{Sym}\,V$ is identified with the dual vector space V^* via the correspondence between elements $\phi \in V^*$ and maximal ideals $\mathfrak{m}_\phi \subseteq \mathsf{Sym}\,V$, of elements of $\mathsf{Sym}\,V$ such that evaluating everything in V against ϕ simultaneously yields zero, i.e., $\sum_i v_{i,1} \cdots v_{i,j_i} \in \mathfrak{m}_\phi$ if and only if $\sum_i \phi(v_{i,1}) \cdots \phi(v_{i,j_i}) = 0$, for $v_{i,k} \in V$ and some $j_i \geq 1$.

Next, recall that an action of a discrete group Γ on an affine variety $X = \operatorname{Spec} \mathcal{O}(X)$ is the same as an action of Γ on the commutative algebra $\mathcal{O}(X)$. The following definition is then standard:

DEFINITION 1.2.11. The quotient X/Γ of an affine variety X by a discrete group Γ is the spectrum, $\operatorname{Spec} \mathcal{O}(X)^{\Gamma}$, of the algebra of invariants, $\mathcal{O}(X)^{\Gamma}$.

Therefore, one identifies V^*/Γ with $\operatorname{Spec}(\operatorname{Sym} V)^{\Gamma}$, i.e., $(\operatorname{Sym} V)^{\Gamma} = \mathcal{O}(V^*/\Gamma)$.

Later on we will also have use for the case where Γ is replaced by a non-discrete algebraic group, in which case the same definition is made with a different notation:

DEFINITION 1.2.12. The (categorical) quotient $X /\!/ G$ is $\operatorname{Spec} \mathcal{O}(X)^G$.

The reason for the (standard) double slash $/\!/$ and for "categorical" in parentheses is to distinguish from other types of quotients of X by G, most notably the stack and GIT (geometric invariant theory) quotients, which we will not need in these notes.

Let us return to the setting of $\Gamma < \operatorname{Sp}(V)$ which is relatively prime to the characteristic of $\Bbbk$. One easily sees that $\operatorname{Weyl}(V)^G$ is filtered (using the additive filtration)[1] and that

$$\operatorname{gr}(\operatorname{Weyl}(V)^{\Gamma}) \cong \operatorname{Sym}(V)^{\Gamma} =: \mathcal{O}(V^*/\Gamma).$$

In the exercises, we will see that the algebra $\operatorname{Weyl}(V)^G$ can be further deformed; this yields the so-called spherical symplectic reflection algebras, which are an important subject of Chapter III.

EXAMPLE 1.2.13. The simplest case is already interesting: $V = \Bbbk^2$ and

$$\Gamma = \{\pm \operatorname{Id}\} \cong \mathbb{Z}/2,$$

with $\Bbbk$ not having characteristic two. Then,

$$\operatorname{Sym}(V)^{\Gamma} = \Bbbk[x^2, xy, y^2] \cong \Bbbk[u, v, w]/(v^2 - uw),$$

the algebra of functions on a singular quadric hypersurface in $\mathbb{A}^3$. The noncommutative deformation, $\operatorname{Weyl}(V)^{\Gamma}$, on the other hand, is *homologically smooth* for all V and $\Gamma < \operatorname{Sp}(V)$ finite, i.e., the algebra $A := \operatorname{Weyl}(V)^{\Gamma}$ has a finitely-generated projective A-bimodule resolution. In fact, it is a *Calabi–Yau algebra of dimension* $\dim V$ (see Definition 3.7.9 below).

1.3. Universal enveloping algebras of Lie algebras. Next, we consider the enveloping algebra $U\mathfrak{g}$, which deforms the symmetric algebra $\operatorname{Sym} \mathfrak{g}$. This is the algebra whose representations are the same as representations of the Lie algebra $\mathfrak{g}$ itself.

The enveloping algebra. Let $\mathfrak{g}$ be a Lie algebra with Lie bracket $[-, -]$. Then the representation theory of $\mathfrak{g}$ can be restated in terms of the representation theory of its *enveloping algebra,*

$$U\mathfrak{g} := T\mathfrak{g}/(xy - yx - [x, y] \mid x, y \in \mathfrak{g}), \tag{1.A}$$

---

[1]When G preserves the span of the x_i, then it also preserves the geometric filtration, and in this case $\operatorname{Weyl}(V)^G$ is also filtered with the geometric filtration. More generally, if G acts on an affine variety X, one can consider the G-invariant differential operators $\mathcal{D}(X)^G$.

where $T\mathfrak{g}$ is the tensor algebra of $\mathfrak{g}$ and $(-)$ denotes the two-sided ideal generated by this relation.

PROPOSITION 1.3.1. *A representation V of $\mathfrak{g}$ is the same as a representation of $U\mathfrak{g}$.*

PROOF. If V is a representation of $\mathfrak{g}$, we define the action of $U\mathfrak{g}$ by

$$x_1 \cdots x_n(v) = x_1(x_2(\cdots(x_n(v))\cdots)).$$

We only have to check the relation defining $U\mathfrak{g}$:

$$(xy - yx - [x, y])(v) = x(y(v)) - y(x(v)) - [x, y](v), \tag{1.B}$$

which is zero since the action of $\mathfrak{g}$ was a Lie action.

For the opposite direction, if V is a representation of $U\mathfrak{g}$, we define the action of $\mathfrak{g}$ by restriction from $U\mathfrak{g}$ to $\mathfrak{g}$. This defines a Lie action since the LHS of (1.B) is zero. $\qquad\square$

REMARK 1.3.2. More conceptually, the assignment $\mathfrak{g} \mapsto U\mathfrak{g}$ defines a functor from Lie algebras to associative algebras. Then the above statement says

$$\mathrm{Hom}_{\mathrm{Lie}}(\mathfrak{g}, \mathrm{End}(V)) = \mathrm{Hom}_{\mathrm{Ass}}(U\mathfrak{g}, \mathrm{End}(V)),$$

where Lie denotes Lie algebra homomorphisms, and Ass denotes associative algebra homomorphisms. This statement is a consequence of the statement that $\mathfrak{g} \mapsto U\mathfrak{g}$ is a functor from Lie algebras to associative algebras which is left adjoint to the restriction functor $A \mapsto A^- := (A, [a, b] := ab - ba)$. That is, we can write the above equivalently as

$$\mathrm{Hom}_{\mathrm{Lie}}(\mathfrak{g}, \mathrm{End}(V)^-) = \mathrm{Hom}_{\mathrm{Ass}}(U\mathfrak{g}, \mathrm{End}(V)).$$

$U\mathfrak{g}$ as a filtered deformation of Sym $\mathfrak{g}$. There is a natural increasing filtration on $U\mathfrak{g}$ that assigns degree one to $\mathfrak{g}$: that is, we define

$$(U\mathfrak{g})_{\leq m} = \langle x_1 \cdots x_j \mid x_1, \ldots, x_j \in \mathfrak{g}, j \leq m \rangle.$$

There is a surjection of algebras,

$$\mathrm{Sym}\,\mathfrak{g} \twoheadrightarrow \mathrm{gr}(U\mathfrak{g}), \quad (x_1 \cdots x_m) \mapsto \mathrm{gr}(x_1 \cdots x_m), \tag{1.C}$$

which is well-defined since, in $\mathrm{gr}(U\mathfrak{g})$, one has $xy - yx = 0$. The Poincaré–Birkhoff–Witt theorem states that this is an isomorphism:

THEOREM 1.3.3 (PBW). *The map (1.C) is an isomorphism.*

The PBW is the *key* property that says that the deformations have been deformed in a *flat* way, so that the algebra has not gotten any smaller (the algebra cannot get bigger by a filtered deformation of the relations, only smaller). This is a very special property of the deformed relations: see the following exercise.

EXERCISE 1.3.4. Suppose more generally that $B = TV/(R)$ for $R \subseteq TV$ a homogeneous subspace (i.e., R is spanned by homogeneous elements). Suppose also that $E \subseteq TV$ is an arbitrary filtered deformation of the relations, i.e., $\mathrm{gr}\, E = R$. Let $A := TV/(E)$. Show that there is a canonical surjection

$$B \twoheadrightarrow \mathrm{gr}\, A.$$

So by deforming relations, the algebra A can only get smaller than B (by which we mean that the above surjection is not injective), and cannot get larger. In general, it can get smaller: see Exercise 1.10.1 for an example where the above surjection is not injective; in that example, B is infinite-dimensional and $\mathrm{gr}\, A$ is one-dimensional. In the case when the above surjection is injective, we call the deformation A *flat*, which is equivalent to saying that A is a filtered deformation in the sense of Definition 1.2.9.

1.4. Quantization of Poisson algebras. As we will explain, the isomorphism (1.C) is compatible with the natural Poisson structure on $\mathrm{Sym}\, \mathfrak{g}$.

Poisson algebras.

DEFINITION 1.4.1. A Poisson algebra is a commutative algebra B equipped with a Lie bracket $\{-, -\}$ satisfying the Leibniz identity:

$$\{ab, c\} = a\{b, c\} + b\{a, c\}.$$

Now, let $B := \mathrm{Sym}\, \mathfrak{g}$ for $\mathfrak{g}$ a Lie algebra with bracket $[-, -]$. Then, B has a canonical Poisson bracket which extends the Lie bracket:

$$\{a_1 \cdots a_m, b_1 \cdots b_n\} = \sum_{i,j} [a_i, b_j] a_1 \cdots \hat{a}_i \cdots a_m b_1 \cdots \hat{b}_j \cdots b_n. \qquad (1.D)$$

Here and in the sequel, the hat denotes that the corresponding term is *omitted* from the list of terms: so a_i and b_j do not appear in the products, but all other a_k and b_ℓ do (for $k \neq i$ and $\ell \neq j$).

Poisson structures on associated graded algebras. Generally, let A be an increasingly filtered algebra such that $\mathrm{gr}\, A$ is commutative. This implies that there exists $d \geq 1$ such that that

$$[A_{\leq m}, A_{\leq n}] \subseteq A_{\leq (m+n-d)}, \quad \forall m, n. \qquad (1.E)$$

One can always take $d = 1$, but in general we want to take d maximal so that the above property is satisfied; see Exercise 1.4.3 below. Fix a value of $d \geq 1$ satisfying (1.E).

We claim that $\mathrm{gr}\, A$ is canonically Poisson, with the bracket, for $a \in A_{\leq m}$ and $b \in A_{\leq n}$,

$$\{\mathrm{gr}_m a, \mathrm{gr}_n b\} := \mathrm{gr}_{m+n-d}(ab - ba).$$

EXERCISE 1.4.2. Verify that the above is indeed a Poisson bracket, i.e., it satisfies the Lie and Leibniz identities.

EXERCISE 1.4.3. Suppose that $d \geq 1$ is *not* the maximum possible value such that (1.E) is satisfied. Show that the Poisson bracket on $\mathrm{gr}\, A$ is zero. This explains why we usually will take d to be the maximum possible.

We conclude that $\mathrm{Sym}\, \mathfrak{g}$ is equipped with a Poisson bracket by Theorem 1.3.3, i.e., by the isomorphism (1.C), taking $d := 1$.

EXERCISE 1.4.4. Verify that the Poisson bracket on $\mathsf{Sym}\,\mathfrak{g}$ obtained from (1.C) is the same as the one of (1.D), for $d := 1$.

EXERCISE 1.4.5. Equip $\mathsf{Sym}\,V$ with the unique Poisson bracket such that $\{v, w\} = (v, w)$ for $v, w \in V$. This bracket has degree -2. Show that, with the additive filtration, $\mathrm{gr}\,\mathsf{Weyl}(V) \cong \mathsf{Sym}\,V$ as Poisson algebras, where $d = 2$ in (1.E).

On the other hand, show that, with the geometric filtration, one can take $d = 1$, and then one obtains an isomorphism $\mathrm{gr}\,\mathsf{Weyl}(V) \cong \mathsf{Sym}\,V'$, where V' is the same underlying vector space as V, but placing the x_i in degree zero and the y_i in degree one (so $V' = (V')_0 \oplus (V')_1$ with $\dim(V')_0 = \dim(V')_1 = \frac{1}{2}\dim V$). The Poisson bracket on $\mathsf{Sym}\,V'$ is given by the same formula as for $\mathsf{Sym}\,V$.

Filtered quantizations. The preceding example motivates the definition of a filtered quantization:

DEFINITION 1.4.6. Let B be a graded Poisson algebra, such that the Poisson bracket has negative degree $-d$. Then a *filtered quantization* of B is a filtered associative algebra A such that (1.E) is satisfied, and such that $\mathrm{gr}\,A \cong B$ as Poisson algebras.

Again, the key property here is that $\mathrm{gr}\,A \cong B$. Since, in the case where $A = U\mathfrak{g}$, this property is the PBW theorem, one often refers to this property in general as the "PBW property." In many examples of B and A of interest, proving that this property holds is an important theorem, which is often called a "PBW theorem."

1.5. Algebras of differential operators. As mentioned above, when $\Bbbk$ has characteristic zero, the Weyl algebra $\mathsf{Weyl}_n = \mathsf{Weyl}(\Bbbk^{2n})$ is isomorphic to the algebra of differential operators on $\Bbbk^n$ with polynomial coefficients.

More generally:

DEFINITION 1.5.1 (Grothendieck). Let B be a commutative $\Bbbk$-algebra. We define the space $\mathrm{Diff}_{\leq m}(B)$ of differential operators *of order* $\leq m$ inductively on m. For $a \in B$ and $\phi \in \mathrm{End}_{\Bbbk}(B)$, let $[\phi, a] \in \mathrm{End}_{\Bbbk}(B)$ be the linear operator

$$[\phi, a](b) := \phi(ab) - a\phi(b), \quad \forall b \in B.$$

Then we define

$$\mathrm{Diff}_{\leq 0}(B) := \{\phi \in \mathrm{End}_{\Bbbk}(B) \mid [\phi, a] = 0, \forall a \in B\} = \mathrm{End}_B(B) \cong B; \quad (1.F)$$

$$\mathrm{Diff}_{\leq m}(B) := \{\phi \in \mathrm{End}_{\Bbbk}(B) \mid [\phi, a] \in \mathrm{Diff}_{\leq (m-1)}(B), \forall a \in B\}. \quad (1.G)$$

Let $\mathrm{Diff}(B) := \bigcup_{m \geq 0} \mathrm{Diff}_{\leq m}(B)$.

EXERCISE 1.5.2. Verify that $\mathrm{Diff}(B)$ is a nonnegatively filtered associative algebra whose associated graded algebra is commutative.

Now, suppose that B is finitely generated commutative and $X := \mathrm{Spec}(B)$. Then the global vector fields on X are the same as $\Bbbk$-algebra derivations of the algebra B,

$$\mathrm{Vect}(X) := \mathrm{Der}_{\Bbbk}(\mathcal{O}(X), \mathcal{O}(X))$$

$$:= \{\phi \in \mathrm{Hom}_{\Bbbk}(\mathcal{O}(X), \mathcal{O}(X)) \mid \phi(fg) = \phi(f)g + f\phi(g), \forall f, g \in \mathcal{O}(X)\}.$$

This is naturally a $\mathcal{O}(X)$-module. Note that it can also be viewed as global sections of the tangent sheaf, denoted by T_X (which is in general defined so as to have sections on open affine subsets given by derivations as above).

Recall that the total space of the cotangent bundle can be defined by $T^*X :=$ $\operatorname{Spec} \operatorname{Sym}_{\mathcal{O}(X)} \operatorname{Vect}(X)$. Points of T^*X are in bijection with pairs (x, p) where $x \in X$ and $p \in T_x^* X = \mathfrak{m}_x / \mathfrak{m}_x^2$, where $\mathfrak{m}_x \subseteq \mathcal{O}(X)$ is the maximal ideal of functions vanishing at x. In the case where X is a smooth affine complex variety, this is the same as a pair of a point of the complex manifold X and a cotangent vector at x.

Now we can state the result that, for smooth affine varieties in characteristic zero, $\operatorname{Diff}(\mathcal{O}(X))$ quantizes the cotangent bundle:

PROPOSITION 1.5.3. *If X is a smooth (affine) variety and $\Bbbk$ has characteristic zero, then as Poisson algebras,*

$$\operatorname{gr} \operatorname{Diff}(\mathcal{O}(X)) \cong \operatorname{Sym}_{\mathcal{O}(X)} \operatorname{Vect}(X) \cong \mathcal{O}(T^*X).$$

REMARK 1.5.4. When X is not affine (but still smooth), the above generalizes if we replace the algebras $\mathcal{O}(X)$ and $\mathcal{O}(T^*X)$ by the sheaves $\mathcal{O}_X$ and $\mathcal{O}_{T^*X}$ on X, and the global vector fields $\operatorname{Vect}(X)$ by the tangent sheaf T_X. The material in the remainder of this section also generalizes similarly to the smooth nonaffine context.

The proposition requires some clarifications. First, since $\operatorname{Vect}(X)$ is actually a Lie algebra, we obtain a Poisson structure on $\operatorname{Sym}_{\mathcal{O}(X)} \operatorname{Vect}(X)$ by the formula

$$\{\xi_1 \cdots \xi_m, \eta_1 \cdots \eta_n\} = \sum_{i,j} [\xi_i, \eta_j] \cdot \xi_1 \cdots \hat{\xi}_i \cdots \xi_m \eta_1 \cdots \hat{\eta}_j \cdots \eta_n.$$

Proposition 1.5.3 says that, for smooth affine varieties, the algebra $\operatorname{Diff}(\mathcal{O}(X))$ quantizes $\mathcal{O}(T^*X)$. To prove this result, we will want to have an alternative construction of $\operatorname{Diff}(\mathcal{O}(X))$:

DEFINITION 1.5.5. The universal enveloping algebroid $U_{\mathcal{O}(X)}(\operatorname{Vect}(X))$ is the quotient of the usual enveloping algebra $U(\operatorname{Vect}(X)) = U_{\Bbbk}(\operatorname{Vect}(X))$ by the relations

$$f \cdot \xi = f\xi, \quad \xi \cdot f = \xi(f) + f\xi, \quad f \in \mathcal{O}(X), \quad \xi \in \operatorname{Vect}(X). \tag{1.H}$$

REMARK 1.5.6. More generally, the above definition extends to the setting when we replace $\operatorname{Vect}(X)$ by (global sections of) an arbitrary Lie algebroid L over X. Namely, such an L is a Lie algebra which is an $\mathcal{O}(X)$-module together with a $\mathcal{O}(X)$-linear map $a : L \to T_X$ satisfying the Leibniz rule,

$$[\xi, f\eta] = a(\xi)(f)\eta + f[\xi, \eta], \quad \forall \xi, \eta \in L, \ \forall f \in \mathcal{O}(X).$$

This implies that the anchor map a is a homomorphism of Lie algebras. Note that a simple example of such an L is $\operatorname{Vect}(X)$ itself, with $a = \operatorname{Id}$. One then defines the universal enveloping algebroid of L as $U_{\mathcal{O}(X)}L := UL/(f \cdot \xi - f\xi, \xi \cdot f - a(\xi)(f) - f\xi)$.

REMARK 1.5.7. The universal enveloping algebroid can also be defined in a way which generalizes the definition of the usual enveloping algebra. Namely, if L is as in the previous remark, define a new $\mathcal{O}(X)$-bimodule structure on L by giving the usual left action, and defining the right multiplication by $\xi \cdot f = a(\xi)(f) + f\xi$. Given any bimodule M over a (not-necessarily commutative) associative algebra A, we can define the tensor algebra $T_A^{nc} M := \bigoplus_{m \geq 0} M^{\otimes_m A}$. Then, $U_{\mathcal{O}(X)} L := T_{\mathcal{O}(X)}^{nc} L / (\xi \cdot \eta - \eta \cdot \xi - [\xi, \eta])$. Then, in the case that $\mathcal{O}(X) = \Bbbk$ itself, this recovers the usual definition of the enveloping algebra $UL = U_{\Bbbk} L$.

SKETCH OF PROOF OF PROPOSITION 1.5.3. Filter $U_{\mathcal{O}(X)} \operatorname{Vect}(X)$ by saying that the image of $T^{\leq m}(\operatorname{Vect}(X)) = \bigoplus_{0 \leq i \leq m} \operatorname{Vect}(X)^{\otimes i}$ under the defining quotient is $(U_{\mathcal{O}(X)} \operatorname{Vect}(X))_{\leq m}$. We have a natural map $\operatorname{Sym}_{\mathcal{O}(X)} \operatorname{Vect}(X) \to \operatorname{gr} U_{\mathcal{O}(X)} \operatorname{Vect}(X)$.

Next, if $\operatorname{Vect}(X)$ is free, or more generally projective (equivalently, T_X is a locally free sheaf), as will be true when X is smooth, then the PBW theorem generalizes to show that the natural map $\operatorname{Sym}_{\mathcal{O}(X)} \operatorname{Vect}(X) \to \operatorname{gr} U_{\mathcal{O}(X)} \operatorname{Vect}(X)$ is an isomorphism.

So, the proposition reduces to showing that, in the case X is smooth, the canonical map

$$U_{\mathcal{O}(X)} \operatorname{Vect}(X) \to \operatorname{Diff}(\mathcal{O}(X)) \tag{1.I}$$

is an isomorphism.

To prove this, it suffices to show that the associated graded homomorphism,

$$\operatorname{Sym}_{\mathcal{O}(X)} \operatorname{Vect}(X) \to \operatorname{gr} \operatorname{Diff}(\mathcal{O}(X)),$$

is an isomorphism. This statement can be checked locally, in the formal neighborhood of each point $x \in X$, which is isomorphic to a formal neighborhood of affine space of the same dimension at the origin. More precisely, we can replace $\mathcal{O}(X)$ by a formal power series ring $\Bbbk[[x_1, \ldots, x_n]]$ and require that all derivations and differential operators are continuous in the adic topology (i.e., if ξ is a derivation, we require that every sum $\sum a_{r_1, \ldots, r_n} \xi(x_1^{r_1} \cdots x_n^{r_n})$ converges as a formal power series). But then $\operatorname{Diff}(\mathcal{O}(X))$ becomes $\Bbbk[[x_1, \ldots, x_n]][\partial_1, \ldots, \partial_n]$, a completion of the Weyl algebra, and the statement follows as in Exercises 1.2.1 and 1.2.7. (Alternatively, instead of using formal power series, one can use ordinary localization at x; since $\dim \mathfrak{m}_x / \mathfrak{m}_x^2 = \dim X$, Nakayama's lemma implies that a basis for this vector space lifts to a collection of $\dim X$ algebra generators of the local ring $\mathcal{O}_{X,x}$, and then differential operators of order $\leq m$ are determined by their action on products of $\leq m$ generators, and all possible actions are given by $\operatorname{Sym}_{\mathcal{O}_{X,x}}^{\leq m} T_{X,x}$.)   □

REMARK 1.5.8. Whenever L is (the global sections of) a Lie algebroid over $\mathcal{O}(X)$ as in Remark 1.5.6, then it follows just as in the case $L = \operatorname{Vect}(X)$ that $\operatorname{Sym}_{\mathcal{O}(X)} L$ is a Poisson algebra. Then as before we have a natural map $\operatorname{Sym}_{\mathcal{O}(X)} L \to \operatorname{gr}(U_{\mathcal{O}(X)} L)$ and in the case L is locally free, the PBW theorem generalizes to show that this is an isomorphism. This applies even in the case X is not smooth (and hence $\operatorname{Vect}(X)$ itself is not locally free), since often one can nonetheless define interesting locally free Lie algebroids (and the same applies when X need not be affine, replacing $\mathcal{O}(X)$ by $\mathcal{O}_X$ and $\operatorname{Vect}(X)$ by T_X). In fact, this is the setting of a large body of interesting recent work, such as Calaque–Van den Bergh's analogues of Kontsevich's formality theorem [59, 58] and their proof of Căldăraru's conjecture [57] on the compatibility of the former with cap products with Hochschild homology.

Preview: $\mathcal{D}$-modules. Since Proposition 1.5.3 does not apply to singular varieties, it is less clear how to treat algebras of differential operators and their modules in the singular case (again with $\Bbbk$ having characteristic zero).

One solution, discovered by Kashiwara, is to take a singular variety X and embeds it into a smooth variety V, and define the category of right $\mathcal{D}$-modules on X to be the category of right $\mathcal{D}(V)$-modules supported on X, i.e., right modules M over the ring $\mathcal{D}(V)$ of differential operators on V, with the property that, for all $m \in M$, there exists $N \geq 1$ such that $m \cdot I_X^N = 0$, where I_X is the ideal corresponding to X.

This circumvents the problem that the ring $\mathcal{D}(X)$ of differential operators is not well-behaved, and one obtains a very useful theory. An equivalent definition to Kashiwara's goes under the name of *(right) crystals.* Under some restrictions (when the $\mathcal{D}$-modules are holonomic with regular singularities), one can also replace $\mathcal{D}$-modules by *perverse sheaves,* which are, roughly speaking, gluings of local systems on subvarieties, or more precisely, complexes of such gluings with certain properties.

1.6. Invariant differential operators. It is clear that the group of automorphisms $\mathrm{Aut}(X)$ of the variety X acts by filtered automorphisms also on $\mathrm{Diff}(X)$ and also by graded automorphisms of $\mathrm{gr}\,\mathrm{Diff}(X) = \mathrm{Sym}_{\mathcal{O}(X)}\mathrm{Vect}(X)$. Let us continue to assume $\Bbbk$ has characteristic zero.

Now, suppose that $G < \mathrm{Aut}(X)$ is a finite subgroup of automorphisms of X. Then one can form the algebras $\mathrm{Diff}(X)^G$ and $(\mathrm{Sym}_{\mathcal{O}(X)}\mathrm{Vect}(X))^G$. By Proposition 1.5.3, we conclude that $\mathrm{gr}\,\mathrm{Diff}(X)^G$ is a quantization of $(\mathrm{Sym}_{\mathcal{O}(X)}\mathrm{Vect}(X))^G$.

One example of this is when $X = \mathbb{A}^n$ and $G < \mathrm{GL}(n) < \mathrm{Aut}(\mathbb{A}^n)$. Then we obtain that $\mathrm{Weyl}(\mathbb{A}^{2n})$ is a quantization of $\mathcal{O}(T^*\mathbb{A}^n) = \mathcal{O}(\mathbb{A}^{2n})$, which is the special case of the example of §1.2.10 where $G < \mathrm{GL}(n)$ (note that $\mathrm{GL}(n) < \mathrm{Sp}(2n)$, where explicitly, a matrix A acts by the block matrix

$$\begin{pmatrix} A & 0 \\ 0 & (A^t)^{-1} \end{pmatrix},$$

with A^t denoting the transpose of A.)

REMARK 1.6.1. By deforming $\mathrm{Diff}(X)^G$, one obtains global analogues of the spherical rational Cherednik algebras [97]; see Example 2.7.9 below.

EXAMPLE 1.6.2. Let $X = \mathbb{A}^1 \setminus \{0\}$ (note that this is affine) and let $G = \{1, g\}$ where $g(x) = x^{-1}$. Then $(T^*X)//G$ is a "global" or "multiplicative" version of the variety $\mathbb{A}^2/(\mathbb{Z}/2)$ of Example 1.2.13. A quantization is $\mathcal{D}(X)^G$, and one has, by the above,

$$\mathrm{gr}\,\mathcal{D}(\mathbb{A}^1 \setminus \{0\})^{\mathbb{Z}/2} \cong \mathcal{O}(T^*(\mathbb{A}^1 \setminus \{0\}))^{\mathbb{Z}/2}.$$

Explicitly, the action on tangent vectors is $g(\partial_x) = -x^2\partial_x$, i.e., so that, applying the operator $g(\partial_x)$ to $g(x)$, we get $g(\partial_x)(g(x)) = 1 = -x^2\partial_x(x^{-1})$. Thus, setting $y := \mathrm{gr}\,\partial_x$, we have $g(y) = -x^2y$. So $\mathcal{O}(T^*X//G) = \mathbb{C}[x + x^{-1}, y - x^2y, x^2y^2]$ and

$$\mathcal{D}(X)^G = \mathbb{C}[x + x^{-1}, \partial_x - x^2\partial_x, (x\partial_x)^2] \subseteq \mathcal{D}(X)^G.$$

1.7. Quantization of the nilpotent cone.

Central reductions. Suppose, generally, that A is a filtered quantization of $B = \operatorname{gr} A$. Suppose in addition that there is a central filtered subalgebra $Z \subseteq A$. Then $\operatorname{gr} Z$ is Poisson central in B:

DEFINITION 1.7.1. The center, $Z(B)$, of a Poisson algebra, B, is the subalgebra of elements $z \in B$ such that $\{z, b\} = 0$ for all $b \in B$. An element is called Poisson central if it is in $Z(B)$. A subalgebra $C \subseteq B$ is called central if $C \subseteq Z(B)$.

Next, for every character $\eta : Z \to \Bbbk$, we obtain *central reductions*

$$A^\eta := A/\ker(\eta)A, \quad B^\eta := B/\operatorname{gr}(\ker \eta)B.$$

Note here that $\ker(\eta)A$ is actually a two-sided ideal since Z is central. In the case that $B_0 = \Bbbk$ (or more generally $(\operatorname{gr} Z)_0 = \Bbbk$), which will be the case for us, note that B^η does not actually depend on η, and we obtain $B^\eta = B/(\operatorname{gr} Z)_+ B$ for all η, where $(\operatorname{gr} Z)_+ \subseteq \operatorname{gr} Z$ is the augmentation ideal (the ideal of positively-graded elements).

Then, the category $\operatorname{Rep}(A^\eta)$ can be identified with the category of representations V of A such that $Z(A)$ acts by the character η, i.e., for all $v \in V$ and all $z \in Z(A)$, we have $z \cdot v = \eta(z)v$.

REMARK 1.7.2. The subcategories $\operatorname{Rep}(A^\eta) \subseteq \operatorname{Rep}(A)$ are all *orthogonal* for distinct η, which means that there are no nontrivial homomorphisms or extensions between representations $V \in \operatorname{Rep}(A^\eta)$, $W \in \operatorname{Rep}(A^\xi)$ with distinct central characters $\eta \neq \xi$. This is because the ideal generated by $(z - \eta(z))$ maps to the unit ideal in A^ξ for $\xi \neq \eta$, i.e., there is an element $z \in Z(A)$ which acts by one on all representations in $\operatorname{Rep}(A^\xi)$ and by zero on all representations in $\operatorname{Rep}(A^\eta)$, and this allows one to canonically split any extension of representations in the two categories $\operatorname{Rep}(A^\xi)$ and $\operatorname{Rep}(A^\eta)$: if V is such an extension, then $V = zV \oplus (1 - z)(V)$.

The nilpotent cone. Now let us restrict to our situation of $A = U\mathfrak{g}$ and $B = \operatorname{Sym}\mathfrak{g}$ with $\Bbbk$ of characteristic zero. Let us suppose moreover that $\mathfrak{g}$ is finite-dimensional semisimple (see, for example, [135]; for the reader who is not familiar with this, one can restrict to the most important examples, such as the Lie algebra $\mathfrak{sl}_n(\Bbbk)$ of trace-zero $n \times n$ matrices, or the Lie algebra $\mathfrak{so}_n(\Bbbk)$ of skew-symmetric $n \times n$ matrices). Then, the structure of the center $Z(A)$ is well-known.

EXAMPLE 1.7.3. Let $\mathfrak{g} = \mathfrak{sl}_2$ with basis (e, f, h). Then $Z(U\mathfrak{g}) = \Bbbk[C]$, where the element C is the Casimir element, $C = ef + fe + \frac{1}{2}h^2$. In this case, the central reduction $(U\mathfrak{g})^\eta$ describes those representations on which C acts by a fixed scalar $\eta(C)$. For example, there exists a finite-dimensional representation of $(U\mathfrak{g})^\eta$ if and only if $\eta(C) \in \{m + \frac{1}{2}m^2 \mid m \geq 0\}$, since $\eta(C)$ acts on a highest-weight vector v of h of weight λ, i.e., a vector such that $ev = 0$ and $hv = \lambda v$, by $C \cdot v = (h + \frac{1}{2}h^2)v = (\lambda + \frac{1}{2}\lambda^2)v$. In particular, there are only countably many such characters η that admit a finite-dimensional representation.

Moreover, when $\eta(C) \in \{m + \frac{1}{2}m^2 \mid m \geq 0\}$, there is *exactly one* finite-dimensional representation of $(U\mathfrak{g})^\eta$: the one with highest weight m. So these quantizations $(U\mathfrak{g})^\eta$ of

(Sym $\mathfrak{g}$)/(gr C) have at most one finite-dimensional representation, and only countably many have this finite-dimensional representation.

More generally, if $\mathfrak{g}$ is finite-dimensional semisimple (still with $\Bbbk$ of characteristic zero), it turns out that $Z(U\mathfrak{g})$ is a polynomial algebra, and gr $Z(U\mathfrak{g}) \to Z(\text{Sym } \mathfrak{g})$ is an isomorphism of polynomial algebras.

DEFINITION 1.7.4. Given a Lie algebra $\mathfrak{g}$ and a representation V, the invariants $V^{\mathfrak{g}}$ are defined as $V^{\mathfrak{g}} := \{v \in V \mid x \cdot v = 0, \forall x \in \mathfrak{g}\}$.

Note that, if B is an algebra with a $\mathfrak{g}$-action, i.e., B is an $\mathfrak{g}$ representation and an algebra such that the multiplication map $B \otimes B \to B$ is a map of $\mathfrak{g}$-representations, then $B^{\mathfrak{g}} \subseteq B$ is a subalgebra. In particular, one has (Sym $\mathfrak{g})^{\mathfrak{g}} \subseteq$ Sym $\mathfrak{g}$ which is an algebra. In fact, this is the Poisson center of Sym $\mathfrak{g}$, since $\{z, f\} = 0$ for all $f \in$ Sym $\mathfrak{g}$ if and only if $\{z, x\} = 0$ for all $x \in \mathfrak{g}$: i.e., $Z(\text{Sym } \mathfrak{g}) = (\text{Sym } \mathfrak{g})^{\mathfrak{g}}$.

We then have the following extremely important result, whose history is discussed in more detail in Remark 1.7.13:

THEOREM 1.7.5 (H. Cartan, Chevalley [67, 68], Coxeter, Harish-Chandra [127], Koszul, Shephard and Todd [201], Weil). *Let $\mathfrak{g}$ be finite-dimensional semisimple and $\Bbbk$ of characteristic zero. Then $Z(U\mathfrak{g}) \cong \Bbbk[x_1, \ldots, x_r]$ is a polynomial algebra, with r equal to the semisimple rank of $\mathfrak{g}$. Moreover, the polynomial algebra gr $Z(U\mathfrak{g})$ equals the Poisson center $Z(\text{Sym } \mathfrak{g}) = (\text{Sym } \mathfrak{g})^{\mathfrak{g}}$ of gr $U\mathfrak{g}$, and there is a canonical algebra isomorphism $(\text{Sym } \mathfrak{g})^{\mathfrak{g}} \to Z(U\mathfrak{g})$.*

The final isomorphism $(\text{Sym } \mathfrak{g})^{\mathfrak{g}} \to Z(U\mathfrak{g})$ is called the *Harish-Chandra isomorphism*; it actually generalizes to an isomorphism which is defined for *arbitrary* finite-dimensional $\mathfrak{g}$ and is due to Kirillov and Duflo; we will explain how this latter isomorphism also follows from Kontsevich's formality theorem in Corollary 4.11.7 (and in fact, Kontsevich's result implies that it holds even for finite-dimensional Lie superalgebras).

The degrees d_i of the generators gr x_i are known as the *fundamental degrees*. (In fact, they satisfy $d_i = m_i + 1$ where m_i are the Coxeter exponents of the associated root system, cf. e.g., [136]; we will not need to know anything about Coxeter exponents or precisely what the m_i are below.)

By the theorem, for every character $\eta : Z(U\mathfrak{g}) \to \Bbbk$, one obtains an algebra $(U\mathfrak{g})^{\eta}$ which quantizes $(\text{Sym } \mathfrak{g})/((\text{Sym } \mathfrak{g})^{\mathfrak{g}}_+)$. Here $(\text{Sym } \mathfrak{g})^{\mathfrak{g}}_+$ is the augmentation ideal of Sym $\mathfrak{g}$, which equals gr(ker η) since (Sym $\mathfrak{g})_0 = \Bbbk$ (cf. the comments above).

Recall that the dual vector space $\mathfrak{g}^*$ is canonically a representation of $\mathfrak{g}$, with action, called the coadjoint action, given by $(x \cdot \phi)(y) := -\phi([x, y])$ for $x, y \in \mathfrak{g}$ and $\phi \in \mathfrak{g}^*$. We denote $x \cdot \phi$ also by $\text{ad}(x)(\phi)$.

DEFINITION 1.7.6. The *nilpotent cone* Nil $\mathfrak{g} \subseteq \mathfrak{g}^*$ is the set of elements $\phi \in \mathfrak{g}^*$ such that, for some $x \in \mathfrak{g}$, we have $\text{ad}(x)\phi = \phi$.

REMARK 1.7.7. Note that, in the case where $\mathfrak{g} = \text{Lie } G$ for G a connected Lie (or algebraic) group, then Nil $\mathfrak{g}$ is the set of elements ϕ such that the coadjoint orbit $G \cdot \phi$ contains the line $\Bbbk^{\times} \cdot \phi$ (in the Lie group case, $\Bbbk$ should be $\mathbb{R}$ or $\mathbb{C}$).

REMARK 1.7.8. In the case when $\mathfrak{g}$ is finite-dimensional and semisimple, it is well-known that the Killing form isomorphism $\mathfrak{g}^* \cong \mathfrak{g}$ takes $\mathrm{Nil}(\mathfrak{g})$ to the cone of elements which are ad-nilpotent (i.e., $(\mathrm{ad}\, x)^N = 0$ for some $N \geq 1$, where $\mathrm{ad}\, x(y) := [x, y]$), which explains the terminology. It is perhaps more standard to define the nilpotent cone as the latter cone inside $\mathfrak{g}$, but for us it is more natural to use the above definition.

PROPOSITION 1.7.9. *Let $\mathfrak{g}$ be finite-dimensional semisimple. Then, the algebra B^0 is the algebra of functions on the nilpotent cone* $\mathrm{Nil}\,\mathfrak{g} \subseteq \mathfrak{g}^*$.

We give a proof modulo a number of facts about semisimple Lie algebras and groups (and algebraic groups), so the reader not familiar with them may skip it.

PROOF. We may assume that $\Bbbk$ is algebraically closed (otherwise let $\overline{\Bbbk}$ be an algebraic closure and replace $\mathfrak{g}$ by $\mathfrak{g} \otimes_{\Bbbk} \overline{\Bbbk}$).

Let $\mathfrak{h} \subseteq \mathfrak{g}$ be a Cartan subalgebra and W be the Weyl group. The Chevalley isomorphism states that $(\mathrm{Sym}\,\mathfrak{g})^{\mathfrak{g}} \cong \mathcal{O}(\mathfrak{h}^*/W)$. This isomorphism maps the augmentation ideal $(\mathrm{Sym}\,\mathfrak{g})^{\mathfrak{g}}_+$ to the augmentation ideal of $\mathcal{O}(\mathfrak{h}^*/W)_+$, which is the ideal of the zero element $0 \in \mathfrak{h}^*$.

Now, let G be a connected algebraic group such that $\mathfrak{g} = \mathrm{Lie}\, G$ (this exists by the well-known classification of semisimple Lie algebras, see, e.g., [135], and by the existence theorem for reductive algebraic groups given root data, see, e.g., [212, 16.5]); in the case $\Bbbk = \mathbb{C}$, one can let G be a complex Lie group such that $\mathrm{Lie}\, G = \mathfrak{g}$. Then, $(\mathrm{Sym}\,\mathfrak{g})^{\mathfrak{g}} = (\mathrm{Sym}\,\mathfrak{g})^G = \mathcal{O}(\mathfrak{g}^*//G)$, and the ideal $\mathcal{O}(\mathfrak{h}^*/W)_+ \cdot \mathrm{Sym}\,\mathfrak{g}$ thus defines those elements $\phi \in \mathfrak{g}^*$ such that $\overline{G \cdot \phi} \cap \mathfrak{h}^* = \{0\}$. This set is stable under dilation, so that for all such nonzero ϕ, the coadjoint orbit $G \cdot \phi$ intersects a neighborhood of ϕ in the line $\Bbbk \cdot \phi$, and is hence in the nilpotent cone; conversely, if ϕ is in the nilpotent cone, the ideal vanishes on ϕ. □

COROLLARY 1.7.10. *For $\mathfrak{g}$ finite-dimensional semisimple, the central reductions $(U\mathfrak{g})^{\eta} = U\mathfrak{g}/(\ker\eta) \cdot U\mathfrak{g}$ quantize $\mathcal{O}(\mathrm{Nil}\,\mathfrak{g})$.*

EXAMPLE 1.7.11. In the case $\mathfrak{g} = \mathfrak{sl}_2$, the quantizations $(U\mathfrak{g})^{\eta}$, whose representations are those $\mathfrak{sl}_2$-representations on which the Casimir C acts by a fixed scalar, all quantize the cone of nilpotent 2×2 matrices,

$$\mathrm{Nil}(\mathfrak{sl}_2) = \left\{ \begin{pmatrix} a & b \\ c & -a \end{pmatrix} \,\middle|\, a^2 + bc = 0 \right\},$$

which are identified with functions on matrices via the trace pairing: $X(Y) := \mathrm{tr}(XY)$.

REMARK 1.7.12. The quadric of Example 1.7.11 is isomorphic to the one $v^2 = uw$ of Example 1.2.13, i.e., $\mathbb{A}^2/(\mathbb{Z}/2) \cong \mathrm{Nil}(\mathfrak{sl}_2)$. Thus we have given *two quantizations* of the same variety: one by the invariant Weyl algebra, $\mathrm{Weyl}_1^{\mathbb{Z}/2}$, and the other a family of quantizations given by the central reductions $(U\mathfrak{sl}_2)^{\eta} = U\mathfrak{sl}_2/(C - \eta(C))$. In fact, the latter family is a universal family of quantizations (i.e., all quantizations are isomorphic to one of these via a filtered isomorphism whose associated graded homomorphism is the identity), and one can see that $\mathrm{Weyl}_1^{\mathbb{Z}/2} \cong (U\mathfrak{sl}_2)^{\eta}$ where $\eta(C) = -\frac{3}{8}$ (see Exercise 1.10.2).

This coincidence is the first case of a part of the *McKay correspondence* [173, 46, 205, 206], which identifies, for every finite subgroup $\Gamma < SL_2(\mathbb{C})$, the quotient $\mathbb{C}^2/\Gamma$ with a certain two-dimensional "Slodowy" slice of $Nil(\mathfrak{g})$, where $\mathfrak{g}$ is the Lie algebra whose Dynkin diagram has vertices labeled by the irreducible representations of Γ, and a single edge from V to W if and only if $V \otimes W$ contains a copy of the standard representation $\mathbb{C}^2$. See Chapter IV, § 5. These varieties $\mathbb{C}^2/\Gamma$ are called du Val or Kleinian.

REMARK 1.7.13. Theorem 1.7.5 is stated anachronistically and deserves some explanation. Let us restrict for simplicity to the case $\Bbbk = \mathbb{C}$ (although everything below applies to the case $\Bbbk$ is algebraically closed of characteristic zero, and to deal with the non-algebraically closed case, one can tensor by an algebraic closure). Coxeter originally observed that $(\operatorname{Sym}\mathfrak{h})^W$ is a polynomial algebra for W a Weyl (or Coxeter) group. In the more general situation where W is a complex reflection group, the degrees d_i were computed by Shephard and Todd [201] in a case-by-case study, and shortly after this Chevalley [68] gave a uniform proof that $(\operatorname{Sym}\mathfrak{h})^W$ is a polynomial algebra if and only if W is a complex reflection group.

Chevalley also observed that $(\operatorname{Sym}\mathfrak{g})^{\mathfrak{g}} \cong (\operatorname{Sym}\mathfrak{h})^W$ where $\mathfrak{h} \subseteq \mathfrak{g}$ is a Cartan subalgebra and W is the Weyl group (the Chevalley restriction theorem) (this mostly follows from the more general property of conjugacy of Cartan subgroups and its proof, in [67]). Note here that $(\operatorname{Sym}\mathfrak{g})^{\mathfrak{g}} = \mathcal{O}(\mathfrak{g}^*//G)$, the functions on coadjoint orbits in $\mathfrak{g}^*$. The latter are identified with adjoint orbits of G in $\mathfrak{g}$ by the Killing form $\mathfrak{g}^* \cong \mathfrak{g}$. The closed adjoint orbits all contain points of $\mathfrak{h}$, and their intersections with $\mathfrak{h}$ are exactly the W-orbits.

Harish-Chandra [127] constructed an explicit isomorphism $HC : Z(U\mathfrak{g}) \xrightarrow{\sim} \operatorname{Sym}(\mathfrak{h})^W$ (such an isomorphism was, according to Godement's review on Mathematical Reviews (MR0044515) of this article, independently a consequence of results of H. Cartan, Chevalley, Koszul, and Weil in algebraic topology). Harish-Chandra's isomorphism is defined by the property that, for every highest-weight representation V_λ of $\mathfrak{g}$ with highest weight $\lambda \in \mathfrak{h}^*$ and (nonzero) highest weight vector $v \in V_\lambda$,

$$z \cdot v = HC(z)(\lambda) \cdot v,$$

viewing $HC(z)$ as a polynomial function on $\mathfrak{h}^*$. The Harish-Chandra isomorphism is nontrivial: indeed, the target of HC equips $\mathfrak{h}^*$ not with the usual action of W, but the *affine* action, defined by $w \cdot \lambda := w(\lambda + \delta) - \delta$, where the RHS uses the usual action of W on $\mathfrak{h}$, and δ is the sum of the fundamental weights. This shifting phenomenon is common for the center of a quantization. In this case one can explicitly see why the shift occurs: the center must act by the same character on highest weight representations V_λ and $V_{w \cdot \lambda}$, and computing this character (done in the $\mathfrak{sl}_2$ case in Exercise 1.7.3 below) yields the invariance under the affine action. To see why the center acts by the same character on V_λ and $V_{w \cdot \lambda}$, in the case where λ is dominant (i.e., $\lambda(\alpha) \geq 0$ for all positive roots α), and these are the Verma modules (i.e., $V_\lambda = \operatorname{Ind}_{\mathfrak{b}}^{\mathfrak{g}} \chi_\lambda$ where χ_λ is the corresponding character of a Borel subalgebra $\mathfrak{b}$ containing $\mathfrak{h}$, and the same is true for $V_{w \cdot \lambda}$), one can explicitly check that $V_{w \cdot \lambda} \subseteq V_\lambda$. The general case follows from this one. See [135, §23] for a detailed proof.

1.8. Beilinson–Bernstein localization theorem and global quantization. The purpose of this section is to explain a deep property of the quantization of the nilpotent cone of a semisimple Lie algebra discussed above: this *resolves* to a global (nonaffine) symplectic quantization of a cotangent bundle (namely, the cotangent bundle of the flag variety). We still assume that $\Bbbk$ has characteristic zero.

Return to the example $\mathrm{Nil}(\mathfrak{sl}_2) \cong \mathbb{A}^2/\{\pm \mathrm{Id}\}$. There is another way to view this Poisson algebra, by the *Springer resolution*. Namely, let us view $\mathrm{Nil}(\mathfrak{sl}_2)$ as the locus of nilpotent elements in $\mathfrak{sl}_2$, i.e., the nilpotent two-by-two matrices (this is consistent with Definition 1.7.6 if we identify $\mathfrak{sl}_2 \cong \mathfrak{sl}_2^*$ via the trace pairing $(x, y) = \mathrm{tr}(xy)$, cf. Remark 1.7.8). Then, a nonzero nilpotent element $x \in \mathfrak{sl}_2$, up to scaling, is uniquely determined by the line $\ker(x) = \mathrm{im}(x)$ in $\Bbbk^2$. Similarly, a nonzero element $\phi \in \mathrm{Nil}(\mathfrak{sl}_2) \subseteq \mathfrak{g}^*$ is uniquely determined, up to scaling, by the line $\ell \subseteq \Bbbk^2$ such that $\phi(x) = 0$ whenever $\mathrm{im}(x) \subseteq \ell$. With a slight abuse of notation we will also refer to this line as $\ker(\phi)$.

Consider the locus of pairs

$$X := \{(\ell, x) \in \mathbb{P}^1 \times \mathrm{Nil}(\mathfrak{sl}_2) \mid \ker(x) \subseteq \ell\} \subseteq \mathbb{P}^1 \times \mathrm{Nil}(\mathfrak{sl}_2).$$

This projects to $\mathrm{Nil}(\mathfrak{sl}_2)$. Moreover, the fiber over $x \neq 0$ is evidently a single point, since $\ker(x)$ determines ℓ. Only over the singular point $0 \in \mathrm{Nil}(\mathfrak{sl}_2)$ is there a larger fiber, namely $\mathbb{P}^1$ itself.

LEMMA 1.8.1. $X \cong T^*\mathbb{P}^1$.

PROOF. Fix $\ell \in \mathbb{P}^1$. Note that $T_\ell \mathbb{P}^1$ is naturally $\mathrm{Hom}(\ell, \Bbbk^2/\ell)$. On the other hand, the locus of x such that $\ker(x) \subseteq \ell$ naturally acts linearly on $T_\ell \mathbb{P}^1$: given such an x and given $\phi \in \mathrm{Hom}(\ell, \Bbbk^2/\ell)$, we can take $x \circ \phi \in \mathrm{Hom}(\ell, \ell) \cong \Bbbk$. Since this locus of x is a one-dimensional vector space, we deduce that it is $T_\ell^*\mathbb{P}^1$, as desired.   □

Thus, we obtain a resolution of singularities

$$\rho : T^*\mathbb{P}^1 \twoheadrightarrow \mathrm{Nil}(\mathfrak{sl}_2) = \mathbb{A}^2/\{\pm \mathrm{Id}\}. \tag{1.J}$$

Note that $\mathrm{Nil}(\mathfrak{sl}_2)$ is affine, so that, since the map ρ is birational (as all resolutions must be), $\mathcal{O}(\mathrm{Nil}(\mathfrak{sl}_2))$ is the algebra of global sections $\Gamma(T^*\mathbb{P}^1, \mathcal{O}_{T^*\mathbb{P}^1})$ of functions on $T^*\mathbb{P}^1$.

Moreover, equipping $T^*\mathbb{P}^1$ with its standard symplectic structure, ρ is a Poisson map, i.e., $\rho^* \pi_{\mathrm{Nil}(\mathfrak{sl}_2)} = \pi_{T^*\mathbb{P}^1}$ for $\pi_{\mathrm{Nil}(\mathfrak{sl}_2)}$ and $\pi_{T^*\mathbb{P}^1}$ the Poisson bivectors on $\mathrm{Nil}(\mathfrak{sl}_2)$ and $T^*\mathbb{P}^1$, respectively. Indeed, the Poisson structure on $\mathcal{O}_{\mathrm{Nil}(\mathfrak{sl}_2)}$ is obtained from the one on $\mathcal{O}_{T^*\mathbb{P}^1}$ by taking global sections.

In fact, the map ρ can be *quantized*: let $\mathcal{D}_{\mathbb{P}^1}$ be the sheaf of differential operators with polynomial coefficients on $\mathbb{P}^1$. This quantizes $\mathcal{O}_{T^*\mathbb{P}^1}$, as we explained, since $\mathbb{P}^1$ is smooth (this fact works for nonaffine varieties as well, if one uses sheaves of algebras). Then, there is the deep

THEOREM 1.8.2 (Beilinson–Bernstein for $\mathfrak{sl}_2$).   (i) *There is an isomorphism of algebras* $(U\mathfrak{sl}_2)^{\eta_0} \xrightarrow{\sim} \Gamma(\mathbb{P}^1, \mathcal{D}_{\mathbb{P}^1})$, *where* $\eta_0(C) = 0$;
  (ii) *Taking global sections yields an equivalence of abelian categories*

$$\mathcal{D}_{\mathbb{P}^1}\text{-mod} \xrightarrow{\sim} (U\mathfrak{sl}_2)^{\eta_0}\text{-mod}.$$

This quantizes the Springer resolution (1.J).

REMARK 1.8.3. This implies that the quantization $\mathcal{D}_{\mathbb{P}^1}$ of $\mathcal{O}_{T^*\mathbb{P}^1}$ is, in a sense, *affine*, since its category of representations is equivalent to the category of representations of its global sections. The analogous property holds for the sheaf of algebras $\mathcal{O}(X)$ on an arbitrary variety X if and only if X is affine. But, even though $\mathbb{P}^1$ is projective (the opposite of affine), the noncommutative algebra $\mathcal{D}_{\mathbb{P}^1}$ is still affine, in this sense.

There is a longstanding conjecture that says that, if there is an isomorphism $\mathcal{D}_X\text{-mod} \xrightarrow{\sim} \Gamma(X, \mathcal{D}_X)\text{-mod}$ (i.e., X is "$\mathcal{D}$-affine"), and X is smooth, projective, and connected, then X is of the form $X \cong G/P$ where $P < G$ is a parabolic subgroup of a connected semisimple algebraic group. (The converse is a theorem of Beilinson–Bernstein, which generalizes Theorem 1.8.4 below to the parabolic case G/P instead of G/B.) Similarly, there is also an (even more famous) "associated graded" version of the conjecture, which says that if X is a smooth projective variety and $T^*X \twoheadrightarrow Y$ is a symplectic resolution with Y affine (i.e., $Y = \operatorname{Spec}\Gamma(T^*X, \mathcal{O}(T^*X))$), then $X \cong G/P$ as before.

In fact, this whole story generalizes to arbitrary connected semisimple algebraic groups G with $\mathfrak{g} := \operatorname{Lie} G$ the associated finite-dimensional semisimple Lie algebra. Let $\mathcal{B}$ denote the flag variety of G, which can be defined as the symmetric space G/B for $B < G$ a Borel subgroup. Then, $\operatorname{Nil}(\mathfrak{g}) \subseteq \mathfrak{g}^*$ consists of the union of $\mathfrak{b}^\perp := \{\phi \mid \phi(x) = 0, \forall x \in \mathfrak{b}\}$ over all Borels.

THEOREM 1.8.4.   (i) *(Springer resolution) There is a symplectic resolution* $T^*\mathcal{B} \to \operatorname{Nil}(\mathfrak{g})$, *which is the composition*

$$T^*\mathcal{B} = \{(\mathfrak{b}, x) \mid x \in \mathfrak{b}^\perp\} \subseteq (\mathcal{B} \times \mathfrak{g}^*) \twoheadrightarrow \operatorname{Nil}(\mathfrak{g}),$$

where the last map is the second projection;
  (ii) *(Beilinson–Bernstein) There is an isomorphism* $(U\mathfrak{g})^{\eta_0} \xrightarrow{\sim} \Gamma(\mathcal{B}, \mathcal{D}_\mathcal{B})$, *where* η_0 *is the augmentation character, i.e.,* $\ker(\eta_0)$ *acts by zero on the trivial representation of* $\mathfrak{g}$;
  (iii) *(Beilinson–Bernstein) Taking global sections yields an equivalence of abelian categories,* $\mathcal{D}_\mathcal{B}\text{-mod} \xrightarrow{\sim} (U\mathfrak{g})^{\eta_0}\text{-mod}.$

REMARK 1.8.5. The theorem generalizes in order to replace the augmentation character η_0 by arbitrary characters η, at the price of replacing the category of D-modules $\mathcal{D}_B\text{-mod}$ by the category of twisted D-modules, with twisting corresponding to the character η.

1.9. Quivers and preprojective algebras. There is a very important generalization of tensor algebras, and hence also of finitely presented algebras, that replaces a set of variables $x_1, \ldots, x_n$ by a directed graph, which is typically called a *quiver*; the variables $x_1, \ldots, x_n$ then correspond to the directed edges, which are called *arrows*. Note that it may seem odd to have a special name for a completely ordinary object, the directed graph, but we have P. Gabriel to thank for this suggestive renaming: essentially, whenever you see the word "quiver," you should realize this is merely a directed graph, except that one is probably interested in representations of the quiver, or equivalently of its path algebra, as defined below.

DEFINITION 1.9.1. A quiver is a directed graph, whose directed edges are called arrows. Loops (arrows from a vertex to itself) and multiple edges (multiple arrows with the same endpoints) are allowed.

Typically, a quiver Q has its set of vertices denoted by Q_0 and its set of arrows denoted by Q_1.

DEFINITION 1.9.2. A representation $(\rho, (V_i)_{i \in Q_0})$ of a quiver Q is an assignment to each vertex $i \in Q_0$ a vector space V_i, and to each arrow $a : i \to j$ a linear map $\rho(a) : V_i \to V_j$. The dimension vector of ρ is defined as $d(\rho) := (\dim V_i)_{i \in Q_0} \in \mathbb{Z}_{\geq 0}^{Q_0}$.

DEFINITION 1.9.3. For every dimension vector $d \in \mathbb{Z}_{\geq 0}^{Q_0}$, let $\mathrm{Rep}_d(Q)$ be the set of representations of the form $(\rho, (\Bbbk^{d_i}))$, i.e., with $V_i = \Bbbk^{d_i}$ for all i.

DEFINITION 1.9.4. The path algebra $\Bbbk Q$ of a quiver Q is defined as the vector space with basis the set of paths in the quiver Q (allowing paths of length zero at each vertex), with multiplication of paths given by reverse concatenation: if $p : i \to j$ is a path from i to j and $q : j \to \ell$ is a path from j to ℓ, then $qp : i \to \ell$ is the concatenated path from i to ℓ. The product of two paths that cannot be concatenated (because their endpoints don't match up) is zero.

Given an arrow $a \in Q_1$, let a_h be its head (the incident vertex it points to) and a_t be its tail (the incident vertex the arrow points away from), so that $a : a_t \to a_h$. Note that, viewing a_t and a_h as zero-length paths, we have $a = a_h a a_t$ (because we are using reverse concatenation). We further observe that every zero-length path defines an idempotent in the path algebra: in particular, a_h and a_t are idempotents.

EXERCISE 1.9.5.   (i) Show that a representation of Q is the same as a representation of the algebra $\Bbbk Q$.
  (ii) Show that the set of representations $\mathrm{Rep}_d(Q)$ is canonically isomorphic to the vector space $\bigoplus_{a \in Q_1} \mathrm{Hom}(\Bbbk^{d(a_t)}, \Bbbk^{d(a_h)})$.

EXERCISE 1.9.6. Suppose that Q has only one vertex, so that Q_1 consists entirely of loops from the vertex to itself. Then show that $\Bbbk Q$ is the tensor algebra over the vector space with basis Q_1.

DEFINITION 1.9.7. Given a quiver Q, the double quiver $\bar{Q}$ is defined as the quiver with the same vertex set, $\bar{Q}_0 := Q_0$, and with double the number of arrows, $\bar{Q}_1 := Q_1 \sqcup Q_1^*$, obtained by adding, for each arrow $a \in Q_1$, a reverse arrow, $a^* \in Q_1$, such that if $a : i \to j$ has endpoints i and j, then $a^* : j \to i$ has the same endpoints but with the opposite orientation; thus $Q_1^* := \{a^* \mid a \in Q_1\}$.

EXERCISE 1.9.8. Show that there is a canonical isomorphism between $\mathrm{Rep}_d(\bar{Q})$ and $T^* \mathrm{Rep}_d(Q)$.

DEFINITION 1.9.9. Let $\lambda \in \Bbbk^{Q_0}$. Then the *deformed preprojective algebra* $\Pi_\lambda(Q)$ (defined in [75]) is defined as

$$\Pi_\lambda(Q) := \Bbbk\bar{Q}/\Big(@ \sum_{a \in Q_1} aa^* - a^*a - \sum_{i \in Q_0} \lambda_i\Big). \tag{1.K}$$

The (undeformed) preprojective algebra is $\Pi_0(Q)$ (defined in [108]).

The algebra $\Pi_\lambda(Q)$ is filtered by the length of paths: $(\Pi_\lambda(Q))_{\leq m}$ is the span of paths of length $\leq m$.

EXERCISE 1.9.10.   (i) Show that $\Pi_0(Q)$ is actually graded by path length.

(ii) Show that there is a canonical surjection $\Pi_0(Q) \to \mathrm{gr}\, \Pi_\lambda(Q)$ (Hint: observe that the relations are deformed, and refer to Exercise 1.3.4).

(iii) Give an example to show that this surjection need not be flat in general (i.e., $\Pi_\lambda(Q)$ is not a (flat) filtered deformation in general). Hint: Try a quiver with one arrow and two vertices.

REMARK 1.9.11. It is a deep fact that, whenever Q is not Dynkin (i.e., the underlying graph forgetting orientation is not Dynkin), then $\Pi_\lambda(Q)$ is always a (flat) filtered deformation of $\Pi_0(Q)$, but we will not prove this here (it follows from the fact that $\Pi_0(Q)$ satisfies a quiver analogue of the Koszul property: it is Koszul over the semisimple ring $\Bbbk Q_0$).

DEFINITION 1.9.12. For $d = (d_i) \in \mathbb{Z}_{\geq 0}^{Q_0}$, let

$$\mathfrak{g}(d) := \prod_{i \in Q_0} \mathfrak{gl}(d_i).$$

The *moment map* on $\mathrm{Rep}_d(\bar{Q})$ is the map $\mu : \mathrm{Rep}_d(\bar{Q}) \to \mathfrak{g}(d)$ defined by

$$\mu(\rho) = \sum_{a \in Q_1} \rho(a)\rho(a^*) - \rho(a^*)\rho(a). \tag{1.L}$$

Let $\mathrm{Rep}_d(\Pi_\lambda(Q))$ be the set of representations of $\Pi_\lambda(Q)$ on $(\Bbbk^{d_i})$, i.e., the set of $\Bbbk Q_0$-algebra homomorphisms

$$\Pi_\lambda(Q) \to \mathrm{End}(\bigoplus_{i \in Q_0} \Bbbk^{d_i}),$$

equipping the RHS with the $\Bbbk Q_0$-algebra structure where each vertex $i \in Q_0$ is the projection to $\Bbbk^{d_i}$. In other words, this is the set of representations on $\bigoplus_i \Bbbk^{d_i}$ with each arrow $a \in \bar{Q}_1$ given by a map from the a_t-factor $\Bbbk^{d_{a_t}}$ to the a_h-factor $\Bbbk^{d_{a_h}}$.

EXERCISE 1.9.13. Show that there is a canonical identification $\mathrm{Rep}_d(\Pi_\lambda(Q)) = \mu^{-1}(\lambda \cdot \mathrm{Id})$.

Note that, since $\Pi_0(Q)$ is not commutative, the deformation $\Pi_\lambda(Q)$ (in the case it is flat) does *not* give $\Pi_0(Q)$ a Poisson structure (this would not make sense, since for us Poisson algebras are by definition commutative).

REMARK 1.9.14. The above is closely related to an important example of quantization, namely the quantized quiver varieties (this has been known since the origin of preprojective algebras, but see, e.g., [36] for a recent paper on the subject, where the algebra is denoted $A_\lambda(v)$, setting $v = d$ and $w = 0$). These are defined as follows. Letting $G(d) := \prod_{i \in Q_0} \mathrm{GL}(d_i)$, we can consider the variety $\mathrm{Rep}_d(\Pi_\lambda(Q)) /\!/ G(d)$ parameterizing representations of $\Pi_\lambda(Q)$ of dimension vector d *up to isomorphism*. This turns out to be Poisson, i.e., its algebra of functions, $\mathcal{O}(\mathrm{Rep}_d(\Pi_\lambda(Q)))^{G(d)}$, is Poisson. To quantize it, one replaces $T^* \mathrm{Rep}_d(Q)$ by its quantization $\mathcal{D}(\mathrm{Rep}_d(Q))$, and hence replaces the Poisson algebra $\mathcal{O}(\mathrm{Rep}_d(\Pi_0(Q)))^{G(d)}$, which is a quotient of $\mathcal{O}(T^* \mathrm{Rep}_d(Q)) G(d)$, by the corresponding quotient of $\mathcal{D}(\mathrm{Rep}_d(Q))^{G(d)}$: Let $\mathrm{tr} : \mathfrak{g}(d) \to \Bbbk$ be the sum of the trace functions $\mathrm{GL}(V_i) \to \Bbbk$. Then we define

$$A_\mu(d) := \mathcal{D}(\mathrm{Rep}_d(Q))^{G(d)} \Big/ \Big(\mathrm{tr}\Big(x \Big(\lambda \, \mathrm{Id} - \sum_{a \in Q_1} \rho(a)\rho(a^*) - \rho(a^*)\rho(a) \Big) \Big)_{x \in \mathfrak{g}(d)} \Big),$$

where the ideal we quotient by is two-sided (although actually equals the same as the one-sided ideal on either side with the same generators), and we consider the trace to be a function of ρ for every $x \in \mathfrak{g}(d)$, and hence an element of $\mathcal{O}(\mathrm{Rep}_d(Q)) \subseteq \mathcal{D}(\mathrm{Rep}_d(Q))$, which is evidently $G(d)$-invariant.

Dynkin and extended Dynkin quivers. Finally, as motivation for further study, we briefly explain how the behavior of quivers and preprojective algebras falls into three distinct classes, depending on whether the quiver is Dynkin, extended Dynkin, or otherwise. Readers not familiar with Dynkin diagrams can safely skip this subsection. We restrict to the case of a connected quiver, i.e., one whose underlying graph is connected. A quiver is Dynkin if its underlying graph is a type ADE Dynkin diagram, and it is extended Dynkin if its underlying graph is a simply-laced extended Dynkin diagram, i.e., of types $\tilde{A}_n$, $\tilde{D}_n$, or $\tilde{E}_n$ for some n. Note that we already saw one way in which the behavior changes between the Dynkin and non-Dynkin case, in Remark 1.9.11.

Let us assume $\Bbbk$ is algebraically closed of characteristic zero.

THEOREM 1.9.15. *Let Q be a connected quiver. Let $\lambda \in \Bbbk^{Q_0}$ be arbitrary.*

(1) *The following are equivalent:* (a) *Q is Dynkin;* (b) *The algebra $\Pi_\lambda(Q)$ is finite-dimensional;* and (c) *the quiver Q has finitely many indecomposable representations up to isomorphism;*

(2) *The following are equivalent:* (a) *the quiver Q is extended Dynkin;* (b) *the algebra $\Pi_0(Q)$ has an infinite-dimensional center;* (c) *the center of $\Pi_0(Q)$ is of the form $\mathcal{O}(\mathbb{A}^2)^\Gamma = \mathcal{O}(\mathbb{A}^2/\Gamma)$ for a finite subgroup $\Gamma < \mathrm{SL}_2(\Bbbk)$.*

Recall that the singularities $\mathbb{A}^2/\Gamma$ for $\Gamma < SL_2(\Bbbk)$ finite are called du Val or Kleinian singularities. The correspondence between extended Dynkin quivers Q and finite subgroups $\Gamma < SL_2(\Bbbk)$ is called the *McKay correspondence*: when Q is of type $\tilde{A}_n$, then $\Gamma = \mathbb{Z}/n$ is cyclic; when Q is of type $\tilde{D}_n$, then Γ is a double cover of a dihedral group; and when Q is of type $\tilde{E}_n$, then it is a double cover of the tetrahedral, octahedral, or icosahedral rotation group. See also Remark 1.7.12 and Chapter IV, §5.

REMARK 1.9.16. The various parts of the theorem appeared as follows: Equivalence (1)(a) $\Leftrightarrow$ (1)(b) is due to Gelfand and Ponomarev [108] (at least for $\lambda = 0$) and the equivalence (1)(a) $\Leftrightarrow$ (1)(c) is due to Peter Gabriel [107]. Equivalences (2)(a) $\Leftrightarrow$ (2)(b) and (2)(a) $\Leftrightarrow$ (2)(c) follow from the stronger statements that, when Q is neither Dynkin nor extended Dynkin, then the center of $\Pi_0(Q)$ is just $\Bbbk$ ([74, Proposition 8.2.2]; see also [100, Theorem 1.3.1]), whereas when Q is extended Dynkin, then the center is $\mathcal{O}(\mathbb{A}^2)^\Gamma$ [75]. The theorem also extends to characteristic p without much modification (only (2)(c) needs to be modified); see [197, Theorem 10.1.1].

In the next remark and later on, we will need to use skew product algebras:

DEFINITION 1.9.17. Let Γ be a finite (or discrete) group acting by automorphisms on an algebra A. The skew (or smash) product $A \rtimes \Gamma$ is defined as the algebra which, as a vector space, is the tensor product $A \otimes \Bbbk[\Gamma]$, with the multiplication

$$(a_1 \otimes g_1)(a_2 \otimes g_2) := a_1 g_1(a_2) \otimes g_1 g_2.$$

REMARK 1.9.18. In the extended Dynkin case, when λ lies in a particular hyperplane in $\Bbbk^{Q_0}$, then Π_λ also has an infinite center, and this center $Z(\Pi_\lambda)$ gives a commutative deformation of $\mathcal{O}(\mathbb{A}^2/\Gamma)$ which is the versal deformation. For general λ, Π_λ is closely related to a symplectic reflection algebra $H_{t,c}(\Gamma)$ deforming $\mathcal{O}(\mathbb{A}^2) \rtimes \Gamma$ (see Chapter III or Example 2.7.8 below for the definition of this algebra): there is an idempotent $f \in \mathbb{C}[\Gamma]$ such that $\Pi_\lambda = f H_{t,c}(\Gamma) f$ for t, c determined by λ, and this makes Π_λ Morita equivalent to the symplectic reflection algebra; this follows from [75]. Similarly, $\mathcal{O}(\mathbb{A}^2/\Gamma) = e\Pi_0 e$ for $e \in \Bbbk Q$ the idempotent corresponding to the extending vertex, and $e\Pi_\lambda e$ yields the spherical symplectic reflection algebra, $e H_{t,c}(\Gamma) e$.

REMARK 1.9.19. There is also an analogue of (1)(c) for part two (due to [177] and [87], see also [84]): Continue to assume that $\Bbbk$ has characteristic zero (or, it suffices for this statement for it to be infinite). Then, a quiver Q is extended Dynkin if and only if, for each dimension vector, the isomorphism classes of representations can be parametrized by finitely many curves and points, i.e., the variety of representations modulo equivalence has dimension ≤ 1, and there exists a dimension vector for which this has dimension one (i.e., it does not have finitely many indecomposable representations for every dimension vector, which would imply it is Dynkin by the theorem). In fact, the only dimension vectors for which there are infinitely many indecomposable representations are the imaginary roots (the dimension vectors that are in the kernel of the Cartan matrix associated to the extended Dynkin diagram), and for these all but finitely many indecomposable representations are parametrized by the projective line $\mathbb{P}^1$.

1.10. Additional exercises. (See also 1.2.1, 1.2.3, 1.2.7, 1.2.8, 1.3.4, 1.4.2, 1.4.3, 1.4.4, 1.4.5, and 1.5.2; and the exercises on quivers: 1.9.5, 1.9.6, 1.9.8, 1.9.10, and 1.9.13.)

EXERCISE 1.10.1. We elaborate on the final point of Exercise 1.3.4, giving an example where a filtered deformation of homogeneous relations can yield an algebra of smaller dimension. Consider the quadratic algebra $B = \Bbbk\langle x, y\rangle/(xy, yx)$ (by quadratic, we mean presented by quadratic relations, i.e., homogeneous relations of degree two). First show that B has a basis consisting of monomials in either x or y but not both, and hence it is infinite-dimensional. Now consider the family of deformations parametrized by pairs $(a, b) \in \Bbbk^2$, given by

$$A_{a,b} := \Bbbk\langle x, y\rangle/(xy - a, yx - b),$$

i.e., this is a family of filtered algebras obtained by deforming the relations of B. Show that, for $a \neq b$, we get $A_{a,b} = \{0\}$, the zero ring. Hence $A_{a,b}$ is not a (flat) filtered deformation of $A_{0,0}$ for $a \neq b$; equivalently, the surjection of Exercise 1.3.4 is not an isomorphism for $a \neq b$.

On the other hand, show that, for $a = b \neq 0$, then

$$A_{a,a} \cong \Bbbk[x, x^{-1}],$$

and verify that the basis we obtained for B (monomials in x or y but not both) gives also a basis for $A_{a,a}$. So the family $A_{a,b}$ is flat along the diagonal $\{(a, a)\} \subseteq \Bbbk^2$.

EXERCISE 1.10.2. (a) Verify, using Singular, that

$$\mathsf{Weyl}_1^{\mathbb{Z}/2} \cong (U\mathfrak{sl}_2)^\eta,$$

where

$$\eta(C) = \eta\left(ef + fe + \tfrac{1}{2}h^2\right) = -\tfrac{3}{8}.$$

How to do this: Type the commands

```
LIB "nctools.lib";
def a = makeWeyl(1);
setring a;
a;
```

Now you can play with the Weyl algebra with variables D and x; D corresponds to ∂_x. Now you need to figure out some polynomials $e(x, D)$, $f(x, D)$, and $h(x, D)$ so that

$$[e, f] = h, \quad [h, e] = 2e, \quad [h, f] = -2f.$$

Then, once you have done this, compute the value of the Casimir,

$$C = ef + fe + \tfrac{1}{2}h^2,$$

and verify it is $-\tfrac{3}{8}$.

Hint: the polynomials $e(x, D)$, $f(x, D)$, and $h(x, D)$ should be homogeneous quadratic polynomials (that way the bracket is linear).

For example, try first

```
def e=D^2;
def f=x^2;
def h=e*f-f*e;
h*e-e*h;
2*e;
h*f-f*h;
2*f;
```

and you see that, defining e and f as above and h to be $ef - fe$ as needed, we don't quite get $he - eh = 2e$ or $hf - fh = 2f$. But you can correct this...

(b) Show that the highest weight of a highest weight representation of $(U\mathfrak{sl}_2)^\eta$, for this η, is either $-\frac{3}{2}$ or $-\frac{1}{2}$. Here, a highest weight representation is one generated by a vector v such that $e \cdot v = 0$ and $h \cdot v = \mu v$ for some $\mu \in \Bbbk$. Then, μ is called its highest weight.

The value $-\frac{1}{2}$ is halfway between the value 0 of the highest weight of the trivial representation of $\mathfrak{sl}_2$ and the value -1 of the highest weight of the Verma module for the unique χ such that $(U\mathfrak{sl}_2)^\chi$ has infinite Hochschild dimension (see Remark 3.7.8 for the definition of Hochschild dimension).

REMARK 1.10.3. Here, a Verma module of $\mathfrak{sl}_2$ is a module of the form $U\mathfrak{sl}_2/(e, h-\mu)$ for some μ; this Verma module of highest weight -1 is also the unique one that has different central character from all other Verma modules, since the action of the Casimir on a Verma with highest weight μ is by $\frac{1}{2}\mu^2 + \mu$, so C acts by multiplication by $-\frac{1}{2}$ on this Verma and no others. Therefore, the Bernstein–Gelfand–Gelfand [25] category $\mathcal{O}$ of modules for $U\mathfrak{sl}_2$ (these are defined as modules which are finitely generated, have semisimple action of the Cartan subalgebra, and are such that, for every vector v, $e^N v = 0$ for some $N \geq 0$) which factor through the central quotient $U\mathfrak{sl}_2/(C + \frac{1}{2})$ is equivalent to the category of vector spaces, with this Verma as the unique simple object. This is the only central quotient $U\mathfrak{sl}_2/(C - (\frac{1}{2}\mu^2 + \mu))$, with μ integral, having this property. Also, the corresponding Cherednik algebra $H_{1,c}(\mathbb{Z}/2)$ considered in Chapter III, of which $U\mathfrak{sl}_2/(C + \frac{1}{2})$ is the spherical subalgebra, has semisimple category $\mathcal{O}$ as defined there (this algebra has two simple Verma modules: one of them is killed by symmetrization $V \mapsto eVe$, so only one yields a module over the spherical subalgebra).

(c) Identify the representation $\Bbbk[x^2]$ over $\mathrm{Weyl}_1^{\mathbb{Z}/2}$ with a highest weight representation of $(U\mathfrak{sl}_2)^\eta$ (using an appropriate choice of e, h, and f in $\mathrm{Weyl}_1^{\mathbb{Z}/2}$).

(d) Use the isomorphism $\mathrm{Weyl}_1^{\mathbb{Z}/2} \cong (U\mathfrak{sl}_2)^\eta$ and representation theory of $\mathfrak{sl}_2$ to show that $\mathrm{Weyl}_1^{\mathbb{Z}/2}$ admits no finite-dimensional irreducible representations. Hence it admits no finite-dimensional representations at all. (Note: in Exercise 1.10.7 below, we will see that $\mathrm{Weyl}_1^{\mathbb{Z}/2}$ is in fact simple, which strengthens this, and we will generalize it to $\mathrm{Weyl}(V)^{\mathbb{Z}/2}$ for arbitrary V and $\mathbb{Z}/2$ acting by $\pm \mathrm{Id}$.)

We will often need the notion of *Hilbert series*:

DEFINITION 1.10.4. Let B be a graded algebra, $B = \bigoplus_{m \in \mathbb{Z}} B_m$, with all B_m finite-dimensional. Then the Hilbert series is defined as $h(B; t) = \sum_{m \in \mathbb{Z}} \dim B_m t^m$.

EXERCISE 1.10.5. In this exercise, we will learn another characterization of flatness of deformations via Hilbert series. Let A and B be as in Exercise 1.3.4. Suppose first that B is finite-dimensional. Show that A is a (flat) filtered deformation of B, i.e., the surjection $B \to \mathrm{gr}(A)$ is an isomorphism, if and only if $\dim A = \dim B$. More generally, suppose that B has finite-dimensional weight spaces. Show then that $\mathrm{gr}(A)$ automatically also has finite-dimensional weight spaces, and that A is a (flat) filtered deformation if and only if $h(\mathrm{gr}\, A; t) := h(B; t)$.

EXERCISE 1.10.6. Now we use GAP to try to play with flat deformations. In these examples, you may take the definition of a flat filtered deformation of B to be an algebra A given by a filtered deformation of the relations of B such that $h(\mathrm{gr}\, A; t) = h(B; t)$.

One main technique to use is that of Gröbner bases. See Chapter I for the necessary background on these.

Here is code to get you started for the universal enveloping algebra of $\mathfrak{sl}_2$:

```
LoadPackage("GBNP");
A:=FreeAssociativeAlgebraWithOne(Rationals, "e", "f", "h");
e:=A.e;; f:=A.f;; h:=A.h;; o:=One(A);;
uerels:=[f*e-e*f+h,h*e-e*h-2*e,h*f-f*h+2*f];
uerelsNP:=GP2NPList(uerels);;
PrintNPList(uerelsNP);
GBNP.ConfigPrint(A);
GB:=SGrobner(uerelsNP);;
PrintNPList(GB);
```

This computes the Gröbner basis for the ideal generated by the relations. You can also get explicit information about how each Gröbner basis element is obtained from the original relations:

```
GB:=SGrobnerTrace(uerelsNP);;
PrintNPListTrace(GB);
PrintTracePol(GB[1]);
```

Here the second line gives you the list of Gröbner basis elements, and for each element $GB[i]$, the third line will tell you how it is expressed in terms of the original relations.

(a) Verify that, with the relations, $[x, y] = z$, $[y, z] = x$, $[z, x] = y$, one gets a flat filtered deformation of $\Bbbk[x, y, z]$, by looking at the Gröbner basis. Observe that this is the enveloping algebra of a three-dimensional Lie algebra isomorphic to $\mathfrak{sl}_2$.

(b) Now play with modifying those relations and see that, for most choices of filtered deformations, one need not get a flat deformation.

(c) Now play with the simplest Cherednik algebra: the deformation of $\mathrm{Weyl}_1 \rtimes \mathbb{Z}/2$ (cf. Definition 1.9.17). The algebra $\mathrm{Weyl}_1 \rtimes \mathbb{Z}/2$ itself is defined by relations $x * y - y * x - 1$ (the Weyl algebra relation), together with $z^2 - 1$ (so z generates $\mathbb{Z}/2$) and $x * z + z * x$, $y * z + z * y$, so z anticommutes with x and y. Show that this is a flat deformation of $\mathsf{Sym}(\Bbbk^2) \rtimes \mathbb{Z}/2$, i.e., $\Bbbk[x, y] \rtimes \mathbb{Z}/2$. Equivalently, show that the latter is the associated graded algebra of $\mathsf{Weyl}_1 \rtimes \mathbb{Z}/2$.

Then, the deformation of $\mathsf{Weyl}_1 \rtimes \mathbb{Z}/2$ is given by replacing the relation $x * y - y * x - 1$ with the relation $x * y - y * x - 1 - \lambda \cdot z$, for $\lambda \in \Bbbk$ a parameter. Show that this is also a flat deformation of $\Bbbk[x, y] \rtimes \mathbb{Z}/2$ for all λ (again you can just compute the Gröbner basis).

(d) Now modify the action of $\mathbb{Z}/2$ so as to not preserve the symplectic form: change $y * z + z * y$ into $y * z - z * y$. What happens to the algebra defined by these relations now?

EXERCISE 1.10.7. Prove that, over a field $\Bbbk$ of characteristic zero, V a symplectic vector space, and $\mathbb{Z}/2 = \{\pm \mathsf{Id}\} < \mathsf{Sp}(V)$, the skew product $\mathsf{Weyl}(V) \rtimes \mathbb{Z}/2$ is a simple algebra, and similarly show the same for $(\mathsf{Weyl}(V))^{\mathbb{Z}/2}$. Hint: For the latter, you can use the former, together with the symmetrizer element $e = \frac{1}{2}(1 + \sigma)$, for $\mathbb{Z}/2 = \{1, \sigma\}$, satisfying $e^2 = e$ and $e(\mathsf{Weyl}(V) \rtimes \mathbb{Z}/2)e = \mathsf{Weyl}(V)^{\mathbb{Z}/2}$.

(In fact, one can show that the same is true for $\mathsf{Weyl}(V) \rtimes \Gamma$ for arbitrary finite $\Gamma < \mathsf{Sp}(V)$: this is a much more difficult exercise you can try if you are ambitious. From this one can deduce that $(\mathsf{Weyl}(V))^{\Gamma}$ is also simple by the following Morita equivalence: if $e \in \Bbbk[\Gamma]$ is the symmetrizer element, then $\mathsf{Weyl}(V)^{\Gamma} = e(\mathsf{Weyl}(V) \rtimes \Gamma)e$ and $(\mathsf{Weyl}(V) \rtimes \Gamma)e(\mathsf{Weyl}(V) \rtimes \Gamma) = \mathsf{Weyl}(V) \rtimes \Gamma$.)

For the next exercises, we need to recall the notions of graded and filtered modules.

Recall that for a graded algebra $B = \bigoplus_m B_m$, a graded module is a module $M = \bigoplus_m M_m$ such that $B_i \cdot M_j \subseteq M_{i+j}$ for all $i, j \in \mathbb{Z}$. As before, we will call these gradings the weight gradings. We say that $\phi : M \to N$ has weight k if $\phi(M_m) \subseteq N_{m+k}$ for all m. If M is a graded right B-module and N is a graded left B-module, then $M \otimes_B N$ is a graded vector space, placing $M_m \otimes_B N_n$ in weight $m + n$.

This induces gradings also on the corresponding derived functors, i.e., $M \otimes_B N$ becomes canonically graded and $\mathsf{Ext}^i_B(M, N)$ has a notion of weight k elements. Explicitly, if M is a graded left B-module and $P_\bullet \twoheadrightarrow M$ is a graded projective resolution, then weight k cocycles of the complex $\mathsf{Hom}_B(P_\bullet, N)$ induce weight k elements of $\mathsf{Ext}^\bullet_B(M, N)$. Similarly if M is a graded right B-module and $P_\bullet \twoheadrightarrow M$ is a graded projective resolution, then the complex $P_\bullet \otimes_B N$ computing $\mathsf{Tor}_\bullet(M, N)$ is a graded complex.

Given a nonnegatively filtered algebra A, a (good) nonnegatively filtered module M is one such that $M_{\leq -1} = 0 \subseteq M_{\leq 0} \subseteq \cdots$, with $M = \bigcup_m M_{\leq m}$, and $A_{\leq m} M_{\leq n} \subseteq M_{\leq (m+n)}$. (Note that everything generalizes to the case where B is $\mathbb{Z}$-graded and M is $\mathbb{Z}$-filtered, if we replace this condition by $\bigcap_m M_{\leq m} = \{0\}$.)

EXERCISE 1.10.8. Given an algebra B, to understand its Hochschild cohomology (which governs deformations of associative algebras!) as well as many things, we need to construct resolutions. A main tool then is *deformations* of resolutions.

Suppose that B is nonnegatively graded, i.e., $B = \bigoplus_{m \geq 0} B_m$ for B_m the degree m part of B. Let A be a filtered deformation (i.e., $\mathsf{gr}\, A = B$) and M be a nonnegatively filtered A-module. (We could also work for this problem in the $\mathbb{Z}$-graded and filtered setting, but do not do so for simplicity.)

Suppose that $Q_\bullet \to M$ is a nonnegatively filtered complex with Q_i projective such that $\operatorname{gr} Q_\bullet \to \operatorname{gr} M$ is a projective resolution of $\operatorname{gr} M$. Show that $Q_\bullet \to M$ is a projective resolution of M.

Hint: Show that $\dim H_i(\operatorname{gr} C_\bullet) \geq \dim H_i(C_\bullet)$ for an arbitrary filtered complex $C_\bullet$, because $\ker(\operatorname{gr} C_i \to \operatorname{gr} C_{i-1}) \supseteq \operatorname{gr} \ker(C_i \to C_{i-1})$ and $\operatorname{im}(\operatorname{gr} C_{i+1} \to \operatorname{gr} C_i) \subseteq \operatorname{gr} \operatorname{im}(C_{i+1} \to C_i)$. Conclude, for an arbitrary filtered complex, that exactness of $\operatorname{gr}(C_\bullet)$ implies exactness of $C_\bullet$ (the converse is not true).

PROPOSITION 1.10.9. *Suppose that $B = TV/(R)$ for some homogeneous relations $R \subseteq R$, and that B has a nonnegatively graded free B-bimodule resolution, $P_\bullet \twoheadrightarrow B$, which is finite-dimensional in each weighted degree. If $E \subseteq TV$ is a filtered deformation of the homogeneous relations R, i.e., $\operatorname{gr} E = R$, then the following are equivalent:*

(i) *$A := TV/(E)$ is a flat filtered deformation of B, i.e., the canonical surjection $B \twoheadrightarrow \operatorname{gr} A$ is an isomorphism;*

(ii) *The resolution $P_\bullet \to B$ deforms to a filtered free resolution $Q_\bullet \to A$.*

(ii') *The resolution $P_\bullet$ deforms to a filtered complex $Q_\bullet \to A$.*

In this case, the deformed complex in (ii) *is a free resolution of A.*

We remark that the existence of $P_\bullet$ is actually automatic, since B always admits a unique (up to isomorphism) minimal free resolution (where minimal means that, for each i, $P_i = B_i \otimes V_i$ for V_i a nonnegatively graded vector space with minimum possible Hilbert series, i.e., for any other choice V_i', we have that $h(V_i'; t) - h(V_i; t)$ has nonnegative coefficients). For the minimal resolution, each P_i is in degrees $\geq i$ for all i, and is finite-dimensional in each degree, which implies that $\bigoplus_i P_i$ is also finite-dimensional in each degree.

SKETCH OF PROOF. It is immediate that (ii) implies (ii'), and also that (ii') implies (i), since part of the statement is that $\operatorname{gr}(A) \cong B$. Also, by the exercise, (ii') implies (ii).

To show (i) implies (ii) — it does not really help to try to show only (ii') — inductively construct a deformation of $P_\bullet$ to a filtered free resolution of A. It suffices to let $P_\bullet$ be the minimal resolution as above, since any other resolution is obtained from this one by summing with a split exact complex of free graded modules. First of all, we know that $P_0 \twoheadrightarrow B$ deforms to $Q_0 \twoheadrightarrow A$, since we can arbitrarily lift the map $P_0 = B^{r_0} \to B$ to a filtered map $Q_0 := A^{r_0} \to A$, and this must be surjective since $B \twoheadrightarrow \operatorname{gr} A$ is surjective. Since in fact it is an isomorphism, the Hilbert series (Definition 1.10.4) of the kernels are the same, and so we can arbitrarily lift the surjection $P_1 = B^{r_1} \twoheadrightarrow \ker(P_0 \to B)$ to a filtered map $Q_1 := A^{r_1} \twoheadrightarrow \ker(Q_0 \to A)$, etc. Fill in the details! $\square$

In the case B is Koszul, we can do better using the form of the Koszul complex. This is divided over the next three exercises. In the first exercise we explain augmented algebras. In the next exercise, we explain quadratic and Koszul algebras. In the third exercise, we improve the above result in the case B is Koszul.

We begin with the definition and an exercise on augmented algebras:

DEFINITION 1.10.10. An augmented algebra is an associative $\Bbbk$-algebra B together with an algebra homomorphism $B \to \Bbbk$. The *augmentation ideal* is the kernel, $B_+ \subseteq B$.

EXERCISE 1.10.11. (a) Show that an augmented algebra is equivalent to the data of an algebra B and a codimension-one ideal $B_+ \subseteq B$.

(b) Show that TV is augmented with augmentation ideal $(TV)_+ := \bigoplus_{m \geq 1} V^{\otimes m}$.

(c) Show that an augmented algebra B is always of the form $TV/(R)$ for V a vector space and $R \subseteq (TV)_+$.

(d) Suppose now that $R \subseteq ((TV)_+)^2$.[2] Then we have an isomorphism $V \cong \mathrm{Tor}_1(\Bbbk, \Bbbk)$ (for $B = TV/(R)$). Hint: one has a projective B-module resolution of $\Bbbk$ of the form

$$\cdots \to B \otimes R \to B \otimes V \to B \twoheadrightarrow \Bbbk. \tag{1.M}$$

Here the maps are given by

$$b \otimes \sum_i (v_{i,1} \cdots v_{i,k_i}) \mapsto \sum_i b(v_{i,1} \cdots v_{i,k_i-1}) \otimes v_{i,k_i}, \quad b \otimes v \mapsto bv.$$

In other words, the first map is obtained by restricting to $B \otimes R$ the splitting map $B \otimes (TV)_+ \to B \otimes V$, given by linearly extending to all of $B \otimes (TV)_+$ the assignment

$$b \otimes v_1 \cdots v_k \mapsto bv_1 \cdots v_{k-1} \otimes v_k.$$

Then show that, when $R \subseteq ((TV)_+)^2$, then after applying the functor $M \mapsto M \otimes_B \Bbbk$ to the sequence, the last two differentials (before the map $B \twoheadrightarrow \Bbbk$) become zero.

(e) Continue to assume $R \subseteq ((TV)_+)^2$. Assume that $R \cap (RV + VR) = \{0\}$, where $(RV + VR)$ denotes the ideal, i.e., $(TV)_+ R(TV) + (TV)R(TV)_+$; this is a minimality condition. Show in this case that $\mathrm{Tor}_2(\Bbbk, \Bbbk) \cong R$. In particular, if R is spanned by homogeneous elements, show that, replacing R with a suitable subspace, it satisfies this condition, and then $\mathrm{Tor}_2(\Bbbk, \Bbbk) \cong R$. Hint: Show that the projective resolution (1.M) can be extended to

$$\cdots \to B \otimes S \to B \otimes R \to B \otimes V \to B \twoheadrightarrow \Bbbk, \tag{1.N}$$

where $S := (TV \cdot R) \cap (TV \cdot R \cdot TV_+)$ (or any subspace which generates this as a left TV-module). Show that the multiplication map map $TV \otimes R \to TV \cdot R$ is injective. Then, the map $B \otimes S \to B \otimes R$ is given by the restriction of

$$B \otimes (TV \otimes R) \to B \otimes R, \quad b \otimes (f \otimes r) \mapsto bf \otimes r.$$

Check that this gives the exact sequence (1.N) (beginning with $B \otimes S$).

We proceed to the definition and an exercise on quadratic algebras.

DEFINITION 1.10.12. A quadratic algebra is an algebra of the form $B = TV/(R)$ where $R \subseteq T^2 V = V \otimes V$.

---

[2]One can show that, if we complete B and TV with respect to the augmentation ideals, we can always assume this; for simplicity here we will not take the completion.

In particular, such an algebra is nonnegatively graded, $B = \bigoplus_{m \geq 0} B_m$, with $B_0 = \Bbbk$. For the rest of the problem, whenever B is graded, we will call the grading on B the *weight grading*, to avoid confusion with the *homological grading* on complexes of (graded) B-modules.

EXERCISE 1.10.13. Let $B = \bigoplus_{m \geq 0} B_m$ be a nonnegatively graded algebra with $B_0 = \Bbbk$. Show that B is quadratic if and only if $\mathrm{Tor}_2(\Bbbk, \Bbbk)$ is concentrated in weight two. In this case, conclude that $B \cong TV/(R)$ where $V \cong \mathrm{Tor}_1(\Bbbk, \Bbbk)$ and $R \cong \mathrm{Tor}_2(\Bbbk, \Bbbk)$.

Now we define and study Koszul algebras:

DEFINITION 1.10.14. A Koszul algebra is a nonnegatively graded algebra B with $B_0 = \Bbbk$ such that $\mathrm{Tor}_i^B(\Bbbk, \Bbbk)$ is concentrated in weight i for all $i \geq 1$.

We remark that the above is often stated dually as: $\mathrm{Ext}_B^i(\Bbbk, \Bbbk)$ is concentrated in weight $-i$.

EXERCISE 1.10.15. Show that the following are equivalent:

(1) B is Koszul.

(2) $\Bbbk$ has a projective resolution of the form

$$\cdots \to B \otimes V_i \to \cdots \to B \otimes V_1 \to B \twoheadrightarrow \Bbbk, \tag{1.O}$$

where V_i are graded vector spaces placed in degree i for all i, and the differentials preserve the weight grading (or alternatively, you can view V_i as vector spaces in degree zero, and then the differentials all increase weights by 1).

We now proceed to discuss a deep property of Koszul algebras which makes them very useful in deformation theory, the *Koszul deformation principle*, due to Drinfeld.

EXERCISE 1.10.16. (a) For a quadratic algebra $B = TV/(R)$, let $S := R \otimes V \cap V \otimes R$, taking the intersection in the tensor algebra. Let $E \subseteq T^{\leq 2}$ be a deformation of R, in the sense that the composite map $E \to T^{\leq 2}V \twoheadrightarrow V^{\otimes 2}$ is an isomorphism onto R. Let $A := TV/(E)$. Then we can consider $T := E \otimes V \cap (V + \Bbbk) \otimes E$, again taking the intersection in the tensor algebra (where $(V + \Bbbk)$ does *not* denote an ideal).

Show first that the composition $T \to T^{\leq 3}V \twoheadrightarrow V^{\otimes 3}$ is an injection to S.

(b) Verify that the following are complexes:

$$B \otimes S \to B \otimes R \to B \otimes V \to B \twoheadrightarrow \Bbbk,$$

and

$$A \otimes T \to A \otimes E \to A \otimes V \to A \twoheadrightarrow \Bbbk.$$

The maps in the above are all obtained by restriction from splitting maps, $x \otimes (\sum_i v_i \otimes y_i) \mapsto x v_i \otimes y_i$ for x in either A or B, $v_i \in V$, and y_i in one of the spaces T, E, S, or R; the map $A \otimes V \to A$ is the multiplication map.

Show also that: they are exact at $B \otimes V$ and B, and at $A \otimes V$ and A, respectively, and that the first complex is exact in degrees ≤ 3.

(c) Prove that, if A is a flat deformation, then the map in (a) must be an isomorphism, and hence the second sequence in (b) deforms the first (in the sense that the first is obtained from the second by taking the associated graded sequence, $\text{gr}(T) = S$, $\text{gr}(E) = R$). Note the similarity to Proposition 1.10.9.

(d)(*, but with the solution outlined) Finally, we state the Koszul deformation principle, which is a converse of (c) in the Koszul setting (see Theorem 2.8.5 below for the usual version). In this case, by hypothesis, the first sequence in (b) is exact (in all degrees).

THEOREM 1.10.17 (Koszul deformation principle: filtered version [89, 40, 26, 181]). *Suppose $B = TV/(R)$ is a Koszul algebra and $E \subseteq T^{\leq 2}V$ is a deformation of R, i.e., $\text{gr}(E) = R$. Then $A = TV/(E)$ is a flat deformation of B if and only if $\dim T = \dim S$, i.e., the injection $\text{gr}(T) \to S$ is an isomorphism.*

That is, in the Koszul case, A is a flat deformation if and only if T is a flat deformation of S. In this case, the second sequence in (b) deforms the first one, and hence is also exact.

We will give the more standard version of this using formal deformations in Exercise 2.8.4.

Prove that, in the situation of the theorem, the whole resolution (1.O) *deforms to a resolution of A.*

We outline how to do this: First, you can form a minimal resolution of A analogously to (1.O), of the form

$$\cdots \to A \otimes W_i \to \cdots \to A \otimes W_1 \to A, \qquad (1.P)$$

where W_i are now filtered vector spaces with $W_{\leq(i-1)} = 0$. Note that $W_1 = V$, $W_2 = E$, and $W_3 = T$, and the assumption of the theorem already gives that $\text{gr}(W_i) = V_i$ for $i \leq 3$.

If we take the associated graded of (1.P), since $\text{gr}(A) = B$, we obtain a complex which is exact beginning with $B \otimes S = B \otimes V_3$. We want to show it is exact. Suppose it is not, and that the first nonzero homology is in degree $m \geq 3$, i.e., $\ker(B \otimes \text{gr}(W_m) \to B \otimes \text{gr}(W_{m-1})) \neq \text{im}(B \otimes \text{gr}(W_{m+1}) \to B \otimes \text{gr}(W_m))$.

First verify that, in this case, $\text{gr}(W_i) = V_i$ for all $i \leq m$, but $\text{gr}_{m+1}(W_{m+1}) \subsetneq V_{m+1}$.

Next, taking the Hilbert series

$$h(\text{gr } A; t) + \sum_{i \geq 1}(-1)^i h(\text{gr}(W_i); t)h(\text{gr } A; t),$$

we must get zero, since the sequence is exact. The same fact says that $h(B; t) + \sum_{i \geq 1}(-1)^i h(V_i; t)h(B; t) = 0$. On the other hand, flatness is saying that $\text{gr } A = B$. Conclude that

$$\sum_{i \geq 1}(-1)^i h(V_i; t) = \sum_{i \geq 1}(-1)^i h(\text{gr}(W_i); t).$$

Finally, by construction, $(W_i)_{\leq(i-1)} = 0$ for all i, so that $h(\text{gr}(W_i); t)$ is zero in degrees $\leq (i-1)$. Moreover, $\text{gr}(W_i) = V_i$ for $i \leq m$. Taking the degree $m + 1$ part of the above equation, we then get $h(V_{m+1}; t) = h(\text{gr}_{m+1}(W_{m+1}); t)$. This contradicts the fact that $\text{gr}_{m+1}(W_{m+1}) \subsetneq V_{m+1}$ (as noted above by our assumption).

2. Formal deformation theory and Kontsevich's theorem

In this section, we state and study the question of deformation quantization of Poisson structures on the polynomial algebra $\mathbb{k}[x_1, \ldots, x_n] = \mathcal{O}(\mathbb{A}^n)$, or more generally on $\mathcal{O}(X)$ for X a smooth affine variety, with $\mathbb{k}$ of characteristic zero. We defined the notion of Poisson structure in the previous section: such a structure on $B := \mathcal{O}(X)$ means a Poisson bracket $\{-, -\}$ on B. In this section we define the notion of deformation quantization, which essentially means an associative product $\star$ on $B[[\hbar]]$ which, modulo $\hbar$, reduces to the usual multiplication, and modulo $\hbar^2$, satisfies $[a, b] \equiv \hbar\{a, b\}$, for $a, b \in \mathcal{O}(X)$. As proved by Kontsevich for $\mathbb{k} = \mathbb{R}$, all Poisson structures on $X = \mathbb{A}^n$ admit a quantization, and his proof extends to the case of smooth affine (and some nonaffine) varieties, as fleshed out by Yekutieli and others (and his proof can even be extended to the case of general characteristic zero fields, as shown recently in [85]). Moreover, Kontsevich gives an explicit formula for the quantization in terms of operators associated to graphs. The coefficients of these operators are given by certain, very interesting, explicit integrals over configuration spaces of points in the upper-half plane, which unfortunately cannot be explicitly evaluated in general.

Our goal is to develop enough of the definitions and background in order to explain Kontsevich's answer, omitting the integral formulas for the coefficients. To our knowledge, this result itself is not proved on its own in the literature: Kontsevich proves it as a corollary of a more general result, his formality theorem, which led to an explosion of literature on refinements, analogues, and related results.

2.1. Differential graded algebras. In this section, we will often work with dg (differential graded) algebras. These algebras have *homological* grading, which means that one should always think of the permutation of tensors $v \otimes w \mapsto w \otimes v$ as carrying the additional sign $(-1)^{|v||w|}$. Precisely, this means the following:

REMARK 2.1.1. We will use superscripts for the grading on dg algebras since the differential increases degree by one (i.e., this is cohomological grading). If one uses subscripts to indicate the grading (homological grading) then the differential should rather decrease degree by one. We will typically refer to this grading as "homological" even when it is really cohomological.

DEFINITION 2.1.2. A dg vector space, or complex of vector spaces, is a graded vector space $V = \bigoplus_{m \in \mathbb{Z}} V^m$ equipped with a linear differential $d : V^\bullet \to V^{\bullet+1}$ satisfying $d(d(b)) = 0$ for all $b \in V$. A morphism of dg vector spaces is a linear map $\phi : V \to W$ such that $\phi(V^m) \subseteq W^m$ for all $m \in \mathbb{Z}$ and $\phi \circ d = d \circ \phi$.

DEFINITION 2.1.3. A dg associative algebra (or dg algebra) is a dg vector space $A = \bigoplus_{m \in \mathbb{Z}} A^m$ which is also a graded associative algebra, i.e., equipped with an associative multiplication satisfying $A^m A^n \subseteq A^{m+n}$, such that the differential is a graded derivation:

$$d(ab) = d(a) \cdot b + (-1)^{|a|} a \cdot d(b).$$

A dg algebra morphism $A \to B$ is a homomorphism of associative algebras which is also a morphism of dg vector spaces.

DEFINITION 2.1.4. A dg commutative algebra B is a dg associative algebra satisfying the graded commutativity rule, for homogeneous $a, b \in B$,

$$ab = (-1)^{|a||b|} ba. \tag{2.A}$$

A dg commutative algebra morphism is the same thing as a dg (associative) algebra morphism (that is, commutativity adds no constraint on the homomorphism).

DEFINITION 2.1.5. A dg Lie algebra (dgla) L is a dg vector space equipped with a bracket $[-, -] : L \otimes L \to L$ which is a morphism of complexes (i.e., $[L^m, L^n] \subseteq L^{m+n}$ and $d[a, b] = [da, b] + (-1)^{|a|}[a, db]$) and satisfies the graded skew-symmetry and Jacobi identities:

$$[v, w] = -(-1)^{|v||w|}[w, v], \tag{2.B}$$

$$[u, [v, w]] + (-1)^{|u|(|v|+|w|)}[v, [w, u]] + (-1)^{(|u|+|v|)|w|}[w, [u, v]] = 0. \tag{2.C}$$

A dg Lie algebra morphism is a homomorphism of Lie algebras which is also a morphism of dg vector spaces.

REMARK 2.1.6. If you are comfortable with the idea of the category of dg vector spaces (i.e., complexes) as equipped with a tensor product (precisely, a symmetric monoidal category where the permutation is, as indicated above, the signed one $v \otimes w \mapsto (-1)^{|v||w|} w \otimes v$), then a dg (associative, commutative, Lie) algebra is exactly an (associative, commutative, Lie) algebra in the category of dg vector spaces.

2.2. Definition of Hochschild (co)homology. Let A be an associative algebra. The Hochschild (co)homology is the natural (co)homology theory attached to associative algebras. We give a convenient definition in terms of Ext and Tor. Let $A^e := A \otimes_{\Bbbk} A^{\mathrm{op}}$, where A^{op} is the opposite algebra, defined to be the same underlying vector space as A, but with the opposite multiplication, $a \cdot b := ba$. Note that A^e-modules are the same as A-bimodules (where, by definition, $\Bbbk$ acts the same on the right and the left, i.e., by the fixed $\Bbbk$-vector space structure on the bimodule).

DEFINITION 2.2.1. Define the Hochschild homology and cohomology, respectively, of A, with coefficients in an A-bimodule M, by

$$\mathsf{HH}_i(A, M) := \mathrm{Tor}_i^{A^e}(A, M), \quad \mathsf{HH}^i(A, M) := \mathrm{Ext}_{A^e}^i(A, M).$$

Without specifying the bimodule M, we are referring to $M = A$, i.e., $\mathsf{HH}_i(A) := \mathsf{HH}_i(A, A)$ and $\mathsf{HH}^i(A) := \mathsf{HH}^i(A, A)$.

In fact, $\mathsf{HH}^\bullet(A)$ is a *ring* under the Yoneda product of extensions: see §3.9 below for more details (we will not use this until then). It moreover has a Gerstenhaber bracket (which is a shifted version of a Poisson bracket), which we will introduce and use beginning in §4.1 below.

The most important object above for us will be $\mathsf{HH}^2(A)$, in accordance with the (imprecise) principle:

The space $\mathsf{HH}^2(A)$ parametrizes all (infinitesimal, filtered, or formal) deformations of A, up to obstructions (in $\mathsf{HH}^3(A)$) and equivalences (in $\mathsf{HH}^1(A)$). $\tag{2.D}$

We will give a first definition of infinitesimal and formal deformations in the next subsection, leaving the general notion to §2.5 below.

More generally, given any type of algebra structure, $\mathcal{P}$, on an ordinary (not dg) vector space B:

The space $H_{\mathcal{P}}^2(B, B)$ parametrizes (infinitesimal, filtered, or formal) deformations of the $\mathcal{P}$-algebra structure on B,
up to obstructions (in $H_{\mathcal{P}}^3(B, B)$) and equivalences (in $H_{\mathcal{P}}^1(B, B)$).

When $\mathcal{P}$ indicates associative algebras, $H_{\mathcal{P}}^\bullet(B, B) = \mathrm{HH}^\bullet(B, B)$, and for Lie, commutative, and Poisson algebras, one gets Chevalley–Eilenberg, Harrison, and Poisson cohomology, respectively. For example, as we will use later, when $\mathcal{P}$ indicates Poisson algebras, we denote the Poisson cohomology by $\mathrm{HP}^\bullet(B, B)$, so that $\mathrm{HP}^2(B, B)$ parametrizes Poisson deformations of a Poisson algebra B, up to the aforementioned caveats.

REMARK 2.2.2. One way of making the above assertion precise is to make $\mathcal{P}$ an *operad* and B an algebra over this operad (see, e.g., [169, 166]): for example, there are commutative, Lie, Poisson, and associative operads, and algebras over each of these are commutative, Lie, Poisson, and associative algebras, respectively. In some more detail, a $\Bbbk$-linear *operad* $\mathcal{P}$ is an algebraic structure which consists of, for every $m \geq 0$, a vector space $\mathcal{P}(m)$ equipped with an action of the symmetric group S_m, together with composition operations $\mathcal{P}(m) \times \big(\mathcal{P}(n_1) \times \mathcal{P}(n_2) \times \cdots \times \mathcal{P}(n_m)\big) \to \mathcal{P}(n_1 + n_2 + \cdots + n_m)$, for all $m \geq 1$ and $n_1, n_2, \ldots, n_m \geq 0$, together with a unit $1 \in \mathcal{P}(1)$, satisfying certain associativity and unitality conditions. Then, a $\mathcal{P}$-algebra is a vector space V together with, for all $m \geq 0$ and $\xi \in \mathcal{P}(m)$, an m-ary operation $\mu_\xi : V^{\otimes m} \to V$, satisfying certain associativity and unitality conditions. Then, for any operad $\mathcal{P}$ and any $\mathcal{P}$-algebra A, there is a natural notion of the $\mathcal{P}$-(co)homology of A, which recovers (in degrees at least two) in the associative case the Hochschild (co)homology; in the Lie case over characteristic zero, the Chevalley–Eilenberg (co)homology; and in the commutative case over characteristic zero, the Harrison or André–Quillen (co)homology. Moreover, one has the notion of an A-module M, which in the aforementioned cases recover bimodules, Lie modules, and modules, respectively, and one has the notion of $\mathcal{P}$-(co)homology of A valued in M, which recovers (in degrees at least two) the Hochschild, Chevalley–Eilenberg, and Harrison (co)homology valued in M.

In the case of Lie algebras $\mathfrak{g}$, one can explicitly describe the Chevalley–Eilenberg cohomology of $\mathfrak{g}$, owing to the fact that the category of $\mathfrak{g}$-modules is equivalent to the category of modules over the associative algebra $U\mathfrak{g}$. Let $H_{\mathrm{Lie}}^\bullet(\mathfrak{g}, -)$ denote the Lie (or Chevalley–Eilenberg) cohomology of $\mathfrak{g}$ with coefficients in Lie modules. One then has (cf. [239, Exercise 7.3.5]) $\mathrm{Ext}_{U\mathfrak{g}}^\bullet(M, N) \cong H_{\mathrm{Lie}}^\bullet(\mathfrak{g}, \mathrm{Hom}_{\Bbbk}(M, N))$. So $H_{\mathrm{Lie}}^\bullet(\mathfrak{g}, N) \cong \mathrm{Ext}_{U\mathfrak{g}}^\bullet(\Bbbk, N)$, the extensions of $\Bbbk$ by N as $\mathfrak{g}$-modules; this differs from the associative case where one considers the bimodule extensions of A, rather than the Lie module extensions of $\Bbbk$. (Note that, since A need not be augmented, one does not necessarily have a module $\Bbbk$; indeed one need not have a one-dimensional module at all, such as in the case where A is the algebra of n by n matrices for $n \geq 2$, or when A is the Weyl algebra in characteristic zero, which has no finite-dimensional modules).

2.3. Formality theorems. We will also state results for C^∞ manifolds, both for added generality, and also to help build intuition. In the C^∞ case we will take $\Bbbk = \mathbb{C}$ and let $\mathcal{O}(X)$ be the algebra of smooth complex-valued functions on X; there is, however, one technicality: we restrict in this case to the "local" part of Hochschild cohomology as defined in Remark 3.2.1 (this difference can be glossed over in a first reading). In the affine case, in this section, we will assume that $\Bbbk$ is a field of characteristic zero.

THEOREM 2.3.1 (Hochschild–Kostant–Rosenberg). *Let X be a either smooth affine variety over a field $\Bbbk$ of characteristic zero or a C^∞ manifold. Then, the Hochschild cohomology ring* $\mathsf{HH}^\bullet(\mathcal{O}(X), \mathcal{O}(X)) \cong \bigwedge^\bullet_{\mathcal{O}(X)} \mathrm{Vect}(X)$ *is the ring of **polyvector fields** on X.*

Explicitly, a polyvector field of degree d is a sum of elements of the form

$$\xi_1 \wedge \cdots \wedge \xi_d,$$

where each $\xi_i \in \mathrm{Vect}(X)$ is a vector field on X, i.e., when X is an affine algebraic variety, $\mathrm{Vect}(X) = \mathrm{Der}(\mathcal{O}(X), \mathcal{O}(X))$.

Thus, the deformations are classified by certain *bivector fields* (those whose obstructions vanish). As we will explain, the first obstruction is that the bivector field be Poisson. Moreover, deforming along this direction is the same as quantizing the Poisson structure. So the question reduces to: which Poisson structures on $\mathcal{O}(X)$ can be quantized? It turns out, by a very deep result of Kontsevich, that *they all can.*

To make this precise, we need to generalize the notion of quantization: we spoke about quantizing graded $\mathcal{O}(X)$, but usually this is not graded when X is smooth. (Indeed, a grading would mean that X has a $\mathbb{G}_m$-action, and in general the fixed point(s) will be singular. For example, if $\mathcal{O}(X)$ is nonnegatively graded with $(\mathcal{O}(X))_0 = \Bbbk$, i.e., the action is contracting with a single fixed point, then X is singular unless $X = \mathbb{A}^n$ is an affine space.)

Instead, we introduce a formal parameter $\hbar$.[3] Given a vector space V, let $V[\![\hbar]\!] := \left\{ \sum_{m \geq 0} v \hbar^m \right\}$ be the vector space of formal power series with coefficients in V. If $V = B$ is an algebra, we obtain an algebra $B[\![\hbar]\!]$. We will be interested in assigning this a deformed associative multiplication $\star$, i.e., a $\Bbbk[\![\hbar]\!]$-bilinear associative multiplication $\star : B[\![\hbar]\!] \otimes B[\![\hbar]\!] \to B[\![\hbar]\!]$. Moreover, we will require that $\star$ be continuous in the $\hbar$-adic topology, i.e., that, for all $b_m, c_n \in B$,

$$\left(\sum_{m \geq 0} b_m \hbar^m \right) \star \left(\sum_{n \geq 0} c_n \hbar^n \right) = \sum_{m,n \geq 0} (b_m \star c_n) \hbar^{m+n}. \tag{2.E}$$

The following exercise explains why, in general, the continuity assumption (while in some sense a technicality) is necessary when B is infinite-dimensional:

---

[3]The use of $\hbar$ to denote the formal parameter originates from quantum physics, where it is actually a constant, given by Planck's constant divided by 2π.

EXERCISE 2.3.2. Show that, if B is finite-dimensional, then (2.E) is automatic, given that $\star$ is $\Bbbk[[\hbar]]$-bilinear. Explain why this need not hold if B is infinite-dimensional (where one can multiply two series of the form $\sum_{m \geq 0} b_m \hbar^m$ for $\{b_0, b_1, b_2, \ldots\}$ linearly independent).

DEFINITION 2.3.3. A *(one-parameter) formal deformation* of an associative algebra B is an associative algebra $A_\hbar = (B[[\hbar]], \star)$ such that

$$a \star b \equiv ab \quad (\text{mod } \hbar), \quad \forall a, b \in B. \tag{2.F}$$

We require that $\star$ be associative, $\Bbbk[[\hbar]]$-bilinear, and continuous (2.E).

The general definition of formal deformation will be given in §2.5 below; in contexts where it is clear we are speaking about one-parameter formal deformations, we may omit "one-parameter."

REMARK 2.3.4. Equivalent to the above, and often found in the literature, is the following alternative formulation: A one-parameter formal deformation is a $\Bbbk[[\hbar]]$-algebra $A_\hbar$, whose multiplication is continuous in the $\hbar$-adic topology[4] together with an algebra isomorphism $A_\hbar/\hbar A_\hbar \xrightarrow{\sim} B$, such that $A_\hbar$ is topologically free (i.e., isomorphic to $V[[\hbar]]$ for some V, which in this case we can take to be any section of the map $A_\hbar \twoheadrightarrow B$). The equivalence is given by taking any vector space section $\tilde{B} \subseteq A_\hbar$ of $A_\hbar \twoheadrightarrow B$, and writing $A_\hbar = (\tilde{B}[[\hbar]], \star)$ for a unique binary operation $\star$, which one can check must be continuous, bilinear, and satisfy (2.F). Conversely, given $A_\hbar = (\tilde{B}[[\hbar]], \star)$, one has a canonical isomorphism $A_\hbar/\hbar A_\hbar = B$.

Modulo $\hbar^2$, we get the notion of infinitesimal deformation:

DEFINITION 2.3.5. An *infinitesimal deformation* of an associative algebra is an algebra $(B[\varepsilon]/(\varepsilon^2), \star)$ such that $a \star b \equiv ab$ (mod ε).

Now we let B be a Poisson algebra (which is automatically commutative).

DEFINITION 2.3.6. A *deformation quantization* is a one-parameter formal deformation of a Poisson algebra B which satisfies the identity

$$a \star b - b \star a \equiv \hbar\{a, b\} \quad (\text{mod } \hbar^2), \quad \forall a, b \in B. \tag{2.G}$$

We will often use the fact that, for X a smooth affine variety in odd characteristic or a smooth manifold, a Poisson structure on $\mathcal{O}(X)$ is the same as a bivector field $\pi \in \bigwedge^2_{\mathcal{O}(X)} \text{Vect}(X)$, via

$$\{f, g\} = i_\pi(df \wedge dg) := \pi(f \otimes g),$$

satisfying the Jacobi identity (note here and in the sequel that $i_\eta(\alpha)$ denotes the contraction of a polyvector field η with a differential form α). When X is affine space $\mathbb{A}^n$ or a smooth manifold, this is clear. For the general case of a smooth affine variety, this can be shown as follows (the reader uncomfortable with the necessary

---

[4]This means explicitly that (2.E) holds allowing b_i and c_j to be arbitrary elements of $A_\hbar$ such that the sums on the LHS converge; by topological freeness they actually converge for all b_i and c_j.

algebraic geometry can skip it and take $X = \mathbb{A}^n$): One needs to show, more generally, that $\bigwedge^2_{\mathcal{O}(X)} \text{Vect}(X)$ is canonically isomorphic to the vector space, $\text{SkewBiDer}(\mathcal{O}(X))$, of skew-symmetric biderivations $\mathcal{O}(X) \otimes_{\Bbbk} \mathcal{O}(X) \to \mathcal{O}(X)$, i.e., skew-symmetric brackets satisfying the Leibniz identity (but not necessarily the Jacobi identity). Then, one has a canonical map $\bigwedge^2 \text{Vect}(X) \to \text{SkewBiDer}(\mathcal{O}(X))$, and this is a map of $\mathcal{O}(X)$-modules. Then, the fact that it is an isomorphism is a local statement. However, if X is smooth, then $\text{Vect}(X)$ is a projective $\mathcal{O}(X)$-module, i.e., the tangent sheaf T_X is locally free. On an open affine subset $U \subseteq X$ such that $\text{Vect}(U)$ is free as a $\mathcal{O}(U)$-module, it is clear that the canonical map is an isomorphism. This implies the statement.

In terms of π, the Jacobi identity says $[\pi, \pi] = 0$, using the Schouten–Nijenhuis bracket (see Proposition 4.5.4), defined as follows:

DEFINITION 2.3.7. The Schouten–Nijenhuis Lie bracket on $\bigwedge^{\bullet}_{\mathcal{O}(X)} \text{Vect}(X)$ is given by the formula

$$[\xi_1 \wedge \cdots \wedge \xi_m, \eta_1 \wedge \cdots \wedge \eta_n]$$
$$= \sum_{i,j} (-1)^{i+j+m-1} [\xi_i, \eta_j] \wedge \xi_1 \wedge \cdots \hat{\xi}_i \cdots \wedge \xi_m \wedge \eta_1 \wedge \cdots \hat{\eta}_j \cdots \wedge \eta_n. \quad (2.\text{H})$$

As before, the hat indicates that the given terms are *omitted* from the product.

REMARK 2.3.8. Alternatively, the Schouten–Nijenhuis bracket is the Lie bracket uniquely determined by the conditions that $[\xi, \eta]$ is the ordinary Lie bracket for $\xi, \eta \in \text{Vect}(X)$, that $[\xi, f] = \xi(f)$ for $\xi \in \text{Vect}(X)$ and $f \in \mathcal{O}(X)$, and such that the graded Leibniz identity is satisfied, for all homogeneous $\theta_1, \theta_2, \theta_3 \in \bigwedge^{\bullet}_{\mathcal{O}(X)} \text{Vect}(X)$ (see also Definition 4.1.2 below):

$$[\theta_1, \theta_2 \wedge \theta_3] = [\theta_1, \theta_2] \wedge \theta_3 + (-1)^{|\theta_2||\theta_3|} [\theta_1, \theta_3] \wedge \theta_2. \quad (2.\text{I})$$

EXAMPLE 2.3.9. The simplest example is the case $X = \mathbb{A}^n$ with a constant Poisson bivector field, which can always be written up to choice of coordinates — which we denote by $(x_1, \ldots, x_m, y_1, \ldots, y_m, z_1, \ldots, z_{n-2m})$ — as

$$\pi = \sum_{i=1}^m \partial_{x_i} \wedge \partial_{y_i}, \quad \text{i.e.,} \quad \{f, g\} = \sum_{i=1}^m \frac{\partial f}{\partial x_i} \frac{\partial g}{\partial y_i} - \frac{\partial f}{\partial y_i} \frac{\partial g}{\partial x_i},$$

for a unique $m \leq n/2$. Then, for $\Bbbk$ of characteristic zero, there is a well-known deformation quantization, called the Moyal–Weyl star product:

$$f \star g = \mu \circ e^{\frac{1}{2}\hbar\pi}(f \otimes g), \quad \mu(a \otimes b) := ab.^5$$

When $2m = n$, so that the Poisson structure is symplectic, this is actually isomorphic to the usual Weyl quantization: see the next exercise.

---

[5]In physics, over $\Bbbk = \mathbb{C}$, often one sees an $i = \sqrt{-1}$ also in the exponent, so that $a \star b - b \star a \equiv i\hbar\{a, b\}$ (mod $\hbar^2$), but according to our definition, which works over arbitrary $\Bbbk$, we don't have it.

EXERCISE 2.3.10. We consider the Moyal–Weyl star product for $X = \mathbb{A}^n$ with coordinates $(x_1, \ldots, x_m, y_1, \ldots, y_m, z_1, \ldots, z_{n-2m})$ as above, for $\Bbbk$ of characteristic zero.

(a) Show that, for the Moyal–Weyl star product on $\mathcal{O}(X)[[\hbar]]$ with $X = \mathbb{A}^n$ as above, in the above basis,

$$x_i \star y_j - y_j \star x_i = \hbar \delta_{ij}, \quad x_i \star x_j = x_j \star x_i, \quad y_i \star y_j = y_j \star y_i.$$

whereas $z_1, \ldots, z_{n-2m}$ are central: $z_i \star f = f \star z_i$ for all $1 \le i \le n - 2m$ and all $f \in \mathcal{O}(X)$.

(b) Show that this star product is actually defined over polynomials in $\hbar$, i.e., on $\mathcal{O}(X)[\hbar]$. That is, if $f, g \in \mathcal{O}(X)[\hbar]$, so is $f \star g$, and hence we get an associative algebra $(\mathcal{O}(X)[\hbar], \star)$. Note that this is homogeneous with respect to the grading where $|x_i| = |y_i| = 1$ and $|\hbar| = 2$.

(c) Note that $(\hbar - 1)$ is an ideal in $\Bbbk[\hbar]$ and hence in $(\mathcal{O}(X)[\hbar], \star)$. Taking the quotient, get a filtered quantization $(\mathcal{O}(X)[\hbar], \star)/(\hbar - 1)$, which is in other words obtained by setting $\hbar = 1$ above.

(d) Now we get to the goal of the exercise: to relate the quantization

$$(\mathcal{O}(X)[\hbar], \star)/(\hbar - 1)$$

to the Weyl algebra Weyl_m. Show first that there is a unique isomorphism of algebras,

$$\mathsf{Weyl}_m \otimes \Bbbk[z_1, \ldots, z_{n-2m}] \to (\mathcal{O}(X)[\hbar], \star)/(\hbar - 1),$$

satisfying $x_i \mapsto x_i$, $y_i \mapsto y_i$, and $z_i \mapsto z_i$.

(e)(*) The main point of the exercise is to give the explicit *inverse* of (d). To begin, for any vector space V, with $\Bbbk$ of characteristic zero, we can consider the symmetrization map, $\mathsf{Sym}\, V \to TV$, given by

$$v_1 \cdots v_k \mapsto \frac{1}{k!} \sum_{\sigma \in S_k} v_{\sigma(1)} \otimes \cdots \otimes v_{\sigma(k)}$$

where $v_i \in V$ for all i. (Caution: this is *not* an algebra homomorphism.)

Now let $V = \mathsf{Span}\{x_1, \ldots, x_m, y_1, \ldots, y_m, z_1, \ldots, z_{n-2m}\}$, and consider the composition of the above linear map $\mathsf{Sym}\, V = \mathcal{O}(X) \to TV$ with the obvious quotient $TV \to \mathsf{Weyl}_m \otimes \Bbbk[z_1, \ldots, z_{n-2m}]$. Show that the result yields an algebra homomorphism

$$(\mathcal{O}(X)[\hbar], \star)/(\hbar - 1) \to \mathsf{Weyl}_m \otimes \Bbbk[z_1, \ldots, z_{n-2m}],$$

which inverts the homomorphism of (d).

EXERCISE 2.3.11. In this exercise, we explain a uniqueness statement for the Moyal–Weyl quantization. Suppose that $\star'$ is any other star product formula of the form

$$f \star' g = \mu \circ F(\pi)(f \otimes g), \tag{2.J}$$

for $F = 1 + \frac{1}{2}\hbar\pi + \sum_{i \ge 2} \hbar^i F_i(\pi)$, with each F_i a polynomial in π. Then, if $(\mathcal{O}(\mathbb{A}^n)[[\hbar]], \star')$ quantizes the Poisson bracket $\{-, -\}$ given by π, show that we still have the relations of part (a) of the previous exercise, for $\star'$ instead of $\star$. Similarly to part (d) above, conclude that we have an isomorphism

$$(\mathcal{O}(\mathbb{A}^n)[\![\hbar]\!], \star) \to (\mathcal{O}(\mathbb{A}^n)[\![\hbar]\!], \star'),$$

given uniquely by

$$v_1 \star \cdots \star v_m \mapsto v_1 \star' \cdots \star' v_m,$$

for all linear functions $v_1, \ldots, v_m$ in $\mathrm{Span}\{x_1, \ldots, x_m, y_1, \ldots, y_m, z_1, \ldots, z_{n-2m}\}$. Moreover, show that this isomorphism is the identity modulo $\hbar$.

(It is a much deeper fact that, when $n = 2m$, we can drop the assumption on the formula (2.J).)

We will also need the notion of a *formal Poisson deformation* of a Poisson algebra $\mathcal{O}(X)$:

DEFINITION 2.3.12. A formal Poisson deformation of a Poisson algebra $\mathcal{O}(X)$ is a continuous $\Bbbk[\![\hbar]\!]$-linear Poisson bracket on $\mathcal{O}(X)[\![\hbar]\!]$ which reduces modulo $\hbar$ to the original Poisson bracket on $\mathcal{O}(X)$.

Here, continuous as before means continuous in the $\hbar$-adic topology, which means that the identity obtained from (2.E) by replacing the star product by the deformed Poisson bracket holds.

THEOREM 2.3.13 ([156, 155, 245, 86], among others). *Every Poisson structure on a smooth affine variety over a field of characteristic zero, or on a C^∞ manifold, admits a canonical deformation quantization. In particular, every graded Poisson algebra with Poisson bracket of degree $-d < 0$ admits a filtered quantization.*

Moreover, there is a canonical bijection, up to isomorphisms equal to the identity modulo $\hbar$, between deformation quantizations and formal Poisson deformations.

REMARK 2.3.14. As shown by O. Mathieu [171], in general there are obstructions to the existence of a quantization. We explain this following §1.4 of www.math.jussieu. fr/~keller/emalca.pdf, which more generally forms a really nice reference for much of the material discussed in these notes!

Let $\mathfrak{g}$ be a Lie algebra over a field $\Bbbk$ of characteristic zero such that $\mathfrak{g} \otimes_\Bbbk \bar{\Bbbk}$ is simple and not isomorphic to $\mathfrak{sl}_n(\bar{\Bbbk})$ for any n, where $\bar{\Bbbk}$ is the algebraic closure of $\Bbbk$. (In the cited references, one takes $\Bbbk = \mathbb{R}$ and thus $\bar{\Bbbk} = \mathbb{C}$, but this assumption is not needed.) For instance, one could take $\mathfrak{g} = \mathfrak{so}(n)$ for $n = 5$ or $n \geq 7$.

One then considers the Poisson algebra $B := \mathrm{Sym}\,\mathfrak{g}/(\mathfrak{g}^2)$, equipped with the Lie bracket on $\mathfrak{g}$. In other words, this is the quotient of $\mathcal{O}(\mathfrak{g}^*)$ by the square of the augmentation ideal, i.e., the maximal ideal of the origin. Then, $\mathrm{Spec}\,B$, which is known as the first infinitesimal neighborhood of the origin in $\mathfrak{g}^*$, is nonreduced and set-theoretically a point, albeit with a nontrivial Poisson structure.

We claim that B does not admit a deformation quantization. For a contradiction, suppose it did admit one, $B_\hbar := (B[\![\hbar]\!], \star)$. Since $B = \Bbbk \oplus \mathfrak{g}$ has the property that $B \otimes_\Bbbk \bar{\Bbbk} \cong (\bar{\Bbbk} \oplus (\mathfrak{g} \otimes_\Bbbk \bar{\Bbbk}))$ with $\mathfrak{g} \otimes_\Bbbk \bar{\Bbbk}$ a semisimple Lie algebra, it follows from basic Lie cohomology (e.g., [239, §7]) that $H^2_{\text{Lie}}(B \otimes_{\Bbbk} \bar{\Bbbk}, B \otimes_{\Bbbk} \bar{\Bbbk}) = 0$ and hence also $H^2_{\text{Lie}}(B, B) = 0$, i.e., B has no deformations as a Lie algebra. Hence, $B_\hbar \cong B[[\hbar]]$ as a Lie algebra; in fact there is a continuous $\Bbbk[[\hbar]]$-linear isomorphism of Lie algebras which is the identity modulo $\hbar$. Now let $K := \bar{\Bbbk}((\hbar))$, an algebraically closed field, and set $\tilde{B} := B_\hbar \otimes_{\Bbbk[[\hbar]]} K$. Since this is finite-dimensional over the algebraically closed field K, Wedderburn theory implies that, if $J \subseteq \tilde{B}$ is the nilradical, then $M := \tilde{B}/J$ is a product of matrix algebras over K of various sizes. Then J is also a Lie ideal in $\tilde{B}$ which is, as a Lie algebra, a sum $K \oplus (\mathfrak{g} \otimes_{\Bbbk} K)$ (using that $\mathfrak{g}$ is finite-dimensional). As J cannot be the unit ideal and J is nilpotent, it must be zero, so $\tilde{B}$ itself is a direct sum of matrix algebras. But this would imply that $\mathfrak{g} \otimes_{\Bbbk} K \cong \mathfrak{sl}_n(K)$, and hence $\mathfrak{g} \otimes_{\Bbbk} \bar{\Bbbk} \cong \mathfrak{sl}_n(\bar{\Bbbk})$, contradicting our assumptions.

EXAMPLE 2.3.15. In the case $X = \mathbb{A}^n$ with a constant Poisson bivector field, Kontsevich's star product coincides with the Moyal–Weyl one.

EXAMPLE 2.3.16. Next, let $\{-, -\}$ be a linear Poisson bracket on $\mathfrak{g}^* = \mathbb{A}^n$, i.e., a Lie bracket on the vector space $\mathfrak{g} = \Bbbk^n$ of linear functions. As explained in Exercise 4.14.1 below (following [156]), if $(\mathcal{O}(\mathbb{A}^n)[[\hbar]], \star)$ is Kontsevich's canonical quantization, then

$$x \star y - y \star x = \hbar[x, y],$$

so that, as in Exercise 2.3.10, the map which is the identity on linear functions yields an isomorphism

$$U_\hbar(\mathfrak{g}) \to (\mathcal{O}(\mathfrak{g}^*)[[\hbar]], \star).$$

Modulo $\hbar$, the inverse to this is the symmetrization map of Exercise 2.3.10. Thus, if we apply a gauge equivalence to Kontsevich's star-product, then the inverse really is the symmetrization map. This is explained in detail in [82], where the gauge equivalence is also explicitly computed. The resulting star product on $\mathcal{O}(\mathfrak{g}^*)$ is called the Gutt product and dates to [126]; for a description of this product, see, e.g., [82, (13)]. Moreover, as first noticed in [3] (see also [82]), there is no gauge equivalence required when $\mathfrak{g}$ is nilpotent, i.e., in this case the Kontsevich star-product equals the Gutt product (this is essentially because nothing else can happen in this case).

These theorems all rest on the basic statement that the Hochschild cohomology of a smooth affine variety in characteristic zero or C^∞ manifold is *formal*, i.e., the Hochschild cochain complex (which computes its cohomology), is equivalent to its cohomology not merely as a vector space, but as dg Lie algebras, up to homotopy. (In fact, this statement can be made to be equivalent to the bijection of Theorem 2.3.13 if one extends to deformations over dg commutative rings rather than merely formal power series.)

We will explain what formality of dg Lie algebras means precisely later: we note only that it is a completely different use of the term "formal" than we have used it before for formal deformations. For now, we only give some motivation via the analogous concept of formality for vector spaces, modules, etc. First of all, note that all *complexes of vector spaces are always equivalent to their cohomology*, i.e., they are all formal.

This follows because, given a complex $C^\bullet$, one can always find an *isomorphism of complexes*

$$C^\bullet \xrightarrow{\sim} H^\bullet(C^\bullet) \oplus S^\bullet,$$

where $H^\bullet$ is the homology of $C^\bullet$, and $S^\bullet$ is a contractible complex. Here, contractible means that there exists a linear map $h : C^\bullet \to C^{\bullet-1}$, decreasing degree by one, so that $dh + hd = \mathrm{Id}$; such an h is called a *contracting homotopy*. In particular, a contractible complex is acyclic (and for complexes of vector spaces, the converse is also true).

But if we consider modules over a more general ring that is not a field, it is no longer true that all complexes are isomorphic to a direct sum of their homology and a contractible complex. Consider, for example, the complex

$$0 \to \mathbb{Z} \xrightarrow{\cdot 2} \mathbb{Z} \to 0.$$

The homology is $\mathbb{Z}/2$, but it is impossible to write the complex as a direct sum of $\mathbb{Z}/2$ with a contractible complex, since $\mathbb{Z}$ has no torsion. That is, the above complex is *not* formal.

Now, the subtlety with the formality of Hochschild cohomology is that, even though the underlying Hochschild cochain complex is automatically formal as a *complex of vector spaces*, it is *not* automatically formal as a *dg Lie algebra*. For example, it may not necessarily be isomorphic to a direct sum of its cohomology and another dg Lie algebra (although being formal does not require this, but only that the dg Lie algebra be "homotopy equivalent" to its cohomology).

The statement that the Hochschild cochain complex is formal is *stronger* than merely the existence of deformation quantizations. It implies, for example:

THEOREM 2.3.17. *If (X, ω) is either an affine symplectic variety over a field of characteristic zero or a symplectic C^∞ manifold, and $A = (\mathcal{O}(X)[\![\hbar]\!], \star)$ is a deformation quantization of X, then $\mathrm{HH}^\bullet(A[\hbar^{-1}]) \cong H_{DR}^\bullet(X, \Bbbk(\!(\hbar)\!))$, and there is a versal formal deformation of $A[\hbar^{-1}]$ over the base $\hat{\mathcal{O}}(H_{DR}^2(X))$ (the completion of $\mathrm{Sym}(H_{DR}^2(X))^*$).*

Here $H_{DR}^\bullet$ denotes the algebraic de Rham cohomology, which in the case $\Bbbk = \mathbb{C}$ coincides with the ordinary topological de Rham cohomology.

We will define the notion of versal formal deformation more precisely later; roughly speaking, such a deformation U over a base $\Bbbk[\![t_1, \ldots, t_n]\!]$ is one so that every formal deformation $(A[\hbar^{-1}][\![t]\!], \star)$ of $A[\hbar^{-1}]$ with deformation parameter t is isomorphic, as a $\Bbbk[\![t]\!]$-algebra, to $U \otimes_{\Bbbk[\![t_1, \ldots, t_n]\!]} \Bbbk[\![t]\!]$ for some continuous homomorphism $\Bbbk[\![t_1, \ldots, t_n]\!] \to \Bbbk[\![t]\!]$ (i.e., some assignment of each t_i to a power series in t without constant term) and that this isomorphism is the identity modulo t. In the situation of the theorem, $t_1, \ldots, t_n$ should be a basis of $(H_{DR}^2)^*$. We will sketch how the theorem follows from Kontsevich's theorem in §4.11 below.

2.4. Description of Kontsevich's deformation quantization for $\mathbb{R}^d$. It is worth explaining the general form of the star-products given by Kontsevich's theorem for $X = \mathbb{R}^d$, considered either as a smooth manifold or an affine algebraic variety over $\Bbbk = \mathbb{R}$. This is taken from [156, §2]; the interested reader is recommended to look there for more details.

Suppose we are given a Poisson bivector $\pi \in \bigwedge^2_{\mathcal{O}(X)} \mathrm{Vect}(X)$. Then Kontsevich's star product $f \star g$ is a linear combination of all possible ways of applying π multiple times to f and g, but with very sophisticated weights.

The possible ways of applying π are easy to describe using directed graphs. Namely, the graphs we need to consider are placed in the closed upper-half plane $\{(x, y) \mid y \geq 0\} \subseteq \mathbb{R}^2$, satisfying the following properties:

(1) There are exactly two vertices along the x-axis, labeled by L and R. The other vertices are labeled $1, 2, \ldots, m$.

(2) L and R are sinks, and all other vertices have exactly two outgoing edges.

(3) At every vertex $j \in \{1, 2, \ldots, m\}$, the two outgoing edges should be labeled by the symbols e_1^j and e_2^j. That is, we fix an ordering of the two edges and denote them by e_1^j and e_2^j.

Examples of such graphs are given in Figures 1 and 2.

Write our Poisson bivector π in coordinates as

$$\pi = \sum_{i<j} \pi^{i,j} \partial_i \wedge \partial_j.$$

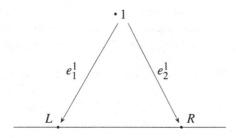

Figure 1. The graph corresponding to $f \otimes g \mapsto \{f, g\}$

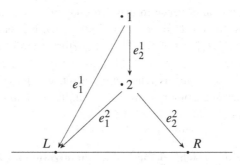

Figure 2. The graph corresponding to $f \otimes g \mapsto \sum_{i,j,k,\ell} \pi^{i,j} \partial_j (\pi^{k,\ell}) \partial_i \partial_k(f) \partial_\ell(g)$

Let $\pi^{j,i} := -\pi^{i,j}$. We attach to the graph Γ a bilinear differential operator $B_{\Gamma,\pi}$: $\mathcal{O}(X)^{\otimes 2} \to \mathcal{O}(X)$, as follows. Let $E_\Gamma = \{e_1^1, e_1^2, \ldots, e_m^1, e_m^2\}$ be the set of edges. Given an edge $e \in E_\Gamma$, let $t(e)$ denote the target vertex of e.[6] Then

$$B_{\Gamma,\pi}(f \otimes g) := \sum_{I:E_\Gamma \to \{1,\ldots,d\}} \left[\prod_{i=1}^m \left(\prod_{e \in E_\Gamma | t(e)=i} \partial_{I(e)} \right) \pi^{I(e_i^1),I(e_i^2)} \right]$$

$$\cdot \left(\prod_{e \in E_\Gamma | t(e)=L} \partial_{I(e)} \right)(f) \cdot \left(\prod_{e \in E_\Gamma | t(e)=R} \partial_{I(e)} \right)(g).$$

Now, let V_Γ denote the set of vertices of Γ other than L and R, so that in the above formula, $m = |V_\Gamma|$. Then the star product is given by

$$f \star g = \sum_\Gamma \hbar^{|V_\Gamma|} w_\Gamma B_{\Gamma,\pi}(f \otimes g),$$

where we sum over isomorphism classes of graphs satisfying conditions (1)–(3) above (we only need to include one for each isomorphism class forgetting the labeling, since the operator is the same up to a sign). The $w_\Gamma \in \mathbb{R}$ are weights given by very explicit integrals (which in general are impossible to evaluate). Note that, in order to have $f \star g \equiv fg$ (mod $\hbar$) (i.e., (2.F)), then if Γ_0 is the graph with no edges (and thus only vertices L and R), $w_{\Gamma_0} = 1$. Similarly, if we let Γ_1 be the graph in Figure 1 with only three vertices L, R, and 1 and we include this graph in the sum but not the isomorphic graph obtained by swapping the labels of the edges, then the relation $f \star g - g \star f \equiv \hbar\{f, g\}$ (mod $\hbar^2$) (i.e., (2.G)) forces $w_{\Gamma_1} = 1/2$ as well.

The **main observation** which motivates the operators $B_{\Gamma,\pi}$ is the following: linear combinations of operators $B_{\Gamma,\pi}$ as above are exactly all of bilinear operators obtainable by contracting tensor powers of π with f and g.

2.5. Formal deformations of algebras. Now we define more precisely formal deformations. These generalize star products to deformations of arbitrary associative algebras, and also to the setting where more than one deformation parameter is allowed. Still more generally, we are interested in deformations over a base commutative augmented $\mathbb{k}$-algebra $R \supseteq R_+$ such that R is complete with respect to the R_+-adic topology, i.e., such that $R = \lim_{m \to \infty} R/R_+^m$, taking the inverse limit under the system of surjections $\cdots \to R/R_+^m \to R/R_+^{m-1} \to \cdots \to R/R_+ = \mathbb{k}$. We call such rings *complete augmented commutative $\mathbb{k}$-algebras*. We will need the completed tensor product,

$$A \hat{\otimes} R := \lim_{m \to \infty} A \otimes R/R_+^m. \tag{2.K}$$

EXAMPLE 2.5.1. When R is a formal power series ring $R = \mathbb{k}[[t_1, \ldots, t_n]]$,

$$A \hat{\otimes} R = A[[t_1, \ldots, t_n]] = \left\{ \sum_{i_1,\ldots,i_n \geq 0} a_{i_1,\ldots,i_n} t_1^{i_1} \cdots t_n^{i_n} \,\middle|\, a_{i_1,\ldots,i_n} \in A \right\}.$$

---

[6] This corresponds to what we called the *head*, e_h in the quiver context: we caution that there e_t was the tail, which is *not* the same as the target vertex $t(e)$ here!

DEFINITION 2.5.2. A formal deformation of A over a commutative complete augmented $\Bbbk$-algebra R is an R-algebra A' isomorphic to $A \hat{\otimes}_{\Bbbk} R$ as an R-module such that $A' \otimes_R (R/R_+) \cong A$ as a $\Bbbk$-algebra. Moreover, we require that $A' = \lim_{m \to \infty} A' \otimes_{\Bbbk} R/R_+^m$ (i.e., the product on A' is continuous).

Equivalently, a formal deformation is an algebra $(A \hat{\otimes}_{\Bbbk} R, \star)$ such that $a \star b \equiv ab$ (mod R_+) and the product $\star$ is continuous in the (R_+)-adic topology.

EXAMPLE 2.5.3. If $R = \Bbbk[\![t_1, \dots, t_n]\!]$, then a formal deformation of A over R is the same as an algebra $(A[\![t_1, \dots, t_n]\!], \star)$ such that $a \star b \equiv ab$ (mod $t_1, \dots, t_n$). In the case $n = 1$, we often denote the parameter by $\hbar$, and then we recover the notion of one-parameter formal deformations (Definition 2.3.3).

As a special case of formal deformations, we also have deformations over augmented $\Bbbk$-algebras where the augmentation ideal is nilpotent: $R_+^n = 0$ for some $n \geq 1$. Such deformations are often simply called deformations, since one can also think of R as an ordinary (abstract) $\Bbbk$-algebra without a topology, and then R is already complete: $R = \lim_{m \to \infty} R/R_+^m$. (Actually, it is enough to study only these, and this is frequently done in some literature, since formal deformations are always limits of such deformations; we are nonetheless interested in formal deformations in these notes and Kontsevich's theorem as well as many nice examples are tailored to them.)

In particular, such deformations include infinitesimal deformations (Definition 2.3.5) as well as n-th order deformations:

DEFINITION 2.5.4. An n-th order deformation is a deformation over $R = \Bbbk[\varepsilon]/(\varepsilon^{n+1})$, with $R_+ := \varepsilon R$.

Note that a first-order deformation is the same thing as an infinitesimal deformation.

2.6. Formal vs. filtered deformations. Main idea: if A is a graded algebra, we can consider filtered deformations on the same underlying filtered vector space A. These are equivalent to homogeneous formal deformations of A (over $\Bbbk[\![\hbar]\!]$), by replacing relations of degree $\leq m$, $\sum_{i=0}^{m} p_i = 0$ with $|p_i| = i$, by $\sum_i \hbar^{m-i} p_i = 0$, which are now homogeneous with the sum of the grading on A and $|\hbar| = 1$. We can also do the same thing with $|\hbar| = d \geq 1$ and $d \mid m$, replacing $\sum_{i=0}^{m/d} p_{di}$ for $|p_{di}| = di$ by $\sum_i \hbar^{m-di} p_{di}$.

We begin with two motivating examples:

EXAMPLE 2.6.1. We can form a homogenized version of the Weyl algebra with $|\hbar| = 2$:

$$\mathrm{Weyl}_\hbar(V) = TV[\![\hbar]\!]/(vw - wv - \hbar(v, w)).$$

EXAMPLE 2.6.2. The homogenized universal enveloping algebra $U_\hbar \mathfrak{g}$ is given by

$$U_\hbar \mathfrak{g} = T\mathfrak{g}[\![\hbar]\!]/(xy - yx - \hbar[x, y]),$$

with $|\hbar| = 1$.

We now proceed to precise definitions and statements, for the case $|\hbar| = 1$:

DEFINITION 2.6.3. Let A be an increasingly filtered associative algebra. The *Rees algebra* of A is the graded algebra

$$RA := \bigoplus_{i \in \mathbb{Z}} A_{\leq i} \cdot \hbar^i,$$

with $|\hbar| = 1$ and $|A| = 1$, equipped with the multiplication

$$(a \cdot \hbar^i) \cdot (b \cdot \hbar^j) = ab \cdot \hbar^{i+j}.$$

Similarly, let the *completed* Rees algebra be

$$\hat{A} := \left\{ \sum_{i=m}^{\infty} a_i \cdot \hbar^i \;\middle|\; a_i \in A_{\leq i}, m \in \mathbb{Z} \right\}.$$

In other words, A is graded by powers of the ideal $(\hbar)$ and $\hat{A}$ is the $\hbar$-adic completion.

The Rees algebra construction defines an equivalence between filtered deformations and homogeneous formal deformations:

LEMMA 2.6.4. *The functor $A \mapsto RA$ defines an equivalence of categories from increasingly filtered $\Bbbk$-algebras to nonnegatively graded $\Bbbk[\hbar]$-algebras (with $|\hbar| = 1$) which are free as $\Bbbk[\hbar]$-modules. A quasi-inverse is given by $C \mapsto C/(\hbar - 1)$, assigning the filtration which lets $(C/(\hbar-1))_{\leq m}$ be the image of those elements of degree $\leq m$ in C.*

The proof is left as an exercise. In other words, the equivalence replaces $A_{\leq m}$ with the homogeneous part $(RA)_m$.

REMARK 2.6.5. The Rees algebra can be viewed as interpolating between A and $\mathrm{gr}\, A$, in that $RA/(\hbar - 1) = A$ and $RA/(\hbar) = \mathrm{gr}\, A$; actually for any $c \in \Bbbk \setminus \{0\}$ we have a canonical isomorphism $RA/(\hbar - c) \cong A$ (sending $[a_{\leq i} \cdot \hbar^i]$, in the image of $A_{\leq i} \cdot \hbar^i \subseteq RA$, to $c^i a_{\leq i} \in A$). Similarly, $\hat{A}[\hbar^{-1}] = A(\!(\hbar)\!)$ and $\hat{A}/(\hbar) = \mathrm{gr}\, A$. So over $1 \in \mathrm{Spec}\, \Bbbk[\hbar]$ (or any point other than the origin), or over the generic point of $\mathrm{Spec}\, \Bbbk[\![\hbar]\!]$, we recover A or $A(\!(\hbar)\!)$, respectively, and at the origin, or the special point of $\mathrm{Spec}\, \Bbbk[\![\hbar]\!]$, we obtain $\mathrm{gr}\, A$.

Given a graded algebra A, we canonically obtain a filtered algebra by $A_{\leq n} = \bigoplus_{m \leq n} A_m$. In this case, we can consider filtered deformations of A of the form $(A, \star_f)$, where $\star_f$ reduces to the usual product on $\mathrm{gr}\, A = A$, i.e., $\mathrm{gr}(A, \star_f) = A$.

COROLLARY 2.6.6. *Let A be a graded algebra. Then there is an equivalence between filtered deformations $(A, \star_f)$ and formal deformations $(A[\![\hbar]\!], \star)$ which are graded with respect to the sum of the grading on A and $|\hbar| = 1$, given by $(A, \star_f) \mapsto \widehat{(A, \star_f)}$. For the opposite direction, because of the grading, $(A[\![\hbar]\!], \star) = (A[\hbar], \star) \hat{\otimes}_{\Bbbk[\hbar]} \Bbbk[\![\hbar]\!]$, so one can take $(A[\hbar], \star)/(\hbar - 1)$.*

REMARK 2.6.7. One can similarly give an analogue in the case $|\hbar| = d$: if one begins with a graded Poisson algebra B with Poisson bracket of degree $-d$, then filtered quantizations $(B, \star_f)$ of the form $a \star_f b = ab + \sum_{i \geq 1} f_i(a, b)$ with $|f_i(a, b)| = |ab| - di$ are equivalent to formal deformations $(B[\![\hbar]\!], \star)$ which are homogeneous for $|\hbar| = d$.

In the case $d = 2$, this includes the important Example 2.6.1, which we have also discussed earlier.

2.7. Universal deformations and gauge equivalence. We begin with the proper notion of equivalence of two deformation quantizations:

DEFINITION 2.7.1. Given two formal deformations $(A[[\hbar]], \star)$ and $(A[[\hbar]], \star')$, a gauge equivalence $\star \sim \star'$ means a continuous $\Bbbk[[\hbar]]$-linear automorphism Φ of $A[[\hbar]]$ which is the identity modulo $\hbar$, such that

$$\Phi(a \star b) = \Phi(a) \star' \Phi(b). \tag{2.L}$$

Similarly, given two formal deformations $(A \hat{\otimes} R, \star)$ and $(A \hat{\otimes} R, \star')$ over R, a gauge equivalence is a continuous R-linear automorphism of $A \hat{\otimes} R$ which is the identity modulo R_+, such that (2.L) holds.

EXERCISE 2.7.2. Similarly to Exercise 2.3.2, show that the assumption that Φ be continuous is automatic in the case that A is finite-dimensional.

We are interested in finding a family which parametrizes all formal deformations up to gauge equivalence. To do so, we define two notions: a versal deformation exhausts all formal deformations, and a universal deformation exhausts the formal deformations uniquely. In general, one cannot expect to obtain a (nice) explicit (uni)versal deformation, but there are cases where one can, as we explain in Theorem 2.7.7 below.

DEFINITION 2.7.3. Given a continuous homomorphism $p : R \to R'$ and a formal deformation $(A \hat{\otimes} R, \star)$, the formal deformation defined by base-change from p is $(A \hat{\otimes} R', p(\star))$, with

$$a \, p(\star) \, b := (\mathrm{Id} \otimes p)(a \star b).$$

DEFINITION 2.7.4. A *versal* formal deformation $(A \hat{\otimes} R, \star)$, is one such that, for every other formal deformation $(A \hat{\otimes} R', \star')$, there exists a continuous homomorphism $p : R \to R'$ such that $(A \hat{\otimes} R', \star')$ is gauge-equivalent to the base-change deformation $(A \hat{\otimes} R', p(\star))$.

The deformation is *universal* if the homomorphism p is always unique.

In particular, if $R' = \Bbbk[[\hbar]]$, we see that, given a versal deformation $(A \hat{\otimes} R, \star)$, all formal deformations $(A[[\hbar]], \star)$ are obtained from continuous homomorphisms $R \to \Bbbk[[\hbar]]$.

REMARK 2.7.5. The relationship between $(A[[\hbar]], \star)$ and the (uni)versal $(A \hat{\otimes} R, \star_u)$ can be stated geometrically as follows: p is a formal Spf $\Bbbk[[\hbar]]$-point p of Spf R, and $(A[[\hbar]], \star)$ is gauge-equivalent to the pullback of $(A \hat{\otimes} R, \star_u)$, namely, $(A[[\hbar]], \star) = (A \hat{\otimes} R, \star_u) \otimes_R p$.

REMARK 2.7.6. One can also define similarly the notion of versal and universal filtered deformations. These are not actually filtered deformations of A, but rather filtered deformations of $A \otimes R$ for some graded $\Bbbk$-algebra R, such that all filtered deformations are obtained by base-change by a character $R \to \Bbbk$, up to $\Bbbk$-linear filtered automorphisms of A whose associated graded automorphism is the identity. We will not need this notion, although we will give some examples (see Examples 2.7.8 and 2.7.9 below).

The principle that $HH^3(A)$ classifies obstructions (which will be made precise in §3.8 below) leads to the following important result, which is provable using the Maurer–Cartan formalism discussed in §4. Given a finite-dimensional vector space V, let $\hat{\mathcal{O}}(V)$ be the completion of the algebra $\mathcal{O}(V)$ at the augmentation ideal. Explicitly, if $v_1, \ldots, v_n$ is a basis of V and $c_1, \ldots, c_n$ the dual basis of V^*, so that $\mathcal{O}(V) = \Bbbk[f_1, \ldots, f_n]$, then we set $\hat{\mathcal{O}}(V) := \Bbbk[\![c_1, \ldots, c_n]\!]$, the formal power series algebra.

THEOREM 2.7.7. *If* $HH^3(A) = 0$, *then there exists a versal formal deformation of* A *over* $\hat{\mathcal{O}}(HH^2(A))$. *If, furthermore,* $HH^1(A) = 0$, *then this is a universal deformation.*

At the end of the exercises from Section 4, we will include an outline of a proof of a slightly weaker version of the first statement of the theorem.

In the case that $HH^3(A) \neq 0$, then in general there may not exist a versal formal deformation over $\hat{\mathcal{O}}(HH^2(A))$; see, however, Theorem 2.3.17 for another situation where this exists.

EXAMPLE 2.7.8. If $A = \mathsf{Weyl}(V) \rtimes G$ for $G < \mathsf{Sp}(V)$ a finite subgroup (see Definition 1.9.17), then, by [1], $HH^i(A)$ is the space of conjugation-invariant functions on the set of group elements $g \in G$ such that $\mathrm{rk}(g - \mathrm{Id}) = i$. In particular, $HH^i(A) = 0$ when i is odd, and $HH^2(A)$ is the space of conjugation-invariant functions on the set of *symplectic reflections*: those elements fixing a codimension-two symplectic hyperplane. Thus, there is a universal formal deformation over $\hat{\mathcal{O}}(HH^2(V))$.

Moreover, $HH^2(A) = HH^2(A)_{\leq -2}$ is all in degree -2 (the degree also of the Poisson bracket on $\mathcal{O}(V)$). This actually implies that there is a universal filtered deformation of A parametrized by elements $c \in HH^2(A)$ as above, in the sense that every filtered deformation is isomorphic to one of these via a filtered isomorphism whose associated graded isomorphism is the identity. (In fact, the universal formal deformation is obtained from this one by completing at the augmentation ideal of $\mathcal{O}(HH^2(V))$.)

This universal filtered deformation admits an explicit description, known as the *symplectic reflection algebra* [99], first constructed by Drinfeld, which is defined as follows. Let S be the set of symplectic reflections, i.e., elements such that $\mathrm{rk}(g - \mathrm{Id}) = 2$. Then $HH^2(A) \cong \Bbbk[S]^G$. Let $c \in \mathrm{Fun}_G(S, \Bbbk)$ be a conjugation-invariant $\Bbbk$-valued function on S. Then the corresponding symplectic reflection algebra $H_{1,c}(G)$ is presented as

$$H_{1,c}(G) := TV \Big/ \Big(xy - yx - \omega(x, y) + 2 \sum_{s \in S} c(s) \omega_s(x, y) \cdot s \Big),$$

where ω_s is the composition of ω with the projection to the sum of the nontrivial eigenspaces of s (which is a two-dimensional symplectic vector space). In other words, ω_s is the restriction of the symplectic form ω to the two-dimensional subspace orthogonal to the symplectic reflecting hyperplane of s. The algebra $H_{1,c}(G)$ above is the case $t = 1$ of the more general symplectic reflection algebra $H_{t,c}(G)$ studied in Chapter III, which is obtained by replacing the $\omega(x, y)$ by $t \cdot \omega(x, y)$ in the relation.

EXAMPLE 2.7.9. If X is a smooth affine variety or smooth C^∞ manifold, and G is a group acting by automorphisms on X, then by [97], $\mathrm{HH}^2(\mathcal{D}(X) \rtimes G) \cong H^2_{DR}(X)^G \oplus \Bbbk[S]^G$, where S is the set of pairs (g, Y) where $g \in G$ and $Y \subseteq X^g$ is a connected (hence irreducible) subvariety of codimension one. Then $\Bbbk[S]^G$ is the space of $\Bbbk$-valued functions on S which are invariant under the action of G, $h \cdot (g, Y) = (hgh^{-1}, h(Y))$. Furthermore, by [97], all deformations of $\mathcal{D}(X) \rtimes G$ are *unobstructed* and there exists a universal filtered deformation $H_{1,c,\omega}(X)$ parametrized by $c \in \Bbbk[S]^G$ and $\omega \in H^2_{DR}(X)^G$. More precisely, to each such parameters we have a filtered deformation of $\mathcal{D}(X) \rtimes G$, and these exhaust all filtered deformations up to filtered isomorphism.

2.8. Additional exercises. (See also 2.3.2, 2.3.10, 2.3.11, and 2.7.2.)

EXERCISE 2.8.1. We introduce Koszul complexes, which are one of the main tools for computing Hochschild cohomology.

First consider, for the algebra $\mathrm{Sym}\, V$, the Koszul resolution of the augmentation module $\Bbbk$:

$$0 \to \mathrm{Sym}\, V \otimes \textstyle\bigwedge^{\dim V} V \to \mathrm{Sym}\, V \otimes \textstyle\bigwedge^{\dim V-1} V \to \cdots \to \mathrm{Sym}\, V \otimes V \to \mathrm{Sym}\, V \twoheadrightarrow \Bbbk, \tag{2.M}$$

$$f \otimes (v_1 \wedge \cdots \wedge v_i) \mapsto \sum_{j=1}^{i} (-1)^{j-1} (f v_j) \otimes (v_1 \wedge \cdots \hat{v}_j \cdots \wedge v_i), \tag{2.N}$$

where $\hat{v}_j$ means that v_j was omitted from the wedge product. We remark that this complex itself cannot deform to a complex of $\mathrm{Weyl}(V)$-modules, since the $\Bbbk$ itself does not deform (since $\mathrm{Weyl}(V)$ is simple, it has no finite-dimensional modules). Nonetheless, we will be able to deform a bimodule analogue of the above.

(a) Construct analogously to the Koszul resolution a bimodule resolution of $\mathrm{Sym}\, V$, of the form

$$\mathrm{Sym}\, V \otimes \textstyle\bigwedge^\bullet V \otimes \mathrm{Sym}\, V \twoheadrightarrow \mathrm{Sym}\, V.$$

Using this complex, show that

$$\mathrm{HH}^i(\mathrm{Sym}\, V, \mathrm{Sym}\, V \otimes \mathrm{Sym}\, V) := \mathrm{Ext}^i_{(\mathrm{Sym}\, V)^e}(\mathrm{Sym}\, V, \mathrm{Sym}\, V \otimes \mathrm{Sym}\, V)$$
$$\cong \mathrm{Sym}\, V[-\dim V].$$

(As we will see in Exercise 3.7.11, this implies that $\mathrm{Sym}\, V$ is a Calabi–Yau algebra of dimension $\dim V$.)

(b) Now replace $\mathrm{Sym}\, V$ by the Weyl algebra, $\mathrm{Weyl}(V)$. Show that the Koszul complex above deforms to give a complex whose zeroth homology is $\mathrm{Weyl}(V)$:

$$0 \to \mathrm{Weyl}(V) \otimes \textstyle\bigwedge^{\dim V} V \otimes \mathrm{Weyl}(V) \to \mathrm{Weyl}(V) \otimes \textstyle\bigwedge^{\dim V-1} V \otimes \mathrm{Weyl}(V)$$
$$\to \cdots \to \mathrm{Weyl}(V) \otimes V \otimes \mathrm{Weyl}(V) \to \mathrm{Weyl}(V) \otimes \mathrm{Weyl}(V) \twoheadrightarrow \mathrm{Weyl}(V), \tag{2.O}$$

using the same formula. Hint: You only need to show it is a complex, by Exercise 1.10.8. Note also the fact that such a deformation exists is a consequence of the corollaries of Exercise 1.10.8.

Deduce that $HH^\bullet(Weyl(V), Weyl(V) \otimes Weyl(V)) \cong Weyl V[-d]$ (so, as we will see in Exercise 3.7.11, it is also Calabi–Yau of dimension dim V).

(c) Suppose that $V = \mathfrak{g}$ is a (finite-dimensional) Lie algebra. Deform the complex to a resolution of the universal enveloping algebra using the Chevalley–Eilenberg complex:

$$U\mathfrak{g} \otimes \textstyle\bigwedge^{\dim \mathfrak{g}} \mathfrak{g} \otimes U\mathfrak{g} \to U\mathfrak{g} \otimes \textstyle\bigwedge^{\dim \mathfrak{g}-1} \mathfrak{g} \otimes U\mathfrak{g} \to \cdots \to U\mathfrak{g} \otimes \mathfrak{g} \otimes U\mathfrak{g} \to U\mathfrak{g} \otimes U\mathfrak{g} \twoheadrightarrow U\mathfrak{g},$$

where the differential is the sum of the preceding differential for Sym $V = $ Sym $\mathfrak{g}$ and the additional term,

$$x_1 \wedge \cdots \wedge x_k \mapsto \sum_{i<j} [x_i, x_j] \wedge x_1 \wedge \cdots \hat{x}_i \cdots \hat{x}_j \cdots \wedge x_k. \tag{2.P}$$

That is, verify this is a complex, and conclude from (a) and Exercise 1.10.8 that it must be a resolution. Conclude as before that $HH^\bullet(U\mathfrak{g}, U\mathfrak{g} \otimes U\mathfrak{g}) \cong U\mathfrak{g}[-d]$. (As we will see in Exercise 3.10.1, $U\mathfrak{g}$ is twisted Calabi–Yau, but not Calabi–Yau in general.)

(d) Recall the usual Chevalley–Eilenberg complexes computing Lie algebra (co)homology with coefficients in a $\mathfrak{g}$-module M: if $C_\bullet^{CE}(\mathfrak{g}) = \textstyle\bigwedge^\bullet \mathfrak{g}$ is the complex inside of the two copies of $U\mathfrak{g}$ above (with the differential as in (2.P)), these are

$$C_\bullet^{CE}(\mathfrak{g}, M) := C_\bullet^{CE}(\mathfrak{g}) \otimes M, \qquad C_{CE}^\bullet(\mathfrak{g}, M) := \mathrm{Hom}_\Bbbk(C_\bullet^{CE}(\mathfrak{g}), M).$$

Conclude that (cf. [165, Theorem 3.3.2]), if M is a $U\mathfrak{g}$-bimodule,

$$HH^\bullet(U\mathfrak{g}, M) \cong H_{CE}^\bullet(\mathfrak{g}, M^{\mathrm{ad}}), \qquad HH_\bullet(U\mathfrak{g}, M) \cong H_\bullet^{CE}(\mathfrak{g}, M^{\mathrm{ad}}),$$

where M^{ad} is the $\mathfrak{g}$-module obtained from M by the adjoint action,

$$\mathrm{ad}(x)(m) := xm - mx,$$

where the LHS gives the Lie action of x on m, and the RHS uses the $U\mathfrak{g}$-bimodule action. Conclude that $HH^\bullet(U\mathfrak{g}) \cong H_{CE}^\bullet(\mathfrak{g}, U\mathfrak{g}^{\mathrm{ad}})$.

Next, recall (or accept as a black box) that, when $\mathfrak{g}$ is finite-dimensional semisimple (or more generally reductive) and $\Bbbk$ has characteristic zero, then $U\mathfrak{g}^{\mathrm{ad}}$ decomposes into a direct sum of finite-dimensional irreducible representations, and $H_{CE}^\bullet(\mathfrak{g}, V) = \mathrm{Ext}_{U\mathfrak{g}}^\bullet(\Bbbk, V) = 0$ if V is finite-dimensional irreducible and *not* the trivial representation.

Conclude that, in this case, $HH^\bullet(U\mathfrak{g}) \cong \mathrm{Ext}_{U\mathfrak{g}}^\bullet(\Bbbk, U\mathfrak{g}^{\mathrm{ad}}) \cong Z(U\mathfrak{g}) \otimes \mathrm{Ext}_{U\mathfrak{g}}^\bullet(\Bbbk, \Bbbk)$.

Bonus: Compute that, still assuming that $\mathfrak{g}$ is finite-dimensional semisimple (or reductive) and $\Bbbk$ has characteristic zero, $\mathrm{Ext}_{U\mathfrak{g}}(\Bbbk, \Bbbk) \cong H_{CE}^\bullet(\mathfrak{g}, \Bbbk)$ is isomorphic to $(\textstyle\bigwedge^\bullet \mathfrak{g}^*)^\mathfrak{g}$. To do so, show that the inclusion $C_{CE}^\bullet(\mathfrak{g}, \Bbbk)^\mathfrak{g} \hookrightarrow C_{CE}^\bullet(\mathfrak{g}, \Bbbk)$ is a quasi-isomorphism. This follows by defining an operator d^* with opposite degree of d, so that the Laplacian $dd^* + d^*d$ is the quadratic Casimir element C. Thus this Laplacian operator defines a contracting homotopy onto the part of the complex on which the quadratic Casimir acts by zero. Since C acts by a positive scalar on all nontrivial finite-dimensional irreducible representations, this subcomplex is $C_{CE}^\bullet(\mathfrak{g}, \Bbbk)^\mathfrak{g}$. But the latter is just $(\textstyle\bigwedge \mathfrak{g}^*)^\mathfrak{g}$ with zero differential.

Conclude that $HH^\bullet(U\mathfrak{g}) = Z(U\mathfrak{g}) \otimes (\textstyle\bigwedge^\bullet \mathfrak{g}^*)^\mathfrak{g}$. In particular, if $\mathfrak{g}$ is semisimple, we have $HH^2(U\mathfrak{g}) = HH^1(U\mathfrak{g}) = 0$. As we will see, this implies that $U\mathfrak{g}$ has no nontrivial formal deformations.

REMARK 2.8.2. For $\mathfrak{g} = \mathfrak{sl}_2$, $U\mathfrak{sl}_2$ is a Calabi–Yau algebra of dimension three, since it is a Calabi–Yau deformation of the Calabi–Yau algebra $\mathcal{O}(\mathfrak{sl}_2^*) \cong \mathbb{A}^3$, which is Calabi–Yau with the usual volume form. So $\mathrm{HH}^3(U\mathfrak{sl}_2) \cong \mathrm{HH}_0(U\mathfrak{sl}_2)$ by Van den Bergh duality, and the latter is $U\mathfrak{sl}_2/[U\mathfrak{sl}_2, U\mathfrak{sl}_2]$, which you should be able to prove is isomorphic to $(U\mathfrak{sl}_2)_{\mathfrak{sl}_2} \cong (U\mathfrak{sl}_2)^{\mathfrak{sl}_2}$, i.e., this is a rank-one free module over $\mathrm{HH}^0(U\mathfrak{sl}_2)$.

In other words, $H^3_{CE}(\mathfrak{g}, \Bbbk) \cong \Bbbk$, which is also true for a general simple Lie algebra. Since this group controls the *tensor category deformations* of $\mathfrak{g}$-mod, or the Hopf algebra deformations of $U\mathfrak{g}$, this is saying that there is a one-parameter deformation of $U\mathfrak{g}$ as a Hopf algebra which gives the quantum group $U_q\mathfrak{g}$. But since $\mathrm{HH}^2(U\mathfrak{g}) = 0$, as an associative algebra, this deformation is trivial, i.e., equivalent to the original algebra; this is *not* true as a Hopf algebra!

EXERCISE 2.8.3. Using the Koszul deformation principle (Theorem 1.10.17), show that the symplectic reflection algebra (Example 2.7.8) is a flat filtered deformation of $\mathrm{Sym}(V) \rtimes G$. Hint: This amounts to the "Jacobi" identity

$$[\omega_s(x, y), z] + [\omega_s(y, z), x] + [\omega_s(z, x), y] = 0,$$

for all symplectic reflections $s \in S$, and all $x, y, z \in V$, as well as the fact that the relations are G-invariant. Compare with [98, §2.3], where it is also shown that (the Rees algebras of these) yield all formal deformations of $\mathrm{Weyl}(V) \rtimes G$.

EXERCISE 2.8.4. In this exercise, we deduce the filtered version of the Koszul deformation principle, Theorem 1.10.17, from the more standard deformation version (Theorem 2.8.5), using the correspondence between filtered deformations and homogeneous formal deformations.

Let $R \subseteq TV$ and $B := TV/(R)$. Let $E \subseteq TV[[\hbar]]$ be a subspace which lifts R modulo $\hbar$, i.e., such that the composition $E \to TV[[\hbar]] \to TV$ is an isomorphism onto R. Let $A := TV[[\hbar]]/(E)$. Then we have a canonical surjection $B \to A/\hbar A$. We call A a (flat) formal deformation of B if this surjection is an isomorphism.

(a) A priori, the above is a slightly stronger version of formal deformation than the usual one, because it requires a deformation E of the relations. Show, however, that such an E always exists given a (one-parameter) formal deformation A of $B := TV/(R)$, so the notion is equivalent to the usual one. Hint: Recall that, by Remark 2.3.4, a one-parameter formal deformation A can be alternatively viewed as a topologically free $\Bbbk[[\hbar]]$-algebra A together with an isomorphism $\phi : A/\hbar A \xrightarrow{\sim} B$. Show that there exists $E \subseteq V^{\otimes 2}[[\hbar]]$ deforming R such that $A = TV[[\hbar]]/(E)$ and $\phi : A/\hbar A \to B$ is the identity on V.

(b)(*, but with the solution outlined) We are interested below in the case that $R \subseteq V^{\otimes 2}$, i.e., B is a quadratic algebra, and that $E \subseteq V^{\otimes 2}[[\hbar]]$, i.e., B is also quadratic (as a $\Bbbk[[\hbar]]$-algebra). (The proof of (a) then adapts to show that such an E exists for every graded formal deformation A of B; moreover, the span $\Bbbk[[\hbar]] \cdot E \cong R[[\hbar]]$ is in fact canonical, and independent of the choice of E above, since it is the degree-two part of the kernel of $TV[[\hbar]] \to A$.)

The usual version of the Koszul deformation principle is as follows:

THEOREM 2.8.5 (Koszul deformation principle [89, 40, 26, 181]). *Assume that B is Koszul. If $E \subseteq V^{\otimes 2}[\![\hbar]\!]$ is a formal deformation, then $A := TV[\![\hbar]\!]/(E)$ is a (flat) formal deformation of B if and only if it is flat in weight three, i.e., the surjection $B_3 \twoheadrightarrow A_3/\hbar A_3$ is an isomorphism.*

Use this to prove Theorem 1.10.17, as follows. Given a Koszul algebra $B = TV/(R)$ and a filtered deformation $E \subseteq T^{\leq 2}V$ of R, i.e., such that the composition $E \to T^{\leq 2} \twoheadrightarrow V \otimes V$ is an isomorphism onto R, we can form a corresponding formal deformation as follows. Let $A := TV/(E)$. First we homogenize A by forming the Rees algebra over the parameter t:

$$R_t A := \bigoplus_{i \geq 0} A_{\leq i} t^i.$$

This algebra is graded by degree in t, and deforms B in the sense that $R_t A/t R_t A \cong B$. Thus $R_t A$ is a quadratic algebra. Set $\tilde{V} := V \oplus \Bbbk \cdot t$. Then we in fact have $R_t A = T\tilde{V}/\tilde{E}$, where $\tilde{E}$ is the span of $tv - vt$ for $v \in V$ together with the homogenized versions of the elements of E: given a relation $r = r_2 + r_1 + r_0 \in E$ for $r_i \in V^{\otimes i}$, we have an associated element $\tilde{r} \in \tilde{E}$ given by $\tilde{r} = r_2 + r_1 t + r_0 t^2$. This $\tilde{E}$ is spanned by the $\tilde{r}$ and the $tv - vt$.

Moreover, $R_t A$ can also be viewed as a filtered deformation of $B[t]$: put the filtration on $R_t A$ by the filtration in A (ignoring the t), i.e., $(R_t A)_{\leq m} = A_{\leq m} t^m \otimes \Bbbk[t]$ (thus, this is *not* the filtration given by the grading on $R_t A$). Then $\mathrm{gr}(R_t A) = B[t]$.

Using the preceding filtration, form the completed Rees algebra in $\hbar$,

$$\hat{A} := \widehat{R}_\hbar R_t A := \prod_{i \geq 0} (R_t A)_{\leq i} \cdot \hbar^i.$$

The result is an algebra $\hat{A}$ such that $\hat{A}/\hbar\hat{A} \cong B[t]$ (rather than $R_t A$), since we used the filtration on $R_t A$ coming from A only. Moreover, $\hat{A}$ is a quadratic formal deformation of $R_t A$, using the grading on $R_t A$, placing $\hbar$ in degree zero.

Finally, A satisfies the assumption of Theorem 1.10.17, i.e., that $\mathrm{gr}_3(E \otimes V \cap (V + \Bbbk) \otimes E) = (R \otimes V \cap V \otimes R)$ (where $\mathrm{gr}_3 : T^{\leq 3}V \to V^{\otimes 3}$ is the projection modulo $V^{\otimes \leq 2}$), if and only if $\hat{A}$ satisfies the condition that the natural map

$$(R \otimes V \cap V \otimes R) \to (\tilde{E} \otimes \tilde{V} \cap \tilde{V} \otimes \tilde{E})/\hbar(\tilde{E} \otimes \tilde{V} \cap \tilde{V} \otimes \tilde{E})$$

is an isomorphism.

Check that the latter condition is equivalent to the condition that, in degree three, $\hat{A}$ is a flat deformation of $B[t]$, i.e., that $(B[t])_3 \cong (\hat{A})_3/\hbar(\hat{A})_3$. (Note that $(B[t])_3 = B_3 \oplus B_2 t \oplus B_1 t^2 \oplus \Bbbk t^3$, and the difficulty is in dealing with the B_3 part).

Thus, Theorem 2.8.5 applies and yields that, if B is Koszul, then for every filtered deformation E of the relations R of B, the algebra $A = TV/(E)$ is a (flat) filtered deformation of B if and only if $\hat{A}$ is a (flat) formal graded deformation, which is true if and only if the assumption of Theorem 1.10.17 is satisfied, which proves that theorem.

EXERCISE 2.8.6. Parallel to Exercise 1.10.15, show that, if A is a (flat) graded formal deformation of a Koszul algebra B as in Theorem 2.8.5, then A is also Koszul over $\Bbbk[[\hbar]]$ (meaning, $\text{Tor}_i^A(\Bbbk[[\hbar]], \Bbbk[[\hbar]])$ is concentrated in degree i, or alternatively A admits a free graded $A \otimes_{\Bbbk[[\hbar]]} A^{\text{op}}$-module resolution $P_\bullet^\hbar \to A$ with $P_i^\hbar$ generated in degree i).

3. Hochschild cohomology and infinitesimal deformations

In this section, we return to a more basic topic, that of Hochschild (co)homology of algebras. We first describe the explicit complexes used to compute these, and how they arise from the bar resolution of an algebra as a bimodule over itself. Using this description, we give the standard interpretation of the zeroth, first, second, and third Hochschild cohomology groups as the center, the vector space of outer derivations, the vector space of infinitesimal deformations up to equivalence, and the vector space of obstructions to lifting an infinitesimal deformation to a second-order deformation. We also explain the definition of Calabi–Yau algebras, both to illustrate an application of Hochschild cohomology, and to further understand our running examples of Weyl algebras and universal enveloping algebras.

3.1. The bar resolution. To compute Hochschild (co)homology using Definition 2.2.1, one can resolve A as an A^e-module, i.e., A-bimodule. The standard way to do this is via the *bar resolution*:

DEFINITION 3.1.1. The bar resolution, $C^{\text{bar}}(A)$, of an associative algebra A over $\Bbbk$ is the complex

$$\cdots \longrightarrow A \otimes A \otimes A \longrightarrow A \otimes A \longrightarrow A,$$

$$(a \otimes b \otimes c) \longmapsto ab \otimes c - a \otimes bc, \tag{3.A}$$

$$a \otimes b \longmapsto ab.$$

More conceptually, we can define the complex as

$$T_A(A \cdot \epsilon \cdot A), \quad d = \partial_\epsilon,$$

viewing $A \cdot \epsilon \cdot A$ as an A-bimodule, assigning degrees by $|\epsilon| = 1, |A| = 0$, and viewing the differential $d = \partial_\epsilon$ is a *graded derivation* of degree -1, i.e.,

$$\partial_\epsilon(x \cdot y) = \partial_\epsilon(x) \cdot y + (-1)^{|y|} x \cdot \partial_\epsilon(y).$$

Finally, $\partial_\epsilon(\epsilon) = 1$ and $\partial_\epsilon(A) = 0$.

3.2. The Hochschild (co)homology complexes. Using the bar resolution, we conclude that

$$\text{HH}_i(A, M) = \text{Tor}_i^{A^e}(A, M) = H_i(C_\bullet^{\text{bar}}(A) \otimes_{A^e} M) = H_i(C_\bullet(A, M)), \tag{3.B}$$

where

$$C_\bullet(A, M) = (M \otimes_A C_\bullet^{\text{bar}}(A))/[A, M \otimes_A C_\bullet^{\text{bar}}(A)] \tag{3.C}$$

$$= \cdots \longrightarrow M \otimes A \otimes A \longrightarrow M \otimes A \longrightarrow M,$$

$$(m \otimes b \otimes c) \longmapsto mb \otimes c - m \otimes bc + cm \otimes b,$$

$$m \otimes a \longmapsto ma - am.$$

Similarly,

$$\mathrm{HH}^i(A, M) = \mathrm{Ext}^i_{A^e}(A, M) = H^i(C^\bullet(A, M)), \tag{3.D}$$

where

$$C^\bullet(A, M) = \mathrm{Hom}_{A^e}(C^{\mathrm{bar}}_\bullet(A), M) \tag{3.E}$$

$$= \cdots \longleftarrow \mathrm{Hom}_{\Bbbk}(A \otimes A, M) \longleftarrow \mathrm{Hom}_{\Bbbk}(A, M) \longleftarrow M$$

$$(d\phi)(a \otimes b) = a\phi(b) - \phi(ab) + \phi(a)b \longleftarrow \phi$$

$$(dx)(a) = ax - xa \longleftarrow x.$$

As for cohomology, we will use the notation $C^\bullet(A) := C^\bullet(A, A)$.

REMARK 3.2.1. To extend Hochschild cohomology $\mathrm{HH}^\bullet(A)$ to the C^∞ context, i.e., where $A = \mathcal{O}(X) := C^\infty(X)$ for X a C^∞ manifold, we do not make the same definition as above (i.e., we do not treat A as an ordinary associative algebra). Instead, we define the Hochschild cochain complex so as to place in degree m the subspace of $\mathrm{Hom}_{\Bbbk}(A^{\otimes m}, A)$ of *smooth polydifferential operators*, i.e., linear maps spanned by tensor products of smooth (rather than algebraic) differential operators $A \to A$, or equivalently, operators which are spanned by polynomials in smooth vector fields (which are not the same as algebraic derivations in the smooth setting: a smooth vector field is always a derivation of A as an abstract algebra, but not conversely). This can be viewed as restricting to the "local" part of Hochschild cohomology (because polydifferential operators are local in the sense that the result at a point x only depends on the values in a neighborhood of x).

3.3. Zeroth Hochschild homology. We see from the above that $\mathrm{HH}_0(A) = A/[A, A]$, where the quotient is taken as a vector space.

EXAMPLE 3.3.1. Suppose that $A = TV$ is a free algebra on a vector space V. Then $\mathrm{HH}_0(A) = TV/[TV, TV]$ is the vector space of *cyclic words* in V.

Similarly, if $A = \Bbbk Q$ is the path algebra of a quiver Q (see §1.9), then $\mathrm{HH}_0(A) = \Bbbk Q/[\Bbbk Q, \Bbbk Q]$ is the vector space of *cyclic paths* in the quiver Q (which by definition do not have an initial or terminal vertex).

EXAMPLE 3.3.2. If A is commutative, then $\mathrm{HH}_0(A) = A$, since $[A, A] = 0$.

3.4. Zeroth Hochschild cohomology. Note that $\mathrm{HH}^0(A) = \{a \in A \mid ab - ba = 0, \ \forall b \in A\} = Z(A)$, the center of the algebra A.

EXAMPLE 3.4.1. If $A = TV$ is a tensor algebra and $\dim V \geq 2$, then $\mathrm{HH}^0(A) = \Bbbk$: only scalars are central. The same is true if $A = \Bbbk Q$ for Q a connected quiver which has more than one edge or more than one vertex (see §1.9).

EXAMPLE 3.4.2. If A is commutative, then $\mathrm{HH}^0(A) = A$, since every element is central.

3.5. First Hochschild cohomology. A Hochschild one-cocycle is an element $\phi \in \mathrm{End}_{\Bbbk}(A)$ such that $a\phi(b) + \phi(a)b = \phi(ab)$ for all $a, b \in A$. These are the derivations of A. A Hochschild one-coboundary is an element $\phi = d(x)$, $x \in A$, and this has the form $\phi(a) = ax - xa$ for all $a \in A$. Therefore, these are the inner derivations. We conclude that

$$\mathrm{HH}^1(A) = \mathrm{Out}(A) := \mathrm{Der}(A)/\mathrm{Inn}(A),$$

the vector space of outer derivations of A, which is by definition the quotient of all derivations $\mathrm{Der}(A)$ by the inner derivations $\mathrm{Inn}(A)$. We remark that $\mathrm{Inn}(A) \cong A/Z(A) = A/\mathrm{HH}^0(A)$, since an inner derivation $a \mapsto ax - xa$ is zero if and only if x is central.

EXAMPLE 3.5.1. If $A = TV$ for dim $V \geq 2$, then

$$\mathrm{HH}^1(A) = \mathrm{Out}(TV) = \mathrm{Der}(TV)/\mathrm{Inn}(TV) = \mathrm{Der}(TV)/\overline{TV},$$

where $\overline{TV} = TV/\Bbbk$, since $Z(TV) = \Bbbk$. Explicitly, derivations of TV are uniquely determined by their restrictions to linear maps $V \to TV$, i.e., $\mathrm{Der}(TV) \cong \mathrm{Hom}_{\Bbbk}(V, TV)$. So we get $\mathrm{HH}^1(A) \cong \mathrm{Hom}_{\Bbbk}(V, TV)/\overline{TV}$. (Note that the inclusion $\overline{TV} \to \mathrm{Hom}_{\Bbbk}(V, TV)$ is explicitly given by $f \mapsto \mathrm{ad}(f)$, where $\mathrm{ad}(f)(v) = fv - vf$.)

EXAMPLE 3.5.2. If A is commutative, then $\mathrm{Inn}(A) = 0$, so $\mathrm{HH}^1(A) = \mathrm{Der}(A)$. If, moreover, $A = \mathcal{O}(X)$ is the commutative algebra of functions on an affine variety, then this is also known as the global vector fields $\mathrm{Vect}(X)$, as we discussed.

EXAMPLE 3.5.3. In the case $A = C^\infty(X)$, restricting our Hochschild cochain complex to differential operators in accordance with Remark 3.2.1, then $\mathrm{HH}^1(A)$ is the space of smooth vector fields on X.

EXAMPLE 3.5.4. In the case that V is smooth affine (or more generally normal), $\Bbbk$ has characteristic zero, and Γ is a finite group of automorphisms of V which acts freely outside of a codimension-two subset of V, then all vector fields on the smooth locus of V/Γ extend to derivations of $\mathcal{O}(V/\Gamma) = \mathcal{O}(V)^\Gamma$, since V/Γ is normal. Moreover, vector fields on the smooth locus of V/Γ are the same as Γ-invariant vector fields on the locus, call it $U \subseteq V$, where Γ acts freely. These all extend to all of V since $V \setminus U$ has codimension two. Thus we conclude that $\mathrm{HH}^1(\mathcal{O}(V)^\Gamma) = \mathrm{Der}(\mathcal{O}(V)^\Gamma) = \mathrm{Vect}(V)^\Gamma$ is the space of Γ-invariant vector fields on V (the first equality here is because $\mathcal{O}(V)^\Gamma$ is commutative). In particular this includes the case where V is a symplectic vector space and $\Gamma < \mathrm{Sp}(V)$, since the hyperplanes with nontrivial stabilizer group must be symplectic and hence of codimension at least two.

3.6. Infinitesimal deformations and second Hochschild cohomology. We now come to a key point. A Hochschild two-cocycle is an element $\gamma \in \mathrm{Hom}_{\Bbbk}(A \otimes A, A)$ satisfying

$$a\gamma(b \otimes c) - \gamma(ab \otimes c) + \gamma(a \otimes bc) - \gamma(a \otimes b)c = 0. \tag{3.F}$$

This has a nice interpretation in terms of infinitesimal deformations (Definition 2.3.5). Explicitly, an infinitesimal deformation is given by a linear map $\gamma : A \otimes A \to A$, by the formula

$$a \star_\gamma b = ab + \varepsilon\gamma(a \otimes b).$$

Then, the associativity condition is exactly (3.F).

Moreover, we can interpret two-coboundaries as *trivial* infinitesimal deformations. More generally, we say that two infinitesimal deformations γ_1, γ_2 are *equivalent* if there is a $\Bbbk[\varepsilon]/(\varepsilon^2)$-module automorphism of A_ε which is the identity modulo ε which takes γ_1 to γ_2. Such a map has the form $\phi := \mathrm{Id} + \varepsilon \cdot \phi_1$ for some linear map $\phi_1 : A \to A$, i.e., $\phi_1 \in C^1(A)$. We then compute that

$$\phi^{-1}\big(\phi(a) \star_\gamma \phi(b)\big) = a \star_{\gamma + d\phi_1} b.$$

We conclude:

PROPOSITION 3.6.1. $HH^2(A)$ *is the vector space of equivalence classes of infinitesimal deformations of A.*

EXAMPLE 3.6.2. For $A = TV$, a tensor algebra, we claim that $HH^2(A) = 0$, so there are no nontrivial infinitesimal deformations. Indeed, one can construct a short bimodule resolution of A,

$$0 \longrightarrow A \otimes V \otimes A \longrightarrow A \otimes A \twoheadrightarrow A,$$
$$a \otimes v \otimes b \mapsto av \otimes b - a \otimes vb, \quad a \otimes b \mapsto ab.$$

Since this is a projective resolution of length one, we conclude that $\mathrm{Ext}^2_{A^e}(A, M) = 0$ for all bimodules M, i.e., $HH^2(A, M) = 0$ for all bimodules M.

EXAMPLE 3.6.3. If $A = \mathrm{Weyl}(V)$ for a symplectic vector space V, then $HH^\bullet(A) = \Bbbk$ (in degree zero), and hence $HH^2(\mathrm{Weyl}(V))$ and $HH^1(\mathrm{Weyl}(V))$ are zero, so that there are no nontrivial infinitesimal deformations nor outer derivations. Moreover, $HH_\bullet(A) = \Bbbk[-\dim V]$ (i.e., $\Bbbk$ in degree $\dim V$ and zero elsewhere). To see this, we can use the resolution of Exercise 2.8.1 (see also the solution).

3.7. Remarks on Calabi–Yau algebras. The Koszul resolutions above imply that $\mathrm{Weyl}(V)$ and $\mathrm{Sym}\, V$ are *Calabi–Yau algebras of dimension* $\dim V$, which we will define as follows (and which was also mentioned in Chapter I). One can also conclude from this that $\mathrm{Weyl}(V) \rtimes \Gamma$ and $\mathrm{Sym}\, V \rtimes \Gamma$ are Calabi–Yau of dimension $\dim V$ for $\Gamma < \mathrm{Sp}(V)$ finite, as we will do in the exercises.

DEFINITION 3.7.1. An A-bimodule U is right invertible if there exists an A-bimodule R such that $U \otimes_A R \cong A$ as A-bimodules. It is left invertible if there exists an A-bimodule L such that $L \otimes_A U \cong A$ as A-bimodules. It is invertible if it is both left and right invertible.

Note that, if U is invertible, then any left inverse is a right inverse (just as in the case of monoids), so this is also the unique (two-sided) inverse of U up to isomorphism.

QUESTION 3.7.2. Can you find an example of a left but not right invertible bimodule?

The following basic example will be our typical invertible bimodule:

EXAMPLE 3.7.3. Given an algebra automorphism $\sigma : A \to A$, we can define the A-bimodule A^σ, which as a vector space is merely A itself, but with the bimodule action given by the formula, for $a, b \in A$ and $m \in A^\sigma$,

$$a \cdot m \cdot b = (am\sigma(b)).$$

EXERCISE 3.7.4.   (a) Prove that, if M is an A-bimodule which is free of rank one as a left and as a right module (i.e., isomorphic to A as a left and as a right module) using the same generator $m \in M$, then $M \cong A^\sigma$ for some algebra automorphism $\sigma : A \to A$. Hint: Let $1 \in M$ be a left module generator, realizing $A \cong A \cdot 1 = M$. Consider the map $\sigma : A \to A$ given by $1 \cdot h = \sigma(h) \cdot 1$. Show that σ is a $\Bbbk$-algebra homomorphism, with inverse σ^{-1} given by $h \cdot 1 = 1 \cdot \sigma^{-1}(h)$.

  (b) Now assume A is graded and M is a graded bimodule obeying the assumptions of (a), with common generator m is in degree zero. Conclude that $M \cong A^\sigma$ as a graded bimodule, where σ is a graded automorphism.

In order to define the notion of Calabi–Yau and twisted Calabi–Yau algebras, we will need the following exercise:

EXERCISE 3.7.5. Verify that $\mathrm{HH}^\bullet(A, A \otimes A)$ is a canonically an A-bimodule using the *inner* action, i.e., the action obtained from the A-bimodule structure on $M = A \otimes A$ given, for $x \otimes y \in M$ and $a, b \in A$, by the formula $a(x \otimes y)b = xb \otimes ay$. The most explicit way to do this is to use (3.E), where the action is inner multiplication on the output, i.e., $(a \cdot \phi \cdot b)(x_1 \otimes \cdots \otimes x_n) := a \cdot \phi(x_1 \otimes \cdots \otimes x_n) \cdot b$.

DEFINITION 3.7.6. An algebra A is *homologically smooth* if A has a finitely-generated projective A-bimodule resolution.

Note that finitely-generated projective, for a complex of (bi)modules, means that the complex is of finite length and consists of finitely-generated projective (bi)modules.

REMARK 3.7.7. More generally, a complex of modules M over an algebra B is called perfect if it is a finite complex of finitely-generated projective modules. If one works in the derived category $D^b(B)$ of B-modules, then a complex $P_\bullet$ is quasi-isomorphic to a perfect complex if and only if it is *compact*, i.e., $\mathrm{Hom}(P_\bullet, -)$ commutes with arbitrary direct sums (it is automatic that $\mathrm{Hom}(M, -)$ commutes with finite direct sums for any M, but the condition to commute with arbitrary direct sums is much more subtle). Compact objects of derived categories are extremely important.

REMARK 3.7.8. In particular, homological smoothness implies that A has finite Hochschild dimension: recall that this means that, for some $N \geq 0$, we have $\mathrm{HH}^i(A, M) = 0$ for all $i > N$ and all bimodules M (the minimal such N is then called the Hochschild dimension of A). This is equivalent to the condition that A have a bounded projective resolution, i.e., a resolution by a complex with finitely many nonzero terms, each of which are projective. To be homologically smooth requires that these terms be finitely-generated as well (this is a subtle strengthening of the condition in general). When A is homologically smooth, one can show that it has a finitely-generated projective bimodule resolution whose length (i.e., number of terms) is the Hochschild dimension of A.

DEFINITION 3.7.9. A is a Calabi–Yau algebra of dimension d if A is homologically smooth and $\mathsf{HH}^\bullet(A, A \otimes A) = A[-d]$ as a graded A-bimodule (i.e., $\mathsf{HH}^i(A, A \otimes A) = 0$ for $i \neq d$, and $\mathsf{HH}^d(A, A \otimes A) = A$ as a graded A-bimodule).

More generally, A is *twisted Calabi–Yau* of dimension d if A is homologically smooth and $\mathsf{HH}^\bullet(A, A \otimes A) = U[-d]$ as a graded A-bimodule, where U is an invertible A-bimodule.

Here, in $\mathsf{HH}^i(A, A \otimes A)$, $A \otimes A$ is considered as a bimodule with the *outer* bimodule structure, and the remaining inner structure induces a bimodule structure on A, as shown in Exercise 3.7.5 above. Typically, in the twisted Calabi–Yau case, $U \cong A^\sigma$ as in Example 3.7.3, which as we know from Exercise 3.7.4, is equivalent to saying that $\mathsf{HH}^d(A, A \otimes A) \cong A$ as both left and right modules individually, with a common generator (for example, A could be graded, and the common generator of $\mathsf{HH}^d(A, A \otimes A)$ could be the unique nonzero element up to scaling in degree zero).

REMARK 3.7.10. By the Van den Bergh duality theorem (Theorem 5.3.1), if A is Calabi–Yau of dimension d, then A has Hochschild dimension d (since $\mathsf{HH}^i(A, M) = 0$ for all $i > d$ and all bimodules M, and $\mathsf{HH}^d(A, A \otimes A) \neq 0$). It follows (as pointed out in Remark 3.7.8) that A has a finitely-generated projective bimodule resolution of length d. Thus, we could have equivalently assumed the latter condition in Definition 3.7.9 rather than homological smoothness.

EXERCISE 3.7.11.   (a) Prove from the Koszul resolutions from Exercise 1.10.15 that $\mathsf{Sym}\, V$ and $\mathsf{Weyl}(V)$ are Calabi–Yau of dimension $\dim V$.

(b) Let $\Gamma < \mathsf{Sp}(V)$ be finite and $\Bbbk$ have characteristic zero. Take the Koszul resolutions and apply $M \mapsto M \otimes \Bbbk[\Gamma]$ to all terms, considered as bimodules over $A \rtimes \Gamma$ where A is either $\mathsf{Sym}\, V$ or $\mathsf{Weyl}(V)$. This bimodule structure is given by

$$(a \otimes g)(m \otimes h)(a' \otimes g') = (a \cdot g(m) \cdot gh(a')) \otimes (ghg').$$

Prove that the result are resolutions of $A \rtimes \Gamma$ as a bimodule over itself. Conclude that $A \rtimes \Gamma$ is also Calabi–Yau of dimension $\dim V$. In the case $A = \mathsf{Sym}\, V$, show that $A \rtimes \Gamma$ is actually Calabi–Yau given only that $\Gamma < \mathsf{SL}(V)$ (not $\mathsf{Sp}(V)$), still assuming Γ is finite.

REMARK 3.7.12. Similarly, one can show that, if A is twisted Calabi–Yau, $\Bbbk$ has characteristic zero, and Γ is a finite group acting by automorphisms on A, then $A \rtimes \Gamma$ is also twisted Calabi–Yau. In the case that A is Calabi–Yau, Γ must obey a unimodularity condition for $A \rtimes \Gamma$ to also be Calabi–Yau.

3.8. Obstructions to second-order deformations and third Hochschild cohomology.
Suppose now that we have an infinitesimal deformation given by $\gamma_1 : A \otimes A \to A$. To extend this to a second-order deformation, we require $\gamma_2 : A \otimes A \to A$, such that

$$a \star b := ab + \varepsilon \gamma_1(a \otimes b) + \varepsilon^2 \gamma_2(a \otimes b)$$

defines an associative product on $A \otimes \Bbbk[\varepsilon]/(\varepsilon^3)$.

Looking at the new equation in second degree, this can be written as

$$a\gamma_2(b \otimes c) - \gamma_2(ab \otimes c) + \gamma_2(a \otimes bc) - \gamma_2(a \otimes b)c = \gamma_1(\gamma_1(a \otimes b) \otimes c) - \gamma_1(a \otimes \gamma_1(b \otimes c)).$$

The LHS is $d\gamma_2(a \otimes b \otimes c)$, so the condition for γ_2 to exist is exactly that the RHS is a Hochschild coboundary. Moreover, one can check that the RHS is always a Hochschild three-cocycle (we will give a more conceptual explanation when we discuss the Gerstenhaber bracket). So the element on the RHS defines a class of $HH^3(A)$ which is the *obstruction* to extending the above infinitesimal deformation to a second-order deformation:

COROLLARY 3.8.1. $HH^3(A)$ *is the space of* obstructions *to extending first-order deformations to second-order deformations. If* $HH^3(A) = 0$, *then all first-order deformations extend to second-order deformations.*

We next consider general n-th order deformations. By definition, such a deformation is a deformation over $\Bbbk[\varepsilon]/(\varepsilon^{n+1})$.

EXERCISE 3.8.2. (*) Show that the obstruction to extending an n-th order deformation $\sum_{i=1}^{n} \varepsilon^i \gamma_i$ (where here $\varepsilon^{n+1} = 0$) to an $(n+1)$-st order deformation $\sum_{i=1}^{n+1} \varepsilon^i \gamma_i$ (now setting $\varepsilon^{n+2} = 0$), i.e., the existence of a γ_{n+1} so that this defines an associative multiplication on $A \otimes \Bbbk[\varepsilon]/(\varepsilon^{n+2})$, is also a class in $HH^3(A)$.

Moreover, if this class vanishes, show that two different choices of γ_{n+1} differ by Hochschild two-cocycles, and that two are equivalent (by applying a $\Bbbk[\varepsilon]/(\varepsilon^{n+2})$-module automorphism of $A \otimes \Bbbk[\varepsilon]/(\varepsilon^{n+2})$ of the form $\mathrm{Id} + \varepsilon^{n+1} \cdot f$) if and only if the two choices of γ_{n+1} differ by a Hochschild two-coboundary. Hence, when the obstruction in $HH^3(A)$ vanishes, the set of possible extensions to a $(n+1)$-st order deformation (modulo gauge transformations which are the identity modulo ε^n) form a set isomorphic to $HH^2(A)$ (more precisely, it forms a *torsor* over the vector space $HH^2(A)$, i.e., an affine space modeled on $HH^2(A)$ without a chosen zero element). We will give a more conceptual explanation when we discuss formal deformations.

Note that, when $HH^3(A) \neq 0$, it can still happen that infinitesimal deformations extend to all orders. For example, by Theorem 2.3.13, this happens for Poisson structures on smooth manifolds (a Poisson structure yields an infinitesimal deformation by, e.g., $a \star b = ab + \frac{1}{2}\{a, b\} \cdot \varepsilon$; this works for arbitrary skew-symmetric biderivations $\{-, -\}$, but only the Poisson ones, i.e., those satisfying the Jacobi identity, extend to all orders).

However, finding this quantization is *nontrivial*: even though Poisson bivector fields are those classes of $HH^2(A)$ whose obstruction in $HH^3(A)$ to extending to second order vanishes, if one does not pick the extension correctly, one *can* obtain an obstruction to continuing to extend to third order, etc. In fact, the proof of Theorem 2.3.13 describes the space of *all* quantizations: as we will see, deformation quantizations are equivalent to formal deformations of the Poisson structure.

3.9. Deformations of modules and Hochschild cohomology. Let A be an associative algebra and M a module over A. Recall that Hochschild (co)homology must take coefficients in an A-*bimodule*, not an A-module. Given M, there is a canonical associated bimodule, namely $\text{End}_{\Bbbk}(M)$ (this is an A-bimodule whether M is a left or right module; the same is true for $\text{Hom}_{\Bbbk}(M, N)$ where M and N are both left modules, or alternatively both right modules).

LEMMA 3.9.1. $\text{HH}^i(A, \text{End}_{\Bbbk}(M)) \cong \text{Ext}_A^i(M, M)$ for all $i \geq 0$. More generally, $\text{HH}^i(A, \text{Hom}_{\Bbbk}(M, N)) \cong \text{Ext}_A^i(M, N)$ for all A-modules M and N.

PROOF. We prove the second statement. First of all, for $i = 0$,

$$\text{HH}^0(A, \text{Hom}_{\Bbbk}(M, N)) = \{\phi \in \text{Hom}_{\Bbbk}(M, N) \mid a \cdot \phi = \phi \cdot a, \ \forall a \in A\} = \text{Hom}_A(M, N).$$

Then the statement for higher i follows because they are the derived functors of the same bifunctors $((A\text{-mod})^{\text{op}} \times A\text{-mod}) \to \Bbbk\text{-mod}$.

Explicitly, if $P_\bullet \twoheadrightarrow A$ is a projective A-bimodule resolution of A, then $P_\bullet \otimes_A M \twoheadrightarrow M$ is a projective A-module resolution of M, and

$$\text{RHom}_A^\bullet(M, N) \cong \text{Hom}_A(P_\bullet \otimes_A M, N) = \text{Hom}_{A^e}(P_\bullet, \text{Hom}_{\Bbbk}(M, N))$$
$$\cong \text{RHom}_{A^e}^\bullet(A, \text{Hom}_{\Bbbk}(M, N)),$$

where for the middle equality, we used the adjunction

$$\text{Hom}_B(X \otimes_A Y, Z) = \text{Hom}_{B \otimes A^{\text{op}}}(X, \text{Hom}_{\Bbbk}(Y, Z)),$$

with X a (B, A)-bimodule, Y a left A-module, and Z a left B-module. □

In particular, this gives the most natural interpretation of $\text{HH}^0(A, \text{End}_{\Bbbk}(M))$: this is just $\text{End}_A(M)$. For the higher groups we recall the following standard descriptions of $\text{Ext}_A^1(M, M)$ and $\text{Ext}_A^2(M, M)$, which are convenient to see using Hochschild cochains valued in M.

DEFINITION 3.9.2. A deformation of an A-module M over an augmented commutative $\Bbbk$ algebra $R = \Bbbk \oplus R_+$ is an A-module structure on $M \otimes_{\Bbbk} R$, commuting with the R action, such that $(M \otimes_{\Bbbk} R) \otimes_R (R/R_+) \cong M$ as an A-module.

Let M be an A-module and let $\rho : A \to \text{End}_{\Bbbk}(M)$ be the original (undeformed) module structure.

PROPOSITION 3.9.3. (i) The space of Hochschild one-cocycles valued in $\text{End}_{\Bbbk}(M)$ is the space of infinitesimal deformations of the module M over $R = \Bbbk[\varepsilon]/(\varepsilon^2)$;

(ii) Two such deformations are equivalent up to an R-module automorphism of $M \otimes_{\Bbbk} R$ which is the identity modulo ε if and only if they differ by a Hochschild one-coboundary.

Thus $\text{HH}^1(A, \text{End}_{\Bbbk}(M)) \cong \text{Ext}_A^1(M, M)$ classifies infinitesimal deformations of M.

(iii) The obstruction to extending an infinitesimal deformation with class $\gamma \in \text{Ext}_A^1(M, M)$ to a second-order deformation, i.e., over $\Bbbk[\varepsilon]/(\varepsilon^3)$, is the element

$$\gamma \cup \gamma \in \text{Ext}_A^2(M, M) \cong \text{HH}^2(A, \text{End}_{\Bbbk}(M)),$$

where $\cup$ is the Yoneda cup product of extensions.

Here and below we make use of the cup product on $HH^\bullet(A, \mathrm{End}_{\Bbbk}(M)) = \mathrm{Ext}^\bullet_A(M, M)$ and on $HH^\bullet(A, A)$. We give an explicit description using the standard cochain complexes (3.E). More generally, let N be an A-algebra, i.e., an algebra equipped with an algebra homomorphism $A \to N$; this makes N also an A-bimodule. Then, if $\gamma_1 \in C^i(A, N)$ and $\gamma_2 \in C^j(A, N)$, set $\gamma_1 \cup \gamma_2 \in C^{i+j}(A, N)$ by the formula

$$(\gamma_1 \cup \gamma_2)(x_1, \ldots, x_{i+j}) = \gamma_1(x_1, \ldots, x_i)\gamma_2(x_{i+1}, \ldots, x_{i+j}),$$

and similarly define $\gamma_2 \cup \gamma_1$. Hence also $HH^\bullet(A, N)$ is an associative algebra. (Note that, while $HH^\bullet(A, A)$ is graded commutative as explained in Exercise 4.1.6 below, this is *not* true for $C^\bullet(A, N)$ in general, and in particular for $\mathrm{Ext}^\bullet(M, M)$, where $N = \mathrm{End}_{\Bbbk}(M)$.)

PROOF OF PROPOSITION 3.9.3. (i) Hochschild one-cocycles are precisely $\gamma \in \mathrm{Hom}_{\Bbbk}(A, \mathrm{End}_{\Bbbk}(M))$ such that $\gamma(ab) = a\gamma(b) + \gamma(a)b$, which are also known as A-bimodule derivations valued in $\mathrm{End}_{\Bbbk}(M)$. Infinitesimal deformations of the A-bimodule M are given by algebra homomorphisms $A \to \mathrm{End}_{\Bbbk}(M)[\varepsilon]/(\varepsilon^2)$ which reduce to the usual action $\rho : A \to \mathrm{End}_{\Bbbk}(M)$ modulo ε. Given a homomorphism $\rho + \varepsilon\phi$ of the latter type, we see that ϕ is an A-bimodule derivation valued in M, and conversely.

(ii) If we apply an automorphism $\phi = \mathrm{Id} + \varepsilon \cdot \phi_1$ of $M \otimes_{\Bbbk} R$, for $\phi_1 \in \mathrm{End}_{\Bbbk}(M)$, then the infinitesimal deformation γ is taken to γ', where

$$(\rho + \varepsilon\gamma')(a) = \phi \circ (\rho + \varepsilon\gamma)(a) \circ \phi^{-1}$$
$$= (\rho + \varepsilon\gamma)(a) + \varepsilon(\phi_1 \circ \rho(a) - \rho(a) \circ \phi_1) = (\rho + \varepsilon(\gamma + d\phi_1))(a).$$

This proves that $\gamma' - \gamma = d\phi_1$, as desired. The converse is similar and is left to the reader.

(iii) Working over $\tilde{R} := \Bbbk[\varepsilon]/(\varepsilon^3)$, given a Hochschild one-cocycle γ_1, and an arbitrary element $\gamma_2 \in C^1(A, \mathrm{End}_{\Bbbk}(M))$,

$$(\rho + \varepsilon\gamma_1 + \varepsilon^2\gamma_2)(ab) - (\rho + \varepsilon\gamma_1 + \varepsilon^2\gamma_2)(a)(\rho + \varepsilon\gamma_1 + \varepsilon^2\gamma_2)(b)$$
$$= \varepsilon^2\big(\gamma_2(ab) - \gamma_1(a)\gamma_1(b) - \gamma_2(a)\rho(b) - \rho(a)\gamma_2(b)\big),$$

and the last expression equals $-\varepsilon^2 \cdot (\gamma_1 \cup \gamma_1 + d\gamma_2)(ab)$. Thus the obstruction to extending the module structure is the class $[\gamma_1 \cup \gamma_1] \in HH^2(A, \mathrm{End}_{\Bbbk}(M))$.   □

Finally, we can study general formal deformations:

DEFINITION 3.9.4. Given an A-module M and a formal deformation A_R of A over $R = \Bbbk \oplus R_+$, a formal deformation of M to an A_R-module is an A_R-module structure on $M \hat{\otimes}_{\Bbbk} R$ whose tensor product over R with $R/R_+ = \Bbbk$ recovers M.

In the case that A_R is the trivial deformation over R, we also call this a formal deformation of the A-module M over R.

Recall in the above definition the notation $\hat{\otimes}$ from (2.K).

Analogously to the above, one can study (uni)versal formal deformations of M; when the space of obstructions, $\mathrm{Ext}^2_A(M, M)$, is zero, parallel to Theorem 2.7.7, there exists a versal formal deformation over the base $\hat{\mathcal{O}}(\mathrm{Ext}^1_A(M, M))$, and in the case that $\mathrm{End}_A(M, M) = \Bbbk$ (or more generally, the map $Z(A) = \mathrm{HH}^0(A, A) \to \mathrm{HH}^0(A, \mathrm{End}_{\Bbbk}(M)) = \mathrm{End}_A(M, M)$ is surjective), then this is universal.

More generally, consider a formal deformation $A_\hbar$ of A and ask not for a formal deformation $M_\hbar$ of M as an A-module, but rather for an $A_\hbar$-module $M_\hbar$ deforming M, i.e., satisfying $M_\hbar/\hbar M_\hbar \cong M$ as A-modules, and such that $M_\hbar \cong M[\![\hbar]\!]$ as $\Bbbk[\![\hbar]\!]$-modules. This recovers formal deformations of A-modules in the case $A_\hbar = A[\![\hbar]\!]$ is the trivial deformation. For general $A_\hbar$, however, $M_\hbar$ need not exist (as one no longer has the trivial deformation $M[\![\hbar]\!]$), so one can also ask when it does. In this generality, the calculations of Proposition 3.9.3 generalize to show that, if $\theta \in \hbar \cdot C^2(A)[\![\hbar]\!]$ gives a formal deformation $A_\hbar$ of A, then the condition for $\gamma \in \hbar \cdot C^1(A, \mathrm{End}_{\Bbbk} M)[\![\hbar]\!]$ to give a formal deformation $M_\hbar$ of M to a module over $A_\hbar$ is

$$(\rho + \gamma) \circ \theta + d\gamma + \gamma \cup \gamma = 0, \tag{3.G}$$

where here $(\gamma \cup \gamma)(a \otimes b) := \gamma(a)\gamma(b)$.

EXAMPLE 3.9.5. In the presence of a multiparameter formal deformation

$$(A[\![t_1, \ldots, t_n]\!], \star)$$

of A, this can be used to show the existence of a deformation $M_\hbar$ over some restriction of the parameter space. Let $U = \mathrm{Span}\{t_1, \ldots, t_n\}$ and let $\eta : U \to \mathrm{HH}^2(A)$ be the map which gives the class of infinitesimal deformation of A. We will need the composition $\rho \circ \eta : U \to \mathrm{HH}^2(A, \mathrm{End}_{\Bbbk} M) = \mathrm{Ext}^2_A(M, M)$. Then one can deduce from the above

PROPOSITION 3.9.6. (see, e.g., [103, Proposition 4.1]) Suppose that the map $\rho \circ \eta$ is surjective with kernel K. Then there exists a formal deformation $M_S := (M \hat{\otimes} \mathcal{O}(S), \rho_S)$ of M over a formal subscheme S of the formal neighborhood of the origin of U, with tangent space K at the origin, which is a module over $(A \hat{\otimes} \mathcal{O}(S), \star|_S)$. Moreover, if $\mathrm{Ext}^1_A(M, M) = 0$, then S is unique and M_S is unique up to continuous $\mathcal{O}(S)$-linear isomorphisms which are the identity modulo $\mathcal{O}(S)_+$.

Note that the condition $\mathrm{Ext}^1_A(M, M) = 0$ for uniqueness of the formal definition is consistent with the case where $A_\hbar = A[\![\hbar]\!]$ is the trivial deformation, since then, as above, $\mathrm{Ext}^1_A(M, M)$ classifies infinitesimal (and ultimately formal) deformations of M.

In [103], this was used to show the existence of a unique family of irreducible representations of wreath product Cherednik algebras $H_{1,(k,c)}(\Gamma^n \rtimes S_n)$ for $\Gamma < \mathrm{SL}_2(\mathbb{C})$ finite deforming a module of the form $Y^{\otimes n} \otimes V$ for Y an irreducible finite-dimensional representation of $H_{1,c_0}(\Gamma)$ and V a particular irreducible representation of S_n (whose Young diagram is a rectangle). Here $c_0 \in \mathrm{Fun}_\Gamma(\mathcal{R}, \mathbb{C})$ a conjugation-invariant function on the set $\mathcal{R} \subseteq \Gamma$ of symplectic reflections (here $\mathcal{R} = \Gamma \setminus \{\mathrm{Id}\}$), and $k \in \mathbb{C}$, and there is a unique formal subscheme $S \subseteq \mathbb{C} \times \mathrm{Fun}_\Gamma(\mathcal{R}, \mathbb{C})$ containing $(0, c_0)$ such that (k, c) is restricted to lie in S. Note here that $\mathbb{C} \times \mathrm{Fun}_\Gamma(\mathcal{R}, \mathbb{C})$ is viewed as $\mathrm{Fun}_{\Gamma^n \rtimes S_n}(\mathcal{R}', \mathbb{C})$ where $\mathcal{R}' \subseteq (\Gamma^n \rtimes S_n)$ is the set of symplectic reflections (they are all conjugate to reflections in $\Gamma = (\Gamma \times \{1\}^{n-1}) \rtimes \{1\}$ except for the conjugacy class of the transposition in S_n).

3.10. Additional exercises. (See also 3.7.4, 3.7.5, 3.7.11, and 3.8.2.)

EXERCISE 3.10.1. Prove that $U\mathfrak{g}$ is twisted Calabi–Yau, with $\mathrm{HH}^d(U\mathfrak{g}, U\mathfrak{g} \otimes U\mathfrak{g}) \cong U\mathfrak{g}^\sigma$, where $\sigma(x) = x - \mathrm{tr}(\mathrm{ad}(x))$. Therefore, it is Calabi–Yau if (and only if) $\mathfrak{g}$ is *unimodular*, i.e., $\mathrm{tr}(\mathrm{ad}(x)) = 0$ for all x. This will follow also from the Koszul resolutions from Exercise 2.8.1. Observe that every semisimple Lie algebra, i.e., one satisfying $\mathfrak{g} = [\mathfrak{g}, \mathfrak{g}]$, is unimodular (and hence the same is true for a reductive Lie algebra, i.e., one which is the direct sum of a semisimple and an abelian Lie algebra). The same is true for every nilpotent Lie algebra.

REMARK 3.10.2. More generally, one can consider, for any finite group Γ acting on the Lie algebra $\mathfrak{g}$ by automorphisms, the skew product algebra $U\mathfrak{g} \rtimes \Gamma$. For $\Bbbk$ of characteristic zero, generalizing the above (by tensoring the complexes with $\Bbbk[\Gamma]$ and suitably modifying the differentials) one can show that $U\mathfrak{g} \rtimes \Gamma$ is also twisted Calabi–Yau (cf. Remark 3.7.12). Then, [130] computes that this is Calabi–Yau if and only if $\Gamma < \mathsf{SL}(V)$ and $\mathfrak{g}$ is unimodular. (This extension should not be too surprising, since the skew-product algebra $\mathcal{O}(V) \rtimes \Gamma$ itself is Calabi–Yau if and only if $\Gamma < \mathsf{SL}(V)$, which is the condition for Γ to preserve the volume form giving the Calabi–Yau structure on V.)

EXERCISE 3.10.3. (*, but with many hints) In this exercise we compute the Hochschild (co)homology of a skew group ring.

Let A be an associative algebra over a field $\Bbbk$ of characteristic zero, and Γ a finite group acting on A by automorphisms. Form the algebra $A \rtimes \Gamma$, which as a vector space is $A \otimes \Bbbk[\Gamma]$, with the multiplication

$$(a_1 \otimes g_1)(a_2 \otimes g_2) = (a_1 g_1(a_2) \otimes g_1 g_2).$$

Next, given any Γ-module N, let $N^\Gamma := \{n \in N \mid g \cdot n = n \text{ for all } g \in \Gamma\}$ and $N_\Gamma := N/\{n - g \cdot n \mid n \in N, g \in \Gamma\}$ be the invariants and coinvariants, respectively.

(a) Let M be an $A \rtimes \Gamma$-bimodule. Prove that

$$\mathrm{HH}^\bullet(A \rtimes \Gamma, M) \cong \mathrm{HH}^\bullet(A, M)^\Gamma, \quad \mathrm{HH}_\bullet(A \rtimes \Gamma, M) \cong \mathrm{HH}_\bullet(A, M)_\Gamma,$$

where in the RHS, Γ acts on A and M via the adjoint action, $g \cdot_{\mathrm{Ad}} m = (gmg^{-1})$.
   Hint: Write the first one as $\mathrm{Ext}^\bullet_{A^e \rtimes (\Gamma \times \Gamma)}(A \rtimes \Gamma, M)$, using that $\Bbbk[\Gamma] \cong \Bbbk[\Gamma^{\mathrm{op}}]$ via the map $g \mapsto g^{-1}$. Notice that $A \rtimes \Gamma = \mathrm{Ind}^{A^e \rtimes (\Gamma \times \Gamma)}_{A^e \rtimes \Gamma_\Delta} A$, where $\Gamma_\Delta := \{(g, g) \mid g \in \Gamma\} \subseteq \Gamma \times \Gamma$ is the diagonal subgroup. Then, there is a general fact called Shapiro's lemma, for $H < K$ a subgroup,

$$\mathrm{Ext}^\bullet_{\Bbbk[K]}(\mathrm{Ind}^K_H M, N) \cong \mathrm{Ext}^\bullet_{\Bbbk[H]}(M, N).$$

Similarly, we have $\mathrm{Ext}^\bullet_{A \rtimes K}(\mathrm{Ind}^K_H M, N) \cong \mathrm{Ext}^\bullet_{A \rtimes H}(M, N)$. Using the latter isomorphism, show that

$$\mathrm{Ext}^\bullet_{A^e \rtimes (\Gamma \times \Gamma)}(A \rtimes \Gamma, M) \cong \mathrm{Ext}^\bullet_{A^e \rtimes \Gamma_\Delta}(A, M) = \mathrm{Ext}^\bullet_{A^e \rtimes \Gamma}(A, M^{\mathrm{Ad}}), \qquad (3.\mathrm{H})$$

where M^{Ad} means that Γ acts by the adjoint action from the $A \rtimes \Gamma$-bimodule structure. Since taking Γ-invariants is an exact functor (as $\Bbbk$ has characteristic zero and Γ is finite), this says that the RHS above is isomorphic to

$$\text{Ext}_{A^e}^{\bullet}(A, M^{\text{Ad}})^{\Gamma} = \text{HH}^{\bullet}(A, M)^{\Gamma}.$$

The proof for Hochschild homology is essentially the same, using Tor.

(b) Now we apply the formula in part (a) to the special case $M = A \rtimes \Gamma$ itself.

Let C be a set of representatives of the conjugacy classes of Γ: that is, $C \subseteq \Gamma$ and for every element $g \in \Gamma$, there exists a unique $h \in C$ such that g is conjugate to h. For $g \in \Gamma$, let $Z_g(\Gamma) < \Gamma$ be the centralizer of g, i.e., the collection of elements that commute with g. Prove that

$$\text{HH}^{\bullet}(A \rtimes \Gamma) \cong \bigoplus_{h \in C} \text{HH}^{\bullet}(A, A \cdot h)^{Z_h(\Gamma)},$$

Here, the bimodule action of A on $A \cdot h$ is by

$$a(b \cdot h) = ab \cdot h, \quad (b \cdot h)a = (bh(a)) \cdot h,$$

and $Z_h(\Gamma)$ acts by the adjoint action.

(c) Now specialize to the case that $A = \text{Sym } V$ and $\Gamma < \text{GL}(V)$. We will prove here that

$$\text{HH}^{\bullet}(A \rtimes G) \cong \bigoplus_{h \in C} \left(\left(\textstyle\bigwedge_{\text{Sym } V^h} \text{Vect}((V^h)^*) \right) \otimes \left(\textstyle\bigwedge^{\dim(V^h)^{\perp}} \langle \partial_{\phi} \rangle_{\phi \in (V^h)^{\perp}} \right) \right)^{Z_h(\Gamma)}.$$

The perpendicular space $(V^h)^{\perp}$ here is the subspace of V^* annihilating V^h. The degree $\bullet$ on the LHS is the total degree of polyvector field on the RHS, i.e., the sum of the degree in the first exterior algebra with $\dim(V^h)^{\perp} = \dim V - \dim V^h$.

The same argument shows that (for $\Omega^{\bullet}(X)$ denoting the algebraic differential forms on X),

$$\text{HH}_{\bullet}(A \rtimes G) \cong \bigoplus_{h \in C} \left(\Omega^{\bullet}((V^h)^*) \otimes \left(\textstyle\bigwedge^{\dim(V^h)^{\perp}} d(((V^*)^h)^{\perp}) \right) \right)_{Z_h(\Gamma)}.$$

Hints: first, up to conjugation, we can always assume h is diagonal (since Γ is finite). Suppose that $\lambda_1, \ldots, \lambda_n$ are the eigenvalues of h on the diagonal. Then let $h_1, \ldots, h_n \in \text{GL}_1$ be the one-by-one matrices $h_i = (\lambda_i)$. Show that

$$A = \Bbbk[x_1] \otimes \cdots \otimes \Bbbk[x_n], \quad A \cdot h = (\Bbbk[x_1] \cdot h_1) \otimes \cdots \otimes (\Bbbk[x_n] \cdot h_n).$$

Conclude using the Künneth formula, $\text{HH}^{\bullet}(A \otimes B, M \otimes N) = \text{HH}^{\bullet}(A, M) \otimes \text{HH}^{\bullet}(B, N)$ (for M an A-bimodule and N a B-bimodule, under suitable hypotheses on A, B, M, and N which hold here), that

$$\text{HH}^{\bullet}(A, A \cdot h)^{Z_h(\Gamma)} \cong \bigotimes_{i=1}^{n} \text{HH}^{\bullet}(\Bbbk[x_i], \Bbbk[x_i] \cdot h_i)^{Z_{h_i}(\Gamma)}. \tag{3.I}$$

Since $\Bbbk[x]$ has Hochschild dimension one (as it has a projective bimodule resolution of length one: see Remark 3.7.8), conclude that $\text{HH}^j(\Bbbk[x], \Bbbk[x] \cdot h) = 0$ unless

$j \leq 1$. Using the explicit description as center and outer derivations of the module, show that, if $h \in GL_1$ is not the identity,

$$HH^0(\Bbbk[x], \Bbbk[x] \cdot h) = 0, \quad HH^1(\Bbbk[x], \Bbbk[x] \cdot h) = \Bbbk.$$

Note for the second equality that you must remember to mod by inner derivations. On the other hand, recall that

$$HH^0(\Bbbk[x], \Bbbk[x]) = \Bbbk[x], \, HH^1(\Bbbk[x], \Bbbk[x]) = \Bbbk[x],$$

since $HH^\bullet(\Bbbk[x]) = \bigwedge^\bullet_{\Bbbk[x]} \mathrm{Der}(\Bbbk[x])$.

Now suppose in (3.I) that $h_i \neq \mathrm{Id}$ for $1 \leq i \leq j$, and that $h_i = \mathrm{Id}$ for $i > j$ (otherwise we can conjugate everything by a permutation matrix). Conclude that (3.I) implies

$$HH^\bullet(A, A \cdot h)^{Z_h(\Gamma)} \cong \left((\partial_{x_1} \wedge \cdots \wedge \partial_{x_j}) \otimes \bigwedge_{\mathrm{Sym}(V^h)} \mathrm{Der}(\mathrm{Sym}(V^h)) \right)^{Z_h(\Gamma)}. \quad (3.J)$$

Note that, without having to reorder the x_i, we could write

$$\partial_{x_1} \wedge \cdots \wedge \partial_{x_j} = \bigwedge^{\dim(V^h)^\perp} \langle \partial_\phi \rangle_{\phi \in (V^h)^\perp}.$$

Put together, we get the statement. A similar argument works for Hochschild homology.

(d) Use the same method to prove the main result of [1] for V symplectic and $\Gamma < Sp(V)$ finite:

$$HH^i(\mathrm{Weyl}(V) \rtimes \Gamma) \cong \Bbbk[S_i]^\Gamma, \quad HH_i(\mathrm{Weyl}(V) \rtimes \Gamma) \cong \Bbbk[S_{\dim V - i}]^\Gamma$$

where

$$S_i := \{ g \in \Gamma \mid \mathrm{rk}(g - \mathrm{Id}) = i \}.$$

Observe also that $S_i = \varnothing$ if i is odd, and $S_2 =$ the set of symplectic reflections (as defined in Example 2.7.8).

Hint: Apply the result of part (b) and the method of part (c). This reduces the result to the case $\dim V = 2$, and to computing $HH^\bullet(\mathrm{Weyl}_1, \mathrm{Weyl}_1)$ and $HH^\bullet(\mathrm{Weyl}_1, \mathrm{Weyl}_1 \cdot g)$ for nontrivial $g \in SL_2(\Bbbk)$. Then you can see from the Koszul complexes that the first is $\Bbbk$ (or, this can be deduced from Theorem 2.3.17 in the special case $X = \mathbb{A}^2$, or you can explicitly compute it using the Van den Bergh duality $HH^2(\mathrm{Weyl}_1, \mathrm{Weyl}_1) \cong HH_0(\mathrm{Weyl}_1, \mathrm{Weyl}_1)$ since Weyl_1 is Calabi–Yau). The second you can see must be $\Bbbk[-2]$ since this is already true for $HH^\bullet(\Bbbk[x, y], \Bbbk[x, y] \cdot g)$, and this surjects to $\mathrm{gr}\, HH^\bullet(\mathrm{Weyl}_1, \mathrm{Weyl}_1 \cdot g)$.

4. Dglas, the Maurer–Cartan formalism, and proof of formality theorems

Now the distinction between dg objects and ungraded objects becomes important (especially for the purpose of signs): we will recall in particular the notion of dg Lie algebras (dglas), which have homological grading, and hence parity (even or odd degree).

4.1. The Gerstenhaber bracket on Hochschild cochains. We turn first to a promised fundamental structure of Hochschild cochains: the Lie bracket, which is called its *Gerstenhaber bracket*:

DEFINITION 4.1.1. The *circle product* of Hochschild cochains $\gamma \in C^m(A)$, $\eta \in C^n(A)$ is the element $\gamma \circ \eta \in C^{m+n-1}(A)$ given by

$$\gamma \circ \eta(a_1 \otimes \cdots \otimes a_{m+n-1}) \qquad\qquad (4.A)$$

$$:= \sum_{i=1}^{m} (-1)^{(i-1)(n+1)} \gamma(a_1 \otimes \cdots \otimes a_{i-1} \otimes \eta(a_i \otimes \cdots \otimes a_{i+n-1}) \otimes a_{i+n} \otimes \cdots \otimes a_{m+n-1}).$$

DEFINITION 4.1.2. The *Gerstenhaber bracket* $[\gamma, \eta]$ of $\gamma \in C^m(A)$, $\eta \in C^n(A)$ is

$$[\gamma, \eta] := \gamma \circ \eta - (-1)^{(m+1)(n+1)} \eta \circ \gamma.$$

DEFINITION 4.1.3. Given a cochain complex C, let $C[m]$ denote the shifted complex, so $(C[m])^i = C^{i+m}$.

In other words, letting C^m denote the ordinary vector space obtained as the degree m part of $C^\bullet$, so C^m by definition is a graded vector space in degree zero, we have

$$C = \bigoplus_{m \in \mathbb{Z}} C^m[-m].$$

REMARK 4.1.4. The circle product also defines a natural structure on $\mathfrak{g} := C^\bullet(A)[1]$ viewed as a graded vector space with zero differential: that of a graded *right pre-Lie* algebra. This means that it satisfies the graded pre-Lie identity

$$\gamma \circ (\eta \circ \theta) - (\gamma \circ \eta) \circ \theta = (-1)^{|\theta||\eta|} \big(\gamma \circ (\theta \circ \eta) - (\gamma \circ \theta) \circ \eta \big).$$

Given any graded (right) pre-Lie algebra, the obtained bracket

$$[x, y] = x \circ y - (-1)^{|x||y|} y \circ x$$

defines a graded Lie algebra structure.

EXERCISE 4.1.5. (*) Verify the assertions of the remark! (The second assertion is easy, but the first is a long computation.)

EXERCISE 4.1.6. In fact, the circle product was originally defined by Gerstenhaber in order to prove that the cup product is graded commutative on cohomology. Prove the following identity of Gerstenhaber, for $\gamma_1, \gamma_2 \in C^\bullet(A)$:

$$\gamma_1 \cup \gamma_2 - (-1)^{|\gamma_1||\gamma_2|} \gamma_2 \cup \gamma_1 = d(\gamma_1 \circ \gamma_2) - ((d\gamma_1) \circ \gamma_2) - (-1)^{|\gamma_1|} (\gamma_1 \circ (d\gamma_2)).$$

Conclude from this identity that (a) the cup product is graded commutative, and (b) the Gerstenhaber bracket is compatible with the differential, i.e., it is a morphism of complexes $C^\bullet(A)[1] \otimes C^\bullet(A)[1] \to C^\bullet(A)[1]$.

The remark and exercises immediately imply

PROPOSITION 4.1.7. *The Gerstenhaber bracket defines a dg Lie algebra structure on the shifted complex* $\mathfrak{g} := C^\bullet(A)[1]$.

4.2. The Maurer–Cartan equation. We now come to the key description of formal deformations:

DEFINITION 4.2.1. Let $\mathfrak{g}$ be a dgla over a field of characteristic not equal to two. The Maurer–Cartan equation is

$$d\xi + \tfrac{1}{2}[\xi, \xi] = 0, \quad \xi \in \mathfrak{g}^1. \tag{4.B}$$

A solution of this equation is called a *Maurer–Cartan element*. Denote the space of solutions by $\mathrm{MCE}(\mathfrak{g})$.

The equation can be written suggestively as $[d + \xi, d + \xi] = 0$, if one defines $[d, d] = d^2 = 0$ and $[d, \xi] := d\xi$. In this form the equation is saying that the "connection" $d + \xi$ is flat:

EXAMPLE 4.2.2. Here is one of the original instances and motivation of the Maurer–Cartan equation. Let $\mathfrak{g}$ be a Lie algebra and X a manifold or affine algebraic variety X. Then we can consider the dg Lie algebra $(\Omega^\bullet(X, \mathfrak{g}), d) := (\Omega^\bullet(X) \otimes \mathfrak{g}, d)$, which is the de Rham complex of X tensored with the Lie algebra $\mathfrak{g}$. The grading is given by the de Rham grading, with $|\mathfrak{g}| = 0$. Then, associated to this is a notion of *connection*, defined as a formal expression $\nabla^\alpha := d + \alpha$, where $\alpha \in \Omega^1(X, \mathfrak{g})$; thus connections are in bijection with $\mathfrak{g}$-valued one forms. (We will explain below for the relationship with the standard notion of connections on principal bundles.) Associated to ∇^α is the endomorphism of $\Omega^\bullet(X, \mathfrak{g})$, given by $\beta \mapsto d\beta + [\alpha, \beta]$.

The curvature of ∇^α, denoted $(\nabla^\alpha)^2$ or $\tfrac{1}{2}[\nabla^\alpha, \nabla^\alpha]$, is formally defined as

$$(\nabla^\alpha)^2 = (d + \alpha)^2 = d\alpha + \tfrac{1}{2}[\alpha, \alpha]. \tag{4.C}$$

Then the Maurer–Cartan equation for α says that this is zero. This is clearly equivalent to the assertion that the corresponding endomorphism to ∇^α has square zero, i.e., it is a differential on $\Omega^\bullet(X, \mathfrak{g})$. In other words, Maurer–Cartan elements give *deformations of the differential* on the de Rham complex valued in $\mathfrak{g}$ (where α acts via the Lie bracket). In general, this is a good way to think about the Maurer–Cartan equation, as we will formalize following this example.

We explain the relationship with the standard terminology: If G is an algebraic or Lie group such that $\mathfrak{g} = \mathrm{Lie}\, G$, then ∇^α as above is equivalent to a connection on the trivial principal G-bundle on X. Precisely, the connection on $\pi : G \times X \to X$ associated to ∇^α is the one-form $\omega + \pi^*\alpha \in \Omega^1(G \times X, \mathfrak{g})$, with ω the canonical connection on the trivial bundle, and the curvature of $\omega + \pi^*\alpha$ is the pullback $\pi^*(d\alpha + \tfrac{1}{2}[\alpha, \alpha])$ of the curvature as defined above.

Closely related to Example 4.2.2 is the following very important observation.

PROPOSITION 4.2.3. *Suppose* $\xi \in \mathrm{MCE}(\mathfrak{g})$.

(i) *The map* $d^\xi : y \mapsto dy + [\xi, y]$ *defines a new differential on* $\mathfrak{g}$. *Moreover,* $(\mathfrak{g}, d^\xi, [-, -])$ *is also a dgla.*

(ii) *Maurer–Cartan elements of $\mathfrak{g}$ are in bijection with those of the twist of a Maurer–Cartan element $\mathfrak{g}^\xi$ by the correspondence*

$$\xi + \eta \in \mathfrak{g} \leftrightarrow \eta \in \mathfrak{g}^\xi.$$

DEFINITION 4.2.4. We call the dg Lie algebra $(\mathfrak{g}, d^\xi, [-, -])$ given by the above proposition the *twist by* ξ, and denote it by $\mathfrak{g}^\xi$.

PROOF OF PROPOSITION 4.2.3. (i) This is an explicit verification: $(d^\xi)^2(y) = [\xi, dy] + [\xi, [\xi, y]] + d[\xi, y] = [d\xi + \frac{1}{2}[\xi, \xi], y]$, and

$$d^\xi[x, y] - [d^\xi x, y] - (-1)^{|x|}[x, d^\xi y] = [\xi, [x, y]] - [[\xi, x], y] - (-1)^{|x|}[x, [\xi, y]] = 0,$$

where the first equality uses that d is a (graded) derivation for $[-, -]$, and the second equality uses the (graded) Jacobi identity for $[-, -]$.

(ii) One immediately sees that $d^\xi(\eta) + \frac{1}{2}[\eta, \eta] = d(\xi + \eta) + \frac{1}{2}[\xi + \eta, \xi + \eta]$, using that $d\xi + \frac{1}{2}[\xi, \xi] = 0$.   □

4.3. General deformations of algebras.

PROPOSITION 4.3.1. *One-parameter formal deformations $(A[[\hbar]], \star)$ of an associative algebra A are in bijection with Maurer–Cartan elements of the dgla $\mathfrak{g} := \hbar \cdot (C^\bullet(A)[1])[[\hbar]]$.*

PROOF. Let $\gamma := \sum_{m \geq 1} \hbar^m \gamma_m \in \mathfrak{g}^1$. Here $\gamma_m \in C^2(A)$ for all m, since $\mathfrak{g}$ is shifted. To $\gamma \in \mathfrak{g}^1$ we associate the star product $f \star g = fg + \sum_{m \geq 1} \hbar^m \gamma_m(f \otimes g)$. We need to show that $\star$ is associative if and only if γ satisfies the Maurer–Cartan equation. This follows from a direct computation (see Remark 4.3.2 for a more conceptual explanation):

$$f \star (g \star h) - (f \star g) \star h$$
$$= \sum_{m \geq 1} \hbar^m \cdot \left(f \gamma_m(g \otimes h) - \gamma_m(fg \otimes h) + \gamma_m(f \otimes gh) - \gamma_m(f \otimes g)h \right)$$
$$+ \sum_{m,n \geq 1} \hbar^{m+n} \left(\gamma_m(f \otimes \gamma_n(g \otimes h)) - \gamma_m(\gamma_n(f \otimes g) \otimes h) \right)$$
$$= d\gamma + \gamma \circ \gamma = d\gamma + \frac{1}{2}[\gamma, \gamma].   □ \qquad (4.D)$$

REMARK 4.3.2. For a more conceptual explanation of the proof, note that, if we let A_0 be an algebra with the zero multiplication, so that $C^\bullet(A_0)$ is a dgla with zero differential, then associative multiplications are the same as elements $\mu \in C^2(A_0) = \mathfrak{g}^1$ satisfying $\frac{1}{2}[\mu, \mu] = 0$, where $\mathfrak{g} = C^\bullet(A_0)[1]$ as before. (This is the Maurer–Cartan equation for $\mathfrak{g}$.) If we now take an arbitrary algebra A, we can set A_0 to be A but viewed as an algebra with the *zero* multiplication. Let $\mu \in C^2(A_0)$ represent the multiplication on A, hence $[\mu, \mu] = 0$ by associativity. Then, given $\gamma := \sum_{m \geq 1} \hbar^m \gamma_m \in \hbar \mathfrak{g}^1 = \hbar C^2(A)[[\hbar]]$, the product $\mu + \gamma$ is associative if and only if, working in $(C^\bullet(A_0)[1])[[\hbar]]$, we have

$$0 = [\mu + \gamma, \mu + \gamma] = [\mu, \mu] + 2[\mu, \gamma] + [\gamma, \gamma].$$

Now, $[\mu, \gamma] = d_A(\gamma)$, with d_A the (Hochschild) differential on $(C^\bullet(A)[1])[[\hbar]]$. More conceptually, this is saying that $C^\bullet(A)[1] = C^\bullet(A_0)[1]^\mu$, the twist by μ; cf. Proposition

4.2.3 and Definition 4.2.4, as well as Lemma 4.11.1 below. Then, by Proposition 4.2.3, Maurer–Cartan elements $\xi \in \mathrm{MCE}((C^{\bullet}(A)[1])[\![\hbar]\!])$ are the same as associative multiplications $\mu + \xi$. They are μ modulo $\hbar$ if and only if $\xi \in \hbar C^{\bullet}(A)[1][\![\hbar]\!]$.

REMARK 4.3.3. The above formalism works, with the same proof, for formal deformations over arbitrary complete augmented commutative $\Bbbk$-algebras. Namely, associative multiplications on $A \hat{\otimes}_{\Bbbk} R$ deforming the associative multiplication μ on A are the same as Maurer–Cartan elements of the dgla $C(A)[1] \hat{\otimes}_{\Bbbk} R_{+}$.

4.4. Gauge equivalence. Recall from Example 4.2.2 the example of flat connections with values in $\mathfrak{g} = \mathrm{Lie}\, G$ as solutions of the Maurer–Cartan equation. In that situation, one has a clear notion of equivalence of connections, namely gauge equivalence: for $\gamma : X \to G$ a map, and $\iota : G \to G$ the inversion map,

$$\nabla \mapsto (\mathrm{Ad}\,\gamma)(\nabla); \quad (d + \alpha) \mapsto d + (\mathrm{Ad}\,\gamma)(\alpha) + \gamma \cdot d(\iota \circ \gamma).$$

Here, the meaning of $\gamma \cdot d(\iota \circ \gamma)$ is as follows: the derivative $d(\iota \circ \gamma)$ is defined, at each $x \in X$, as a map $d(\iota \circ \gamma)|_x : T_x X \to T_{\gamma(x)^{-1}} G$, and then we apply the derivative of the left multiplication by $\gamma(x)$, $dL_{\gamma(x)} : T_{\gamma(x)^{-1}} G \to T_e G$, to obtain an operator $\gamma \cdot d(\iota \circ \gamma)|_x : T_x X \to T_e G = \mathfrak{g}$. We obtain in this way a one-form $\gamma \cdot d(\iota \circ \gamma) \in \Omega^1(X, \mathfrak{g})$. (We may think of $d + \gamma \cdot d(\iota \circ \gamma)$ formally as $\mathrm{Ad}(\gamma)(d)$; see also below for the case $\gamma = \exp(\beta)$.)

Now, restrict to the case $\gamma = \exp(\beta)$ for $\beta \in \mathcal{O}(X) \otimes \mathfrak{g}$, assuming that $\Bbbk = \mathbb{R}$ or $\mathbb{C}$ so exp is the usual exponential map (if we restrict to the case where G is connected, then such elements γ generate G, so generate all gauge equivalences). By taking a faithful representation, we may even assume without loss of generality that $G < \mathrm{GL}_n$ and $\mathfrak{g} < \mathfrak{gl}_n$, so γ and β are matrix-valued functions. We can then rewrite the above formula in a way not requiring G or the definition of $\gamma \cdot d(\iota \circ \gamma)$ as:

$$\alpha \mapsto \exp(\mathrm{ad}\,\beta)(\alpha) + \frac{1 - \exp(\mathrm{ad}\,\beta)}{\mathrm{ad}\,\beta}(d\beta), \tag{4.E}$$

where $(\mathrm{ad}\,\beta)(\alpha) := [\beta, \alpha]$, using the Lie bracket on $\mathfrak{g}$. The last term above can be thought of as $\exp(\mathrm{ad}\,\beta)(d) - d$, where we set $[d, \beta] = d(\beta)$, as explained in the following exercise:

EXERCISE 4.4.1. Verify (4.E). To do so, replace $\exp : \mathfrak{g} \to G$ by its Taylor series, and use the standard identity $\mathrm{Ad}(\exp(\beta)) = \exp(\mathrm{ad}\,\beta) := \sum_{m \geq 0} (\mathrm{ad}\,\beta)^m$, which holds formally (setting $\mathrm{Ad}(\exp(\beta))(f) = \exp(\beta) \cdot f \cdot \exp(-\beta)$ and $\mathrm{ad}(\beta)(f) = \beta \cdot f - f \cdot \beta$), and follows from the basic theory of Lie groups. Note also that, for $\alpha \in \mathcal{O}(X) \otimes \mathfrak{g}$ arbitrary, we have $([d, \beta])\alpha := d(\beta\alpha) - \beta d\alpha = (d\beta)\alpha$, so we can formally write $[d, \beta] = d(\beta)$ as above. Then, use all of this to expand and simplify $\gamma \cdot d(\iota \circ \gamma)$. Hint: write the latter, formally, as $\mathrm{Ad}(\exp(\beta))(d) - d$, then apply all of the above.

The above discussion motivates the following general definition, where now $\mathfrak{g}$ can be an arbitrary dgla (no longer a finite-dimensional Lie algebra as above). **From now until the end of Section 4, $\Bbbk$ should be a characteristic zero field.**

DEFINITION 4.4.2. Two Maurer–Cartan elements $\alpha, \alpha' \in \mathfrak{g}^1$ of a dgla are called gauge equivalent by an element $\beta \in \mathfrak{g}^0$ if $\alpha' = \exp(\mathrm{ad}\,\beta)(\alpha) + \frac{1 - \exp(\mathrm{ad}\,\beta)}{\beta}(d\beta)$, when this formula makes sense: for us, we take either (a) $\Bbbk = \mathbb{R}$ or $\mathbb{C}$ and $\mathfrak{g}$ is finite-dimensional as above; (b) $\mathfrak{g} = \mathfrak{h} \otimes_{\Bbbk} R_+$ with $\mathfrak{h}$ an arbitrary dgla and R a complete augmented (dg) commutative algebra; or (c) with R as in (b), we can also take $\mathfrak{g} = \mathfrak{h} \otimes_{\Bbbk} R$ and $\beta \in (R_+ \mathfrak{g})^0$.

This definition is consistent with Definition 2.7.1:

PROPOSITION 4.4.3. *Two formal deformations* $(A[[\hbar]], \star)$ *and* $(A[[\hbar]], \star')$ *are gauge equivalent, i.e., isomorphic via a continuous* $\Bbbk[[\hbar]]$-*linear automorphism of* $A[[\hbar]]$ *which is the identity modulo* $\hbar$, *if and only if the corresponding Maurer–Cartan elements of* $\mathfrak{g} = \hbar \cdot C(A)[[\hbar]]$ *are gauge equivalent.*

Here, the automorphism of $A[[\hbar]]$ does not respect the algebra structure on A: it is just a continuous $\Bbbk[[\hbar]]$-linear automorphism. Being the identity modulo $\hbar$ means that the automorphism Φ satisfies the property that $\Phi - \mathrm{Id}$ is a multiple of $\hbar$ as an endomorphism of the vector space $A[[\hbar]]$.

PROOF. This is an explicit verification: Let ϕ be a continuous automorphism of $A[[\hbar]]$ which is the identity modulo $\hbar$. We can write $\phi = \exp(\alpha)$ where $\alpha \in \hbar\,\mathrm{End}_{\Bbbk}(A)[[\hbar]]$; one can check that $\exp(\alpha) = 1 + \alpha + \alpha^2/2! + \cdots$ makes sense since we are using power series in $\hbar$. Let $\gamma, \gamma' \in \mathfrak{g}^1$ be the Maurer–Cartan elements corresponding to $\star$ and $\star'$. Let $\mu : A \otimes A \to A$ be the undeformed multiplication. Then

$$\exp(\alpha)(\exp(-\alpha)(a) \star \exp(-\alpha)(b))$$
$$= \exp(\mathrm{ad}\,\alpha)(\mu + \gamma)(a \otimes b)$$
$$= \Big(\mu + \exp(\mathrm{ad}\,\alpha)(\gamma) + \frac{1 - \exp(\mathrm{ad}\,\alpha)}{\mathrm{ad}\,\alpha}(d\alpha)\Big)(a \otimes b), \quad (4.\text{F})$$

where the final equality follows because $[\mu, \alpha] = d\alpha$. $\square$

4.5. The dgla of polyvector fields, Poisson deformations, and Gerstenhaber algebra structures.

Let X again be a smooth affine algebraic variety over a characteristic zero field or a C^∞ manifold. By the Hochschild–Kostant–Rosenberg theorem (Theorem 2.3.1), the Hochschild cohomology $\mathrm{HH}^\bullet(\mathcal{O}(X))$ is isomorphic to the algebra of polyvector fields, $\bigwedge_{\mathcal{O}(X)}^\bullet \mathrm{Vect}(X)$. Since, as we now know, $C^\bullet(\mathcal{O}(X))[1]$ is a dgla, one concludes that $\bigwedge_{\mathcal{O}(X)}^\bullet \mathrm{Vect}(X)[1]$ is also a dg Lie algebra (with zero differential). In fact, this structure coincides with the Schouten–Nijenhuis bracket:

PROPOSITION 4.5.1. *The Lie bracket on* $\bigwedge_{\mathcal{O}(X)}^\bullet \mathrm{Vect}(X)[1]$ *induced by the Gerstenhaber bracket is the Schouten–Nijenhuis bracket, as defined in Definition 2.3.7.*

Such a structure is called a *Gerstenhaber algebra*:

DEFINITION 4.5.2. A (dg) Gerstenhaber algebra is a dg commutative algebra B equipped with a dg Lie algebra structure on the shift $B[1]$, such that (2.I) is satisfied.

Note that, by definition, a Gerstenhaber algebra has to be (cohomologically) graded; sometimes when the adjective "dg" is omitted one means a dg Gerstenhaber algebra with zero differential. This is the case for $\bigwedge^\bullet_{\mathcal{O}(X)} \text{Vect}(X)$.

REMARK 4.5.3. Note that the definition of a Gerstenhaber algebra is very similar to that of a Poisson algebra: the difference is that the Lie bracket on a Gerstenhaber algebra is *odd*: it has homological degree -1.

We easily observe:

PROPOSITION 4.5.4. *A bivector field* $\pi \in \bigwedge^2 \text{Vect}(X)$ *defines a Poisson bracket if and only if* $[\pi, \pi] = 0$. *That is,* Poisson bivectors π are solutions of the Maurer–Cartan equation in $\bigwedge^\bullet_{\mathcal{O}(X)} \text{Vect}(X)[1]$.

EXERCISE 4.5.5. Prove Proposition 4.5.4!

The same proof implies:

COROLLARY 4.5.6. *Formal Poisson structures in* $\hbar \cdot \bigwedge^2_{\mathcal{O}(X)} \text{Vect}(X)[[\hbar]]$ *are the same as Maurer–Cartan elements of the dgla* $\hbar \cdot \left(\bigwedge^\bullet_{\mathcal{O}(X)} \text{Vect}(X)[1] \right)[[\hbar]]$.

4.6. Kontsevich's formality and quantization theorems. We can now make a precise statement of Kontsevich's formality theorem. As before, we need $\Bbbk$ to be a characteristic zero field for the remainder of Section 4.

REMARK 4.6.1. Kontsevich proved this result for $\mathbb{R}^n$ or smooth C^∞ manifolds; for the general smooth affine setting, when $\Bbbk$ contains $\mathbb{R}$, one can extract this result from [155]; for more details see [245], and also, e.g., [233]. These proofs also yield a sheaf-level version of the statement for the nonaffine algebraic setting. For a simpler proof in the affine algebraic setting, which works over arbitrary fields of characteristic zero, see [86]. We remark also that, recently in [85], Dolgushev showed that there actually exists a "correction" of Kontsevich's formulas which involve only rational weights, which replaces Kontsevich's proof by one that works over $\mathbb{Q}$.

The one parameter version of the theorem is this:

THEOREM 4.6.2 ([156, 155, 245, 86]). *There is a map*

Formal Poisson bivectors in $\hbar \cdot \bigwedge^2 \text{Vect}(X)[[\hbar]] \to$ *Formal deformations of* $\mathcal{O}(X)$

which induces a bijection modulo continuous automorphisms of $\mathcal{O}(X)[[\hbar]]$ *which are the identity modulo* $\hbar$, *and sends a formal Poisson structure* $\hbar\pi_\hbar$ *to a deformation quantization of the ordinary Poisson structure* $\pi \equiv \pi_\hbar \pmod{\hbar}$.

REMARK 4.6.3. By dividing the formal Poisson structure by $\hbar$, we also get a bijection modulo gauge equivalence from *all* formal Poisson structures to formal deformations, now sending $\pi_\hbar$ to a deformation quantization of π; the way it is stated above generalizes better to the full (multiparameter) version below.

We can state the full version of the theorem as follows:

THEOREM 4.6.4 ([156, 155, 245, 86]). *There is a map, functorial in dg commutative complete augmented* $\Bbbk$-*algebras* $R = \Bbbk \oplus R_+$,

$$\mathcal{U} : Poisson \ bivectors \ in \ \textstyle\bigwedge^2_{\mathcal{O}(X)} \mathrm{Vect}(X) \hat{\otimes}_\Bbbk R_+ \rightarrow Formal \ deformations \ (\mathcal{O}(X) \hat{\otimes}_\Bbbk R, \star)$$

which induces a bijection modulo continuous automorphisms of $\mathcal{O}(X) \hat{\otimes}_\Bbbk R$. *Moreover, modulo* R_+^2, *this reduces to the identity on bivectors valued in* R_+/R_+^2.

To explain what we mean by "the identity" in the end of the theorem, we note that, working modulo R_+^2, the Jacobi and associativity constraints become trivial. Similarly, formal deformations over R/R_+^2 are given by (not necessarily skew-symmetric) biderivations $\mathcal{O}(X) \otimes_\Bbbk \mathcal{O}(X) \rightarrow \mathcal{O}(X) \otimes R_+/R_+^2$. Just as in the case where $R = \Bbbk[\varepsilon]$, up to equivalence, these are given by their skew-symmetrization, a bivector $\bigwedge^2_{\mathcal{O}(X)} \mathrm{Vect}(X) \otimes R_+/R_+^2$. Thus, up to equivalence, both the domain and target reduce modulo R_+^2 to bivectors valued in R_+/R_+^2, and we can ask that the map reduce to the identity in this case.

4.7. Restatement in terms of morphisms of dglas.

We would like to restate the theorems above without using coefficients in R, just as a statement relating the two dglas in question. Let us name these: $T_{\mathrm{poly}} := \left(\bigwedge^{\bullet}_{\mathcal{O}(X)} \mathrm{Vect}(X) \right)[1]$ is the dgla of (shifted) polyvector fields on X, and $D_{\mathrm{poly}} := C^{\bullet}(\mathcal{O}(X))[1]$ is the dgla of (shifted) Hochschild cochains on X, which in the C^∞ setting are required to be differential operators.

These dglas are clearly not isomorphic on the nose, since T_{poly} has zero differential and not D_{poly}. They have isomorphic cohomology, by the Hochschild–Kostant–Rosenberg theorem. In this section we will explain how they are quasi-isomorphic, which is equivalent to the functorial equivalence of Theorem 4.6.4.

First, the Hochschild–Kostant–Rosenberg (HKR) theorem (Theorem 2.3.1) in fact gives a quasi-isomorphism of complexes $\mathrm{HKR} : T_{\mathrm{poly}} \rightarrow D_{\mathrm{poly}}$, defined by

$$\mathrm{HKR}(\xi_1 \wedge \cdots \wedge \xi_m)(f_1 \otimes \cdots \otimes f_m) = \frac{1}{m!} \sum_{\sigma \in S_m} \mathrm{sign}(\sigma) \xi_{\sigma(1)}(f_1) \cdots \xi_{\sigma(m)}(f_m).$$

This clearly sends $T_{\mathrm{poly}}^{m-1} = \bigwedge^m_{\mathcal{O}(X)} \mathrm{Vect}(X)$ to $D_{\mathrm{poly}}^{m-1} = C^m(\mathcal{O}(X), \mathcal{O}(X))$, since the target is an $\mathcal{O}(X)$-multilinear differential operator. Moreover, it is easy to see that the target is closed under the Hochschild differential. By the proof of the HKR theorem, one in fact sees that HKR is a quasi-isomorphism of complexes.

However, HKR is *not* a dgla morphism, since it does not preserve the Lie bracket. It does preserve it when restricted to vector fields, but already does not on bivector fields (which would be needed to apply it in order to take a Poisson bivector field and produce a star product). For example, $[\mathrm{HKR}(\xi_1 \wedge \xi_2), \mathrm{HKR}(\eta_1 \wedge \eta_2)]$, for vector fields ξ_1, ξ_2, η_1, and η_2, is not, in general, in the image of HKR: it is not skew-symmetric, as one can see by Definition 4.1.2.

The fundamental idea of Kontsevich was to correct this deficiency by adding higher order terms to HKR. The result will not be a morphism of dglas (this cannot be done), but it will be a more general type of morphism called an L_∞ morphism, which we introduce in the next subsection. The idea behind an L_∞ morphism is as follows: If we know that $\phi : \mathfrak{g} \to \mathfrak{h}$ has the property that $\phi[a, b] - [\phi(a), \phi(b)]$ is a boundary, say equal to dc, then we try to incorporate the data of the c into the morphism, by defining a map $\phi_2 : \mathfrak{g} \wedge \mathfrak{g} \to \mathfrak{h}$ sending $a \wedge b \to c$, and more generally such that $\phi[x, y] - [\phi(x), \phi(y)] = d\phi_2(x \wedge y)$ for all x, y. Then, we also need to define $\phi_3 : \wedge^3 \mathfrak{g} \to \mathfrak{h}$ as well, and so on. A full L_∞ morphism is then a sequence of linear maps $\phi_m : \wedge^m \mathfrak{g} \to \mathfrak{h}$ satisfying certain axioms.

Kontsevich therefore constructs an explicit sequence of linear maps

$$\mathcal{U}_m : \mathrm{Sym}^m(T_{\mathrm{poly}}[1]) \to D_{\mathrm{poly}}[1]$$

which satisfy these axioms, and hence yield an L_∞ morphism. Kontsevich constructs the $\mathcal{U}_m$ using graphs as in §2.4, except that now we must allow an arbitrary number of vertices on the real axis, not merely two (the number of vertices corresponds to two more than the degree of the target in $D_{\mathrm{poly}}[1]$), and the outgoing valence of vertices above the real axis can be arbitrary as well. As before, the vertices on the real axis are sinks. Note that $\mathcal{U}_1 = \mathsf{HKR}$ is just the sum of all graphs with a single vertex above the real axis, and all possible numbers of vertices on the real axis.

Then, if we plug in a formal Poisson bivector $\pi_\hbar$, we obtain the star product described in §2.4,

$$f \star g = \sum_{m \geq 1} \frac{1}{m!} \mathcal{U}_m(\pi_\hbar^m)(f \otimes g),$$

i.e., the star product is $\mathcal{U}(\exp(\pi_\hbar))$, where $\mathcal{U} = \sum_{m \geq 1} \mathcal{U}_m$.

4.8. L_∞ morphisms. One way to motivate L_∞ morphisms is to study what we require to obtain a functor on Maurer–Cartan elements. We will study this generally for two arbitrary dglas, $\mathfrak{g}$ and $\mathfrak{h}$.

Given two augmented algebras (A, A_+) and (B, B_+), an augmented algebra morphism is an algebra morphism $\phi : A \to B$ such that $\phi(A_+) \subseteq B_+$. We will always require our maps of augmented algebras be augmented algebra morphisms. The following then follows from definitions:

PROPOSITION 4.8.1. *Any dgla morphism* $F : \mathfrak{g} \to \mathfrak{h}$ *induces a functorial map in complete augmented dg commutative* $\Bbbk$-*algebras* $R = \Bbbk \oplus R_+$,

$$F : \mathsf{MCE}(\mathfrak{g} \hat{\otimes}_\Bbbk R_+) \to \mathsf{MCE}(\mathfrak{h} \hat{\otimes}_\Bbbk R_+).$$

However, it is not true that all functorial maps are obtained from dgla morphisms; in particular, if they were, then all functorial maps as above would define functorial maps if we replace R_+ by arbitrary (unital or nonunital) rings, by the remark below. But this is not true: with general coefficients the infinite sums in, e.g., §2.4 need not converge.

REMARK 4.8.2. In fact, dgla morphisms also induce functorial maps in ordinary (not necessarily complete augmented or even augmented) dg commutative $\Bbbk$-algebras R, taking the ordinary tensor product. However, the generalization to L_∞ morphisms below requires complete augmented $\Bbbk$-algebras.

It turns out that there is a complete augmented dg commutative $\Bbbk$-algebra B which represents the functor $R \mapsto \mathsf{MCE}(\mathfrak{g}\hat{\otimes}_{\Bbbk}R_+)$. This means that R-points of Spf B, i.e., continuous augmented dg algebra morphisms $B \to R$, are functorially in bijection with Maurer–Cartan elements of $\mathfrak{g}\hat{\otimes}_{\Bbbk}R$. To see what B is, first consider the case where $\mathfrak{g}$ is abelian with zero differential. If R is concentrated in degree zero, then Maurer–Cartan elements with coefficients in R_+ are elements of $\mathfrak{g}^1\hat{\otimes}R$. For any ungraded vector space V, the algebra of polynomial functions on it is $\mathsf{Sym}(V^*)$; so here one can consider $B = \mathsf{Sym}((\mathfrak{g}^1)^*)$, and then elements in $\mathfrak{g}^1$ are the same as continuous algebra homomorphisms $B \to \Bbbk$. Taking the completion $\hat{B}$ of B at the augmentation ideal (V^*), continuous augmented algebra homomorphisms $\hat{B} \to R$ are the same as elements of $\mathfrak{g}\hat{\otimes}R_+$. But if we take coefficients in a graded ring R, then we need to incorporate all of $\mathfrak{g}$, not just $\mathfrak{g}^1$. Observing that $\mathfrak{g}^1 = (\mathfrak{g}[1])^0$, the natural choice of graded algebra is $\mathsf{Sym}((\mathfrak{g}[1])^*)$. We wanted a completed algebra, so we take the completion $\hat{S}(\mathfrak{g}[1])^*$, which has the same continuous augmented maps to complete augmented rings R. Then, the points of $\hat{S}(\mathfrak{g}[1])^*$ valued in an ordinary (non-dg) complete augmented algebra R are in bijection with elements of $\mathfrak{g}^1\hat{\otimes}R$, as desired.

Thus, we consider the completed symmetric algebra $\hat{S}(\mathfrak{g}[1])^*$. In the case $\mathfrak{g}$ is nonabelian, we can account for the Lie bracket by deforming the differential on $\hat{S}(\mathfrak{g}[1])^*$, so that the spectrum consists of Maurer–Cartan elements rather than all of $H^1(\mathfrak{g})$.

The result is the *Chevalley–Eilenberg complex of* $\mathfrak{g}$, which you may already know as the complex computing the Lie algebra cohomology of $\mathfrak{g}$.

REMARK 4.8.3. We need to consider here $(\mathfrak{g}[1])^*$ as the *topological* dual to $\mathfrak{g}[1]$. Since $\mathfrak{g}$ is considered as discrete (hence $\mathfrak{g}^i = \lim_{\substack{\to \\ V \subseteq \mathfrak{g}^i \text{ f. d.}}} V$ is an inductive limit of its finite-dimensional subspaces), the dual $\mathfrak{g}^*$ is the topological dg vector space $\bigoplus_{m\in\mathbb{Z}}(\mathfrak{g}^m)^*[m]$, where each $(\mathfrak{g}^m)^*$ is equipped with a not-necessarily discrete topology, given by the *inverse* limit of the finite-dimensional *quotients* $(\mathfrak{g}^m)^* \twoheadrightarrow V^*$, which are the duals of the finite-dimensional subspaces $V \subseteq \mathfrak{g}^m$:

$$(\mathfrak{g}^m)^* := \lim_{\substack{\leftarrow \\ (\mathfrak{g}^m)^* \twoheadrightarrow V^* \text{ f.d.}}} V^*.$$

This is an inverse limit of finite-dimensional vector spaces.[7] It is discrete if and only if V is finite-dimensional.

DEFINITION 4.8.4. The Chevalley–Eilenberg complex is the complete dg commutative algebra $C_{CE}(\mathfrak{g}) := (\hat{S}(\mathfrak{g}[1])^*, d_{CE})$, where d is the derivation such that $-d_{CE}(x) = d_{\mathfrak{g}}^*(x) + \frac{1}{2}\delta_{\mathfrak{g}}(x)$, where the degree one map $\delta_{\mathfrak{g}} : \mathfrak{g}[1]^* \to \mathsf{Sym}^2 \mathfrak{g}[1]^*$ is the dual of the Lie bracket $\bigwedge^2\mathfrak{g} \to \mathfrak{g}$.

PROPOSITION 4.8.5. *Let* $R = \Bbbk \oplus R_+$ *be a dg commutative complete augmented* $\Bbbk$-*algebra. Then there is a canonical bijection between continuous dg commutative*

---

[7]Such a vector space is sometimes a *pseudocompact vector space*. Another term that appears in some literature is *formal vector space*, not to be confused, however, with the notion of formality we are discussing as in Kontsevich's theorem!

augmented algebra morphisms $C_{CE}(\mathfrak{g}) \to R$ *and Maurer–Cartan elements of* $\mathfrak{g}\hat{\otimes}R$, *given by restricting to* $\mathfrak{g}[1]^*$.

PROOF. It is clear that, if we do not consider the differential, continuous algebra homomorphisms $C_{CE}(\mathfrak{g}) \to R$ are in bijection with continuous graded maps χ : $\mathfrak{g}[1]^* \to R_+$. Such elements, because of the continuity requirement, are the same as elements $x_\chi \in \mathfrak{g}[1]\hat{\otimes}R_+$. Then, χ commutes with the differential if and only if $\chi \circ d_\mathfrak{g}^* + \frac{1}{2}(\chi \otimes \chi) \circ \delta_\mathfrak{g} + d \circ \chi = 0$, i.e., if and only if $d(x_\chi) + \frac{1}{2}[x_\chi, x_\chi] = 0$. ☐

COROLLARY 4.8.6. *There is a canonical bijection*

{Functorial in R maps $F : MCE(\mathfrak{g}\hat{\otimes}R_+) \to \mathrm{MCE}(\mathfrak{h}\hat{\otimes}R_+)\}$

$\leftrightarrow$ *{continuous dg commutative augmented morphisms* $F^* : C_{CE}(\mathfrak{h}) \to C_{CE}(\mathfrak{g})\}$, (4.G)

where R ranges over dg commutative complete augmented $\Bbbk$*-algebras.*

PROOF. This is a Yoneda type result: given a continuous dg commutative augmented morphism $C_{CE}(\mathfrak{h}) \to C_{CE}(\mathfrak{g})$, the pullback defines a map $MCE(\mathfrak{g}\hat{\otimes}R_+) \to \mathrm{MCE}(\mathfrak{h}\hat{\otimes}R_+)$ for every R as described, which is functorial in R. Conversely, given the functorial map $\mathrm{MCE}(\mathfrak{g}\hat{\otimes}R_+) \to \mathrm{MCE}(\mathfrak{h}\hat{\otimes}R_+)$, we apply it to $R = C_{CE}(\mathfrak{g})$ itself. Then, by Proposition 4.8.5, the identity map $C_{CE}(\mathfrak{g}) \to R$ yields a Maurer–Cartan element $I \in \mathfrak{g}\hat{\otimes}C_{CE}(\mathfrak{g})_+$ (the "universal" Maurer–Cartan element). Its image in $\mathrm{MCE}(\mathfrak{h}\hat{\otimes}C_{CE}(\mathfrak{g})_+)$ yields, by Proposition 4.8.5, a continuous dg commutative augmented morphism $C_{CE}(\mathfrak{h}) \to C_{CE}(\mathfrak{g})$. It is straightforward to check that these maps are inverse to each other. ☐

DEFINITION 4.8.7. An L_∞ morphism $F : \mathfrak{g} \to \mathfrak{h}$ is a continuous dg commutative augmented morphism $F^* : C_{CE}(\mathfrak{h}) \to C_{CE}(\mathfrak{g})$.

REMARK 4.8.8. If we remove the requirement "augmented," then one obtains so-called *curved* L_∞ morphisms.

EXERCISE 4.8.9. Show that a dgla morphism is an L_∞ morphism. More precisely, show that a dgla morphism $\mathfrak{g} \to \mathfrak{h}$ induces a canonical continuous dg commutative morphism $C_{CE}(\mathfrak{h}) \to C_{CE}(\mathfrak{g})$ (note that one can define a canonical linear map owing to the dual in the definition of C_{CE} (Definition 4.8.4); you need to show it is actually a morphism of dg commutative algebras).

We will refer to F^* as the *pullback* of F. Thus, Corollary 4.8.6 can be alternatively stated as

{Functorial in R maps $F : \mathrm{MCE}(\mathfrak{g}\hat{\otimes}R_+) \to \mathrm{MCE}(\mathfrak{h}\hat{\otimes}R_+)\}$

$\leftrightarrow \{L_\infty$ morphisms $F : \mathfrak{g} \to \mathfrak{h}\}$.   (4.H)

Finally, the above extends to describe quasi-isomorphisms:

DEFINITION 4.8.10. An L_∞ quasi-isomorphism is a L_∞ morphism which is an isomorphism on homology, i.e., a dg commutative augmented quasi-isomorphism $F^* : C_{CE}(\mathfrak{h}) \to C_{CE}(\mathfrak{g})$.

This can also be called a homotopy equivalence of dglas.

THEOREM 4.8.11. *If F is a quasi-isomorphism, then the above functorial map is a bijection on gauge equivalence classes.*

The theorem is part of [156, Theorem 4.6], but is older and considered standard. (For example, to see that quasi-isomorphisms admit quasi-inverses, see [132]; this is the dg version of the statement that a map of formal neighborhoods of the origin of two vector spaces is an isomorphism if and only if it is an isomorphism on tangent spaces (the formal inverse function theorem). Using this, the statement reduces to showing that a quasi-isomorphism which is the identity on cohomology is also the identity on gauge equivalence classes.)

REMARK 4.8.12. In fact, the above theorem can be significantly strengthened: the Maurer–Cartan set $MCE(\mathfrak{g}\hat\otimes R_+)$ is not just a set with gauge equivalences, but in fact a simplicial complex. The statement that $MCE(\mathfrak{g}\hat\otimes R_+) \to MCE(\mathfrak{h}\hat\otimes R_+)$ is a bijection on gauge equivalences is the same as the statement that it induces an isomorphism on π_0. In fact, a quasi-isomorphism $\mathfrak{g} \to \mathfrak{h}$ induces a homotopy equivalence $MCE(\mathfrak{g}\hat\otimes R_+) \simeq MCE(\mathfrak{h}\hat\otimes R_+)$ [110, Proposition 4.9], which is stronger.

REMARK 4.8.13. In fact, everything we have discussed above is an instance of Koszul duality: the dgla $\mathfrak{g}$ is (derived) Koszul dual to its Chevalley–Eilenberg complex $C_{CE}(\mathfrak{g})$, a dg commutative algebra; in general, if A and B are algebras of any type (e.g., algebras over an operad) and $A^!$ is the (derived) Koszul dual of A (an algebra of the Koszul dual type, e.g., an algebra over the Koszul dual operad), then Proposition 4.8.5 generalizes to the statement: Homotopy (i.e., infinity) morphisms $A \to B$ identify with Maurer–Cartan elements in $A^!\hat\otimes B$ (note that $A^!\hat\otimes B$, properly defined, always has a dgla or at least an L_∞-algebra structure). Similarly, Corollary 4.8.6 and (4.H) generalize to: Homotopy morphisms $A \to B$ are in bijection with functorial maps $MCE(A\hat\otimes C) \to MCE(B\hat\otimes C)$ where C is an algebra of the Koszul dual type. (Note that, taking $C = A^!$, the element in $MCE(B\hat\otimes A^!)$ corresponding to the original morphism is the image of the canonical element in $MCE(A\hat\otimes A^!)$ corresponding to the identity.)

4.9. Explicit definition of L_∞ morphisms. Let us write out explicitly what it means to be an L_∞ morphism. Let $\mathfrak{g}$ and $\mathfrak{h}$ be dglas. Then an L_∞ morphism is a continuous commutative dg algebra morphism $F^* : C_{CE}(\mathfrak{h}) \to C_{CE}(\mathfrak{g})$. Since $C_{CE}(\mathfrak{h})$ is the symmetric algebra on $\mathfrak{h}[1]^*$, this map is uniquely determined by its restriction to $\mathfrak{h}[1]^*$. We then obtain a sequence of maps

$$F_m^* : \mathfrak{h}[1]^* \to \mathrm{Sym}^m \mathfrak{g}[1]^*,$$

or dually,

$$F_m : \mathrm{Sym}^m \mathfrak{g}[1] \to \mathfrak{h}[1].$$

The F_m^* are the Taylor coefficients of F^*, since they are the parts of F^* of polynomial degree m, i.e., the order-m Taylor coefficients of the map on Maurer–Cartan elements.

The condition that F^* commute with the differential says that

$$F^*(d_{\mathfrak{h}}^*(x)) + F^*(\tfrac{1}{2}\delta_{\mathfrak{h}}(x)) = d_{\mathfrak{g}}^* F^*(x) + \tfrac{1}{2}\delta_{\mathfrak{g}} F^*(x).$$

Writing this in terms of the F_m, we obtain that, for all $m \geq 1$,

$$d_{\mathfrak{h}} \circ F_m + \frac{1}{2} \sum_{i+j=m} [F_i, F_j]_{\mathfrak{h}} = F_m \circ d_{CE,\mathfrak{g}}, \tag{4.I}$$

where $d_{CE,\mathfrak{g}}$ is the Chevalley–Eilenberg differential for $\mathfrak{g}$.

Now, let us specialize to $\mathfrak{g} = T_{\text{poly}}$ and $\mathfrak{h} = D_{\text{poly}}$, with $F_m = \mathcal{U}_m$ for all $m \geq 1$. In terms of Kontsevich's graphs, the second term of (4.I) involves a sum over all ways of combining two graphs together by a single edge to get a larger graph, multiplying the weights for those graphs. The last term (the RHS) of (4.I) involves summing over all ways of expanding a graph by adding a single edge. The first term on the LHS says to apply the Hochschild differential to the result of all graphs, and this can be suppressed in exchange for adding to Kontsevich's map $\mathcal{U}$ a term $\mathcal{U}_0 : \mathbb{R} = \text{Sym}^0(T_{\text{poly}}[1]) \to D_{\text{poly}}[1]$ which sends $1 \in \mathbb{R}$ to $\mu_A \in C^2(A)$. (Recall that $\Bbbk = \mathbb{R}$ for Kontsevich's construction; this is needed to define the weights of the graphs.)

4.10. Formality in terms of a L_∞ quasi-isomorphism. We deduce from the preceding material that the formality theorem, Theorem 4.6.4, can be restated as

THEOREM 4.10.1 ([156, 155, 245, 86]). *There is an L_∞ quasi-isomorphism,*

$$T_{\text{poly}}(X) \xrightarrow{\sim} D_{\text{poly}}(X).$$

The proof is accomplished for $X = \mathbb{R}^n$ in [156] by finding weights w_Γ to attach to all of the graphs Γ described above, so that the explicit equations of the preceding section are satisfied. These explicit equations are quadratic in the weights, and are of the form, for certain graphs Γ, denoting by $|\Gamma|$ the number of edges of Γ,

$$\sum_{|\Gamma_1|+|\Gamma_2|=|\Gamma|} c(\Gamma, \Gamma_1, \Gamma_2) w_{\Gamma_1} w_{\Gamma_2} + \sum_{|\Gamma'|=|\Gamma|-1} c(\Gamma, \Gamma') w_{\Gamma'} = 0,$$

for suitable coefficients $c(\Gamma, \Gamma_1, \Gamma_2)$ and $c'(\Gamma, \Gamma')$. In fact, the graphs that appear are all of the following form: in the first summation, $\Gamma_1 \subseteq \Gamma$ is a subgraph which is incident to the real line, and $\Gamma_2 = \Gamma/\Gamma_1$ is obtained by contracting Γ_1 to a point on the real line. In the second summation, Γ' is obtained by contracting a single edge in Γ. The coefficient $c(\Gamma, \Gamma_1, \Gamma_2)$ is a signed sum of the ways of realizing $\Gamma_1 \subseteq \Gamma$ such that $\Gamma_2 = \Gamma/\Gamma_1$, and the coefficient $c(\Gamma, \Gamma')$ is a signed sum over edges e in Γ such that $\Gamma/e = \Gamma'$.

To find weights w_Γ satisfying these equations, Kontsevich defines the w_Γ as certain integrals over partially compactified configuration spaces of vertices of the graph, such that the above sum follows from Stokes' theorem for the configuration space C_Γ associated to Γ, whose boundary strata are of the form $C_{\Gamma_1 \times \Gamma_2}$ or $C_{\Gamma'}$ (the technical part of the resulting proof involves showing that the terms from Stokes' theorem not appearing in the desired identity vanish).

REMARK 4.10.2. In fact, T_{poly} and D_{poly} are quasi-isomorphic not merely as L_∞ algebras but in fact as homotopy Gerstenhaber ("G_∞") algebras, i.e., including the structure of cup product (and additional "brace algebra" structures in the case of D_{poly}). Although, by Theorem 2.3.1, the HKR morphism is compatible with cup product on cohomology,[8] the preceding statement is much stronger than this. In [86] the main result actually constructs a homotopy Gerstenhaber equivalence between $T_{\text{poly}}(X)$ and $D_{\text{poly}}(X)$, in the general smooth affine algebraic setting over characteristic zero; in [242], Willwacher completes Kontsevich's L_∞ quasi-isomorphism to a homotopy Gerstenhaber quasi-isomorphism (which requires adding additional Taylor series terms). In fact, in [242], Willwacher shows that Kontsevich's morphism lifts to a "KS_∞ quasi-isomorphism" of pairs

$$(\text{HH}^\bullet(\mathcal{O}(X)), \text{HH}_\bullet(\mathcal{O}(X))) \overset{\sim}{\to} (C^\bullet(\mathcal{O}(X)), C_\bullet(\mathcal{O}(X))),$$

where we equip the Hochschild homology with the natural operations by the contraction and Lie derivative operations from Hochschild cohomology (i.e., the calculus structure), and similarly equip Hochschild chains with the analogous natural operations by Hochschild cochains.

4.11. Twisting the L_∞-morphism; Poisson and Hochschild cohomology. Given a formal Poisson structure $\pi_\hbar \in \text{MCE}(\hbar \cdot T_{\text{poly}}(X)[[\hbar]])$ and its image star product $\star$, corresponding to the element $\mathcal{U}(\pi_\hbar) \in \text{MCE}(\hbar \cdot D_{\text{poly}}(X)[[\hbar]])$, we obtain a quasi-isomorphism of twisted dglas,

$$T_{\text{poly}}(X)((\hbar))^{\pi_\hbar} \overset{\sim}{\to} D_{\text{poly}}(X)((\hbar))^{\mathcal{U}(\pi_\hbar)}. \tag{4.J}$$

This has an important meaning in terms of Poisson and Hochschild cohomology. Namely, the RHS computes the Hochschild cohomology of the algebra $(\mathcal{O}(X)((\hbar)), \mathcal{U}(\pi_\hbar))$, which follows from the following result (cf. Remark 4.3.2):

LEMMA 4.11.1. *Let A_0 be a vector space, viewed as an algebra with the zero multiplication, and let $A = (A_0, \mu)$ be an associative algebra with multiplication map $\mu \in C^2(A_0)$. Then $C^\bullet(A) = C^\bullet(A_0)^\mu$.*

The proof follows, as in Remark 4.3.2, because $C^\bullet(A)$ and $C^\bullet(A_0)$ have the same underlying graded vector space; the differential on $C^\bullet(A_0)$ is zero, and one checks that the differential on $C^\bullet(A)$ is the operation $[\mu, -]$ of taking the Gerstenhaber bracket with μ.

COROLLARY 4.11.2. *If $\mu, \mu' \in C^2(A_0)$ are two different associative multiplications on A_0, then setting $A = (A_0, \mu)$ and $A' = (A_0, \mu')$, we have $C^2(A') = C^2(A)^{\mu'-\mu}$.*

Thus, applying the corollary to the situation where $A_0 = \mathcal{O}(X)[[\hbar]]$, with μ the undeformed multiplication and μ' the one corresponding to $\star$ (i.e., $\mu + \mathcal{U}(\pi_\hbar)$), we obtain the promised

---

[8]The analogous sheaf-theoretic statement for nonaffine smooth varieties is no longer true: see, e.g., [59, 58].

COROLLARY 4.11.3. $\text{HH}^*(\mathcal{O}(X)[[\hbar]], \star) \cong H^*(D_{\text{poly}}(X)((\hbar))^{\mathcal{U}(\pi_\hbar)})$.

Similarly, we can interpret the first term as the Poisson cohomology of the Poisson algebra $(\mathcal{O}(X)((\hbar)), \pi_\hbar)$:

DEFINITION 4.11.4 ([55]). Let X be a smooth affine variety with a Poisson bivector π. Then, the Poisson cohomology of X is the cohomology of the dgla $T_{\text{poly}}(X)^\pi$.

REMARK 4.11.5. The definition of Poisson cohomology in the general nonsmooth affine context is quite different, and is expressed as the cohomology of a canonical differential on the Lie algebra of derivations of the free Poisson algebra generated by $\mathcal{O}(X)^*[1]$, taking the topological dual as in Remark 4.8.3. Note that this is completely analogous to the definition of Hochschild cohomology, which can also be defined as the cohomology of a canonical differential on the Lie algebra of derivations of the free associative algebra generated by $A^*[1]$ (in general, the analogous cohomology for an algebra A of any type is given as the cohomology of a free algebra of the Koszul dual type generated by $A^*[1]$ with a canonical differential, cf. Remark 4.8.13). In the smooth affine case, one can show this coincides with the above definition.

COROLLARY 4.11.6. Working over $\Bbbk((\hbar))$, the Poisson cohomology of $(\mathcal{O}(X)((\hbar)), \pi_\hbar)$ is $H^*(T_{\text{poly}}(X)((\hbar))^{\pi_\hbar})$.

We conclude:

COROLLARY 4.11.7. There is an isomorphism of graded vector spaces,

$$\text{HP}^\bullet(\mathcal{O}(X)((\hbar)), \pi_\hbar) \xrightarrow{\sim} \text{HH}^\bullet(\mathcal{O}(X)((\hbar)), \star). \tag{4.K}$$

In particular, applied to degrees 0, 1, and 2, there are canonical $\Bbbk((\hbar))$-linear isomorphisms:

(i) from the Poisson center of $(\mathcal{O}(X)((\hbar)), \pi_\hbar)$ to the center of $(\mathcal{O}(X)((\hbar)), \star)$;

(ii) from the outer derivations of $(\mathcal{O}(X)((\hbar)), \pi_\hbar)$ to those of $(\mathcal{O}(X)((\hbar)), \star)$;

(iii) from infinitesimal Poisson deformations of $(\mathcal{O}(X)((\hbar)), \star)$ to infinitesimal algebra deformations of $(\mathcal{O}(X)((\hbar)), \star)$.

REMARK 4.11.8. In fact, the above isomorphism (4.K) is an isomorphism of $\Bbbk((\hbar))$-algebras, and hence also the morphism (i) is an isomorphism of algebras, and the morphisms (ii)–(iii) are compatible with the module structures over the Poisson center on the LHS and the center of the quantization on the RHS. This is highly nontrivial and was proved by Kontsevich in [156, §8]. More conceptually, the reason why this holds is that Kontsevich's L_∞ quasi-isomorphism lifts to a homotopy Gerstenhaber isomorphism, as we pointed out in Remark 4.10.2.

We can now sketch a proof of Theorem 2.3.17. Suppose that $X = (X, \omega)$ is an symplectic affine variety or C^∞ manifold equipped with its canonical Poisson bracket. Recall that the symplectic condition means that ω is a closed two-form such that the map $\xi \mapsto i_\xi(\omega)$ defines an isomorphism $\omega^\sharp : \text{Vect}(X) \to \Omega^1(X)$ between vector fields and one-forms. The corresponding Poisson structure is given by $(\omega^\sharp)^{-1} : \Omega^1(X) \to \text{Vect}(X)$, namely, $(\omega^\sharp)^{-1} = \pi^\sharp : \alpha \mapsto i_\pi \alpha$. For ease of notation, we will write $\pi = \omega^{-1}$.

Suppose that X is symplectic as above. Then, [55] shows that the Poisson cohomology $\text{HP}^\bullet(\mathcal{O}(X))$ is isomorphic to the de Rham cohomology of X. Now, let $\pi_\hbar := \hbar\pi$ be the obtained Poisson structure on $\mathcal{O}(X)((\hbar))$; this corresponds to the symplectic structure $\omega_\hbar := \hbar^{-1}\omega$. Then, working over $\Bbbk((\hbar))$, [55] implies the first statement of the theorem.

For the second statement, we first caution that Theorem 2.7.7 *cannot* be applied in general, since $\text{HH}^3(\mathcal{O}(X)((\hbar)), \star) \cong H^3_{DR}(X)((\hbar))$, which need not be zero for general X. Nonetheless, a versal family can be explicitly constructed: by (4.J), it suffices to construct a versal family of deformations of the Poisson structure on $(\mathcal{O}(X)((\hbar)), \pi_\hbar)$. Let $\alpha_1, \ldots, \alpha_k$ be closed two-forms on X which map to a basis of $H^2_{DR}(X)$. Then we obtain the versal family of symplectic structures $\hbar^{-1}(\omega + c_1\alpha_1 + \cdots + c_k\alpha_k)$ over $\Bbbk((\hbar))[[c_1, \ldots, c_k]]$, and similarly the versal family $\hbar(\omega + c_1\alpha_1 + \cdots + c_k\alpha_k)^{-1}$ of Poisson deformations by inverting the elements of this family.

4.12. Explicit twisting of L_∞ morphisms. We caution that, unlike in the untwisted case, (4.K) is **not** obtained merely from the HKR morphism. Indeed, even in degree zero, an element which is Poisson central for $\pi_\hbar$ need not correspond in any obvious way to an element which is central in the quantized algebra. The only obvious statement one could make is that we have a map modulo $\hbar$, where $\pi_\hbar = \hbar \cdot \pi + $ higher order terms, with π an ordinary Poisson structure,

$$Z(\mathcal{O}(X), \pi) \leftarrow Z(\mathcal{O}(X)[[\hbar]], \star)/(\hbar)$$

which is quite different (and weaker) than the above statement.

To write the correct formula for the isomorphism (4.K), we need to discuss functoriality for twisting. Namely, given an L_∞ morphism $F : \mathfrak{g} \to \mathfrak{h}$ and a Maurer–Cartan element $\xi \in \text{MCE}(\mathfrak{g})$, we need to define an L_∞ morphism $F^\xi : \mathfrak{g}^\xi \to \mathfrak{h}^{F(\xi)}$. Given this, we obtain the twisted L_∞ quasi-isomorphism (4.J), then pass to cohomology to obtain (4.K).

For Kontsevich's morphism $\mathcal{U}$, which is explicitly defined by $\mathcal{U}^*$, this produces the HKR isomorphism $H^\bullet(T_{\text{poly}}(X)) = \wedge^{\bullet+1}_{\mathcal{O}(X)} \text{Vect}(X) \to H^\bullet(D_{\text{poly}}(X)) \cong \text{HH}^{\bullet+1}(\mathcal{O}(X))$. However, to apply this to the twisted versions $\mathcal{U}^{\pi_\hbar} : T_{\text{poly}}(X)^{\pi_\hbar} \to D_{\text{poly}}(X)^{\mathcal{U}(\pi_\hbar)}$, we need an explicit formula for the pullback $(F^\xi)^*$. This is nontrivial by the observation at the beginning of the subsection—on cohomology one does *not* obtain the HKR morphism.

By Proposition 4.2.3.(ii), there is a canonical map on Maurer–Cartan elements $F^\xi : \text{MCE}(\mathfrak{g}) \to \text{MCE}(\mathfrak{h})$:

DEFINITION 4.12.1. Set $F^\xi(\eta) := F(\eta + \xi) - F(\xi)$.

We extend this to a map $\text{MCE}(\mathfrak{g} \hat{\otimes} R_+) \to \text{MCE}(\mathfrak{h} \hat{\otimes} R_+)$ for all complete augmented R functorially. Explicitly, this implies that the pullback $(F^\xi)^*$ can be defined as follows: $(F^\xi)^* = T_\xi^* \circ F^* \circ T^*_{-F(\xi)}$, where $T_\xi(\eta) = \eta + \xi$, so $T_\xi^*(x) = x - \xi(x)$ for $x \in \mathfrak{g}[1]^*$

(and $\xi \in \mathrm{MCE}(\mathfrak{g}) \subseteq \mathfrak{g}^1$, so $\xi(x) = 0$ if $x \in (\mathfrak{g}[1]^*)^m$ with $m \neq 0$). This extends uniquely to a continuous augmented dg morphism.

Explicitly, the formula we obtain on Taylor coefficients F_m^ξ is, for $\eta_1, \dots, \eta_m \in \mathfrak{g}$,

$$F_m^\xi(\eta_1 \wedge \cdots \wedge \eta_m) = \sum_{k \geq 0} \binom{m+k}{k} F_{m+k}^\xi(\xi^{\wedge k} \wedge \eta_1 \wedge \cdots \wedge \eta_m). \tag{4.L}$$

Thus, the composition of $(F^\xi)^*$ with the projection yields the map on cohomology,

$$H^\bullet(\mathfrak{g}^\xi) \to H^\bullet(\mathfrak{h}^{F(\xi)}), \quad x \mapsto \sum_{m \geq 1} m F_m(x \wedge \xi^{\wedge(m-1)}). \tag{4.M}$$

Note that this formula is not so obvious from the simple definition above: one has to be careful to interpret that formula functorially so as to be given by conjugating F^* by a dg algebra isomorphism.

Let us now explain a geometric interpretation, which was used by Kontsevich in [156, §8] (this also gives an alternative way to derive (4.M)). He observes that the cohomology of the twisted dgla $\mathfrak{g}^\xi$ is the *tangent space* in the moduli space $\mathrm{MCE}(\mathfrak{g})/\sim$ to the Maurer–Cartan element ξ. Here $\sim$ denotes gauge equivalence. This is because, if we fix ξ, and differentiate the Maurer–Cartan equation $d(\xi + \eta) + \frac{1}{2}[\xi + \eta, \xi + \eta]$ with respect to η, then the tangent space is the kernel of $d + \mathrm{ad}\,\xi$ on $\mathfrak{g}^1$, and two tangent vectors η and η' are gauge equivalent if and only if they differ by $(d + \mathrm{ad}\,\xi)(z)$ for $z \in \mathfrak{g}^0$. Functorially, this says that the dg vector space $(\mathfrak{g}, d + \mathrm{ad}\,\xi)$ is the dg tangent space to the moduli space of Maurer–Cartan elements of $\mathfrak{g}$, taken with coefficients in arbitrary complete augmented dg commutative algebras: this is because, if you use such a complete augmented dg commutative algebra R which is not concentrated in degree zero, then $\mathrm{MCE}(\mathfrak{g} \hat{\otimes} R_+)$ will detect cohomology of $\mathfrak{g}$ which is not merely in degree one. So its cohomology is the (cohomology of the) tangent space.

Therefore, we will adopt Kontsevich's terminology and refer to (4.M) as the *tangent map* (in [156, §8], it is denoted by I_T, at least in the situation of (4.J) with X the dual to a finite-dimensional Lie algebra, equipped with its standard Poisson structure).

4.13. The algebra isomorphism $(\mathrm{Sym}\,\mathfrak{g})^\mathfrak{g} \xrightarrow{\sim} Z(U\mathfrak{g})$ and Duflo's isomorphism. Now

let $\mathfrak{g}$ be a finite-dimensional Lie algebra and $X = \mathfrak{g}^* = \mathrm{Spec}\,\mathrm{Sym}\,\mathfrak{g}$ the associated affine Poisson variety. The Poisson center $\mathrm{HP}^0(\mathcal{O}(X)) = Z(\mathcal{O}(X))$ is equal to $(\mathrm{Sym}\,\mathfrak{g})^\mathfrak{g}$. By Remark 4.11.8, Kontsevich's morphism induces an isomorphism of algebras

$$(\mathrm{Sym}\,\mathfrak{g})^\mathfrak{g}[\![\hbar]\!] \xrightarrow{\sim} Z(\mathrm{Sym}\,\mathfrak{g}[\![\hbar]\!], \star). \tag{4.N}$$

Moreover, by Example 2.3.16 and Exercise 4.14.1, $(\mathrm{Sym}\,\mathfrak{g}[\![\hbar]\!], \star) \cong U_\hbar \mathfrak{g}$, so we obtain from the above an isomorphism

$$(\mathrm{Sym}\,\mathfrak{g})^\mathfrak{g}[\![\hbar]\!] \xrightarrow{\sim} Z(U_\hbar \mathfrak{g}). \tag{4.O}$$

We can check that this is actually defined over polynomials in $\hbar$, since the Poisson bracket has polynomial degree -1 (i.e., $|\{f, h\}| = |f| + |h| - 1$ for homogeneous f and h), so that the target of an element of $(\mathrm{Sym}\,\mathfrak{g})^\mathfrak{g}$ is polynomial in $\hbar$. Therefore we get an isomorphism $(\mathrm{Sym}\,\mathfrak{g})^{\mathfrak{g}}[\hbar] \overset{\sim}{\to} Z(T\mathfrak{g}[\hbar]/(xy - yx - \hbar\{x, y\})$, and further modding by $(\hbar - 1)$ we obtain an isomorphism

$$(\mathrm{Sym}\,\mathfrak{g})^{\mathfrak{g}} \overset{\sim}{\to} Z(U\mathfrak{g}).$$

That such an isomorphism exists, for *arbitrary* finite-dimensional $\mathfrak{g}$, is a significant generalization of the Harish-Chandra isomorphism for the semisimple case; it was first noticed by Kirillov and then proved by Duflo, using a highly nontrivial formula. By partially computing his isomorphism (4.N), Kontsevich was able to show that his isomorphism (4.O) coincides with the Duflo–Kirillov isomorphism. The latter isomorphism is given by the following explicit formula. Let symm : $\mathrm{Sym}\,\mathfrak{g} \to U\mathfrak{g}$ be the symmetrization map, $x_1 \cdots x_m \mapsto \frac{1}{n!} \sum_{\sigma \in S_n} x_{\sigma(i)} \cdots x_{\sigma(m)}$. This is a vector space isomorphism (even an isomorphism of $\mathfrak{g}$-representations via the adjoint action) but *not* an algebra isomorphism. However, it can be corrected to an isomorphism I_{DK} : $(\mathrm{Sym}\,\mathfrak{g})^{\mathfrak{g}} \overset{\sim}{\to} Z(U\mathfrak{g})$ given by $I_{DK} = \mathrm{symm} \circ I_{\mathrm{strange}}$, where $I_{\mathrm{strange}} : \mathrm{Sym}\,\mathfrak{g} \to \mathrm{Sym}\,\mathfrak{g}$ is given by an (infinite-order) constant-coefficient differential operator. Such operators can be viewed as formal power series functions of $\mathfrak{g}$, i.e., elements of $\hat{S}\mathfrak{g}^*$, via the inclusions

$$\mathrm{Sym}^m \mathfrak{g}^* \hookrightarrow \mathrm{Hom}_{\Bbbk}(\mathrm{Sym}^\bullet \mathfrak{g}, \mathrm{Sym}^{\bullet - m} \mathfrak{g}).$$

In other words, if $\mathfrak{g}$ has a basis x_i with dual basis $\partial_i \in \mathfrak{g}^*$, then an element of $\mathrm{Sym}\,\mathfrak{g}^*$ is a polynomial in the ∂_i, i.e., a constant-coefficient differential operator, and $\hat{S}\mathfrak{g}^*$ is a power series in the ∂_i.

Then, the element $I_{\mathrm{strange}} \in \hat{S}\mathfrak{g}^*$, as a power series function of $\mathfrak{g}$, is expressed as

$$I_{\mathrm{strange}} = \left(x \mapsto \exp\left(\sum_{k \geq 1} \frac{B_{2k}}{4k \cdot (2k)!} \mathrm{tr}(\mathrm{ad}(x)^{2k}) \right) \right), \tag{4.P}$$

where the B_{2k} are Bernoulli numbers. Unpacking all this, we see that $I_{\mathrm{strange}}(x^m) - x^m$ is a certain linear combination of elements of the form

$$x^{m-2(i_1 + \cdots + i_k)} \mathrm{tr}((\mathrm{ad}\,x)^{2i_1}) \cdots \mathrm{tr}((\mathrm{ad}\,x)^{2i_k}),$$

for $i_1, \ldots, i_k$ positive integers such that $2(i_1 + \cdots + i_k) \leq m$ and $k \geq 1$.

Let us explain in more detail how to unpack Kontsevich's isomorphism and see in the process why it might coincide with the Duflo–Kirillov isomorphism. As explained in the previous subsection, (4.N) is a *nontrivial* map, given by

$$f \mapsto f + \sum_{m \geq 1} (m+1)\hbar^m \mathcal{U}_{m+1}(f \wedge \pi^{\wedge m}).$$

We are interested in the case that f is an element of the Poisson center of $\mathcal{O}(X)[\![\hbar]\!]$, which means that f corresponds to a vertex (in the upper-half plane) with no outgoing edges (it is a polyvector field of degree zero). Since π is a bivector field, the term $\pi^{\wedge m}$ corresponds to m vertices in the upper-half plane with two outgoing edges. Thus, the above sum re-expresses as a sum over all graphs in the upper half plane $\{(x, y) \mid y \geq 0\} \subseteq \mathbb{R}^2$, up to isomorphism, which have a single vertex labeled by f in the upper half plane with no outgoing edges and m vertices labeled by π with two outgoing edges each. Moreover, we can discard all the graphs where the two outgoing edges of a given vertex labeled by π have the same target, i.e., we can assume the graph has no *multiple edges*, since π is skew-symmetric, and thus the resulting bilinear operation (cf. §2.4) would be zero.

As a result, we conclude by counting that each vertex labeled by π has exactly one incoming edge, and the vertex labeled f has m incoming edges, one from each vertex labeled by π. We can express such a graph as a union of oriented m-gons labeled by π, with each vertex of each m-gon pointing to a single additional vertex labeled by f. In the case where there is a single m-gon, we put the vertex f in the center of the m-gon, and the resulting graph looks like a wheel, so is called a *wheel*. We consider a general graph to be a union of wheels (where the union is taken by gluing the vertices labeled f together to a single vertex).

As we explain in Exercise 4.14.4.(ii), the differential operator corresponding to an m-gon is $x \mapsto \mathrm{tr}((\mathrm{ad}\,x)^m)$, and we explain in part (iii) that the differential operator attached to a union of m_i-gons is the product of these differential operators. Then, in part (iv), we deduce that Kontsevich's isomorphism should be obtained from a linear map $\mathrm{Sym}\,\mathfrak{g} \to U\mathfrak{g}$ sending x^m to a polynomial in x and $\mathrm{tr}((\mathrm{ad}(x))^j)$, as in the Kirillov–Duflo isomorphism.

Kontsevich shows that the weight of a union of wheels is the product of the weights of the wheels. Moreover, by a symmetry argument he uses also elsewhere, he concludes that the weight is zero when a wheel has an odd number of edges. This shows (see Exercise 4.14.4.(v) below) that Kontsevich's isomorphism must be the composition of symm and an operator of the form (4.P) except possibly with different coefficients than $B_{2k}/(4k \cdot (2k)!)$.

Finally, Kontsevich shows that there can only be at most one isomorphism of this form, so (without computing them!) his coefficients must equal the $B_{2k}/(4k \cdot (2k)!)$.

4.14. Additional exercises. (See also 4.1.5, 4.1.6, 4.4.1, 4.5.5, and 4.8.9.)

EXERCISE 4.14.1. Let $\mathfrak{g}$ be a finite-dimensional Lie algebra. Equip $\mathcal{O}(\mathfrak{g}^*)$ with its standard Poisson structure. Show that, for Kontsevich's star product $\star$ on $\mathrm{Sym}\,\mathfrak{g}$ associated to this Poisson structure, one has

$$v \star w - w \star v = \hbar[v, w], \quad \forall v, w \in \mathfrak{g}.$$

Hint: Show that only the graph corresponding to the Poisson bracket can give a nonzero contribution to $v \star w - w \star v$ when $v, w \in \mathfrak{g}$.

EXERCISE 4.14.2. (*, but with the proof outlined) Prove, following the outline below, the following slightly weaker version of the first statement of Theorem 2.7.7: there is a map from formal power series $\gamma = \sum_{m \geq 1} \hbar^m \cdot \gamma_m \in \hbar \cdot \mathrm{HH}^2(A)[\![\hbar]\!]$ to formal deformations of A which exhausts all formal deformations up to gauge equivalence. (With a bit more work, the proof below can be extended to give the first statement of the Theorem.) This is similar to Exercise 3.8.2 and its proof.

Namely, use the Maurer–Cartan formalism, and the fact that, if $C^\bullet$ is an arbitrary complex of vector spaces, there exists a homotopy $H : C^\bullet \to C^{\bullet-1}$ such that $\mathrm{Id} - (Hd +$

dH) is a projection of $C^\bullet$ onto a subspace of $\ker(d)$ which maps isomorphically to $H^\bullet(C)$ (this is called a Hodge decomposition; we remark that it always satisfies $dHd = d$). In this case, let $i : H^\bullet(C) \cong \mathrm{im}(\mathrm{Id} - (Hd + dH)) \hookrightarrow C^\bullet$ be the obtained inclusion; we have $C^\bullet = i(H^\bullet(C)) \oplus (dH + Hd)(C^\bullet)$, with $(dH + Hd)(C^\bullet)$ a contractible complex.

In the case $\mathrm{HH}^3(A) = 0$, let H be a homotopy as above for $C^\bullet := C^\bullet(A)$. Then, on Hochschild three-cocycles $Z^3(A)$, $dH|_{Z^3(A)} = \mathrm{Id}$ is the identity. Now, if $\gamma = \sum_{m \geq 1} \hbar^m \cdot \gamma_m \in \hbar \cdot \mathrm{HH}^2(A)[\![\hbar]\!]$, we can construct a corresponding solution x of the Maurer–Cartan equation $\mathrm{MC}(x) := dx + \frac{1}{2}[x, x] = 0$ as follows. Set $x^{(1)} := i(\gamma)$, so in particular $dx^{(1)} = 0$. Then, also $\mathrm{MC}(x^{(1)}) \in \hbar^2 Z^3(A)$. Set $x^{(2)} := x^{(1)} - H \circ \mathrm{MC}(x^{(1)})$. Then $\mathrm{MC}(x^{(2)}) \in \hbar^3 C^3(A)[\![\hbar]\!]$. We claim that $\mathrm{MC}(x^{(2)}) \in \hbar^3 Z^3 + \hbar^4 C^3$.

LEMMA 4.14.3. *Let $\mathfrak{g}$ be any dgla, and suppose that $z \in \hbar \mathfrak{g}^1[\![\hbar]\!]$ satisfies $\mathrm{MC}(z) \in \hbar^n \mathfrak{g}^2$. Then, $\mathrm{MC}(z) \in \hbar^n Z^2(\mathfrak{g}) + \hbar^{n+1} \mathfrak{g}^3$.*

PROOF. We have

$$d \, \mathrm{MC}(z) = [dz, z] \equiv -\frac{1}{2}[[z, z], z] \pmod{\hbar^{n+1}},$$

but the RHS is zero by the Jacobi identity. $\qquad\square$

Thus $\mathrm{MC}(x^{(2)}) \in \hbar^3 Z^3(A) + \hbar^4 C^3(A)[\![\hbar]\!]$. Inductively for $n \geq 2$, suppose that $x^{(n)}$ is constructed and $\mathrm{MC}(x^{(n)}) \in \hbar^{n+1} Z^3(A) + \hbar^{n+2} C^3(A)[\![\hbar]\!]$. Then, we set

$$x^{(n+1)} := x^{(n)} - H \circ \mathrm{MC}(x^{(n)}), \quad \text{i.e.,} \quad x^{(n+1)} = (\mathrm{Id} - H \circ \mathrm{MC})^n i(\gamma).$$

It follows from construction that $\mathrm{MC}(x^{(n+1)}) \in \hbar^{n+2} C^3(A)[\![\hbar]\!]$. By the above lemma, it is in $\hbar^{n+2} Z^3(A) + \hbar^{n+3} C^3(A)[\![\hbar]\!]$, completing the inductive step. By construction, $x^{(n+1)} \equiv x^{(n)} \pmod{\hbar^{n+1}}$, so that $x := \lim_{n \to \infty} x^{(n)}$ exists. Also by construction, x is a solution of the Maurer–Cartan equation.

One can show that the map $x \mapsto \gamma$ above yields all possible formal deformations of A up to gauge equivalence, by showing inductively on m that it yields all m-th order deformations, i.e., deformations over $\Bbbk[\hbar]/(\hbar^{m+1})$, for all $m \geq 1$. This is because the tangent space to the space of extensions of an m-th order deformation to $(m + 1)$-st order deformations modulo gauge equivalence is given by $\hbar^{m+1} \cdot \mathrm{HH}^2(A)$.

EXERCISE 4.14.4. (*) Here, following [156, §8], we complete the steps from §4.13 above, outlining why Kontsevich's isomorphism $(\mathrm{Sym}\,\mathfrak{g})^{\mathfrak{g}} \to Z(U\mathfrak{g})$ must be of the form (4.P) except with possibly different coefficients than the $B_{2k}/(4k \cdot (2k)!)$ there. In particular, we show here why it is at least a power series in $\mathrm{tr}(\mathrm{ad}(x)^i)$, and how the desired formula follows from certain properties of the weights.

The description in §2.4 of polydifferential operators from graphs with two vertices on the real line generalizes in a straightforward way to define operators

$$B_\Gamma(\gamma) : \mathcal{O}(X)^{\otimes \Bbbk^n} \to \mathcal{O}(X), \quad \gamma : V_\Gamma \to T_{\mathrm{poly}}(X),$$

where Γ is an arbitrary graph in the upper-half plane with n (not necessarily two) vertices on the real line, and $\gamma : V_\Gamma \to T_{\mathrm{poly}}(X)$ is a function sending each vertex v to an element of degree equal to one less than the number of outgoing edges from v (i.e., the corresponding degree in $\bigwedge^\bullet_{\mathcal{O}(X)} \text{Vect}(X)$ is equal to the number of outgoing edges). Define this generalization, or alternatively see [156].

Now, let $X = \mathfrak{g}^*$ equipped with its standard Poisson structure, for $\mathfrak{g}$ a finite-dimensional Lie algebra. Recall the wheels from §4.13. Let W_m denote the wheel with m vertices labeled π and one vertex labeled f. We are interested in the operators

$$B_{W_m}(-, \pi, \ldots, \pi) : \text{Sym}\, \mathfrak{g} \to \text{Sym}\, \mathfrak{g},$$

placing the vertex labeled f first, where π is the Poisson bivector on $\mathfrak{g}^*$.

(i) Show that $B_{W_m}(-, \pi, \ldots, \pi)$ is a constant-coefficient differential operator of order m, i.e., that

$$B_{W_m}(-, \pi, \ldots, \pi) \in \text{Sym}^m \mathfrak{g}^*.$$

(ii) In terms of $\text{Sym}^m \mathfrak{g}^* = \mathcal{O}(\mathfrak{g})$, show that $B_{W_m}(-, \pi, \ldots, \pi)$ corresponds to the polynomial function

$$x \mapsto \text{tr}((\text{ad}\, x)^m), \quad x \in \mathfrak{g}.$$

(iii) Now suppose that Γ is a graph which is a union (glued at the vertex labeled f) of k wheels $W_{m_1}, \ldots, W_{m_k}$. Show that, considered as polynomials in $\mathcal{O}(\mathfrak{g}) = \text{Sym}^m \mathfrak{g}^*$,

$$B_\Gamma(-, \pi, \ldots, \pi) = x \mapsto \prod_{i=1}^{k} \text{tr}((\text{ad}\, x)^{m_i}).$$

(iv) Conclude that Kontsevich's isomorphism

$$(\text{Sym}\, \mathfrak{g})^{\mathfrak{g}}[\![\hbar]\!] \overset{\sim}{\to} Z(\mathcal{O}(\mathfrak{g}^*)[\![\hbar]\!], \star) \tag{4.Q}$$

has the form, for some constants $c_{m_1,\ldots,m_k}$, now viewed as an element of $\hat{S}\mathfrak{g}^*$, i.e., a power series function contained in the completion $\hat{\mathcal{O}}(\mathfrak{g})$,

$$x \mapsto \sum_{k; m_1, \ldots, m_k} c_{m_1,\ldots,m_k} \cdot \prod_{i=1}^{k} \text{tr}((\text{ad}\, x)^{m_i}),$$

viewed as an element of $\hat{S}\mathfrak{g}^*$, i.e., a formal sum of differential operators (with finitely many summands of each order). Note here that we allow $m_1, \ldots, m_k$ to all be independent (and do not require, for example, $m_1 \leq \cdots \leq m_k$).

(v) Now, it follows from Kontsevich's explicit definition of the weights c_Γ associated to graphs Γ, which are the coefficients of B_Γ in the definition of his L_∞ quasi-isomorphism $\mathcal{U}$, that $c_{m_1,\ldots,m_k} = \frac{1}{k!} \prod_{i=1}^{k} c_{m_i}$, and $c_m = 0$ if m is odd. Using these identities (if you want to see why they are true, see [156]), conclude that (4.Q) is given by the (completed) differential operator corresponding to the series (cf. [156, Theorem 8.5])

$$x \mapsto \exp\left(\sum_{m \geq 1} c_{2m} \, \text{tr}((\text{ad}\, x)^{2m}) \right) \in \hat{S}\mathfrak{g}^* = \hat{\mathcal{O}}(\mathfrak{g}).$$

5. Calabi–Yau algebras and isolated hypersurface singularities

The goal of this section, which may seem somewhat of an abrupt departure from previous sections, is to introduce Calabi–Yau algebras (first mathematically defined and studied in [112]) and use them to study deformation theory. Our motivation is twofold: many of the deformations we have studied can be obtained and understood by realizing algebras as quotients of Calabi–Yau algebras and deforming the Calabi–Yau algebras, and second, because Calabi–Yau algebras are a unifying theme appearing throughout this book. We explain each of these motivations in more detail (in §5.1 and 5.2), before discussing first the homological properties of these algebras (§5.3), then the remarkable phenomenon that they are often defined by potentials (§5.4) and how that can be exploited to deform them (§5.5), and finally how to use them to study deformations of isolated hypersurface singularities and del Pezzo surfaces (§5.6), following Etingof and Ginzburg.

5.1. Motivation: deformations of quotients of Calabi–Yau algebras. We have seen that, in many cases, it is much easier to describe deformations of algebras via generators and relations than via star products (and note that, by Exercise 2.8.4, all formal deformations can be obtained by deforming the relations). For example, this was the case for the Weyl and universal enveloping algebras, as well as the symplectic reflection algebras, where it is rather simple to write down the deformed relations, but the star product is complicated. More generally, given a Koszul algebra, one can easily study the formal homogeneous (or filtered) deformations of the relations that satisfy the PBW property, i.e., give a flat deformation, by using the Koszul deformation principle (Theorem 2.8.5, see also Theorem 1.10.17), even though writing down the corresponding star products could be difficult.

In our running example of $\mathbb{C}[x, y]^{\mathbb{Z}/2} = \mathbb{C}[x^2, xy, y^2] = \mathcal{O}(\mathrm{Nil}\,\mathfrak{sl}_2)$, one way we did this was not by directly finding a noncommutative deformation of the singular ring $\mathbb{C}[x, y]^{\mathbb{Z}/2}$ itself, but rather by deforming either $\mathbb{C}[x, y]$ to Weyl_1 and taking $\mathbb{Z}/2$ invariants, or more generally deforming $\mathbb{C}[x, y] \rtimes \mathbb{Z}/2$ to a symplectic reflection algebra $H_{1,c}(\mathbb{Z}/2)$ and then passing to the spherical subalgebra $eH_{1,c}(\mathbb{Z}/2)e$.

Another way we did it was by realizing the ring as $\mathcal{O}(\mathrm{Nil}\,\mathfrak{sl}_2)$ for the subvariety $\mathrm{Nil}\,\mathfrak{sl}_2 \subseteq \mathfrak{sl}_2 \cong \mathbb{A}^3$ inside affine space, then first deforming $\mathcal{O}(\mathfrak{sl}_2) = \mathcal{O}(\mathfrak{sl}_2^*)$ to the noncommutative ring $U\mathfrak{sl}_2$, then finding a central quotient that yields a quantization of $\mathcal{O}(\mathrm{Nil}\,\mathfrak{sl}_2)$.

We would like to generalize this approach to quantizing more general hypersurfaces in $\mathbb{A}^3$. Suppose $f \in \Bbbk[x, y, z]$ is a hypersurface and $\Bbbk$ has characteristic zero. Then we claim that there is a canonical Poisson bivector field on $Z(f)$. This comes from the *Calabi–Yau* structure on $\mathbb{A}^3$, i.e., *everywhere nonvanishing volume form*. Namely, $\mathbb{A}^3$ is equipped with the volume form

$$dx \wedge dy \wedge dz.$$

The inverse of this is the everywhere nonvanishing top polyvector field

$$\partial_x \wedge \partial_y \wedge \partial_z,$$

in the sense that the contraction of the two is the constant function 1. Now, we can contract this with df and obtain a bivector field,

$$\pi_f := (\partial_x \wedge \partial_y \wedge \partial_z)(df) = \partial_x(f)\partial_y \wedge \partial_z + \partial_y(f)\partial_z \wedge \partial_x + \partial_z(f)\partial_x \wedge \partial_y. \quad (5.A)$$

EXERCISE 5.1.1. Show that this is Poisson. Show also that f is Poisson central, so that the quotient $\mathcal{O}(Z(f)) = \mathcal{O}(\mathbb{A}^3)/(f)$ is Poisson, i.e., the surface $Z(f) \subseteq \mathbb{A}^3$ is canonically equipped with a Poisson bivector from the Calabi–Yau structure on $\mathbb{A}^3$.

Moreover, show that the Poisson bivector π_f is *unimodular*: for every Hamiltonian vector field $\xi_g := i_\pi(dg) = \{g, -\}$ with respect to this Poisson bivector, we have $L_{\xi_g}(\mathrm{vol}) = 0$, where $\mathrm{vol} = dx \wedge dy \wedge dz$. Equivalently, $L_\pi(\mathrm{vol}) = 0$.

More generally (but harder), show that, for an arbitrary complete intersection surface $Z(f_1, \ldots, f_{n-2})$ in $\mathbb{A}^n$ with $\Bbbk$ still of characteristic zero, then

$$(\partial_1 \wedge \cdots \wedge \partial_n)(df_1 \wedge \cdots \wedge df_{n-2})$$

is a Poisson structure. Show that $f_1, \ldots, f_{n-2}$ are Poisson central, so that the surface $Z(f_1, \ldots, f_{n-2})$ is a closed Poisson subvariety, and in particular has a canonical Poisson structure. Moreover, show that π is *unimodular*: for every Hamiltonian vector field $\xi_g := i_\pi(dg) = \{g, -\}$, we have $L_{\xi_g}(\mathrm{vol}) = 0$, where $\mathrm{vol} = dx_1 \wedge \cdots \wedge dx_n$. Equivalently, $L_\pi(\mathrm{vol}) = 0$.

From this, we can deduce that the same holds if $\mathbb{A}^n$ is replaced by an arbitrary n-dimensional Calabi–Yau variety X, i.e., a variety X equipped with a nonvanishing volume form, since the Jacobi condition $[\pi, \pi] = 0$ can be checked in the formal neighborhood of a point of X, which is isomorphic to the formal neighborhood of the origin in $\mathbb{A}^n$.

Now, our strategy, following [101], for deforming $X = Z(f) \subseteq \mathbb{A}^3$ is as follows:

(1) First, consider *Calabi–Yau* deformations of $\mathbb{A}^3$, i.e., noncommutative deformations of $\mathbb{A}^3$ as a Calabi–Yau algebra. We should consider these in the direction of the Poisson structure π_f defined above.

(2) Next, inside such a Calabi–Yau deformation A of $\mathcal{O}(\mathbb{A}^3)$, we identify a central (or more generally normal) element Φ corresponding to $f \in \mathcal{O}(\mathbb{A}^3)$.

(3) Then, the quantization of $\mathcal{O}(X)$ is $A/(\Phi)$.

In order to carry out this program, we need to recall the notion of *(noncommutative) Calabi–Yau algebras* and their convenient presentation by relations derived from a single *potential*. Then it turns out that deforming in the direction of π_f is obtained by deforming the potential of $\mathbb{A}^3$ in the "direction of f." Since f is Poisson central, by Kontsevich's theorem (Corollary 4.11.7), f deforms to a central element of the quantization.

5.2. Calabi–Yau algebras as a unifying theme. Recall Definition 3.7.9. These algebras are ubiquitous and in fact they have appeared throughout the entire book:

(1) A commutative Calabi–Yau algebra is the algebra of functions on a Calabi–Yau affine algebraic variety, i.e., a smooth affine algebraic variety equipped with an everywhere nonvanishing volume form;

(2) Most deformations we have considered of Calabi–Yau algebras are still Calabi–Yau. This includes:
  (a) The universal enveloping algebra $U\mathfrak{g}$ of a finite-dimensional Lie algebra $\mathfrak{g}$ which is unimodular ($\mathrm{tr}(\mathrm{ad}\,x) = 0$ for all $x \in \mathfrak{g}$): this is Calabi–Yau of dimension $\dim \mathfrak{g}$ (and without the unimodular condition, is twisted Calabi–Yau of the same dimension);
  (b) The Weyl algebras $\mathsf{Weyl}(V)$ (Calabi–Yau of dimension $\dim V$);
  (c) The skew-group ring $\mathcal{O}(V) \rtimes G$ for a vector space V and $G < \mathsf{SL}(V)$ finite;
  (d) All symplectic reflection algebras;
(3) The invariant subrings $\mathsf{Weyl}(V)^G$ for $G < \mathsf{Sp}(V)$ finite, and more generally, all homologically smooth spherical symplectic reflection algebras;
(4) All NCCRs that resolve a Gorenstein singularity (as discussed in Chapter IV);
(5) All of the regular algebras discussed in Chapter I are either Calabi–Yau or at least twisted Calabi–Yau. In particular, the quantum versions of $\mathbb{A}^n$, with

$$x_i x_j = r_{ij} x_j x_i, \quad i < j,$$

are Calabi–Yau if and only if, setting $r_{ji} := r_{ij}^{-1}$ for $i < j$,

$$r_{ij} r_{ji} = 1, \ \forall i, j, \quad \prod_{j \neq m} r_{mj} = 1, \ \forall m.$$

For instance, in three variables, we have a single parameter $q = r_{12} = r_{23} = r_{31}$ and then $q^{-1} = r_{21} = r_{32} = r_{13}$.

5.3. Van den Bergh duality and the BV differential.

THEOREM 5.3.1 ([228]). *Let A be Calabi–Yau of dimension d. Then, fixing an isomorphism $\mathsf{HH}^d(A, A \otimes A) \cong A$ yields a canonical isomorphism, for every A-bimodule M,*

$$\mathsf{HH}_i(A, M) \xrightarrow{\sim} \mathsf{HH}^{d-i}(A, M). \tag{5.B}$$

More generally, if A is twisted Calabi–Yau with $\mathsf{HH}^d(A, A \otimes A) \cong U$, for U an invertible A-bimodule, then $\mathsf{HH}_i(A, U \otimes_A M) \xrightarrow{\sim} \mathsf{HH}^{d-i}(A, M)$.

We actually only will require that U be a projective right A-module (which is implied if it is only left invertible, since then the functor $U \otimes_A - : A\text{-mod} \to A\text{-mod}$ has a quasi-inverse and hence preserves projectives).

REMARK 5.3.2. The above theorem is easy to prove, as we will show, but it is an extremely important observation.

PROOF. This is a direct computation. For the first statement, using that $\mathsf{HH}^\bullet(A, M) = H^i(\mathsf{RHom}_{A^e}(A, M))$ and similarly $\mathsf{HH}_\bullet(A, M) = A \otimes_{A^e}^{\mathsf{L}} M$,

$$\mathsf{RHom}_{A^e}(A, M) \cong \mathsf{RHom}_{A^e}(A, A \otimes A) \otimes_{A^e}^{\mathsf{L}} M \cong A[-d] \otimes_{A^e}^{\mathsf{L}} M.$$

The homology of the RHS identifies with $\mathsf{HH}_{d-\bullet}(A, M)$ (the degrees were inverted here because $\mathsf{HH}_\bullet$ uses homological grading and $\mathsf{HH}^\bullet$ uses cohomological grading).

For the second statement, replacing $\mathrm{RHom}_{A^e}(A, A \otimes A)$ by $U[-d]$, we get

$$\mathrm{RHom}^\bullet(A, M) \cong U[-d] \otimes^{\mathrm{L}}_{A^e} M = A[-d] \otimes^{\mathrm{L}}_{A^e} (U \otimes^{\mathrm{L}}_A M),$$

so that if U is projective as a right A-module, hence $U \otimes^{\mathrm{L}}_A M = U \otimes_A M$ is an ordinary A-bimodule, the homology of the RHS is $\mathrm{HH}_{d-\bullet}(A, U \otimes_A M)$, as desired (and as remarked above, invertibility implies projectivity as a right A-module). $\qquad \square$

Next, recall the HKR theorem: $\mathrm{HH}^\bullet(\mathcal{O}(X)) \cong \bigwedge^\bullet_{\mathcal{O}(X)} \mathrm{Vect}(X)$ when X is smooth affine. There is a counterpart for Hochschild homology: $\mathrm{HH}_\bullet(\mathcal{O}(X)) \cong \Omega^\bullet(X)$, the algebraic de Rham differential forms. Moreover, there is a *homological* explanation of the de Rham differential: this turns out to coincide with the *Connes differential*, which is defined on the Hochschild homology of an *arbitrary* associative algebra.

In the presence of the Van den Bergh duality isomorphism (5.B), this differential B yields a differential on the Hochschild cohomology of a Calabi–Yau algebra,

$$\Delta : \mathrm{HH}^\bullet(A) \to \mathrm{HH}^{\bullet-1}(A).$$

This is often called the Batalin–Vilkovisky (BV) differential. As pointed out in [101, §2],

PROPOSITION 5.3.3. *The infinitesimal deformations of a Calabi–Yau algebra A within the class of Calabi–Yau algebras are parametrized by* $\ker(\Delta) : \mathrm{HH}^2(A) \to \mathrm{HH}^1(A)$.

This is not necessarily the right question to ask, however: if one asks for the deformation space of A *together with its Calabi–Yau structure*, i.e., isomorphism $\mathrm{HH}^d(A, A \otimes A) \cong A$, one obtains:

THEOREM 5.3.4 ([235]). *The infinitesimal deformations of pairs (A, η) where A is a Calabi–Yau algebra and η an A-bimodule quasi-isomorphism in the derived category $D(A^e)$,*

$$\eta : C^\bullet(A, A \otimes A) \xrightarrow{\sim} A[-d],$$

is given by the negative cyclic homology $\mathrm{HC}^-_{d-2}(A)$. The obstruction to extending to second order is given by the string bracket $[\gamma, \gamma] \in \mathrm{HC}^-_{d-3}(A)$.

You should think of cyclic homology as a noncommutative analogue of de Rham cohomology, i.e., the homology of $(\mathrm{HH}_\bullet(A), B)$ where B is Connes' differential above (although this is not correct in general, and moreover there are three flavors of cyclic homology: ordinary, negative, and periodic; it is the periodic flavor that coincides with the de Rham cohomology for algebras of functions on smooth affine varieties).

5.4. Algebras defined by a potential, and Calabi–Yau algebras of dimension three.

It turns out that "most" Calabi–Yau algebras are presented by a *(super)potential* (as recalled in the introduction, a precise version of the existence for general dimension is given in [234] when the algebra is complete; on the other hand counterexamples in the general case are given in [77]).

In the case of dimension three, potentials have the following form:

DEFINITION 5.4.1. Let V be a vector space. A *potential* is an element $\Phi \in TV/[TV, TV]$.

For brevity, from now on we write $TV_{\mathrm{cyc}} := TV/[TV, TV]$.

DEFINITION 5.4.2. Let $\xi \in \mathrm{Der}(TV)$ be a constant-coefficient vector field, e.g., $\xi = \partial_i$. We define an action $\xi : TV_{\mathrm{cyc}} \to TV$ as follows: for cyclic words $[v_1 \cdots v_m]$ with $v_i \in V$,

$$\xi[v_1 \cdots v_m] := \sum_{i=1}^{m} \xi(v_i) v_{i+1} \cdots v_m v_1 \cdots v_{i-1}.$$

Then we extend this linearly to TV_{cyc}.

DEFINITION 5.4.3. The algebra A_Φ defined by a potential $\Phi \in TV_{\mathrm{cyc}}$ is

$$A_\Phi := TV/(\partial_1 \Phi, \ldots, \partial_n \Phi).$$

DEFINITION 5.4.4. A potential Φ is called a Calabi–Yau potential (of dimension three) if A_Φ is a Calabi–Yau algebra of dimension three.

EXAMPLE 5.4.5. For $V = \Bbbk^3$, $\mathbb{A}^3$ is defined by the potential

$$\Phi = [xyz] - [xzy] \in TV_{\mathrm{cyc}}, \tag{5.C}$$

since

$$\partial_x \Phi = yz - zy, \quad \partial_y \Phi = zx - xz, \quad \partial_z \Phi = xy - yx.$$

EXAMPLE 5.4.6. The universal enveloping algebra $U\mathfrak{sl}_2$ of a Lie algebra is defined by the potential

$$[efh] - [ehf] - \tfrac{1}{2}[h^2] - 2[ef].$$

EXAMPLE 5.4.7. The Sklyanin algebra with relations

$$xy - tyx + cz^2 = 0, \quad yz - tzy + cx^2 = 0, \quad zx - xz + cy^2 = 0,$$

from Chapter I is given by the potential

$$[xyz] - t[xzy] + \tfrac{1}{3}c[x^3 + y^3 + z^3].$$

EXAMPLE 5.4.8. As mentioned above, NCCRs of Gorenstein singularities are Calabi–Yau; in Chapter IV comments are made throughout that the examples given can be presented by a potential (for example, see Exercise 2.5.4 there). In fact, by [234], any complete Calabi–Yau algebra of dimension three can be presented by a potential; this means that, for any of the examples in Chapter IV, at least after completing at the ideal generated by all the arrows, the relations can be expressed by a potential. In most cases there, the singularity is either already complete (a quotient of $\mathbb{C}[\![x_1, \ldots, x_n]\!]$), or else it is a cone (i.e., the relations are all homogeneous with respect to some assignment of each variable to a weight) and the potential is as well, hence defined without completing.

5.5. Deformations of potentials and PBW theorems. The first part of the following theorem was proved in the filtered case in [33] and in the formal case in [101]. I have written informal notes proving the converse (the second part). The theorem can also be viewed as a consequence of the main result of [234].

THEOREM 5.5.1. *Let A_Φ be a graded Calabi–Yau algebra defined by a (Calabi–Yau) potential Φ. Then for any filtered or formal deformation $\Phi + \Phi'$ of Φ, the algebra $A_{\Phi+\Phi'}$ is a filtered or formal Calabi–Yau deformation of A_Φ.*

Conversely, all filtered or formal deformations of A_Φ are obtained in this way.

Now, let us return to the setting of $\mathcal{O}(\mathbb{A}^3) = A_\Phi$, which is Calabi–Yau using the potential (5.C). Let $f \in \mathcal{O}(\mathbb{A}^3)$ be a hypersurface. Then one obtains as above the Poisson structure π_f on $\mathbb{A}^3$ for which f is Poisson central. As above, all quantizations are given by Calabi–Yau deformations $\Phi + \Phi'$. Moreover, by Corollary 4.11.7, f deforms to a central element, call it $f_{\Phi'}$, of each such quantization. Therefore, one obtains a quantization of the hypersurface $(\mathcal{O}(\mathbb{A}^3)/(f), \pi_f)$, namely $A_{\Phi+\Phi'}/(f_{\Phi'})$.

Next, what we would like to do is to extend this to *graded* deformations, in the case that Φ is homogeneous, i.e., cubic: this will yield the ordinary three-dimensional Sklyanin algebras. More generally, we will consider the *quasihomogeneous* case, which means that it is homogeneous if we assign each of x, y, and z certain weights, which need not be equal. This will recover the "weighted Sklyanin algebras." Note that producing actual graded deformations is not an immediate consequence of the above theorem, and they are not immediately provided by Kontsevich's theorem either. In fact, their existence is a special case (the nicest nontrivial one!) of the following broad conjecture of Kontsevich:

CONJECTURE 5.5.2. [155, Conjecture 1] Suppose that π is a quadratic Poisson bivector on $\mathbb{A}^n$, i.e., $\{-, -\}$ is homogeneous, and $\Bbbk = \mathbb{C}$. Then, the star-product $\star_\hbar$ in the Kontsevich deformation quantization, up to a suitable formal gauge equivalence, actually converges for $\hbar$ in some complex neighborhood of zero, producing an actual deformation $(\mathcal{O}(\mathbb{A}^n), \star_\hbar)$ parametrized by $|\hbar| < \epsilon$, for some $\epsilon > 0$.

As we will see (and was already known, via more indirect constructions), the conjecture holds for π_f with $f \in \mathcal{O}(\mathbb{A}^3)$ quasihomogeneous such that $Z(f)$ has an isolated singularity at the origin.

5.6. Etingof–Ginzburg's quantization of isolated quasihomogeneous surface singularities. In [101], Etingof and Ginzburg constructed a (perhaps universal) family of *graded* quantizations of hypersurface singularities in $\mathbb{A}^3$. These essentially coincide with the Sklyanin algebras associated to types E_6, E_7, and E_8 (the E_6 type is the one of Example 5.4.7). Because these hypersurface singularities also deform to affine surfaces $S \setminus E$ where S is a (projective) del Pezzo surface and $E \subseteq S$ is an elliptic curve, the family yields the universal family of quantizations of these del Pezzo surfaces (in more detail, by [101, Proposition 1.1.3], the Rees algebra of $\mathcal{O}(S \setminus E)$ is the homogeneous coordinate ring of the anti-canonical embedding of S, so the Rees algebras of filtered deformations of $\mathcal{O}(S \setminus E)$ can be viewed as deformations of the projective del Pezzo surface). In this subsection, we will stick to the hypersurface singularities and will not mention del Pezzo surfaces any further.

Let $\Bbbk$ have characteristic zero. Suppose that x, y, and z are assigned positive weights $a, b, c > 0$, not necessarily all one. Let $d := a + b + c$. Suppose $f \in \Bbbk[x, y, z]$ is a polynomial which is weight-homogeneous of degree m. Then the Poisson bracket π_f (5.A) has weight $m - d$. If we want to deform in such a way as to get a *graded* quantization, we will need to have $m = d$, which is called the *elliptic* case. For a filtered but not necessarily graded quantization, we need only that f be a sum of weight-homogeneous monomials of degree $\leq d$. The case where f is weight-homogeneous of degree strictly less than d turns out to yield all the *du Val* (or Kleinian) singularities $\mathbb{A}^2 / \Gamma$ for $\Gamma < SL(2)$ finite, which we discussed in Remark 1.7.12 and Theorem 1.9.15.

So, to look for graded quantizations, suppose f is homogeneous of degree d. We also need to require that $Z(f)$ has an isolated singularity at the origin. In this case, it is well-known that the possible choices of f, up to weight-graded automorphisms of $\Bbbk[x, y, z]$, fall into three possible families, each parametrized by $\tau \in \Bbbk$:

$$\tfrac{1}{3}(x^3 + y^3 + z^3) + \tau \cdot xyz, \quad \tfrac{1}{4}x^4 + \tfrac{1}{4}y^4 + \tfrac{1}{2}z^2 + \tau \cdot xyz, \quad \tfrac{1}{6}x^6 + \tfrac{1}{3}y^3 + \tfrac{1}{2}z^2 + \tau \cdot xyz.$$

Let p, q, and r denote the exponents: so in the first case, $p = q = r = 3$, in the second case $p = q = 4$ and $r = 2$, and in the third case, $p = 6, q = 3$, and $r = 2$. Note that $ap = bq = cr$; so in the first case we could take $a = b = c = 1$; in the second, $a = b = 1, c = 2$; in the third, $a = 1, b = 2$, and $c = 3$. Let $TV_{\text{cyc}}^{\leq m}$ and TV_{cyc}^{m} be the subspaces of TV_{cyc} of weighted degrees $\leq m$ and exactly m, respectively. Similarly define $TV_{\text{cyc}}^{<m} = TV_{\text{cyc}}^{\leq m-1}$. For a filtered algebra A, we similarly define A^m, $A^{\leq m}$, and $A^{<m}$.

We define μ to be the Milnor number of the singularity of the homogeneous f above at the origin, i.e.,

$$\mu = \frac{(a+b)(a+c)(b+c)}{abc} = p + q + r - 1.$$

Finally, we will consider more generally inhomogeneous polynomials replacing f above, of the form $P(x) + Q(y) + R(z)$, such that $\text{gr}(P(x) + Q(y) + R(z)) = f$ and $P(0) = Q(0) = R(0) = 0$. Clearly such f have μ parameters.

Similarly, we will consider potentials $\Phi(P, Q, R; t, c)$ of the form

$$\Phi(P, Q, R; t, c) := [xyz] - t[yxz] + c(P(x) + Q(y) + R(z)). \tag{5.D}$$

If we set $t := e^{\tau \hbar}$, then working over $\Bbbk[[\hbar]]$, with $c \in \Bbbk$ and $P(x), Q(y), R(z)$ polynomials as above, the algebra $A_{\Phi(P,Q,R;t,c)}$ (recall Definition 5.4.3) is a deformation quantization of $\mathcal{O}(\mathbb{A}^3)$, equipped with the Poisson bracket

$$\pi_f, \quad f = \tau \cdot xyz - c(P(x) + Q(y) + R(z)).$$

Thus, for $\Bbbk = \mathbb{C}$, the graded algebras $A_{\Phi(P,Q,R;t,c)}$ indeed give graded quantizations (for actual values of $\hbar$) of the Poisson structures π_f associated to all quasihomogeneous (when P, Q, R are quasihomogeneous), and more generally all filtered polynomials f of weighted degree $\leq d$. This is actually true for arbitrary values of $a, b, c \geq 1$ with

$d = a + b + c$, without requiring that the homogeneous part of f have an isolated singularity at zero. This confirms Kontsevich's conjecture 5.5.2 in these cases, with $\hbar \in \mathbb{C}$ convergent everywhere, up to the fact that $A_{\Phi(P,Q,R;t,c)}$ need not a priori correspond precisely to the Poisson structure $\hbar \cdot \pi_f$, but rather to some formal Poisson deformation thereof.

However, in this case, since the f above are all possible filtered polynomials of degree $\leq d$, we can conclude that, for every $f_\hbar \in \hbar \cdot \mathcal{O}(\mathbb{A}^3)^{\leq d}[\![\hbar]\!]$, letting $\star_{f_\hbar}$ be the Kontsevich quantization of $\pi_{f_\hbar}$, there must exist $\Phi_\hbar \in TV_{\mathrm{cyc}}^{\leq d}[\![\hbar]\!]$ such that, via a continuous isomorphism which is the identity modulo $\hbar$,

$$(\mathcal{O}(\mathbb{A}^3)[\![\hbar]\!], \star_{f_\hbar}) \cong A_{\Phi_\hbar}. \tag{5.E}$$

QUESTION 5.6.1. What is the relation between the parameters $f_\hbar$ and $\Phi_\hbar$ above? In particular:

(1) Does (5.E) send star products that, up to a gauge equivalence, converge for $\hbar$ in some neighborhood of zero to potentials that, up to a continuous automorphism of $TV[\![\hbar]\!]$, also converge in some neighborhood of zero?

(2) Is $A_{\Phi(P,Q,R;t,c)}$ isomorphic (via a continuous $\Bbbk[\![\hbar]\!]$-algebra isomorphism which is the identity modulo $\hbar$) to Kontsevich's quantization of $\hbar \cdot \pi_f$ (or some constant multiple), for $f = \tau \cdot xyz - c(P(x) + Q(y) + R(z))$ (and $t = e^{\tau\hbar}$)?

As stated above, if the second question had a positive answer, this would be enough to confirm Conjecture 5.5.2 in this case.

Returning to the case that the degrees (a, b, c) are in the set $\{(1,1,1), (1,1,2), (1,2,3)\}$, i.e., that generic quasihomogeneous polynomials of degree d have an isolated singularity at the origin, the main result of [101] shows that, for generic A_Φ with Φ filtered of degree $\leq d = a + b + c$, the algebra A_Φ is in the family $A_{\Phi(P,Q,R;t,c)}$ and the family provides a versal deformation; moreover, the center is a polynomial algebra in a single generator, and quotienting by this generator produces a quantization of the original isolated singularity, which also restricts to a versal deformation generically:

THEOREM 5.6.2 ([101]).   (i) *Suppose that $\Phi \in TV_{\mathrm{cyc}}^d$ is a homogeneous Calabi–Yau potential of weighted degree $d = a + b + c$, where $|x| = a$, $|y| = b$, and $|z| = c$ (which can be arbitrary). Then for any potential $\Phi' \in TV_{\mathrm{cyc}}^{<d}$ of degree strictly less than d, $\Phi + \Phi'$ is also a Calabi–Yau potential. Moreover, the Hilbert series of the associated graded algebra of the Calabi–Yau algebra $A_{\Phi+\Phi'}$ is*

$$h(\mathrm{gr}(A_{\Phi+\Phi'}); t) = \frac{1}{(1 - t^a)(1 - t^b)(1 - t^c)}.$$

(ii) *There exists a nonscalar central element $\Psi \in A_{\Phi+\Phi'}^{\leq d}$.*

(iii) *Now suppose that $(a, b, c) \in \{(1, 1, 1), (1, 1, 2), (1, 2, 3)\}$ as above. Then, if Φ is generic, then there exist parameters P, Q, R, t, c, as above, such that for all $\Phi' \in TV_{\mathrm{cyc}}^{<d}$,*

$$A_{\Phi+\Phi'} \cong A_{\Phi(P,Q,R;t,c)}.$$

Moreover, in this case, the center $Z(A_{\Phi+\Phi'}) = \Bbbk[\Psi]$ is a polynomial algebra in one variable, and the Poisson center $Z(\mathrm{gr}\, A_{\Phi+\Phi'}) = \Bbbk[\mathrm{gr}\, \Psi]$.

(iv) *Keep the assumption of (iii). The family $\{A_{\Phi(P',Q',R';t',c')}\}$ restricts to a versal deformation of $A_{\Phi+\Phi'}$ in the formal neighborhood of (P, Q, R, t, c), depending on μ parameters. Moreover, the family $\{A_{\Phi(P',Q',R';t',c')}/\Psi(P', Q', R'; t', c')\}$ restricts to a versal deformation of $A_{\Phi+\Phi'}/(\Psi)$ in the same formal neighborhood.*

QUESTION 5.6.3. Note that we did *not* say anything above about the family $A_{\Phi(P,Q,R;t,c)}/(\Psi(P, Q, R; t, c))$ restricting to a versal quantization of the original singularity $\mathcal{O}(\mathbb{A}^3)/(f)$ for $f = \tau \cdot xyz - c(P(x)+Q(y)+R(z))$. More precisely, does the family of central reductions,

$$A_{\Phi(\hbar \cdot P_\hbar, \hbar \cdot Q_\hbar, \hbar \cdot R_\hbar; t_\hbar, \hbar \cdot c_\hbar)}/(\Psi(\hbar \cdot P_\hbar, \hbar \cdot Q_\hbar, \hbar \cdot R_\hbar; t_\hbar, \hbar \cdot c_\hbar)),$$

produce a versal deformation quantization, for $P_\hbar(x), Q_\hbar(y), R_\hbar(z) \in \mathcal{O}(\mathbb{A}^3)[\![\hbar]\!]$ and $t_\hbar, c_\hbar \in \Bbbk[\![\hbar]\!]$ such that $P = P_0, Q = Q_0, R = R_0, c = c_0$, and $t_\hbar \equiv e^{\tau\hbar} \pmod{\hbar^2}$?

In the case that P, Q, and R are homogeneous, i.e., $f = P(x) + Q(y) + R(z)$, we can explicitly write out the central elements Ψ for the first two possibilities for f ([101, (3.5.1)]). For the first one, $f = \frac{1}{3}(x^3 + y^3 + z^3) + \tau \cdot xyz$, we get

$$\Psi = c \cdot y^3 + \frac{t^3 - c^3}{c^3 + 1}(yzx + c \cdot z^3) - t \cdot zyx. \qquad (5.F)$$

For the second one, $\frac{1}{4}x^4 + \frac{1}{4}y^4 + \frac{1}{2}z^2 + \tau \cdot xyz$, we get

$$\Psi = (t^2 + 1)xyxy - \frac{t^4 + t^2 + 1}{t^2 - c^4}(t \cdot xy^2x + c^2 \cdot y^4) + t \cdot y^2x^2. \qquad (5.G)$$

For the third case, the answer is too long, but you can look it up at

    http://www-math.mit.edu/~etingof/delpezzocenter.

Moreover, there one can obtain the formulas for the central elements Ψ in the filtered cases as well.

CHAPTER III

Symplectic reflection algebras

Gwyn Bellamy

Introduction

The purpose of this chapter is to give the reader a flavour for, and basic grounding in, the theory of symplectic reflection algebras. These algebras, which were introduced by Etingof and Ginzburg in [99], are related to an astonishingly large number of apparently disparate areas of mathematics such as combinatorics, integrable systems, real algebraic geometry, quiver varieties, resolutions of symplectic singularities and, of course, representation theory. As such, their exploration entails a journey through a beautiful and exciting landscape of mathematical constructions. In particular, as we hope to illustrate throughout these notes, studying symplectic reflection algebras involves a deep interplay between geometry and representation theory.

The motivation for introducing and studying symplectic reflection algebras comes from one of the most fundamental problems in mathematics, that of trying to understand the action of a finite group on a vector space. More precisely, from trying to understand the geometry of the space of orbits V/G of this action. This is a fundamental, but difficult, problem in invariant theory, and has a history that stretches back centuries. The modern approach to this problem is to try and apply techniques from noncommutative algebra and the theory of derived categories in order to better understand the geometry of V/G. This "modern era" began with McKay's fundamental observation that the intersection form on the resolution of a Kleinian singularity $\mathbb{C}^2/G$, where G is a finite subgroup of $\mathrm{SL}(2, \mathbb{C})$, is closely related to the representation theory of the skew group algebra $\mathbb{C}[\mathbb{C}^2] \rtimes G$. The precise relationship between the geometry of $\mathbb{C}^2/G$ and the representation theory of $\mathbb{C}[\mathbb{C}^2] \rtimes G$ has been uncovered through the work of many people, e.g., Gonzalez-Sprinberg and Verdier [116] and Kapranov and Vasserot [146]. Based on these examples, it became clear that noncommutative algebras, and in particular the skew group ring $\mathbb{C}[V] \rtimes G$, could provide powerful tools in trying to understand the quotients V/G. This approach to studying the quotients spaces V/G culminated in the generalised McKay correspondences of Bridgeland, King and Reid [44] and Bezrukavnikov and Kaledin [37], which state that, in many examples, the bounded derived category of coherent sheaves on a crepant resolution of V/G is equivalent to the bound derived category of finitely generated $\mathbb{C}[V] \rtimes G$-modules. See Chapter IV on noncommutative resolutions for more on this.

The skew group algebra is only mildly noncommutative, in the sense that it is a finite module over its centre $\mathbb{C}[V]^G$, the algebra of functions on V/G. In such a situation, it is natural to ask if we can go fully noncommutative and *quantize* the algebra $\mathbb{C}[V]^G$, i.e., construct some filtered, noncommutative algebra whose associated graded equals $\mathbb{C}[V]^G$. If this is possible then it implies that the algebra $\mathbb{C}[V]^G$ has some additional structure, namely it is a Poisson algebra. Essentially, this fact forces us to consider the situation where V is a symplectic vector space and G acts by symplectic automorphisms on V. Based on Reid's fundamental work on the birational geometry of threefolds, it has become clear that not all resolutions of V/G are equal — the minimal (in an appropriate sense) ones are called crepant, and these are the ones that are best behaved. When V is symplectic, a crepant resolution Y of V/G is the same as a *symplectic resolution* — one where Y is additionally a symplectic manifold. However, there is a fundamental problem with the notion of a symplectic resolution of singularities - they need not always exist! Thus, it is a fundamental problem in the study of symplectic singularities to decide when a given singularity admits a symplectic resolution.

In the case of V/G, this is where symplectic reflection algebras come to the fore. Deep results of Ginzburg and Kaledin [114] and Namikawa [176] have shown that the existence of a symplectic resolution for V/G is equivalent to the existence of a Poisson deformation of the singularity, which is generically smooth. Being deformations of $\mathbb{C}[V] \rtimes G$, symplectic reflection algebras give rise to a *canonical* Poisson deformation of V/G. Thus, by studying the representation theory of symplectic reflection algebras, we have by now an almost complete classification of those groups $G \subset \mathrm{Sp}(V)$ for which V/G admits a symplectic resolution.

At around the same time that people began to consider the question of existence of symplectic resolutions, M. Haiman was working on his proof of the Macdonald positivity conjecture. The Macdonald polynomials are two-parameter deformations of the classical Schur polynomials of symmetric function theory. Thus, they can be expressed as a linear combination of the Schur polynomials, with coefficients in $\mathbb{C}[q, t]$. The Macdonald positivity conjecture states that these coefficients actually lie in $\mathbb{N}[q, t]$. It was already known by work of Garsia and Haiman that the Macdonald positivity conjecture could be reduced to a purely geometric statement — the $n!$-conjecture, which says that a certain coherent sheaf on the Hilbert scheme of n points in the complex plane is actually a vector bundle (of rank $n!$), called the *Procesi bundle*. Since the Hilbert scheme of n points in the plane is a symplectic resolution of the quotient singularity $\mathbb{C}^{2n}/\mathfrak{S}_n$, it seemed natural to expect that the symplectic reflection algebras associated to the symmetric group might have some close relation to the Hilbert scheme. That this is indeed the case was confirmed in the beautiful work of Gordon and Stafford [123, 124], where it was shown that there is a natural noncommutative analogue of the category of coherent sheaves on the Hilbert scheme, and that this category of quantized sheaves on the Hilbert scheme is (almost always) equivalent to the category of finitely generated modules over the symplectic reflection algebra. Moreover, they showed that the symplectic reflection algebra provides a natural quantization of the Procesi bundle. An equivalent approach using sheaves of deformation-quantization algebras was given by Kashiwara and Rouquier [148]. A vast generalisation of the Macdonald positivity conjecture to the wreath product groups $\mathfrak{S}_n \wr \mathbb{Z}_m$ was later proposed by Haiman. Recently, using symplectic reflection algebras in both zero and positive characteristic, Bezrukavnikov and Finkelberg [35] were able give a prove of this generalised Macdonald conjecture. It is still hoped that the symplectic reflection algebra associated to the symmetric group can be used to give a different proof of the $n!$-conjecture....

When the group G preserves a Lagrangian subspace $\mathfrak{h}$ of V, the associated symplectic reflection algebra is called a *rational Cherednik algebra*. The name comes from the fact that, in those cases where $(\mathfrak{h}, G)$ is a Weyl group, the algebra can also be described as a rational degeneration of the double affine Hecke algebra introduced and studied by I. Cherednik. Rational Cherednik algebras are the most intensely studied class of symplectic reflection algebras since they exhibit many useful features that other symplectic reflection algebras lack. One of these features is a triangular decomposition; see Section 2. The existence of this triangular decomposition implies that there is a natural category $\mathcal{O}$ associated to these algebras, analogous to category $\mathcal{O}$ for a semisimple Lie algebra. In the ground-breaking work by Ginzburg, Guay, Opdam and Rouquier [113], it was shown that category $\mathcal{O}$ is a highest weight category, and thus equivalent to the category of finitely generated modules over a finite-dimensional, quasihereditary algebra. Moreover, it was also shown that there is a very close relationship between category $\mathcal{O}$ and the category of finite-dimensional modules for a Hecke algebra associated to G. Namely, the category of finite-dimensional modules for this Hecke algebra is a quotient of category $\mathcal{O}$ and the quotient functor (called the KZ-functor) is fully faithful on projectives. More recently, it has also been shown, beginning with the work of Shan [199] that category $\mathcal{O}$ can be used to construct categorifications of higher level Fock spaces for affine quantum groups of type A at roots of unity. This illustrates that there is close relationship between rational Cherednik algebras and (quantized) affine Lie algebras. Finally, we note that another problem regarding the representation theory of symplectic reflection algebras, which is currently the subject of intense research, is that of classifying the finite-dimensional simple modules. In the case of rational Cherednik algebras, these finite-dimensional modules (when they exist) belong to category $\mathcal{O}$ and one can use the rich combinatorial structure of these categories in order to better describe the finite-dimensional modules.

Since Poisson geometry plays an important role in the theory of symplectic reflection algebras, it is not so surprising (in retrospect!) that there is a close connection between these algebras and certain integrable systems that appear in mathematical physics. In the original paper by Etingof and Ginzburg it was shown that the centre of the rational Cherednik algebra associated to the symmetric group at $t = 0$ can be identified with the algebra of functions on Wilson's compactification of the Calogero–Moser phase space. Based on this identification, there is a strong interplay between certain symplectic reflection algebras and the Kadomtsev–Petviashvili and Korteweg–de Vries hierarchies. See, for instance, the work of Berest and Chalykh [30, 31].

The theory of symplectic reflection algebras is today viewed as being just one part of a much larger picture of *symplectic representation theory* or the theory of *quantized symplectic singularities*, where one uses sheaves of noncommutative algebras on the symplectic resolution of a symplectic singularity in order to better understand the representation theory of the quantizations of the singularity. The representation theory of semisimple, finite-dimensional Lie algebras also fits into this general framework. Other notable examples include finite W-algebras and hypertoric enveloping algebras.

A brief outline of the content of each section is as follows. In the first section we motivate the definition of symplectic reflection algebras by considering deformations of certain quotient singularities. Once the definition is given, we state the Poincaré–Birkhoff–Witt theorem, which is of fundamental importance in the theory of symplectic reflection algebras. This is the first of many analogies between Lie theory and symplectic reflection algebras. We also introduce a special class of symplectic reflection algebras, the *rational Cherednik algebras*. This class of algebras gives us many interesting examples of symplectic reflection algebras that we can begin to play with. We end the section by describing the double centralizer theorem, which allows us to relate the symplectic reflection algebra with its spherical subalgebra, and also by describing the centre of these algebras.

In the second section, we consider symplectic reflection algebras at $t = 1$. We focus mainly on rational Cherednik algebras and, in particular, on category $\mathcal{O}$ for these algebras. This category of finitely generated $H_c(W)$-modules has a rich, combinatorial representation theory and good homological properties. We show that it is a highest weight category with finitely many simple objects.

Our understanding of category $\mathcal{O}$ is most complete when the corresponding complex reflection group is the symmetric group. In the third section we study this case in greater detail. It is explained how results of Rouquier, Vasserot–Varangolo and Leclerc–Thibon allow us to express the multiplicities of simple modules inside standard modules in terms of the canonical basis of the "level one" Fock space for the quantum affine Lie algebra of type A. A corollary of this result is a character formula for the simple modules in category $\mathcal{O}$. We end the section by stating Yvonne's conjecture which explains how the above mentioned result should be extended to the case where W is the wreath product $\mathfrak{S}_n \wr \mathbb{Z}_m$.

The fourth section deals with the Knizhnik–Zamolodchikov (KZ) functor. This remarkable functor allows one to relate category $\mathcal{O}$ to modules over the corresponding cyclotomic Hecke algebra. In fact, it is an example of a quasihereditary cover, as introduced by Rouquier. The basic properties of the functor are described and we illustrate these properties by calculating explicitly what happens in the rank one case.

The final section deals with symplectic reflection algebras at $t = 0$. For these parameters, the algebras are finite modules over their centres. We explain how the geometry of the centre is related to the representation theory of the algebras. We also describe the Poisson structure on the centre and explain its relevance to representation theory. For rational Cherednik algebras, we briefly explain how one can use the notion of Calogero–Moser partitions, as introduced by Gordon and Martino, in order to decide when the centre of these algebras is regular.

There are several very good lecture notes and survey papers on symplectic reflection algebras and rational Cherednik algebras, including [96, 95, 120, 121, 194]. I would also strongly suggest to anyone interested in learning about symplectic reflection algebras to read the original paper by P. Etingof and V. Ginzburg [99], where symplectic reflection algebras were first defined.[1] It makes a great introduction to the subject and is jam packed with ideas and clever arguments. As noted briefly above, there are strong connections between symplectic reflection algebras and several other areas of mathematics. Due to lack of time and energy, we haven't touched upon those connections here. The interested reader should consult one of the surveys mentioned above. A final remark: to make the sections as readable as possible, there is only a light sprinkling of references in the body of the text. Detailed references can be found at the end of each section.

Acknowledgments. I would like to thank the administrative staff at MSRI for their efforts in organizing and hosting the graduate workshop. My sincerest thanks to all the students who eagerly participated, and showed such enthusiasm over the two weeks. I would also like to extend my thanks to my co-organizers for their hard work at the workshop and for feedback on the lecture notes. Similarly, many thanks to Toby Stafford for his many comments and editorial remarks on earlier drafts, and for taking on the unenviable job of coercing us into completing this work. I would also like to thank Misha Feigin for useful comments on the lecture notes. Finally, my deepest thanks and gratitude to Iain Gordon, who introduced me to symplectic reflection algebras in the first place, and has taught me so much over the years.

1. Symplectic reflection algebras

The action of groups on spaces has been studied for centuries, going back at least to Sophus Lie's fundamental work on transformation groups. In such a situation, one can also study the *orbit space*, i.e., the set of all orbits. This space encodes a lot of the information about the action and provides an effective tool for constructing new spaces out of old ones. The underlying motivation for symplectic reflection algebras is to try and use representation theory to understand a large class of examples of orbit spaces that arise naturally in algebraic geometry.

1.1. Motivation. Let V be a finite-dimensional vector space over $\mathbb{C}$ and let

$$G \subset \mathrm{GL}(V)$$

be a finite group. Fix $\dim V = m$. It is a classical problem in algebraic geometry to try and understand the orbit space $V/G = \mathrm{Spec}\,\mathbb{C}[V]^G$, see Wemyss' lectures in Chapter IV. At the most basic level, we would like to try and answer the questions

QUESTION 1.1.1. Is the space V/G singular?

---

[1] A few years after the publication of [99] it transpired that the definition of symplectic reflection algebras had already appeared in a short paper [88] written by V. Drinfeld in the eighties.

Or, more generally:

QUESTION 1.1.2. How singular is V/G?

The answer to the first question is a classical theorem due to Chevalley and Shephard–Todd. Before I can state their theorem, we need the notion of a complex reflection, which generalises the classical definition of reflection encountered in Euclidean geometry.

DEFINITION 1.1.3. An element $s \in G$ is said to be a *complex reflection* if $1 - s$ has rank 1. Then G is said to be a *complex reflection group* if G is generated by S, the set of all complex reflections contained in G.

Notice that an element $s \in G$ is a complex reflection if and only if all but one of its eigenvalues equals one. Combining the results of Chevalley and Shephard–Todd, we have:

THEOREM 1.1.4. *The space V/G is smooth if and only if G is a complex reflection group. If V/G is smooth then it is isomorphic to $\mathbb{A}^m$, where $\dim V = m$.*

A complex reflection group G is said to be *irreducible* if the reflection representation V is an irreducible G-module. It is an easy exercise (try it!) to show that if $V = V_1 \oplus \cdots \oplus V_k$ is the decomposition of V into irreducible G-modules, then $G = G_1 \times \cdots \times G_k$, where G_i acts trivially on V_j for all $j \neq i$ and (G_i, V_i) is an irreducible complex reflection group. The irreducible complex reflection groups have been classified by Shephard and Todd [201].

EXAMPLE 1.1.5. Let $\mathfrak{S}_n$ the symmetric group act on $\mathbb{C}^n$ by permuting the coordinates. Then, the reflections in $\mathfrak{S}_n$ are exactly the transpositions (i, j), which clearly generate the group. Hence $\mathfrak{S}_n$ is a complex reflection group. If $\mathbb{C}^n = \operatorname{Spec} \mathbb{C}[x_1, \ldots, x_n]$, then the ring of invariants $\mathbb{C}[x_1, \ldots, x_n]^{\mathfrak{S}_n}$ is a polynomial ring with generators $e_1, \ldots, e_n$, where

$$e_k = \sum_{1 \leq i_1 < \cdots < i_k \leq n} x_{i_1} \cdots x_{i_k}$$

is the i-th elementary symmetric polynomial. Notice however that $(\mathfrak{S}_n, \mathbb{C}^n)$ is not an irreducible complex reflection group.

EXAMPLE 1.1.6 (nonexample). Take $m = 2$, i.e., $V = \mathbb{C}^2$ and G a finite subgroup of $\operatorname{SL}_2(\mathbb{C})$. Then it is easy to see that $S = \varnothing$ so G cannot be a complex reflection group. The singular space $\mathbb{C}^2/G$ is called a Kleinian (or Du Val) singularity. The groups G are classified by simply laced Dynkin diagrams, i.e., those diagrams of type ADE, and the singularity $\mathbb{C}^2/G$ is an isolated hypersurface singularity in $\mathbb{C}^3$.

The previous nonexample is part of a large class of groups called symplectic reflection groups. This is the class of groups for which one can try to understand the space V/G using symplectic reflection algebras. In particular, we can try to give a reasonable answer to Question 1.1.2 for these groups. Let (V, ω) be a symplectic vector space, i.e., ω is a nondegenerate, skew symmetric bilinear form on V, and

$$\operatorname{Sp}(V) = \{g \in \operatorname{GL}(V) \mid \omega(gu, gv) = \omega(u, v), \forall u, v \in V\},$$

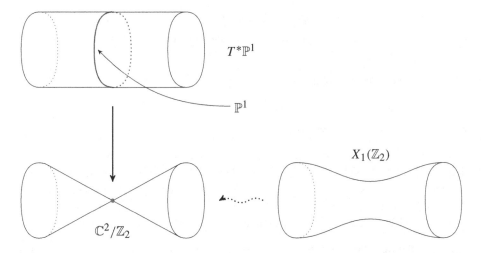

$T^*\mathbb{P}^1$

$\mathbb{P}^1$

$X_1(\mathbb{Z}_2)$

$\mathbb{C}^2/\mathbb{Z}_2$

Figure 1. The resolution and the deformation of the $\mathbb{Z}_2$ quotient singularity.

the symplectic linear group. If $G \subset \mathrm{Sp}(V)$ is a finite subgroup, then G cannot contain any reflections since the determinant of every element in $\mathrm{Sp}(V)$ is equal to one, and hence never a complex reflection. However, we can define $s \in G$ to be a *symplectic reflection* if $\mathrm{rk}(1 - s) = 2$. The idea is that a symplectic reflection is the nearest thing to a genuine complex reflection that one can hope for in a subgroup of $\mathrm{Sp}(V)$.

DEFINITION 1.1.7. The triple (V, ω, G) is a *symplectic reflection group* if (V, ω) is a symplectic vector space and $G \subset \mathrm{Sp}(V)$ is a finite group that is generated by $\mathcal{S}$, the set of all symplectic reflections in G.

Since a symplectic reflection group (V, ω, G) is not "too far" from being a complex reflection group, one might expect V/G to be "not too singular". A measure of the severity of the singularities in V/G is given by how much effort is required to remove them (to "resolve" the singularities). One way to make this precise is to ask whether V/G admits what is called a crepant resolution (the actual definition of a crepant resolution won't be important to us in this course). This is indeed the case for many (but not all!) symplectic reflection groups.[2] In order to classify those groups G for which the space V/G admits a crepant resolution, the first key idea is to try to understand V/G by looking at deformations of the space, i.e., some affine variety $\pi : X \to \mathbb{C}^k$ such that $\pi^{-1}(0) \simeq V/G$ and the map π is *flat*. Intuitively, this is asking that the dimension of the fibres of π don't change. Then it is reasonable to hope that a generic fibre of π is easier to describe, but still tells us something about the geometry of V/G.

However there is a fundamental problem with this idea. We cannot hope to be able to write down generators and relations for the ring $\mathbb{C}[V]^G$ in general. So it seems like a hopeless task to try and write down deformations of this ring. The second key idea is

---

[2]Skip to Section 5.8 for a precise statement.

to try and overcome this problem by introducing noncommutative geometry into the picture. In our case, the relevant noncommutative algebra is the skew group ring.

DEFINITION 1.1.8. The *skew group ring* $\mathbb{C}[V] \rtimes G$ is, as a vector space, equal to $\mathbb{C}[V] \otimes \mathbb{C}G$ and the multiplication is given by

$$g \cdot f = {}^g f \cdot g, \quad \text{for all } f \in \mathbb{C}[V], \ g \in G,$$

where ${}^g f(v) := f(g^{-1}v)$ for $v \in V$.

Exercise 1.9.1 shows that the information of the ring $\mathbb{C}[V]^G$ is encoded in the definition of the skew group ring. On the other hand, the skew group ring has a very explicit, simple presentation. Therefore, we can try to deform $\mathbb{C}[V] \rtimes G$ instead, in the hope that the centre of the deformed algebra is itself a deformation of $\mathbb{C}[V]^G$. We refer the reader to Chapter II for information on the theory of deformations of algebras.

1.2. Symplectic reflection algebras. Thus, symplectic reflection algebras are a particular family of deformations of the skew group ring $\mathbb{C}[V] \rtimes G$, when G is a symplectic reflection group. Fix (V, ω, G), a symplectic reflection group. Let S be the set of symplectic reflections in G. For each $s \in S$, the spaces $\mathrm{Im}(1 - s)$ and $\mathrm{Ker}(1 - s)$ are symplectic subspaces of V with $V = \mathrm{Im}(1 - s) \oplus \mathrm{Ker}(1 - s)$ and $\dim \mathrm{Im}(1 - s) = 2$. We denote by ω_s the 2-form on V whose restriction to $\mathrm{Im}(1 - s)$ is ω and whose restriction to $\mathrm{Ker}(1 - s)$ is zero. Let $c : S \to \mathbb{C}$ be a conjugate invariant function, i.e.,

$$c(gsg^{-1}) = c(s), \quad \text{for all } s \in S, \ g \in G.$$

The space of all such functions equals $\mathbb{C}[S]^G$. Let $TV^* = \mathbb{C} \oplus V^* \oplus (V^* \otimes V^*) \oplus \cdots$ be the tensor algebra on V^*.

DEFINITION 1.2.1. Let $t \in \mathbb{C}$. The *symplectic reflection algebra* $\mathsf{H}_{t,c}(G)$ is defined to be

$$\mathsf{H}_{t,c}(G) =$$
$$TV^* \rtimes G \Big/ \Big\langle u \otimes v - v \otimes u = t\omega(u, v) - 2 \sum_{s \in S} c(s)\omega_s(u, v) \cdot s \mid u, v \in V^* \Big\rangle. \quad (1.\mathrm{A})$$

Notice that the defining relations of the symplectic reflection algebra are trying to tell you how to commute two vectors in V^*. The expression on the right-hand side of (1.A) belongs to the group algebra $\mathbb{C}G$, with $t\omega(u, v) = t\omega(u, v)1_G$, so the price for commuting u and v is that one gets an extra term living in $\mathbb{C}G$.

EXAMPLE 1.2.2. The simplest nontrivial example is $\mathbb{Z}_2 = \langle s \rangle$ acting on $\mathbb{C}^2$. Let $(\mathbb{C}^2)^* = \mathbb{C}\{x, y\}$, where $s \cdot x = -x$, $s \cdot y = -y$ and $\omega(y, x) = 1$. Then $\mathsf{H}_{t,c}(\mathfrak{S}_2)$ is the algebra

$$\mathbb{C}\langle x, y, s \rangle / \langle s^2 = 1, sx = -xs, sy = -ys, [y, x] = t - 2cs \rangle.$$

This example, our "favourite example", will reappear throughout the chapter.

When t and c are both zero, we have $\mathsf{H}_{0,0}(G) = \mathbb{C}[V] \rtimes G$, so that $\mathsf{H}_{t,c}(G)$ really is a deformation of the skew group ring. If $\lambda \in \mathbb{C}^\times$ then $\mathsf{H}_{\lambda t, \lambda c}(G) \simeq \mathsf{H}_{t,c}(G)$ so we normally only consider the cases $t = 0, 1$. The Weyl algebra associated to the symplectic vector space (V, ω) is the noncommutative algebra

$$\mathsf{Weyl}(V, \omega) = TV^* / \langle u \otimes v - v \otimes u = \omega(u, v) \rangle.$$

If $\mathfrak{h}$ is a subspace of V, with $\dim \mathfrak{h} = \frac{1}{2} \dim V$ and $\omega(\mathfrak{h}, \mathfrak{h}) = 0$, then $\mathfrak{h}$ is called a *Lagrangian subspace* of V. Choosing a Lagrangian subspace, we can identify $\mathsf{Weyl}(V, \omega)$ with $\mathcal{D}(\mathfrak{h})$, the ring of differential operators on $\mathfrak{h}$. When $t = 1$ but $c = 0$, we have $\mathsf{H}_{1,0}(G) = \mathsf{Weyl}(V, \omega) \rtimes G$, the skew group ring associated to the Weyl algebra. Hence $\mathsf{H}_{1,c}(G)$ is a deformation of the ring $\mathsf{Weyl}(V, \omega) \rtimes G$.

Let $e = \frac{1}{|G|} \sum_{g \in G} g$ denote the trivial idempotent in $\mathbb{C}G$. The subalgebra $e\mathsf{H}_{t,c}(G)$ of $\mathsf{H}_{t,c}(G)$ is called the *spherical subalgebra* of $\mathsf{H}_{t,c}(G)$. The spherical subalgebra plays a key role in the representation theory of $\mathsf{H}_{t,c}(G)$. It is important to note that the unit in $e\mathsf{H}_{t,c}(G)e$ is e, hence the embedding $e\mathsf{H}_{t,c}(G)e \hookrightarrow \mathsf{H}_{t,c}(G)$ is *not* unital.

EXAMPLE 1.2.3. Again take $V = \mathbb{C}^2$. In this case, $\mathrm{Sp}(V)$ equals $\mathrm{SL}_2(\mathbb{C})$ which implies that every finite subgroup G of $\mathrm{SL}_2(\mathbb{C})$ is a symplectic reflection group since every $g \neq 1$ in G is a symplectic reflection. We have $\omega_g = \omega$. Let x, y be a basis of $(\mathbb{C}^2)^*$ such that $\omega(y, x) = 1$. Then

$$\mathsf{H}_{t,c}(G) = \mathbb{C}\langle x, y \rangle \rtimes G \Big/ \Big\langle [y, x] = t - 2 \sum_{g \in G \setminus \{1\}} c(g)g \Big\rangle.$$

Since c is G-equivariant, the element $z := t - 2 \sum_{g \in G \setminus \{1\}} c(g)g$ belongs to the centre $Z(G)$ of the group algebra of G. Conversely, any element $z \in Z(G)$ can be expressed as $t - 2 \sum_{g \in G \setminus \{1\}} c(g)g$ for some unique t and c. Hence the main relation for the symplectic reflection algebra can simply be expressed more concisely as $[y, x] = z$ for some (fixed) $z \in Z(G)$. For (many) more properties of the algebras $\mathsf{H}_{t,c}(G)$, see [75], where these algebras were first defined and studied.

1.3. Filtrations. To try and study algebras, such as symplectic reflection algebras, that are given in terms of generators and relations, one would like to approximate the algebra by a simpler one, perhaps given by simpler relations, and hope that many properties of the algebra are invariant under this approximation process. An effective way of doing this is by defining a filtration on the algebra and passing to the associated graded algebra, which plays the role of the approximation. Let A be a ring. A *filtration* on A is a nested sequence of abelian subgroups $0 = \mathcal{F}_{-1}A \subset \mathcal{F}_0 A \subset \mathcal{F}_1 A \subset \cdots$ such that $(\mathcal{F}_i A)(\mathcal{F}_j A) \subseteq \mathcal{F}_{i+j}A$ for all i, j and $A = \bigcup_{i \in \mathbb{N}} \mathcal{F}_i A$. The *associated graded* of A with respect to $\mathcal{F}_\bullet$ is $\mathrm{gr}_\mathcal{F} A = \bigoplus_{i \in \mathbb{N}} \mathcal{F}_i A / \mathcal{F}_{i-1} A$. For each $0 \neq a \in A$ there is a unique $i \in \mathbb{N}$ such that $a \in \mathcal{F}_i A$ and $a \notin \mathcal{F}_{i-1} A$. We say that a lies in degree i; $\deg(a) = i$. We define $\sigma(a)$ to be the image of a in $\mathcal{F}_i A / \mathcal{F}_{i-1} A$; $\sigma(a)$ is called the *symbol* of a. This defines a map $\sigma : A \to \mathrm{gr}_\mathcal{F} A$. It's important to note that the map $\sigma : A \to \mathrm{gr}_\mathcal{F} A$ is *not* a morphism of abelian groups even though both domain and target are abelian groups. Let $\bar{a} \in \mathcal{F}_i A / \mathcal{F}_{i-1} A$ and $\bar{b} \in \mathcal{F}_j A / \mathcal{F}_{j-1} A$. One can check that the rule $\bar{a} \cdot \bar{b} := \overline{ab}$, where $\overline{ab}$ denotes the image of ab in $\mathcal{F}_{i+j} A / \mathcal{F}_{i+j-1} A$ extends by additivity to give a well-defined multiplication on $\mathrm{gr}_\mathcal{F} A$, making it into a ring. This multiplication is a "shadow" of that on A and we're reducing the complexity of the situation by forgetting terms of lower order.

There is a natural filtration $\mathcal{F}$ on $\mathsf{H}_{t,c}(G)$, given by putting V^* in degree one and G in degree zero. The crucial result by Etingof and Ginzburg, on which the whole of the theory of symplectic reflection algebras is built, is the Poincaré–Birkhoff–Witt (PBW) theorem (the name comes from the fact that each of Poincaré, Birkhoff and Witt gave proofs of the analogous result for the enveloping algebra of a Lie algebra).

THEOREM 1.3.1. *The map* $\sigma(v) \mapsto v, \sigma(g) \mapsto g$ *defines an isomorphism of algebras*

$$\mathrm{gr}_{\mathcal{F}}(\mathsf{H}_{t,c}(G)) \simeq \mathbb{C}[V] \rtimes G,$$

where $\sigma(D)$ *denotes the symbol, or leading term, of* $D \in \mathsf{H}_{t,c}(G)$ *in* $\mathrm{gr}_{\mathcal{F}}(\mathsf{H}_{t,c}(G))$.

One of the key points of the PBW theorem is that it gives us an explicit basis of the symplectic reflection algebra. Namely, if one fixes an *ordered* basis of V, then the PBW theorem implies that there is an isomorphism of vector spaces $\mathsf{H}_{t,c}(G) \simeq \mathbb{C}[V] \otimes \mathbb{C}G$. One can also think of the PBW theorem as saying that no information is lost in deforming $\mathbb{C}[V] \rtimes G$ to $\mathsf{H}_{t,c}(G)$, since we can recover $\mathbb{C}[V] \rtimes G$ from $\mathsf{H}_{t,c}(G)$.

The proof of this theorem is an application of a general result by Braverman and Gaitsgory [40]. If I is a two-sided ideal of $TV^* \rtimes G$ generated by a space U of (not necessarily homogeneous) elements of degree at most two then [40, Theorem 0.5] gives necessary and sufficient conditions on U so that the quotient $TV^* \rtimes G/I$ has the PBW property. The PBW property immediately implies that $\mathsf{H}_{t,c}(G)$ enjoys some good ring-theoretic properties, for instance:

COROLLARY 1.3.2.   (1) *The algebra* $\mathsf{H}_{t,c}(G)$ *is a prime, (left and right) noether-ian ring.*

  (2) *The algebra* $e\mathsf{H}_{t,c}(G)e$ *is a left and right noetherian integral domain.*

  (3) $\mathsf{H}_{t,c}(G)$ *has finite global dimension (in fact,* gl.dim $\mathsf{H}_{t,c}(G) \le \dim V$*).*

We'll sketch a proof of the corollary, to illustrate the use of filtrations.

PROOF. All three of the results follow from the fact that corresponding statement holds for the skew-group ring $\mathbb{C}[V] \rtimes G$. We'll leave it to the reader to show these statements hold for $\mathbb{C}[V] \rtimes G$. In particular, in part (3) the claim is that the global dimension of $\mathbb{C}[V] \rtimes G$ equals the global dimension of $\mathbb{C}[V]$, which is well-known to equal $\dim V$. So we may assume that $(A, \mathcal{F}_\bullet)$ is a filtered ring such that the above statements hold for $\mathrm{gr}_{\mathcal{F}} A$. Let $I_1 \subset I_2 \subset \cdots$ be a chain of left ideals in A. Then it follows from the definition of multiplication in $\mathrm{gr}_{\mathcal{F}} A$ that $\sigma(I_1) \subset \sigma(I_2) \subset \cdots$ is a chain of left ideals in $\mathrm{gr}_{\mathcal{F}} A$. Therefore, there is some N such that $\sigma(I_{N+i}) = \sigma(I_N)$ for all $i \ge 0$. This implies that $I_{N+i} = I_N$ for all $i \ge 0$, and hence A is left noetherian. The argument for right noetherian is identical. Now assume that I, J are ideals of A such that $I \cdot J = 0$. Then certainly $\sigma(I)\sigma(J) = 0$. Since $\mathrm{gr}_{\mathcal{F}} A$ is assumed to be prime, this implies that either $\sigma(I) = 0$ or $\sigma(J) = 0$. But this can only happen if $I = 0$ or $J = 0$. Hence A is prime.

For part (2), we note that the algebra $e(\mathbb{C}[V] \rtimes W)e$ is isomorphic to

$$\mathbb{C}[V]^G \subset \mathbb{C}[V],$$

and hence is a noetherian integral domain. So we may assume that $(A, \mathcal{F}_\bullet)$ is a filtered ring such that $\mathrm{gr}_{\mathcal{F}} A$ is an integral domain. If $a, b \in A$ such that $a \cdot b = 0$ then certainly $\sigma(a) \cdot \sigma(b) = 0$ in $\mathrm{gr}_{\mathcal{F}} A$. Hence, without loss of generality $\sigma(a) = 0$. But this implies that $a = 0$.

To prove that gl.dim $A \leq$ gl.dim $\mathrm{gr}_{\mathcal{F}} A$ is a bit more involved (but not too difficult) because it involves the notion of filtration on A-modules, compatible with the filtration on A. The proof is given in [172, Section 7.6]. □

1.4. Quantization. In this section we will see how symplectic reflection algebras provide examples of quantizations as described in Schedler's Chapter II. Consider t as a variable and $\mathsf{H}_{t,c}(G)$ a $\mathbb{C}[t]$-algebra. Similarly, we consider $e\mathsf{H}_{t,c}(G)e$ as a $\mathbb{C}[t]$-algebra, where $\mathbb{C}[t]$ is central. Then we can complete $\mathsf{H}_{t,c}(G)$ and $e\mathsf{H}_{t,c}(G)e$ with respect to the two-sided ideal generated by the powers of t:

$$\widehat{\mathsf{H}}_{t,c}(G) = \lim_{\infty \leftarrow n} \mathsf{H}_{t,c}(G)/(t^n), \quad e\widehat{\mathsf{H}}_{t,c}(G)e = \lim_{\infty \leftarrow n} e\mathsf{H}_{t,c}(G)e/(t^n).$$

The Poincaré–Birkhoff–Witt theorem implies that $\mathsf{H}_{t,c}(G)$ and $e\mathsf{H}_{t,c}(G)e$ are free $\mathbb{C}[t]$-modules. Therefore, $\widehat{\mathsf{H}}_{t,c}(G)$ and $e\widehat{\mathsf{H}}_{t,c}(G)e$ are flat, complete $\mathbb{C}[\![t]\!]$-modules. Hence:

PROPOSITION 1.4.1. *The algebra $\widehat{\mathsf{H}}_{t,c}(G)$ is a formal deformation of $\mathsf{H}_{0,c}(G)$ and $e\widehat{\mathsf{H}}_{t,c}(G)e$ is a formal quantization of $e\mathsf{H}_{0,c}(G)e$.*

By formal quantization of $e\mathsf{H}_{0,c}(G)e$, we mean that there is a Poisson bracket on the commutative algebra $e\mathsf{H}_{0,c}(G)e$ such that the first order term of the quantization $e\widehat{\mathsf{H}}_{t,c}(G)e$ is this bracket. This Poisson bracket is described in Section 5.2. The algebras $\mathsf{H}_{t,c}(G)$ are "better" than $\widehat{\mathsf{H}}_{t,c}(G)$ in the sense that one can specialize, in the former, t to any complex number, however in the latter only the specialization $t \to 0$ is well-defined.

1.5. The rational Cherednik algebra. There is a standard way to construct a large number of symplectic reflection groups — by creating them out of complex reflection groups. This class of symplectic reflection algebras is by far the most important and, thus, have been most intensively studied out of all symplectic reflection algebras. So let W be a complex reflection group, acting on its reflection representation $\mathfrak{h}$. Then W acts diagonally on $\mathfrak{h} \times \mathfrak{h}^*$. To be explicit, W acts on $\mathfrak{h}^*$ by $(w \cdot x)(y) = x(w^{-1}y)$, where $x \in \mathfrak{h}^*$ and $y \in \mathfrak{h}$. Then, $w \cdot (y, x) = (w \cdot y, w \cdot x)$. The space $\mathfrak{h} \times \mathfrak{h}^*$ has a natural pairing $(\cdot, \cdot) : \mathfrak{h} \times \mathfrak{h}^* \to \mathbb{C}$ defined by $(y, x) = x(y)$. Using this,

$$\omega((y_1, x_1), (y_2, x_2)) := (y_1, x_2) - (y_2, x_1)$$

defines a W-equivariant symplectic form on $\mathfrak{h} \times \mathfrak{h}^*$. One can easily check that the set of symplectic reflection S in W, consider as a symplectic reflection group $(\mathfrak{h} \times \mathfrak{h}^*, \omega, W)$, is the same as the set of complex reflections S in W, considered as a complex reflection group $(W, \mathfrak{h})$. Therefore, W acts on the symplectic vector space $\mathfrak{h} \times \mathfrak{h}^*$ as a symplectic reflection group if and only if it acts on $\mathfrak{h}$ as a complex reflection group.

The *rational Cherednik algebra*, as introduced by Etingof and Ginzburg [99, p. 250], is the symplectic reflection algebra associated to the triple $(\mathfrak{h} \times \mathfrak{h}^*, \omega, W)$. In this situation, one can simplify somewhat the defining relation (1.A). For each $s \in \mathcal{S}$, fix $\alpha_s \in \mathfrak{h}^*$ to be a basis of the one-dimensional space $\mathrm{Im}(s-1)|_{\mathfrak{h}^*}$ and $\alpha_s^\vee \in \mathfrak{h}$ a basis of the one-dimensional space $\mathrm{Im}(s-1)|_{\mathfrak{h}}$, normalised so that $\alpha_s(\alpha_s^\vee) = 2$. Then the relation (1.A) can be expressed as

$$[x_1, x_2] = 0, \quad [y_1, y_2] = 0, \quad [y_1, x_1] = t(y_1, x_1) - \sum_{s \in \mathcal{S}} c(s)(y_1, \alpha_s)(\alpha_s^\vee, x_1)s, \quad (1.B)$$

for all $x_1, x_2 \in \mathfrak{h}^*$ and $y_1, y_2 \in \mathfrak{h}$. Notice that these relations, together with the PBW Theorem, imply that $\mathbb{C}[\mathfrak{h}]$ and $\mathbb{C}[\mathfrak{h}^*]$ are polynomial subalgebras of $H_{t,c}(W)$.

EXAMPLE 1.5.1. In the previous example we can take $W = \mathfrak{S}_n$, the symmetric group. Choose a basis $x_1, \ldots, x_n$ of $\mathfrak{h}^*$ and dual basis $y_1, \ldots, y_n$ of $\mathfrak{h}$ so that

$$\sigma x_i = x_{\sigma(i)} \sigma, \quad \sigma y_i = y_{\sigma(i)} \sigma, \quad \text{for all } \sigma \in \mathfrak{S}_n.$$

Then $\mathcal{S} = \{s_{i,j} \mid 1 \le i < j \le n\}$ is the set of all transpositions in $\mathfrak{S}_n$. This is a single conjugacy class, so $c \in \mathbb{C}$. Fix

$$\alpha_{i,j} = x_i - x_j, \quad \alpha_{i,j}^\vee = y_i - y_j, \quad \text{for all } 1 \le i < j \le n.$$

Then the relations for $H_{t,c}(\mathfrak{S}_n)$ become $[x_i, x_j] = [y_i, y_j] = 0$ and

$$[y_i, x_j] = cs_{i,j}, \qquad \text{for all } 1 \le i < j \le n,$$

$$[y_i, x_i] = t - c \sum_{j \ne i} s_{i,j}, \qquad \text{for all } 1 \le i \le n.$$

1.6. Double centralizer property.

Being a subalgebra of $H_{t,c}(G)$, the spherical subalgebra $eH_{t,c}(G)e$ inherits a filtration by restriction. It is a consequence of the PBW theorem that $\mathrm{gr}_{\mathcal{F}}(eH_{t,c}(G)e) \simeq \mathbb{C}[V]^G$. Thus, the spherical subalgebra of $H_{t,c}(G)$ is a (not necessarily commutative!) flat deformation of the coordinate ring of V/G; almost exactly what we've been looking for!

The space $H_{t,c}(G)e$ is a $(H_{t,c}(G), eH_{t,c}(G)e)$-bimodule, and is called the Etingof–Ginzburg sheaf. The following result shows that one can recover $H_{t,c}(G)$ from knowing $eH_{t,c}(G)e$ and $H_{t,c}(G)e$.

THEOREM 1.6.1.   (1) *The map* $eh \mapsto (\phi_{eh} : fe \mapsto ehfe)$ *is an isomorphism of left* $eH_{t,c}(G)e$-modules

$$eH_{t,c}(G) \xrightarrow{\sim} \mathrm{Hom}_{eH_{t,c}(G)e}(H_{t,c}(G)e, eH_{t,c}(G)e).$$

(2) $\mathrm{End}_{H_{t,c}(G)}(H_{t,c}(G)e)^{op} \simeq eH_{t,c}(G)e$.

(3) $\mathrm{End}_{(eH_{t,c}e)^{op}}(H_{t,c}(G)e) \simeq H_{t,c}(G)$.

REMARK 1.6.2. As in (1), the natural map of *right* $eH_{t,c}(G)e$-modules

$$H_{t,c}(G)e \to \mathrm{Hom}_{eH_{t,c}(G)e}(eH_{t,c}(G), eH_{t,c}(G)e)$$

is an isomorphism. This, together with (1) imply that $eH_{t,c}(G)$ and $H_{t,c}(G)e$ are *reflexive* left and right $eH_{t,c}(G)e$-modules respectively; see [172, Section 5.1.7].

The above result is extremely useful because, unlike the spherical subalgebra, we have an explicit presentation of $H_{t,c}(G)$. Therefore, we can try to implicitly study $eH_{t,c}(G)e$ by studying instead the algebra $H_{t,c}(G)$. Though the rings $eH_{t,c}(G)e$ and $H_{t,c}(G)$ are never isomorphic, very often the next best thing is true, namely that they are Morita equivalent. This means that the categories of left $eH_{t,c}(G)e$-modules and of left $H_{t,c}(G)$-modules are equivalent. Let A be an algebra. We'll denote by A-mod the category of finitely generated left A-modules. If A is noetherian (which will always be the case for us) then A-mod is abelian.

COROLLARY 1.6.3. *The algebras $H_{t,c}(G)$ and $eH_{t,c}(G)e$ are Morita equivalent if and only if $e \cdot M = 0$ implies $M = 0$ for all $M \in H_{t,c}(G)$-mod.*

PROOF. Theorem 1.6.1, together with a basic result in Morita theory, e.g., [172, Section 3.5], says that the bimodule $H_{t,c}(G)e$ will induce an equivalence of categories $e \cdot - : H_{t,c}(G)\text{-mod} \xrightarrow{\sim} eH_{t,c}(G)e\text{-mod}$ if and only if $H_{t,c}(G)e$ is both a generator of the category $H_{t,c}(G)$-mod and a projective $H_{t,c}(G)$-module. Since $H_{t,c}(G)e$ is a direct summand of $H_{t,c}(G)$ it is projective. Therefore we just need to show that it generates the category $H_{t,c}(G)$-mod. This condition can be expressed as saying that

$$H_{t,c}(G)e \otimes_{eH_{t,c}(G)e} eM \simeq M, \quad \text{for all } M \in H_{t,c}(G)\text{-mod.}$$

Equivalently, we require that $H_{t,c}(G) \cdot e \cdot H_{t,c}(G) = H_{t,c}(G)$. If this is not the case then

$$I := H_{t,c}(G) \cdot e \cdot H_{t,c}(G)$$

is a proper two-sided ideal of $H_{t,c}(G)$. Hence there exists some module M such that $I \cdot M = 0$. But this is equivalent to $e \cdot M = 0$. $\square$

The following notion is very important in the study of rational Cherednik algebras at $t = 1$.

DEFINITION 1.6.4. The parameter (t, c) is said to be *aspherical* for G if there exists a nonzero $H_{t,c}(G)$-module M such that $e \cdot M = 0$.

The value $(t, c) = (0, 0)$ is an example of an aspherical value for G.

1.7. The centre of $H_{t,c}(G)$. One may think of the parameter t as a "quantum parameter". When $t = 0$, we are in the "quasiclassical situation" and when $t = 1$ we are in the "quantum situation" — illustrated by the fact that $H_{0,0}(G) = \mathbb{C}[V] \rtimes G$ and $H_{1,0}(G) = \text{Weyl}(V, \omega) \rtimes G$. The following result gives meaning to such a vague statement. It also shows that the symplectic reflection algebra produces a genuine commutative deformation of the space V/G when $t = 0$.

THEOREM 1.7.1.   (1) *If $t = 0$ then the spherical subalgebra $eH_{t,c}(G)e$ is commutative.*
   (2) *If $t \neq 0$ then the centre of $eH_{t,c}(G)e$ is $\mathbb{C}$.*

REMARK 1.7.2. The word "quasiclassical" appears often in deformation theory. Why are things "quasiclassical" as opposed to "classical"? Well, roughly speaking, quantization is the process of making a commutative algebra (or a space) into a noncommutative algebra (or "noncommutative space"). The word quasi refers to the fact that when it is possible to quantize an algebra, this algebra (or space) has some additional structure. Namely, the fact that an algebra is quantizable implies that it is a Poisson algebra, and not just any old commutative algebra.

One can now use the double centralizer property, Theorem 1.6.1, to lift Theorem 1.7.1 to a result about the centre of $H_{t,c}(G)$.

THEOREM 1.7.3 (the Satake isomorphism). *The map $z \mapsto z \cdot e$ defines an algebra isomorphism $Z(H_{t,c}(G)) \xrightarrow{\sim} Z(eH_{t,c}(G)e)$ for all parameters (t, c).*

PROOF. Clearly $z \mapsto z \cdot e$ is a morphism $Z(H_{t,c}(G)) \to Z(eH_{t,c}(G)e)$. Right multiplication on $H_{t,c}(G) \cdot e$ by an element a in $Z(eH_{t,c}(G)e)$ defines a right $eH_{t,c}(G)e$-linear endomorphism of $H_{t,c}(G) \cdot e$. Therefore Theorem 1.6.1 says that there exists some $\zeta(a) \in H_{t,c}(G)$ such that right multiplication by a equals left multiplication on $H_{t,c}(G) \cdot e$ by $\zeta(a)$. The action of a on the right commutes with left multiplication by any element of $H_{t,c}(G)$ hence $\zeta(a) \in Z(H_{t,c}(G))$. The homomorphism $\zeta : Z(eH_{t,c}(G)e) \to Z(H_{t,c}(G))$ is the inverse to the Satake isomorphism.   □

When $t = 0$, the Satake isomorphism becomes an isomorphism

$$Z(H_{0,c}(G)) \xrightarrow{\sim} eH_{0,c}(G)e,$$

and is in fact an isomorphism of Poisson algebras. Theorems 1.7.1 and 1.7.3 also imply that $H_{0,c}(G)$ is a finite module over its centre. As one might guess, the behaviour of symplectic reflection algebras is very different depending on whether $t = 0$ or 1. It is also a very interesting problem to try and relate the representation theory of the algebras $H_{0,c}(G)$ and $H_{1,c}(G)$ in some meaningful way.

1.8. The Dunkl embedding. In this section, parts of which are designed to be an exercise for the reader, we show how one can use the Dunkl embedding to give easy proofs for rational Cherednik algebras of many of the important theorems described above. In particular, one can give elementary proofs of both the PBW theorem and of the fact that the spherical subalgebra is commutative when $t = 0$. Therefore, we let $(W, \mathfrak{h})$ be a complex reflection group and $H_{t,c}(W)$ the associated rational Cherednik algebra. Let $\mathcal{D}_t(\mathfrak{h})$ be the algebra generated by $\mathfrak{h}$ and $\mathfrak{h}^*$, satisfying the relations

$$[x, x'] = [y, y'] = 0, \quad \text{for all } x, x' \in \mathfrak{h}^*, y, y' \in \mathfrak{h},$$

$$[y, x] = t(y, x), \quad \text{for all } x \in \mathfrak{h}^*, y \in \mathfrak{h}.$$

When $t \neq 0$, $\mathcal{D}_t(\mathfrak{h})$ is isomorphic to $\mathcal{D}(\mathfrak{h})$, the ring of differential operators on $\mathfrak{h}$. But when $t = 0$, the algebra $\mathcal{D}_t(\mathfrak{h}) = \mathbb{C}[\mathfrak{h} \times \mathfrak{h}^*]$ is commutative. Let $\mathfrak{h}_{\text{reg}}$ be the *affine* open subset of $\mathfrak{h}$ on which W acts freely. That $\mathfrak{h}_{\text{reg}}$ is affine is a consequence of the fact that W is a complex reflection group, it is not true in general. We can localize $\mathcal{D}_t(\mathfrak{h})$ to $\mathcal{D}_t(\mathfrak{h}_{\text{reg}})$. Recall from Example 1.5 that we have associated to each $s \in S$ the vector $\alpha_s \in \mathfrak{h}^*$. Define $\lambda_s \in \mathbb{C}^\times$ by $s(\alpha_s) = \lambda_s \alpha_s$. For each $y \in \mathfrak{h}$, the Dunkl operator

$$D_y = y - \sum_{s \in S} \frac{2c(s)}{1 - \lambda_s} \frac{(y, \alpha_s)}{\alpha_s} (1 - s) \tag{1.C}$$

is an element in $\mathcal{D}_t(\mathfrak{h}_{reg}) \rtimes W$, since α_s is invertible on $\mathfrak{h}_{reg}$. Exercise 1.9.3 shows that $x \mapsto x$, $w \mapsto w$ and $y \mapsto D_y$ defines a morphism

$$\phi : \mathsf{H}_{t,c}(W) \to \mathcal{D}_t(\mathfrak{h}_{reg}) \rtimes W.$$

This is enough to prove the Poincaré–Birkhoff–Witt theorem for rational Cherednik algebras. The algebra $\mathcal{D}_t(\mathfrak{h}_{reg}) \rtimes W$ has a natural filtration given by putting $\mathbb{C}[\mathfrak{h}_{reg}] \rtimes W$ in degree zero and $\mathfrak{h} \subset \mathbb{C}[\mathfrak{h}^*]$ in degree one. Similarly, we define a filtration on the rational Cherednik algebra by putting the generators $\mathfrak{h}^*$ and W in degree zero and $\mathfrak{h}$ in degree one.

LEMMA 1.8.1. *The associated graded of* $\mathsf{H}_{t,c}(W)$ *with respect to the above filtration is isomorphic to* $\mathbb{C}[\mathfrak{h} \times \mathfrak{h}^*] \rtimes W$.

PROOF. We begin by showing that the Dunkl embedding (which we have yet to prove is actually an embedding) preserves filtrations, i.e., $\phi(\mathcal{F}_i\mathsf{H}_{t,c}(W))$ is contained in $\mathcal{F}_i(\mathcal{D}_t(\mathfrak{h}_{reg}) \rtimes W)$ for all i. It is clear that ϕ maps $\mathfrak{h}^*$ and W into $\mathcal{F}_0(\mathcal{D}_t(\mathfrak{h}_{reg}) \rtimes W)$ and that $\phi(\mathfrak{h}) \subset \mathcal{F}_1(\mathcal{D}_t(\mathfrak{h}_{reg}) \rtimes W)$. Since the filtration on $\mathsf{H}_{t,c}(W)$ is defined in terms of the generators $\mathfrak{h}^*$, W and $\mathfrak{h}$, the inclusion $\phi(\mathcal{F}_i\mathsf{H}_{t,c}(W)) \subset \mathcal{F}_i(\mathcal{D}_t(\mathfrak{h}_{reg}) \rtimes W)$ follows. Therefore the map ϕ induces a morphism

$$\mathbb{C}[\mathfrak{h} \times \mathfrak{h}^*] \rtimes W \to \mathrm{gr}_{\mathcal{F}}\mathsf{H}_{t,c}(W) \xrightarrow{\mathrm{gr}\,\phi} \mathrm{gr}_{\mathcal{F}}(\mathcal{D}_t(\mathfrak{h}_{reg}) \rtimes W) \to \mathbb{C}[\mathfrak{h}_{reg} \times \mathfrak{h}^*] \rtimes W, \quad (1.D)$$

where the right-hand morphism is the inverse to the map

$$\mathbb{C}[\mathfrak{h}_{reg} \times \mathfrak{h}^*] \rtimes W \to \mathrm{gr}_{\mathcal{F}}(\mathcal{D}_t(\mathfrak{h}_{reg}) \rtimes W),$$

which is well-known to be an isomorphism. The morphism (1.D) maps $\mathfrak{h}$ to $\mathfrak{h}$, $\mathfrak{h}^*$ to $\mathfrak{h}^*$ and W to W. Therefore, it is the natural embedding $\mathbb{C}[\mathfrak{h} \times \mathfrak{h}^*] \rtimes W \hookrightarrow \mathbb{C}[\mathfrak{h}_{reg} \times \mathfrak{h}^*] \rtimes W$. Thus, since the map $\mathbb{C}[\mathfrak{h} \times \mathfrak{h}^*] \rtimes W \to \mathrm{gr}_{\mathcal{F}}\mathsf{H}_{t,c}(W)$ is also surjective, it must be an isomorphism.

Notice that we have also shown that $\mathrm{gr}_{\mathcal{F}}\mathsf{H}_{t,c}(W) \xrightarrow{\mathrm{gr}\,\phi} \mathrm{gr}_{\mathcal{F}}(\mathcal{D}_t(\mathfrak{h}_{reg}) \rtimes W)$ is an embedding. This implies that the Dunkl embedding is also an embedding. As a consequence, $\mathbb{C}[\mathfrak{h}]$ is a faithful $\mathsf{H}_{t,c}(W)$-module. $\square$

For an arbitrary symplectic reflection group, the fact that $e\mathsf{H}_{t,c}(G)e$ is commutative when $t = 0$ relies on a very clever but difficult argument by Etingof and Ginzburg. However, for rational Cherednik algebras, Exercise 1.9.4 shows that this can be deduced from the Dunkl embedding.

A function $f \in \mathbb{C}[\mathfrak{h}]$ is called a *W-semi-invariant* if, for each $w \in W$, we have $w \cdot f = \chi(w)f$ for some linear character $\chi : W \to \mathbb{C}^\times$. The element $\delta := \prod_{s \in \mathcal{S}} \alpha_s \in \mathbb{C}[\mathfrak{h}]$ is a W-semi-invariant. To show this, let $w \in W$ and $s \in \mathcal{S}$. Then $wsw^{-1} \in \mathcal{S}$ is again a reflection. This implies that there is some nonzero scalar β such that $w(\alpha_s) = \beta\alpha_{wsw^{-1}}$. Thus, $w(\delta) = \gamma_w\delta$ for some nonzero scalar γ_w. One can check that $\gamma_{w_1 w_2} = \gamma_{w_1}\gamma_{w_2}$, which implies that δ is a semi-invariant. Therefore, there exists some $r > 0$ such that $\delta^r \in \mathbb{C}[\mathfrak{h}]^W$. The powers of δ^r form an Ore set in $\mathsf{H}_{t,c}(W)$ and we may localize $\mathsf{H}_{t,c}(W)$

at δ^r. Since the tail terms

$$\sum_{s\in\mathcal{S}} \frac{2c(s)}{1-\lambda_s} \frac{(y,\alpha_s)}{\alpha_s}(1-s)$$

of the Dunkl operators D_y belong to $H_{t,c}(W)[\delta^{-r}]$, this implies that each element $y \in \mathfrak{h} \subset \mathcal{D}_t(\mathfrak{h}_{\mathrm{reg}}) \rtimes W$ belongs to $H_{t,c}(W)[\delta^{-r}]$ too. Hence the Dunkl embedding becomes an isomorphism

$$H_{t,c}(W)[\delta^{-r}] \xrightarrow{\sim} \mathcal{D}_t(\mathfrak{h}_{\mathrm{reg}}) \rtimes W.$$

Recall that a ring is said to be simple if it contains no proper two-sided ideals.

PROPOSITION 1.8.2. *The rings $\mathcal{D}(\mathfrak{h})$, $\mathcal{D}(\mathfrak{h})^W$ and $\mathcal{D}(\mathfrak{h}_{\mathrm{reg}})^W$ are simple.*

PROOF. To show that $\mathcal{D}(\mathfrak{h})$ is simple, it suffices to show that $1 \in I$ for any two-sided ideal I. This can be shown by taking the commutator of a nonzero element $h \in I$ with suitable elements in $\mathcal{D}(\mathfrak{h})$. The fact that this implies that $\mathcal{D}(\mathfrak{h})^W$ is simple is standard, but maybe difficult to find in the literature. Firstly one notes that the fact that $\mathcal{D}(\mathfrak{h})$ is simple implies that $\mathcal{D}(\mathfrak{h})^W \simeq e(\mathcal{D}(\mathfrak{h}) \rtimes W)e$ is Morita equivalent to $e(\mathcal{D}(\mathfrak{h}) \rtimes W)e$; we've seen the argument already in the proof of Corollary 1.6.3. This implies that $\mathcal{D}(\mathfrak{h})^W$ is simple if and only if $\mathcal{D}(\mathfrak{h}) \rtimes W$ is simple; see [172, Theorem 3.5.9]. Finally, one can show directly that $\mathcal{D}(\mathfrak{h}), \rtimes W$ is simple: see [172, Proposition 7.8.12] and notice that W acts by outer automorphisms on $\mathcal{D}(\mathfrak{h})$ since the only invertible elements in $\mathcal{D}(\mathfrak{h})$ are the nonzero scalars.

Finally, since $\mathcal{D}(\mathfrak{h}_{\mathrm{reg}})^W$ is the localization of $\mathcal{D}(\mathfrak{h})^W$ at the two-sided Ore set generated by δ^r, a two-sided ideal J in $\mathcal{D}(\mathfrak{h}_{\mathrm{reg}})^W$ is proper if and only if $J \cap \mathcal{D}(\mathfrak{h})^W$ is a proper two-sided ideal. But we have already shown that $\mathcal{D}(\mathfrak{h})^W$ is simple. $\square$

Here is another application of the Dunkl embedding.

COROLLARY 1.8.3. *The centre of $e\mathsf{H}_{1,c}(W)e$ equals $\mathbb{C}$.*

PROOF. By Corollary 1.3.2(2), $e\mathsf{H}_{1,c}(W)e$ is an integral domain. Choose

$$z \in Z(e\mathsf{H}_{1,c}(W)e).$$

The Dunkl embedding defines an isomorphism

$$e\mathsf{H}_{1,c}(W)e[(e\delta^r)^{-1}] \xrightarrow{\sim} e(\mathcal{D}(\mathfrak{h}_{\mathrm{reg}}) \rtimes W)e \simeq \mathcal{D}(\mathfrak{h}_{\mathrm{reg}})^W.$$

Since $\mathcal{D}(\mathfrak{h}_{\mathrm{reg}})^W$ is simple, every nonzero central element is either a unit or zero (otherwise it would generate a proper two-sided ideal). Therefore the image of z in $e\mathsf{H}_{1,c}(W)e[(e\delta^r)^{-1}]$ is either a unit or zero. The fact that the only units in $\mathrm{gr}_{\mathcal{F}} e\mathsf{H}_{1,c}(W)e = \mathbb{C}[\mathfrak{h} \times \mathfrak{h}^*]^W$ are the scalars implies that the scalars are the only units in $e\mathsf{H}_{1,c}(W)e$. If z is a unit then $\alpha(e\delta^r)^a \cdot z = 1$ in $e\mathsf{H}_{1,c}(W)e$, for some $\alpha \in \mathbb{C}^\times$ and $a \in \mathbb{N}$. But $e\delta^r$ is not a unit in $e\mathsf{H}_{1,c}(W)e$ (since the symbol of $e\delta^r$ in $\mathbb{C}[\mathfrak{h} \times \mathfrak{h}^*]^W$ is not a unit). Therefore $a = 0$ and $z \in \mathbb{C}^\times$. On the other hand, if $(e\delta^r)^a \cdot z = 0$ for some a then the fact that $e\mathsf{H}_{1,c}(W)e$ is an integral domain implies that $z = 0$. $\square$

1.9. Exercises.

EXERCISE 1.9.1. Show that the centre $Z(\mathbb{C}[V] \rtimes G)$ of $\mathbb{C}[V] \rtimes G$ equals $\mathbb{C}[V]^G$.

EXERCISE 1.9.2. To see why the PBW theorem is quite a subtle statement, consider the algebra $L(\mathfrak{S}_2)$ defined to be

$$\mathbb{C}\langle x, y, s\rangle / \langle s^2 = 1, sx = -xs, sy = -ys, [y, x] = 1, (y - s)x = xy + s\rangle.$$

Show that $L(\mathfrak{S}_2) = 0$.

EXERCISE 1.9.3. (1) Show that the Dunkl operators, defined in (1.C), act on $\mathbb{C}[\mathfrak{h}]$. Hint: it's strongly recommended that you do the example $W = \mathbb{Z}_2$ first, where

$$D_y = y - \frac{c}{x}(1 - s).$$

(2) Using the fact that

$$s(x) = x - \frac{(\alpha_s^\vee, x)}{2}(1 - \lambda_s)\alpha_s, \qquad \text{for all } x \in \mathfrak{h}^*,$$

show that $x \mapsto x$, $w \mapsto w$ and $y \mapsto D_y$ defines a morphism

$$\phi : \mathsf{H}_{t,c}(W) \to \mathcal{D}_t(\mathfrak{h}_{\text{reg}}) \rtimes W,$$

i.e., show that the commutation relation

$$[D_y, x] = t(y, x) - \sum_{s \in S} c(s)(y, \alpha_s)(\alpha_s^\vee, x)s$$

holds for all $x \in \mathfrak{h}^*$ and $y \in \mathfrak{h}$. Hint: as above, try the case $\mathbb{Z}_2$ first.

EXERCISE 1.9.4. By considering its image under the Dunkl embedding, show that the spherical subalgebra $e\mathsf{H}_{t,c}(W)e$ is commutative when $t = 0$.

EXERCISE 1.9.5. A complex reflection group $(W, \mathfrak{h})$ is said to be *real* if there exists a real vector subspace $\mathfrak{h}^{\text{re}}$ of $\mathfrak{h}$ such that $(W, \mathfrak{h}^{\text{re}})$ is a real reflection group and $\mathfrak{h} = \mathfrak{h}^{\text{re}} \otimes_{\mathbb{R}} \mathbb{C}$. In this case there exists a W-invariant inner product $(-, -)_{\text{re}}$ on $\mathfrak{h}^{\text{re}}$, i.e., an inner product $(-, -)_{\text{re}}$ such that $(wu, wv)_{\text{re}} = (u, v)_{\text{re}}$ for all $w \in W$ and $u, v \in (-, -)_{\text{re}}$. We extend it by linearity to a W-invariant bilinear form $(-, -)$ on $\mathfrak{h}$. The following fact is also very useful when studying rational Cherednik algebras at $t = 0$.

(1) Assume now that W is a real reflection group. Show that the rule $x \mapsto \tilde{x} = (x, -)$, $y \mapsto \tilde{y} = (y, -)$ and $w \mapsto w$ defines an automorphism of $\mathsf{H}_{t,c}(W)$, swapping $\mathbb{C}[\mathfrak{h}]$ and $\mathbb{C}[\mathfrak{h}^*]$.

(2) Show that $\mathbb{C}[\mathfrak{h}]^W$ and $\mathbb{C}[\mathfrak{h}^*]^W$ are central subalgebras of $\mathsf{H}_{0,c}(W)$. Hint: first use the Dunkl embedding to show that $\mathbb{C}[\mathfrak{h}]^W$ is central, then use the automorphism defined above.

1.10. Additional remarks.

In his original paper [68], Chevalley showed that if $(W, \mathfrak{h})$ is a complex reflection group then $\mathbb{C}[\mathfrak{h}]^W$ is a polynomial ring. The converse was shown by Shephard and Todd in [201].

The definition of symplectic reflection algebras first appear in [99].

The PBW theorem, Theorem 1.3.1, and its proof are Theorem 1.3 of [99].

Theorems 1.6.1 and 1.7.3 are also contained in [99], as Theorem 1.5 and Theorem 3.1 respectively.

The first part of Theorem 1.7.1 is due to Etingof and Ginzburg, [99, Theorem 1.6]. The second part is due to Brown and Gordon [51, Proposition 7.2]. Both proofs rely in a crucial way on the Poisson structure of $\mathbb{C}[V]^G$.

2. Rational Cherednik algebras at $t = 1$

In Sections 2 to 4, we focus on rational Cherednik algebras at $t = 1$, and omit t from the notation. We will only be considering rational Cherednik algebras because relatively little is know about general symplectic reflection algebras at $t = 1$. Therefore, we let $(W, \mathfrak{h})$ be a complex reflection group and $\mathsf{H}_c(W)$ the associated rational Cherednik algebra.

As noted in the previous section, the centre of $\mathsf{H}_c(W)$ equals $\mathbb{C}$. Therefore, its behaviour is very different from the case $t = 0$. If we take $c = 0$ then $\mathsf{H}_0(W) = \mathcal{D}(\mathfrak{h}) \rtimes W$ and the category of modules for $\mathcal{D}(\mathfrak{h}) \rtimes W$ is precisely the category of W-equivariant $\mathcal{D}$-modules on $\mathfrak{h}$. In particular, there are *no* finite-dimensional representations of this algebra. In general, the algebra $\mathsf{H}_c(W)$ has very few finite-dimensional representations.

2.1. For rational Cherednik algebras, the PBW theorem implies that, as a vector space, $\mathsf{H}_c(W) \simeq \mathbb{C}[\mathfrak{h}] \otimes \mathbb{C}W \otimes \mathbb{C}[\mathfrak{h}^*]$; there is no need to choose an ordered basis of $\mathfrak{h}$ and $\mathfrak{h}^*$ for this to hold. Since $\mathbb{C}[\mathfrak{h}]$ is in some sense opposite to $\mathbb{C}[\mathfrak{h}^*]$, this is an example of a *triangular decomposition*, just like the triangular decomposition $U(\mathfrak{g}) = U(\mathfrak{n}_-) \otimes U(\mathfrak{h}) \otimes U(\mathfrak{n}_+)$ encountered in Lie theory, where $\mathfrak{g}$ is a finite-dimensional, semisimple Lie algebra over $\mathbb{C}$, $\mathfrak{g} = \mathfrak{n}_- \oplus \mathfrak{h} \oplus \mathfrak{n}_+$ is a decomposition into a Cartan subalgebra $\mathfrak{h}$, the nilpotent radical $\mathfrak{n}_+$ of the Borel $\mathfrak{b} = \mathfrak{h} \oplus \mathfrak{n}_+$, and the opposite $\mathfrak{n}_-$ of the nilpotent radical $\mathfrak{n}_+$. This suggests that it might be fruitful to try and mimic some of the common constructions used in Lie theory. In the representation theory of $\mathfrak{g}$, one of the categories of modules most intensely studied, and best understood, is category $\mathcal{O}$, the abelian category of finitely generated $\mathfrak{g}$-modules that are semisimple as $\mathfrak{h}$-modules and $\mathfrak{n}_+$-locally nilpotent. Therefore, it is natural to try and study an analogue of category $\mathcal{O}$ for rational Cherednik algebras. This is what we will do in this section.

2.2. Category $\mathcal{O}$. Let $\mathsf{H}_c(W)$-mod be the category of all finitely generated left $\mathsf{H}_c(W)$-modules. It is a hopeless task to try and understand in any detail the whole category $\mathsf{H}_c(W)$-mod. Therefore, one would like to try and understand certain interesting, but manageable, subcategories. The PBW theorem suggests the following very natural definition.

DEFINITION 2.2.1. *Category* $\mathcal{O}$ is defined to be the full[3] subcategory of $\mathsf{H}_c(W)$-mod consisting of all modules M such that the action of $\mathfrak{h} \subset \mathbb{C}[\mathfrak{h}^*]$ is locally nilpotent.

REMARK 2.2.2. A module M is said to be locally nilpotent for $\mathfrak{h}$ if, for each $m \in M$ there exists some $N \gg 0$ such that $\mathfrak{h}^N \cdot m = 0$.

As shown in Exercise 2.11.1, every module in category $\mathcal{O}$ is finitely generated as a $\mathbb{C}[\mathfrak{h}]$-module. We will give the proofs of the fundamental properties of category $\mathcal{O}$ since they do not require any sophisticated machinery. However, this does make this section rather formal, so we first outline the key features of category $\mathcal{O}$ so that the reader can get their bearings. Recall that an abelian category is called finite length if every object satisfies the ascending chain condition, and descending chain condition, on subobjects. It is Krull–Schmidt if every module has a unique decomposition (up to permuting summands) into a direct sum of indecomposable modules.

- There are only finitely many simple modules in category $\mathcal{O}$.
- Category $\mathcal{O}$ is a finite length, Krull–Schmidt category.
- Every simple module admits a projective cover, hence category $\mathcal{O}$ contains enough projectives.
- Category $\mathcal{O}$ contains "standard modules", making it a highest weight category.

2.3. Standard objects. One can use induction to construct certain "standard objects" in category $\mathcal{O}$. The skew-group ring $\mathbb{C}[\mathfrak{h}^*] \rtimes W$ is a subalgebra of $\mathsf{H}_c(W)$. Therefore, we can induce to category $\mathcal{O}$ those representations of $\mathbb{C}[\mathfrak{h}^*] \rtimes W$ that are locally nilpotent for $\mathfrak{h}$. Let $\mathfrak{m} = \mathbb{C}[\mathfrak{h}^*]_+$ be the augmentation ideal. Then, for each λ in $\mathrm{Irr}(W)$, we define the $\mathbb{C}[\mathfrak{h}^*] \rtimes W$-module $\tilde{\lambda} = \mathbb{C}[\mathfrak{h}^*] \rtimes W \otimes_W \lambda$. For each $r \in \mathbb{N}$, the subspace $\mathfrak{m}^r \cdot \tilde{\lambda}$ is a proper $\mathbb{C}[\mathfrak{h}^*] \rtimes W$-submodule of $\tilde{\lambda}$ and we set $\lambda_r := \tilde{\lambda}/\mathfrak{m}^r \cdot \tilde{\lambda}$. This is a $\mathfrak{h}$-locally nilpotent $\mathbb{C}[\mathfrak{h}^*] \rtimes W$-module. We set

$$\Delta_r(\lambda) = \mathsf{H}_c(W) \otimes_{\mathbb{C}[\mathfrak{h}^*] \rtimes W} \lambda_r.$$

It follows from Lemma 2.4.1 that $\Delta_r(\lambda)$ is a module in category $\mathcal{O}$. The module $\Delta(\lambda) := \Delta_1(\lambda)$ is called a *standard module* (or, often, a Verma module) of category $\mathcal{O}$. The PBW theorem implies that $\Delta(\lambda) = \mathbb{C}[\mathfrak{h}] \otimes_{\mathbb{C}} \lambda$ as a $\mathbb{C}[\mathfrak{h}]$-module.

EXAMPLE 2.3.1. For our favourite example, $\mathbb{Z}_2$ acting on $\mathfrak{h} = \mathbb{C} \cdot y$ and $\mathfrak{h}^* = \mathbb{C} \cdot x$, we have $\mathrm{Irr}(\mathbb{Z}_2) = \{\rho_0, \rho_1\}$, where ρ_0 is the trivial representation and ρ_1 is the sign representation. Then,

$$\Delta(\rho_0) = \mathbb{C}[x] \otimes \rho_0, \quad \Delta(\rho_1) = \mathbb{C}[x] \otimes \rho_1.$$

The subalgebra $\mathbb{C}[x] \rtimes \mathbb{Z}_2$ acts in the obvious way. The action of y is given as follows:

$$y \cdot f(x) \otimes \rho_i = [y, f(x)] \otimes \rho_i + f(x) \otimes y\rho_i$$

$$= [y, f(x)] \otimes \rho_i, \qquad i = 0, 1.$$

Here $[y, f(x)] \in \mathbb{C}[x] \rtimes \mathbb{Z}_2$ is calculated in $\mathsf{H}_c(\mathbb{Z}_2)$.

---

[3]Recall that a subcategory $\mathcal{B}$ of a category $\mathcal{A}$ is called *full* if $\mathrm{Hom}_{\mathcal{B}}(M, N) = \mathrm{Hom}_{\mathcal{A}}(M, N)$ for all $M, N \in \mathrm{Obj}\mathcal{B}$.

2.4. The Euler element. In Lie theory, the fact that every module M in category $\mathcal{O}$ is semisimple as a $\mathfrak{h}$-module is very important, since M decomposes as a direct sum of weight spaces. We don't have a Cartan subalgebra in $H_c(W)$, but the Euler element is a good substitute (equivalently one can think that the Cartan subalgebras of $H_c(W)$ are one-dimensional).

Let $x_1, \ldots, x_n$ be a basis of $\mathfrak{h}^*$ and $y_1, \ldots, y_n \in \mathfrak{h}$ the dual basis. Define the *Euler element* in $H_c(W)$ to be

$$\mathbf{eu} = \sum_{i=1}^{n} x_i y_i - \sum_{s \in S} \frac{2c(s)}{1 - \lambda_s} s,$$

where λ_s was defined in Section 1.8. The relevance of the element $\mathbf{eu}$ is given by the fact that it satisfies the following fundamental relations:

$$[\mathbf{eu}, x] = x, \quad [\mathbf{eu}, y] = -y, \quad [\mathbf{eu}, w] = 0, \qquad \text{for all } x \in \mathfrak{h}^*, \ y \in \mathfrak{h}, \ w \in W.$$

See Exercise 2.11.2 for the proof. Conjugation by $\mathbf{eu}$ defines a $\mathbb{Z}$-grading on $H_c(W)$, where $\deg(x) = 1$, $\deg(y) = -1$ and $\deg(w) = 0$. The sum

$$-\sum_{s \in S} \frac{2c(s)}{1 - \lambda_s} s$$

belongs to $Z(W)$, the centre of the group algebra. Therefore, if λ is an irreducible W-module, this central element will act by a scalar on λ. This scalar will be denoted by c_λ.

LEMMA 2.4.1. *The modules $\Delta_r(\lambda)$ belong to category $\mathcal{O}$.*

PROOF. We begin by noting that category $\mathcal{O}$ is closed under extensions, i.e., if we have a short exact sequence

$$0 \to M_1 \to M_2 \to M_3 \to 0$$

of $H_c(W)$-modules, where M_1 and M_3 belong to $\mathcal{O}$, then M_2 also belongs to $\mathcal{O}$. The PBW theorem implies that $H_c(W)$ is a free right $\mathbb{C}[\mathfrak{h}^*] \rtimes W$-module. Therefore the short exact sequence $0 \to \lambda_{r-1} \to \lambda_r \to \lambda_1 \to 0$ of $\mathbb{C}[\mathfrak{h}^*] \rtimes W$-modules defines a short exact sequence

$$0 \to \Delta_{r-1}(\lambda) \to \Delta_r(\lambda) \to \Delta(\lambda) \to 0 \qquad (2.A)$$

of $H_c(W)$-modules. Hence, by induction on r, it suffices to show that $\Delta(\lambda)$ belongs to $\mathcal{O}$. We can make $\Delta(\lambda)$ into a $\mathbb{Z}$-graded $H_c(W)$-module by putting $1 \otimes \lambda$ in degree zero. Then, $\Delta(\lambda)$ is actually positively graded with each graded piece finite-dimensional. Since $y \in \mathfrak{h}$ maps $\Delta(\lambda)_i$ into $\Delta(\lambda)_{i-1}$, we have $\mathfrak{h}^{i+1} \cdot \Delta(\lambda)_i = 0$. $\qquad \square$

For each $a \in \mathbb{C}$, the generalised eigenspace of weight a, with respect to $\mathbf{eu}$, of a $H_c(W)$-module M is defined to be

$$M_a = \{m \in M \mid (\mathbf{eu} - 1)^N \cdot m = 0 \text{ for } N \gg 0\}.$$

As for weight modules in Lie theory, we have:

LEMMA 2.4.2. *Each $M \in \mathcal{O}$ is the direct sum of its generalised **eu**-eigenspaces*

$$M = \bigoplus_{a \in \mathbb{C}} M_a,$$

and $\dim M_a < \infty$ *for all* $a \in \mathbb{C}$.

PROOF. Since M is in category $\mathcal{O}$, we can choose a finite-dimensional $\mathbb{C}[\mathfrak{h}^*] \rtimes W$-submodule M' of M that generates M as a $\mathsf{H}_c(W)$-module. Since M' is finite-dimensional, there exists some $r \gg 0$ such that $\mathfrak{h}^r \cdot M' = 0$. Thus we may find $\lambda_1, \ldots, \lambda_k \in \mathrm{Irr}(W)$ such that the sequence

$$\bigoplus_{i=1}^{k} \Delta_r(\lambda_i) \to M \to 0$$

is exact. Each $\Delta(\lambda)_i$ is a generalised **eu**-eigenspace with eigenvalue $i + c_\lambda$. Hence $\Delta(\lambda)$ is a direct sum of its **eu**-eigenspaces. As in the proof of Lemma 2.4.1, one can use this fact together with the short exact sequence (2.A) to conclude that each $\Delta_r(\lambda)$ is a direct sum of its generalised **eu**-eigenspaces, with each eigenspace finite-dimensional. This implies that M has this property too. $\square$

2.5. Characters. Using the Euler operator **eu** we can define the character of a module $M \in \mathcal{O}$ to be

$$\mathrm{ch}(M) = \sum_{a \in \mathbb{C}} (\dim M_a) t^a.$$

The Euler element acts via the scalar c_λ on $1 \otimes \lambda \subset \Delta(\lambda)$. This implies that

$$\mathrm{ch}(\Delta(\lambda)) = \frac{\dim(\lambda) t^{c_\lambda}}{(1 - t)^n}.$$

By Exercise 2.11.4, the character $\mathrm{ch}(M)$ belong to $\bigoplus_{a \in \mathbb{C}} t^a \mathbb{Z}[[t]]$ for any M in category $\mathcal{O}$. In fact one can do even better. As shown in Exercise 2.8.4 below, the standard modules $\Delta(\lambda)$ are a $\mathbb{Z}$-basis of the Grothendieck group $K_0(\mathcal{O})$. The character of M only depends on its image in $K_0(\mathcal{O})$. Therefore if

$$[M] = \sum_{\lambda \in \mathrm{Irr}(W)} n_\lambda [\Delta(\lambda)] \in K_0(\mathcal{O}),$$

for some $n_\lambda \in \mathbb{Z}$, the fact that $\mathrm{ch}(\Delta(\lambda)) = \frac{\dim(\lambda) t^{c_\lambda}}{(1-t)^n}$ implies that $\mathrm{ch}(M) = \frac{f(t)}{(1-t)^n}$, where

$$f(t) = \sum_{\lambda \in \mathrm{Irr}(W)} n_\lambda \dim(\lambda) t^{c_\lambda} \in \mathbb{Z}[x^a \mid a \in \mathbb{C}].$$

Hence, the rule $M \mapsto (1 - t)^n \cdot \mathrm{ch}(M)$ is a morphism of abelian groups $K_0(\mathcal{O}) \to \mathbb{Z}[(\mathbb{C}, +)]$. It is not in general an embedding.

2.6. Simple modules. Two basic problems motivating much of the research in the theory of rational Cherednik algebras are:

(1) Classify the simple modules in $\mathcal{O}$.

(2) Calculate $\mathrm{ch}(L)$ for all simple modules $L \in \mathcal{O}$.

The first problem is easy, but the second is very difficult (and still open in general).

LEMMA 2.6.1. *Let M be a nonzero module in category $\mathcal{O}$. Then, there exists some $\lambda \in \mathrm{Irr}(W)$ and nonzero homomorphism $\Delta(\lambda) \to M$.*

PROOF. Note that the real part of the weights of M are bounded from below, i.e., there exists some $K \in \mathbb{R}$ such that $M_a \neq 0$ implies $\mathrm{Re}(a) \geq K$. Therefore we may choose some $a \in \mathbb{C}$ such that $M_a \neq 0$ and $M_b = 0$ for all $b \in \mathbb{C}$ such that $a - b \in \mathbb{R}_{>0}$. An element $m \in M$ is said to be *singular* if $\mathfrak{h} \cdot m = 0$, i.e., it is annihilated by all y's. Our assumption implies that all elements in M_a are singular. Every weight space M_a is a W-submodule. If λ occurs in M_a with nonzero multiplicity then there is a well defined homomorphism $\Delta(\lambda) \to M$, whose restriction to $1 \otimes \lambda$ injects into M_a. □

LEMMA 2.6.2. *Each standard module $\Delta(\lambda)$ has a simple head $L(\lambda)$ and the set*

$$\{L(\lambda) \mid \lambda \in \mathrm{Irr}(W)\}$$

is a complete set of nonisomorphic simple modules of category $\mathcal{O}$.

PROOF. Let R be the sum of all proper submodules of $\Delta(\lambda)$. It suffices to show that $R \neq \Delta(\lambda)$. The weight subspace $\Delta(\lambda)_{c_\lambda} = 1 \otimes \lambda$ is irreducible as a W-module and generates $\Delta(\lambda)$. If $R_{c_\lambda} \neq 0$ then there exists some proper submodule N of $\Delta(\lambda)$ such that $N_{c_\lambda} \neq 0$. But then $N = \Delta(\lambda)$ since $1 \otimes \lambda$ generates $\Delta(\lambda)$. Thus, $R_{c_\lambda} = 0$, implying that R itself is a proper submodule of $\Delta(\lambda)$. Now let L be a simple module in category $\mathcal{O}$. By Lemma 2.6.1, there exists a nonzero homomorphism $\Delta(\lambda) \to L$ for some $\lambda \in \mathrm{Irr}(W)$. Hence $L \simeq L(\lambda)$. The fact that $L(\lambda) \simeq L(\mu)$ implies $\lambda \simeq \mu$ follows from the fact that L_{sing}, the space of singular vectors in L, is irreducible as a W-module. □

EXAMPLE 2.6.3. Let's consider $\mathsf{H}_c(\mathbb{Z}_2)$ at $c = -\frac{3}{2}$. Then, one can check that

$$\Delta(\rho_1) = \mathbb{C}[x] \otimes \rho_1 \twoheadrightarrow L(\rho_1) = (\mathbb{C}[x] \otimes \rho_1)/(x^3 \mathbb{C}[x] \otimes \rho_1).$$

On the other hand, a direct calculation shows that $\Delta(\rho_0) = L(\rho_0)$ is simple. The composition series of $\Delta(\rho_1)$ is

$$\begin{matrix} L(\rho_1) \\ L(\rho_0) \end{matrix}.$$

COROLLARY 2.6.4. *Every module in category $\mathcal{O}$ has finite length.*

PROOF. Let M be a nonzero object of category $\mathcal{O}$. Choose some real number $K \gg 0$ such that $\mathrm{Re}(c_\lambda) < K$ for all $\lambda \in \mathrm{Irr}(W)$. We write $M^{\leq K}$ for the sum of all weight spaces M_a such that $\mathrm{Re}(a) \leq K$. It is a finite-dimensional subspace. Lemma 2.6.1 implies that $N^{\leq K} \neq 0$ for all nonzero submodules N of M. Therefore, if

$$N_0 \supsetneq N_1 \supsetneq N_2 \supsetneq \cdots$$

is a proper descending chain of submodule of M then

$$N_0^{\leq K} \supsetneq N_1^{\leq K} \supsetneq N_2^{\leq K} \supsetneq \cdots$$

is a proper descending chain of subspaces of $M^{\leq K}$. Hence the chain must have finite length. □

2.7. Projective modules. A module $P \in \mathcal{O}$ is said to be projective if the functor $\mathrm{Hom}_{H_c(W)}(P, -) : \mathcal{O} \to \mathrm{Vect}(\mathbb{C})$ is exact. It is important to note that a projective module $P \in \mathcal{O}$ is *not* projective when considered as a module in $H_c(W)$-mod, i.e., being projective is a relative concept.

DEFINITION 2.7.1. An object Q in $\mathcal{O}$ is said to have a Δ-*filtration* if it has a finite filtration $0 = Q_0 \subset Q_1 \subset \cdots \subset Q_r = Q$ such that $Q_i / Q_{i-1} \simeq \Delta(\lambda_i)$ for some $\lambda_i \in \mathrm{Irr}(W)$ and all $1 \leq i \leq r$.

Let $L \in \mathcal{O}$ be simple. A *projective cover* $P(L)$ of L is a projective module in $\mathcal{O}$ together with a surjection $p : P(L) \to L$ such that any morphism $f : M \to P(L)$ is a surjection whenever $p \circ f : M \to L$ is a surjection. Equivalently, the head of $P(L)$ equals L. Projective covers, when they exist, are unique up to isomorphism. The following theorem, first shown in [125], is of key importance in the study of category $\mathcal{O}$. We follow the proof given in [2].

THEOREM 2.7.2. *Every simple module $L(\lambda)$ in category $\mathcal{O}$ has a projective cover $P(\lambda)$. Moreover, each $P(\lambda)$ has a finite Δ-filtration.*

Unfortunately the proof of Theorem 2.7.2 is rather long and technical. We suggest the reader skips it on first reading. For $a \in \mathbb{C}$ we denote by $\bar{a}$ its image in $\mathbb{C}/\mathbb{Z}$. We write $\mathcal{O}^{\bar{a}}$ for the full subcategory of $\mathcal{O}$ consisting of all M such that $M_b = 0$ for all $b \notin a + \mathbb{Z}$.

PROOF OF THEOREM 2.7.2. Exercise 2.11.5 shows that it suffices to construct a projective cover $P(\lambda)$ for $L(\lambda)$ in $\mathcal{O}^{\bar{a}}$. Fix a representative $a \in \mathbb{C}$ of $\bar{a}$. For each $k \in \mathbb{Z}$, let $\mathcal{O}^{\geq k}$ denote the full subcategory of $\mathcal{O}^{\bar{a}}$ consisting of modules M such that $M_b \neq 0$ implies that $b - a \in \mathbb{Z}_{\geq k}$. Then, for $k \gg 0$, we have $\mathcal{O}^{\geq k} = 0$ and for $k \ll 0$ we have $\mathcal{O}^{\geq k} = \mathcal{O}^{\bar{a}}$. To see this, it suffices to show that such k exist for the finitely many simple modules $L(\lambda)$ in $\mathcal{O}^{\bar{a}}$ — then an arbitrary module in $\mathcal{O}^{\bar{a}}$ has a finite composition series with factors the $L(\lambda)$, which implies that the corresponding statement holds for them too. Our proof of Theorem 2.7.2 will be by induction on k. Namely, for each k and all $\lambda \in \mathrm{Irr}(W)$ such that $L(\lambda) \in \mathcal{O}^{\geq k}$, we will construct a projective cover $P_k(\lambda)$ of $L(\lambda)$ in $\mathcal{O}^{\geq k}$ such that $P_k(\lambda)$ is a *quotient* of $\Delta_r(\lambda)$ for $r \gg 0$. At the end we'll deduce that $P_k(\lambda)$ itself has a Δ-filtration. The idea is to try and lift each $P_k(\lambda)$ in $\mathcal{O}^{\geq k}$ to a corresponding $P_{k-1}(\lambda)$ in $\mathcal{O}^{\geq k-1}$.

Let k_0 be the largest integer such that $\mathcal{O}^{\geq k_0} \neq 0$.

CLAIM 2.7.3. *The category $\mathcal{O}^{\geq k_0}$ is semisimple with $P_{k_0}(\lambda) = \Delta(\lambda) = L(\lambda)$ for all λ such that $L(\lambda) \in \mathcal{O}^{\geq k_0}$.*

PROOF OF THE CLAIM. Note that $\Delta(\lambda)_{c_\lambda} = 1 \otimes \lambda$ and $\Delta(\lambda)_b \neq 0$ implies that $b - c_\lambda \in \mathbb{Z}_{\geq 0}$. If the quotient map $\Delta(\lambda) \to L(\lambda)$ has a nonzero kernel K then choose $L(\mu) \subset K$ a simple submodule. We have $c_\mu - c_\lambda \in \mathbb{Z}_{>0}$, contradicting the minimality of k_0. Thus $K = 0$. Since $\Delta(\lambda)$ is an induced module, adjunction implies that

$$\mathrm{Hom}_{\mathcal{O}^{\geq k_0}}(\Delta(\lambda), M) = \mathrm{Hom}_{\mathbb{C}[\mathfrak{h}^*] \rtimes W}(\lambda, M)$$

for all $M \in \mathcal{O}^{\geq k_0}$. Again, since all the weights of M are at least c_λ, this implies that

$$\mathrm{Hom}_{\mathbb{C}[\mathfrak{h}^*] \rtimes W}(\lambda, M) = \mathrm{Hom}_W(\lambda, M_{c_\lambda}).$$

Since M is a direct sum of its generalised **eu**-eigenspaces, this implies that the functor $\mathrm{Hom}_{\mathcal{O}^{\geq k_0}}(\Delta(\lambda), -)$ is exact, i.e., $\Delta(\lambda)$ is projective. $\square$

Now take $k < k_0$ and assume that we have constructed, for all $L(\lambda) \in \mathcal{O}^{\geq k+1}$, a projective cover $P_{k+1}(\lambda)$ of $L(\lambda)$ in $\mathcal{O}^{\geq k+1}$ with the desired properties. If $\mathcal{O}^{\geq k+1} = \mathcal{O}^{\geq k}$ then there is nothing to do so we may assume that there exist $\mu_1, \ldots, \mu_r \in \mathrm{Irr}(W)$ such that $L(\mu_i)$ belongs to $\mathcal{O}^{\geq k}$, but not to $\mathcal{O}^{\geq k+1}$. Note that $c_{\mu_i} = a + k$ for all i. For all $M \in \mathcal{O}^{\geq k}$, either $M_{a+k} = 0$ (in which case $M \in \mathcal{O}^{\geq k+1}$) or M_{a+k} consists of singular vectors, i.e., $\mathfrak{h} \cdot M_{a+k} = 0$. Therefore, as in the proof of Claim 2.7.3, we have $P_k(\mu_i) = \Delta(\mu_i)$ for $1 \leq i \leq r$. Notice that $P_k(\mu_i)$ is obviously a quotient of $\Delta(\mu_i)$. Thus, we are left with constructing the lifts $P_k(\lambda)$ of $P_{k+1}(\lambda)$ for all those λ such that $L(\lambda) \in \mathcal{O}^{\geq k+1}$. Fix one such λ.

CLAIM 2.7.4. *There exists some integer $N \gg 0$ such that $(\mathbf{eu} - c_\lambda)^N \cdot m = 0$ for all $M \in \mathcal{O}^{\geq k}$ and all $m \in M_{c_\lambda}$.*

By definition, for a given $M \in \mathcal{O}^{\geq k}$ and $m \in M_{c_\lambda}$, there exists some $N \gg 0$ such that $(\mathbf{eu} - c_\lambda)^N \cdot m = 0$. The claim is stating that one can find a particular N that works simultaneously for all $M \in \mathcal{O}^{\geq k}$ and all $m \in M_{c_\lambda}$.

PROOF OF THE CLAIM. Since $\dim(M_{a+k}) < \infty$ there exist $n_i \in \mathbb{N}$ and a morphism

$$\phi : \bigoplus_{i=1}^r \Delta(\mu_i)^{\oplus n_i} \longrightarrow M,$$

such that the cokernel M' of ϕ is in $\mathcal{O}^{\geq k+1}$. Therefore we may construct a surjection

$$\psi : \bigoplus_\eta P_{k+1}(\eta)^{s_\eta} \twoheadrightarrow M',$$

where the sum is over all $\eta \in \mathrm{Irr}(W)$ such that $L(\eta) \in \mathcal{O}^{\geq k+1}$. Since each $\Delta_r(\eta)$ is in category $\mathcal{O}$, Lemma 2.4.2 implies that there is some $N \gg 0$ such that $(\mathbf{eu} - c_\lambda)^{N-1} \cdot q = 0$ for all $q \in \Delta_r(\eta)_{c_\lambda}$. Since $P_{k+1}(\eta)$ is a quotient of $\Delta_r(\eta)$, $(\mathbf{eu} - c_\lambda)^{N-1} \cdot p = 0$ for all $p \in P_{k+1}(\eta)_{c_\lambda}$ too. Therefore $(\mathbf{eu} - c_\lambda)^{N-1} \cdot m$ lies in the image of ϕ for all $m \in M_{c_\lambda}$ and hence $(\mathbf{eu} - c_\lambda)^N \cdot m = 0$. $\square$

Now choose some new integer r such that $r + c_\lambda \gg a + k$ and define

$$R(\lambda) = \frac{\Delta_r(\lambda)}{\mathsf{H}_c(W) \cdot (\mathbf{eu} - c_\lambda)^N (1 \otimes 1 \otimes \lambda)}.$$

Then, for $M \in \mathcal{O}^{\geq k}$,

$$\mathrm{Hom}_{\mathsf{H}_c(W)}(R(\lambda), M) = \{m \in M_{c_\lambda} \mid \mathfrak{h}^r \cdot m = 0\}_\lambda,$$

where the subscript $\{-\}_\lambda$ refers to the λ-isotypic component. Since $c_\lambda - r \ll a + k$ and $M \in \mathcal{O}^{\geq k}$, we have $M_{c_\lambda - r} = 0$. But $\mathfrak{h}^r \cdot m \in M_{c_\lambda - r}$ for all $m \in M_{c_\lambda}$ which means that $\mathrm{Hom}_{\mathsf{H}_c(W)}(R(\lambda), M)$ equals the λ-isotypic component of M_{c_λ}. Thus, $\mathrm{Hom}_{\mathsf{H}_c(W)}(R(\lambda), -)$ is exact on $\mathcal{O}^{\geq k}$ (the functor M maps to the λ-isotypic component of M_{c_λ} being exact). It is nonzero because it surjects onto $\Delta(\lambda)$. The only problem is that it does not necessarily belong to $\mathcal{O}^{\geq k}$. So we let $\tilde{R}(\lambda)$ be the $\mathsf{H}_c(W)$-submodule generated by all weight spaces $R(\lambda)_b$ with $b - a \notin \mathbb{Z}_{\geq k}$ and define $P_k(\lambda) := R(\lambda)/\tilde{R}(\lambda)$. By construction, it belongs to $\mathcal{O}^{\geq k}$ and if $f : R(\lambda) \to M$ is any morphism with $M \in \mathcal{O}^{\geq k}$ then $\tilde{R}(\lambda) \subset \mathrm{Ker}\, f$. Therefore it is the projective cover of $L(\lambda)$ in $\mathcal{O}^{\geq k}$. We have constructed $P_k(\lambda)$ as a quotient of $\Delta_r(\lambda)$, an object equipped with a Δ-filtration.

The only thing left to show is that if $k \ll 0$ such that $\mathcal{O}^{\geq k} = \mathcal{O}^{\bar{a}}$ then $P_k(\lambda)$ has a Δ-filtration. By construction, it is a quotient of an object $M \in \mathcal{O}$ that is equipped with a Δ-filtration. But our assumption on k means that $P_k(\lambda)$ is projective in $\mathcal{O}$. Thus, it is a direct summand of M. Therefore, it suffices to note that if $M = M_1 \oplus M_2$ is an object of $\mathcal{O}$ equipped with a Δ-filtration, then each M_i also has a Δ-filtration (this follows by induction on the length of M from the fact that the modules $\Delta(\lambda)$ are *indecomposable*). This completes the proof of 2.7.2. □

In general, it is very difficult to explicitly construct the projective covers $P(\lambda)$. The object $P = \bigoplus_{\lambda \in \mathrm{Irr}(W)} P(\lambda)$ is a projective generator of category $\mathcal{O}$ i.e for each $M \in \mathcal{O}$ there exists some $N \gg 0$ and surjection $P^N \to M$. Therefore, we have an equivalence of abelian categories

$$\mathcal{O} \simeq A\text{-mod},$$

where $A = \mathrm{End}_{\mathsf{H}_c(W)}(P)$ is a finite-dimensional $\mathbb{C}$-algebra. See [22, Chapter 2], and in particular [22, Corollary 2.6], for details.

2.8. Highest weight categories. Just as for category $\mathcal{O}$ of a semisimple Lie algebra $\mathfrak{g}$ over $\mathbb{C}$, the existence of standard modules in category $\mathcal{O}$ implies that this category has a lot of additional structure. In particular, it is an example of a highest weight (or quasihereditary) category. The abstract notion of a highest weight category was introduced in [70].

DEFINITION 2.8.1. Let $\mathcal{A}$ be an abelian, $\mathbb{C}$-linear and finite length category, and Λ a poset. We say that $(\mathcal{A}, \Lambda)$ is a *highest weight category* if

(1) there is a complete set $\{L(\lambda) \mid \lambda \in \Lambda\}$ of nonisomorphic simple objects labelled by Λ;

(2) there is a collection of *standard* objects $\{\Delta(\lambda) \mid \lambda \in \Lambda\}$ of $\mathcal{A}$, with surjections $\phi_\lambda : \Delta(\lambda) \twoheadrightarrow L(\lambda)$ such that all composition factors $L(\mu)$ of $\mathrm{Ker}\, \phi_\lambda$ satisfy $\mu < \lambda$;

(3) each $L(\lambda)$ has a projective cover $P(\lambda)$ in $\mathcal{A}$ and the projective cover $P(\lambda)$ admits a Δ-filtration $0 = F_0 P(\lambda) \subset F_1 P(\lambda) \subset \cdots \subset F_m P(\lambda) = P(\lambda)$ such that

- $F_m P(\lambda)/F_{m-1} P(\lambda) \simeq \Delta(\lambda)$;
- for $0 < i < m$, $F_i P(\lambda)/F_{i-1} P(\lambda) \simeq \Delta(\mu)$ for some $\mu > \lambda$.

Define a partial ordering on $\mathrm{Irr}(W)$ by setting

$$\lambda \leq_c \mu \iff c_\mu - c_\lambda \in \mathbb{Z}_{\geq 0}.$$

LEMMA 2.8.2. *Choose λ and μ in $\mathrm{Irr}(W)$ such that $\lambda \nleq_c \mu$. Then*

$$\mathrm{Ext}^1_{\mathsf{H}_c(W)}(\Delta(\lambda), \Delta(\mu)) = 0.$$

PROOF. Recall that $\mathrm{Ext}^1_{\mathsf{H}_c(W)}(\Delta(\lambda), \Delta(\mu))$ can be identified with isomorphism classes of short exact sequences $0 \to \Delta(\mu) \to M \to \Delta(\lambda) \to 0$. As such, the group $\mathrm{Ext}^1_{\mathsf{H}_c(W)}(\Delta(\lambda), \Delta(\mu))$ is zero if and only if all such short exact sequences split. Assume that we are given a short exact sequence

$$0 \to \Delta(\mu) \to M \to \Delta(\lambda) \to 0,$$

for some $M \in \mathsf{H}_c(W)$-mod. Then $M \in \mathcal{O}$. Take $0 \neq v \in \Delta(\lambda)_{c_\lambda} = 1 \otimes \lambda$ and $m \in M_{c_\lambda}$ that maps onto v. Since $\lambda \nleq_c \mu$, there is no $a \in \mathbb{C}$ such that $a - c_\lambda \in \mathbb{Z}_{>0}$ and $M_a \neq 0$. Hence $\mathfrak{h} \cdot m = 0$. Then $v \mapsto m$ defines a morphism $\Delta(\lambda) \to M$ which splits the above sequence. $\square$

THEOREM 2.8.3. *Category $\mathcal{O}$ is a highest weight category under the ordering $\leq_c$ given by the Euler element.*

PROOF. The only thing left to check is that we can choose a Δ-filtration on the projective covers $P(\lambda)$ such that the conditions of Definition 2.8.1(3) are satisfied. By Theorem 2.7.2, we can always choose some Δ-filtration $0 = F_0 \subset F_1 \subset \cdots \subset F_m = P(\lambda)$, with $F_i/F_{i-1} \simeq \Delta(\mu_i)$. Since $P(\lambda)$ surjects onto $L(\lambda)$, we must have $\mu_m = \lambda$. I claim that one can always choose the μ_i such that $\mu_i \geq_c \mu_{i+1}$ for all $0 < i < m$. The proof is by induction on i. But first we remark that the fact that $P(\lambda)$ is indecomposable implies that all μ_i are comparable under $<_c$. Assume that $\mu_j \geq_c \mu_{j+1}$ for all $j < i$. If $\mu_i <_c \mu_{i+1}$ then it suffices to show that there is another Δ-filtration of $P(\lambda)$ with composition factors μ'_j such that $\mu'_j = \mu_j$ for all $j \neq i, i+1$ and $\mu'_i = \mu_{i+1}, \mu'_{i+1} = \mu_i$. We have

$$0 \to F_i \to F_{i+1} \to \Delta(\mu_{i+1}) \to 0,$$

which quotienting out by F_{i-1} gives

$$0 \to \Delta(\mu_i) \to F_{i+1}/F_{i-1} \to \Delta(\mu_{i+1}) \to 0.$$

Lemma 2.8.2 implies that the above sequence splits. Hence F_{i+1}/F_{i-1} is isomorphic to $\Delta(\mu_i) \oplus \Delta(\mu_{i+1})$. Thus, we may choose $F_{i-1} \subset F'_i \subset F_i$ such that $F'_i/F_{i-1} \simeq \Delta(\mu_{i+1})$ and $F_{i+1}/F'_i \simeq \Delta(\mu_i)$ as required. Hence the claim is proved. This means that $\mu_{m-1} \geq_c \lambda$. If $\mu_{m-1} =_c \lambda$ then Lemma 2.8.2 implies that

$$P(\lambda)/F_{m-2} \simeq \Delta(\mu_{m-1}) \oplus \Delta(\lambda)$$

and hence the head of $P(\lambda)$ is not simple. This contradicts the fact that $P(\lambda)$ is a projective cover. $\square$

We now describe some consequences of the fact that $\mathcal{O}$ is a highest weight category.

COROLLARY 2.8.4. *The standard modules* $\Delta(\lambda)$ *are a* $\mathbb{Z}$-*basis of the Grothendieck group* $K_0(\mathcal{O})$.

PROOF. Since $\mathcal{O}$ is a finite length, abelian category with finitely many simple modules, the image of those simple modules $L(\lambda)$ in $K_0(\mathcal{O})$ are a $\mathbb{Z}$-basis of $K_0(\mathcal{O})$. Let $k = |\mathrm{Irr}(W)|$ and define the k by k matrix $A = (a_{\lambda,\mu}) \in \mathbb{N}$ by

$$[\Delta(\lambda)] = \sum_{\mu \in \mathrm{Irr}(W)} a_{\lambda,\mu}[L(\mu)].$$

We order $\mathrm{Irr}(W) = \{\lambda_1, \ldots, \lambda_k\}$ so that $i > j$ implies that $\lambda_i \leq_c \lambda_j$. Then property (2) of Definition 2.8.1 implies that A is upper triangular with ones all along the diagonal. This implies that A is invertible over $\mathbb{Z}$ and hence the $[\Delta(\lambda)]$ are a basis of $K_0(\mathcal{O})$. $\square$

A nontrivial corollary of Theorem 2.8.3 is that Bernstein–Gelfand–Gelfand (BGG) reciprocity holds in category $\mathcal{O}$.

COROLLARY 2.8.5 (BGG-reciprocity). *For* $\lambda, \mu \in \mathrm{Irr}W$,

$$(P(\lambda) : \Delta(\mu)) = [\Delta(\mu) : L(\lambda)].$$

We also have the following:

COROLLARY 2.8.6. *The global dimension of* $\mathcal{O}$ *is finite.*

PROOF. As in the proof of Theorem 2.7.2, it suffices to consider modules in the block $\mathcal{O}^{\bar{a}}$ for some $a \in \mathbb{C}$. Recall that we constructed a filtration of this category

$$\mathcal{O}^{\geq k_0} \subset \mathcal{O}^{\geq k_0 - i_1} \subset \cdots \subset \mathcal{O}^{\geq k_0 - i_n} = \mathcal{O}^{\bar{a}},$$

where $0 < i_1 < \cdots < i_n$ are chosen so that each inclusion is proper. For all $\lambda \in \mathrm{Irr}(W)$, define $N(\lambda)$ to be the positive integer such that $L(\lambda) \in \mathcal{O}^{\geq k_0 - i_{N(\lambda)}}$ but $L(\lambda) \notin \mathcal{O}^{\geq k_0 - i_{N(\lambda)-1}}$. We claim that p.d.$(\Delta(\lambda)) \leq n - N(\lambda)$ for all λ. The proof of Theorem 2.7.2 showed that $\Delta(\lambda) = P(\lambda)$ for all λ such that $N(\lambda) = n$. Therefore we may assume that the claim is true for all μ such that $N(\mu) > N_0$. Choose λ such that $N(\lambda) = N_0$. Then, as shown in the proof of Theorem 2.8.3, we have a short exact sequence

$$0 \to K \to P(\lambda) \to \Delta(\lambda) \to 0,$$

where K admits a filtration by $\Delta(\mu)$'s for $\lambda <_c \mu$. The inductive hypothesis, together with standard homological results, e.g., Chapter 4 of [239], imply that p.d.$(K) \leq N_0 - 1$ and hence p.d.$(\Delta(\lambda)) \leq N_0$ as required. Now we show, again by induction, that p.d.$(L(\lambda)) \leq n + N(\lambda)$. Claim 2.7.3 says that $\Delta(\lambda) = L(\lambda)$ for all λ such that $N(\lambda) = 0$. Therefore, we may assume by induction that the claim holds for all μ such that $N(\mu) < N_0$. Assume $N(\lambda) = N_0$. We have a short exact sequence

$$0 \to R \to \Delta(\lambda) \to L(\lambda) \to 0,$$

where R admits a filtration by simple modules $L(\mu)$ with $\mu <_c \lambda$. Hence, as for standard modules, we may conclude that p.d.$(L(\lambda)) \leq n + N(\lambda)$. Note that we have actually shown that the global dimension of $\mathcal{O}^{\bar{a}}$ is at most $2n$. $\square$

COROLLARY 2.8.7. *Category $\mathcal{O}$ is semisimple if and only if $\Delta(\lambda) = L(\lambda)$ for all $\lambda \in \mathrm{Irr}(W)$.*

PROOF. Since $\Delta(\lambda)$ is indecomposable and $L(\lambda)$ a quotient of $\Delta(\lambda)$, it is clear that $\mathcal{O}$ semisimple implies that $\Delta(\lambda) = L(\lambda)$ for all $\lambda \in \mathrm{Irr}(W)$. Conversely, if $\Delta(\lambda) = L(\lambda)$ then BGG reciprocity implies that $P(\lambda) = L(\lambda)$ for all λ which implies that $\mathcal{O}$ is semisimple. $\square$

2.9. Category $\mathcal{O}$ for $\mathbb{Z}_2$. The idea of this section is simply to try and better understand category $\mathcal{O}$ when $W = \mathbb{Z}_2$. Many of the results stated are exercises, which can be found at the end of the chapter. Recall that, in this case, $W = \langle s \rangle$ with $s^2 = 1$ and the defining relations for $\mathsf{H}_c(\mathbb{Z}_2)$ are $sx = -xs$, $sy = -ys$ and

$$[y, x] = 1 - 2cs.$$

Using this defining relation, it is possible to work out when category $\mathcal{O}$ is semisimple and describe the partial ordering on $\mathrm{Irr}(W)$ coming from the highest weight structure on $\mathcal{O}$. See Exercise 2.11.9.

For $W = \mathbb{Z}_2$ one can also relate representations of $\mathsf{H}_c(W)$ to certain representations of $U(\mathfrak{sl}_2)$, the enveloping algebra of $\mathfrak{sl}_2$, using the spherical subalgebra $e\mathsf{H}_c(W)e$. Recall that $\mathfrak{sl}_2 = \mathbb{C}\{E, F, H\}$ with $[H, E] = 2E$, $[H, F] = -2F$ and $[E, F] = H$. See Exercise 2.11.10 for the proof of the following lemma.

LEMMA 2.9.1. *The map $E \mapsto \frac{1}{2}ex^2$, $F \mapsto -\frac{1}{2}ey^2$ and $H \mapsto exy + ec$ is a morphism of Lie algebras $\mathfrak{sl}_2 \to e\mathsf{H}_c(\mathbb{Z}_2)e$, where the right-hand side is thought of as a Lie algebra under the commutator bracket.*

The above morphism extends to a morphism of algebras $\phi_c : U(\mathfrak{sl}_2) \to e\mathsf{H}_c(\mathbb{Z}_2)e$, whose kernel is generated by $\Omega - \alpha$, for some α. Here Ω is the Casimir $\frac{1}{2}H^2 + EF + FE$, which generates the centre of $U(\mathfrak{sl}_2)$. Exercise 2.11.10 shows that ϕ_c descends to a morphism

$$\phi_c' : U(\mathfrak{sl}_2)/(\Omega - \alpha) \to e\mathsf{H}_c(\mathbb{Z}_2)e. \tag{2.B}$$

LEMMA 2.9.2. *The morphism (2.B) is an isomorphism.*

PROOF. This is a filtered morphism, where $e\mathsf{H}_c(\mathbb{Z}_2)e$ is given the filtration as in Section 1.3 and the filtration on $U(\mathfrak{sl}_2)$ is defined by putting E, F and H in degree two. The associated graded of $e\mathsf{H}_c(\mathbb{Z}_2)e$ equals $\mathbb{C}[x, y]^{\mathbb{Z}_2} = \mathbb{C}[x^2, y^2, xy]$ and the associated graded of $U(\mathfrak{sl}_2)/(\Omega - \alpha)$ is a quotient of $\mathbb{C}[E, F, H]/(\frac{1}{2}H^2 + 2EF)$. Since $\mathrm{gr}_{\mathcal{F}} \, e\mathsf{H}_c(\mathbb{Z}_2)e$ is generated by the symbols $x^2 = \sigma(ex^2)$, $y^2 = \sigma(ey^2)$ and $xy = \sigma(exy)$, we see that $e\mathsf{H}_c(\mathbb{Z}_2)e$ is generated by ex^2, ey^2 and exy. Hence ϕ_c' is surjective. On the other hand, the composite

$$\mathbb{C}[E, F, H]/(\tfrac{1}{2}H^2 + 2EF) \to \mathrm{gr}_{\mathcal{F}} \, U(\mathfrak{sl}_2)/(\Omega - \alpha) \xrightarrow{\mathrm{gr}_{\mathcal{F}} \phi_c'} \mathbb{C}[x^2, y^2, xy]$$

is given by $E \mapsto \frac{1}{2}x^2$, $F \mapsto -\frac{1}{2}y^2$ and $H \mapsto xy$ is an isomorphism. This implies that

$$\mathbb{C}[E, F, H]/(\tfrac{1}{2}H^2 + 2EF) \to \mathrm{gr}_{\mathcal{F}} \, U(\mathfrak{sl}_2)/(\Omega - \alpha)$$

is an isomorphism and so too is $\mathrm{gr}_{\mathcal{F}} \, \phi_c'$. Hence ϕ_c' is an isomorphism. $\square$

2.10. Quivers with relations. As noted previously, there exists a finite-dimensional algebra A such that category $\mathcal{O}$ is equivalent to A-mod. In this section, we'll try to construct A in terms of quivers with relations when $W = \mathbb{Z}_2$. This section is included for those who know about quivers and can be skipped if you are not familiar with them. When $W = \mathbb{Z}_2$ one can explicitly describe what the projective covers $P(\lambda)$ of the simple modules in category $\mathcal{O}$ are. Though, as the reader will see, this is a tricky calculation.

The idea is to first use BGG reciprocity to calculate the rank of $P(\lambda)$ as a (free) $\mathbb{C}[x]$-module. We will only consider the case $c = \frac{1}{2} + m$ for some $m \in \mathbb{Z}_{\geq 0}$. The situation $c = -\frac{1}{2} - m$ is completely analogous. We have already seen in Exercise 2.11.9 that

$$[\Delta(\rho_1) : L(\rho_0)] = 0, \qquad [\Delta(\rho_1) : L(\rho_1)] = 1,$$

$$[\Delta(\rho_0) : L(\rho_1)] = 1, \qquad [\Delta(\rho_0) : L(\rho_0)] = 1.$$

Therefore, BGG reciprocity implies that $\Delta(\rho_0) = P(\rho_0)$ and $P(\rho_1)$ is free of rank two over $\mathbb{C}[x]$. Moreover, we have a short exact sequence

$$0 \to \Delta(\rho_0) \to P(\rho_1) \to \Delta(\rho_1) = L(\rho_1) \to 0. \qquad (2.C)$$

As graded $\mathbb{Z}_2$-modules, we write $P(\rho_1) = \mathbb{C}[x] \otimes \rho_0 \oplus \mathbb{C}[x] \otimes \rho_1$, where $\mathbb{C}[x] \otimes \rho_0$ is identified with $\Delta(\rho_0)$. Then the structure of $P(\rho_1)$ is completely determined by the action of x and y on ρ_1:

$$y \cdot (1 \otimes \rho_1) = f_1(x) \otimes \rho_0, \quad x \cdot (1 \otimes \rho_1) = x \otimes \rho_1 + f_0(x) \otimes \rho_0$$

for some $f_0, f_1 \in \mathbb{C}[x]$. For the action to be well-defined we must check the relation $[y, x] = 1 - 2cs$, which reduces to the equation

$$y \cdot (f_0(x) \otimes \rho_0) = 0.$$

Also $s(f_i) = f_i$ for $i = 0, 1$. This implies that $f_0(x) = 1$. The second condition we require is that $P(\rho_1)$ is indecomposable (this will uniquely characterize $P(\rho_1)$ up to isomorphism). This is equivalent to asking that the short exact sequence (2.C) does not split. Choosing a splitting means choosing a vector $\rho_1 + f_2(x) \otimes \rho_0 \in P(\rho_1)$ such that $y \cdot (\rho_1 + f_2(x) \otimes \rho_0) = 0$. One can check that this is always possible, except when $f_1(x) = x^{2m}$. Thus we must take $f_1(x) = x^{2m}$. This completely describes $P(\rho_1)$ up to isomorphism.

Recall that a finite-dimensional $\mathbb{C}$-algebra A is said to be *basic* if the dimension of all simple A-modules is one. Every basic algebra can be described as a quiver with relations. One way to reconstruct A from A-mod is via the isomorphism

$$A = \mathrm{End}_A \left(\bigoplus_{\lambda \in \mathrm{Irr}(A)} P(\lambda) \right),$$

where $\mathrm{Irr}(A)$ is the set of isomorphism classes of simple A-modules and $P(\lambda)$ is the projective cover of λ. Next we will construct a basic A in terms of a quiver with relations such that A-mod $\simeq \mathcal{O}$. The first step in doing this is to use BGG-reciprocity to calculate the dimension of A. Again, we will assume that $c = \frac{1}{2} + m$ for some $m \in \mathbb{Z}_{\geq 0}$. The case $c = -\frac{1}{2} - m$ is similar and all other cases are trivial.

We need to describe $A = \mathrm{End}_{H_c(\mathbb{Z}_2)}(P(\rho_0) \oplus P(\rho_1))$. Using the general formula $\dim \mathrm{Hom}_{H_c(W)}(P(\lambda), M) = [M : L(\lambda)]$, and BGG reciprocity, we see that

$$\dim \mathrm{End}_{H_c(\mathbb{Z}_2)}(P(\rho_0)) = 1, \qquad \dim \mathrm{End}_{H_c(\mathbb{Z}_2)}(P(\rho_1)) = 2 \quad \text{(2.D)}$$

$$\dim \mathrm{Hom}_{H_c(\mathbb{Z}_2)}(P(\rho_0), P(\rho_1)) = 1, \quad \dim \mathrm{Hom}_{H_c(\mathbb{Z}_2)}(P(\rho_1), P(\rho_0)) = 1. \quad \text{(2.E)}$$

Hence $\dim A = 5$. The algebra A will equal $\mathbb{C}Q/I$, where Q is some quiver and I an admissible ideal[4]. The vertices of Q are labelled by the simple modules in $\mathcal{O}$, hence there are two: e_0 and e_1 (corresponding to $L(\rho_0)$ and $L(\rho_1)$ respectively). The number of arrows from e_0 to e_1 equals $\dim \mathrm{Ext}^1_{H_c(\mathbb{Z}_2)}(L(\rho_0), L(\rho_1))$ and the number of arrows from e_1 to e_0 equals $\dim \mathrm{Ext}^1_{H_c(\mathbb{Z}_2)}(L(\rho_1), L(\rho_0))$. Hence there is one arrow $e_1 \leftarrow e_0 : a$ and one arrow $e_0 \leftarrow e_1 : b$. The projective module $P(\rho_0)$ will be a quotient of

$$\mathbb{C}Qe_0 = \mathbb{C}\{e_0, ae_0 = a, ba, aba, \dots\},$$

and similarly for $P(\rho_1)$. Equations (2.D) imply that $ba = (ab)^2 = 0$ in A (note that we cannot have $e_0 - \alpha ba = 0$ etc. because the endomorphism ring of an indecomposable is a local ring). Hence A is a quotient of $\mathbb{C}Q/I$, where $I = \langle ba, (ab)^2 \rangle$. But $\mathbb{C}Q/I$ has a basis given by $\{e_0, e_1, a, b, ab\}$. Hence $\dim \mathbb{C}Q/I = 5$ and the natural map $\mathbb{C}Q/I \to A$ is an isomorphism. Thus,

LEMMA 2.10.1. *Let Q be the quiver with vertices $\{e_0, e_1\}$ and arrows*

$$\{e_1 \xleftarrow{\;a\;} e_0, \; e_0 \xleftarrow{\;b\;} e_1\}.$$

Let I be the admissible ideal $\langle ba, (ab)^2 \rangle$. Then, $\mathcal{O} \simeq \mathbb{C}Q/I$-mod.

2.11. Exercises.

EXERCISE 2.11.1. Show that every module in category $\mathcal{O}$ is finitely generated as a $\mathbb{C}[\mathfrak{h}]$-module.

EXERCISE 2.11.2. Show that $[\mathbf{eu}, x] = x$, $[\mathbf{eu}, y] = -y$ and $[\mathbf{eu}, w] = 0$ for all $x \in \mathfrak{h}^*$, $y \in \mathfrak{h}$ and $w \in W$.

EXERCISE 2.11.3.   (1) Give an example of a module $M \in H_c(W)$-mod that is not the direct sum of its generalised $\mathbf{eu}$-eigenspaces.
(2) Using $\mathbf{eu}$, show that every finite-dimensional $H_c(W)$-module is in category $\mathcal{O}$. Hint: how does $y \in \mathfrak{h}$ act on a generalised eigenspace for $\mathbf{eu}$?

EXERCISE 2.11.4 (harder). Show that $\mathrm{ch}(M) \in \bigoplus_{a \in \mathbb{C}} t^a \mathbb{Z}[[t]]$ for all $M \in \mathcal{O}$.

EXERCISE 2.11.5. Using the fact that all weights of $H_c(W)$ under the adjoint action of $\mathbf{eu}$ are in $\mathbb{Z}$, show that

$$\mathcal{O} = \bigoplus_{\bar{a} \in \mathbb{C}/\mathbb{Z}} \mathcal{O}^{\bar{a}}.$$

---

[4]Recall that an ideal I in a finite-dimensional algebra B is said to be *admissible* if there exists some $m \geq 2$ such that $\mathrm{rad}(B)^m \subset I \subset \mathrm{rad}(B)^2$

EXERCISE 2.11.6. Show that if $c_\lambda - c_\mu \notin \mathbb{Z}_{>0}$ for all $\lambda \neq \mu \in \mathrm{Irr}(W)$ then category $\mathcal{O}$ is semisimple. Conclude that $\mathcal{O}$ is semisimple for generic parameters c.

EXERCISE 2.11.7. Recall that c is said to be *aspherical* if there exists some nonzero $M \in \mathsf{H}_c(W)$-mod such that $e \cdot M = 0$. As in Lie theory, we have a "generalised Duflo theorem":

THEOREM 2.11.8. *Let J be a primitive ideal in $\mathsf{H}_c(W)$. Then there exists some $\lambda \in \mathrm{Irr}(W)$ such that*
$$J = \mathrm{Ann}_{\mathsf{H}_c(W)}(L(\lambda)).$$

Using the generalised Duflo theorem and the arguments as in the proof of Corollary 1.6.3, show that c is aspherical if and only if there exists a simple module $L(\lambda)$ in category $\mathcal{O}$ such that $e \cdot L(\lambda) = 0$.

EXERCISE 2.11.9. In this exercise, we assume $W = \mathbb{Z}_2$.
  (1) For each c, describe the simple modules $L(\lambda)$ as quotients of $\Delta(\lambda)$. For which values of c is category $\mathcal{O}$ semisimple?
  (2) For each c, describe the partial ordering on $\mathrm{Irr}(W)$ coming from the highest weight structure on $\mathcal{O}$.
  (3) Using Exercise 2.11.7, calculate the aspherical values for $W = \mathbb{Z}_2$.

EXERCISE 2.11.10. (1) Show that $E \mapsto \frac{1}{2}ex^2$, $F \mapsto -\frac{1}{2}ey^2$ and $H \mapsto exy + ec$ is a morphism of Lie algebras $\mathfrak{sl}_2 \to e\mathsf{H}_c(\mathbb{Z}_2)e$, where the right-hand side is thought of as a Lie algebra under the commutator bracket.
  (2) Prove that this extends to a morphism of algebras $\phi_c : U(\mathfrak{sl}_2) \to e\mathsf{H}_c(\mathbb{Z}_2)e$.
  (3) The centre of $U(\mathfrak{sl}_2)$ is generated by the Casimir $\Omega = \frac{1}{2}H^2 + EF + FE$. Find $\alpha \in \mathbb{C}$ such that $\Omega - \alpha \in \mathrm{Ker}\ \phi_c$.

2.12. Additional remarks.

The results of this section all come from the papers [91], [125] and [113].

Theorem 2.8.3 is shown in [113].

The fact that BGG reciprocity, Corollary 2.8.5, follows from Theorem 2.8.3 is shown in [113, Proposition 3.3].

The definition given in [70, Definition 3.1] is dual to the one give in Definition 2.8.1. It is also given in much greater generality.

The generalised Duflo theorem, Theorem 2.11.8, is given in [111].

3. The symmetric group

In this section we concentrate on category $\mathcal{O}$ for $W = \mathfrak{S}_n$, the symmetric group. The reason for this is that category $\mathcal{O}$ is much better understood for this group than for other complex reflection groups, though there are still several open problems. For instance, it is known for which parameters c the algebra $\mathsf{H}_c(\mathfrak{S}_n)$ admits finite-dimensional representations, and it turns out that $\mathsf{H}_c(\mathfrak{S}_n)$ admits at most one finite-dimensional, simple module, see [32]. In fact, we now have a good understanding [241] of the "size" (i.e., Gelfand–Kirillov dimension) of all simple modules in category $\mathcal{O}$.

But, in this section, we will concentrate on the main problem mentioned in Section 2, that of calculating the character ch($L(\lambda)$) of the simple modules.

3.1. Outline of the section. The first thing to notice is that, since we can easily write down the character of the standard modules $\Delta(\lambda)$, this problem is equivalent to the problem of calculating the multiplicities of simple modules in a composition series for standard modules; the "multiplicities problem". That is, we wish to find a combinatorial algorithm for calculating the numbers

$$[\Delta(\lambda) : L(\mu)], \quad \text{for all } \lambda, \mu \in \text{Irr}(\mathfrak{S}_n).$$

For the symmetric group, we now also have a complete answer to this question: the multiplicities are given by evaluating at one the transition matrices between standard and canonical basis of a certain Fock space. In order to prove this remarkable result, one needs to introduce a whole host of new mathematical objects, including several new algebras. This can make the journey long and difficult. So we begin by outlining the whole story, so that the reader doesn't get lost along the way. The result relies upon work of several people, namely Rouquier, Varagnolo–Vasserot and Leclerc–Thibon.

Motivated via quantum Schur–Weyl duality, we begin by defining the ν-Schur algebra. This is a finite-dimensional algebra. The category of finite-dimensional modules over the ν-Schur algebra is a highest weight category. Rouquier's equivalence says that category $\mathcal{O}$ for the rational Cherednik algebra of type A is equivalent to the category of modules over the ν-Schur algebra. Thus, we transfer the multiplicity problem for category $\mathcal{O}$ to the corresponding problem for the ν-Schur algebra.

The answer to this problem is known by a result of Vasserot and Varagnolo. However, their answer comes from a completely unexpected place. We forget about ν-Schur algebras for a second and consider instead the Fock space $\mathcal{F}_q$, a vector space over the field $\mathbb{Q}(q)$. This is an infinite-dimensional representation of $\mathcal{U}_q(\widehat{\mathfrak{sl}}_r)$, the quantum group associated to the *affine* Lie algebra $\widehat{\mathfrak{sl}}_r$. On the face of it, $\mathcal{F}_q$ has nothing to do with the ν-Schur algebra, but bare with me!

The Fock space $\mathcal{F}_q$ has a standard basis labelled by partitions. It was shown by Leclerc and Thibon that it also admits a *canonical basis*, in the sense of Lusztig, again labelled by partitions. Hence there is a "change of basis" matrix that relates these two basis. The entries $d_{\lambda,\mu}(q)$ of this matrix are elements of $\mathbb{Q}(q)$. Remarkably, it turns out that they actually belong to $\mathbb{Z}[q]$.

What Varagnolo and Vasserot showed is that the multiplicity of the simple module (for the ν-Schur algebra) labelled by λ in the standard module labelled by μ is given by $d_{\lambda',\mu'}(1)$ (λ' denotes the *transpose* of the partition λ). Thus, to calculate the numbers $[\Delta(\lambda) : L(\mu)]$, and hence the character of $L(\lambda)$, what we really need to do is calculate the change of basis matrix for the Fock space $\mathcal{F}_q$.

3.2. The rational Cherednik algebra associated to the symmetric group. Recall from Example 1.5.1 that the rational Cherednik algebra associated to the symmetric group $\mathfrak{S}_n$ is the quotient of

$$T(\mathbb{C}^{2n}) \rtimes \mathfrak{S}_n = \mathbb{C}\langle x_1, \ldots, x_n, y_1, \ldots, y_n \rangle \rtimes \mathfrak{S}_n$$

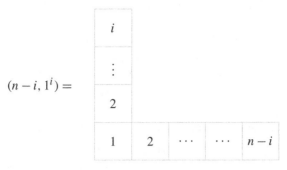

$(n-i, 1^i) =$

Figure 2. The partition $(n-i, 1^i)$ corresponding to the irreducible $\mathfrak{S}_n$-module $\bigwedge^i \mathfrak{h}$.

by the relations

$$[x_i, x_j] = 0,$$
$$[y_i, y_j] = 0, \qquad \qquad \text{for all } i, j,$$
$$[y_i, x_j] = cs_{ij},$$
$$[y_i, x_i] = 1 - c \sum_{j \neq i} s_{ij}, \qquad \text{for all } i \neq j.$$

Since the standard and simple modules in category $\mathcal{O}$ are labelled by the irreducible representations of $\mathfrak{S}_n$, we begin by recalling the parametrization of these representations.

3.3. Representations of $\mathfrak{S}_n$. Recall that a *partition* λ of n is a sequence

$$\lambda_1 \geq \lambda_2 \geq \cdots \geq 0$$

of positive integers such that $\sum_i \lambda_i = n$. It is a classical result, going back to I. Schur, that the irreducible representations of the symmetric group over $\mathbb{C}$ are naturally labelled by partitions of n. Therefore, we can (and will) identify $\mathrm{Irr}(\mathfrak{S}_n)$ with $\mathcal{P}_n$, the set of all partitions of n and denote by λ both a partition of n and the corresponding representation of $\mathfrak{S}_n$. For details on the construction of the representations of $\mathfrak{S}_n$, see [106].

EXAMPLE 3.3.1. The partition (n) labels the trivial representation and (1^n) labels the sign representation. The reflection representation $\mathfrak{h}$ is labelled by $(n-1, 1)$. More generally, each of the representations $\bigwedge^i \mathfrak{h}$ is an irreducible $\mathfrak{S}_n$-module and is labelled by $(n-i, 1^i)$; see Figure 2. Note that the trivial representation is $\bigwedge^0 \mathfrak{h}$ and the sign representation is just $\bigwedge^{n-1} \mathfrak{h}$.

3.4. Partitions. Associated to partitions is a wealth of beautiful combinatorics. We'll need to borrow a little of this combinatorics. Let $\lambda = (\lambda_1, \ldots, \lambda_k)$ be a partition. We visualise λ as a certain array of boxes, called a *Young diagram*, as in the example[5]

---

[5]The numbers in the boxes are the residues of λ modulo 3, see Section 3.8.

$\lambda = (4, 3, 1)$:

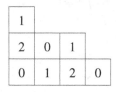

To be precise, the Young diagram of λ is

$$Y(\lambda) := \{(i, j) \in \mathbb{Z}^2 \mid 1 \le j \le k,\ 1 \le i \le \lambda_j\} \subset \mathbb{Z}^2.$$

EXAMPLE 3.4.1. The irreducible $\mathfrak{S}_n$-module labelled by the partition λ, has a natural basis $\{v_\sigma \mid \sigma \in \mathrm{Std}(\lambda)\}$, the *Young basis*, which is labelled by the set $\mathrm{Std}(\lambda)$ of all standard tableau of shape λ. Here a *standard tableau* σ is a filling of the Young diagram of λ by $\{1, \ldots, n\}$ such that the numbers along the row and column, read from left to right, and bottom to top, are increasing. For instance, if $n = 5$ and $\lambda = (3, 2)$, then $\dim \lambda = 5$, and the representation has a basis labelled by all standard tableaux,

4	5	
1	2	3

2	5	
1	3	4

2	4	
1	3	5

3	5	
1	2	4

3	4	
1	2	5

.

3.5. The v-Schur algebra. Recall that the first step on the journey is to translate the multiplicity problem for category $\mathcal{O}$ into the corresponding problem for the v-Schur algebra. By Weyl's complete reducibility theorem, the category $\mathcal{C}_n$ of finite-dimensional, integral representations of the Lie algebra $\mathfrak{gl}_n$, or equivalently of its enveloping algebra $\mathcal{U}(\mathfrak{gl}_n)$, is semisimple. Here a module M is said to be *integral* if the identity matrix $\mathrm{id} \in \mathfrak{gl}_n$ acts on M as multiplication by an integer. The simple modules in this category are the highest weight modules L_λ, where $\lambda \in \mathbb{Z}^n$ such that $\lambda_i - \lambda_{i+1} \ge 0$ for all $1 \le i \le n - 1$. The set $\mathcal{P}(n)$ of all partitions with length at most n can naturally be considered as a subset of this set. Let V denote the vectorial representation of $\mathfrak{gl}_n$. For each $d \ge 1$ there is an action of $\mathfrak{gl}_n$ on $V^{\otimes d}$. The symmetric group also acts on $V^{\otimes d}$ on the right by

$$(v_1 \otimes \cdots \otimes v_d) \cdot \sigma = v_{\sigma^{-1}(1)} \otimes \cdots \otimes v_{\sigma^{-1}(d)}, \qquad \text{for all } \sigma \in \mathfrak{S}_d.$$

It is known that these two actions commute. Thus, we have homomorphisms $\phi_d : \mathcal{U}(\mathfrak{gl}_n) \to \mathrm{End}_{\mathbb{C}\mathfrak{S}_d}(V^{\otimes d})$ and $\psi_d : \mathbb{C}\mathfrak{S}_d \to \mathrm{End}_{\mathcal{U}(\mathfrak{gl}_n)}(V^{\otimes d})^{op}$. Schur–Weyl duality says that

PROPOSITION 3.5.1. *The homomorphisms ϕ_d and ψ_d are surjective.*

Let $S(n, d) = \mathrm{End}_{\mathbb{C}\mathfrak{S}_d}(V^{\otimes d})$ be the image of ϕ_d. It is called the Schur algebra. We denote by $\mathcal{C}_n(d)$ the full subcategory of $\mathcal{C}_n$ consisting of all modules whose composition factors are of the form L_λ for $\lambda \in \mathcal{P}_d(n)$, where $\mathcal{P}_d(n)$ is the set of all partitions of d that belong to $\mathcal{P}(n)$. It is easy to check that $[V^{\otimes d} : L_\lambda] \ne 0$ if and only if $\lambda \in \mathcal{P}_d(n)$. Moreover, it is known that $\mathcal{C}_n(d) \simeq S(n, d)$-mod.

The above construction can be quantized. Let $v \in \mathbb{C}^\times$. Then, the *quantized enveloping algebra* $\mathcal{U}_v(\mathfrak{gl}_n)$ is a deformation of $\mathcal{U}(\mathfrak{gl}_n)$. The quantum enveloping algebra $\mathcal{U}_v(\mathfrak{gl}_n)$ still acts on V. The group algebra $\mathbb{C}\mathfrak{S}_n$ also has a natural flat deformation, the *Hecke algebra* of type A, denoted $\mathcal{H}_v(d)$. This algebra is described in Example 4.6.1. As one might expect, it is also possible to deform the action of $\mathbb{C}\mathfrak{S}_d$ on $V^{\otimes d}$ to an action of $\mathcal{H}_v(d)$ in such a way that this action commutes with the action of $\mathcal{U}_v(\mathfrak{gl}_n)$. The quantum analogue of Schur–Weyl duality, see [90], says

PROPOSITION 3.5.2. *We have surjective homomorphisms*

$$\phi_d : \mathcal{U}_v(\mathfrak{gl}_n) \to \mathrm{End}_{\mathcal{H}_v(d)}(V^{\otimes d}) \quad and \quad \psi_d : \mathcal{H}_v(d) \to \mathrm{End}_{\mathcal{U}_v(\mathfrak{gl}_n)}(V^{\otimes d})^{op}$$

for all $d \geq 1$.

The image $\mathrm{End}_{\mathcal{H}_v(d)}(V^{\otimes d})$ of ϕ_d is the *v-Schur algebra*, denoted $\mathsf{S}_v(n, d)$. The category $\mathcal{C}_{n,v}$ of finite-dimensional integral representations of $\mathcal{U}_v(\mathfrak{gl}_n)$ is no longer semisimple, in general. However, the simple modules in this category are still labelled L_λ, for $\lambda \in \mathbb{Z}^n$ such that $\lambda_i - \lambda_{i+1} \geq 0$ for all $1 \leq i \leq n-1$. Moreover, if we let $\mathcal{C}_{n,v}(d)$ denote the full subcategory of $\mathcal{C}_{n,v}$ consisting of all modules whose composition factors are L_λ, for $\lambda \in \mathcal{P}_d(n)$, then we again have $\mathcal{C}_{n,v}(d) = \mathsf{S}_v(n, d)$-mod. It is known that $\mathsf{S}_v(n, d)$-mod is a highest weight category with standard modules W_λ. When $n = d$, we write $\mathsf{S}_v(n) := \mathsf{S}_v(n, n)$.

3.6. Rouquier's equivalence. As explained at the start of the section, in order to calculate the multiplicities

$$m_{\lambda,\mu} = [\Delta(\lambda) : L(\mu)],$$

one has to make a long chain of connections and reformulations of the question, the end answer relying on several remarkable results. The first of these is Rouquier's equivalence, the proof of which relies in a crucial way on the KZ-functor introduced in Section 4.

THEOREM 3.6.1. *Let[6] $c \in \mathbb{Q}_{\geq 0}$ and set $v = \exp(2\pi\sqrt{-1}c)$. Then there is an equivalence of highest weight categories*

$$\Psi : \mathcal{O} \xrightarrow{\sim} \mathsf{S}_v(n)\text{-mod},$$

such that $\Psi(\Delta(\lambda)) = W_\lambda$ and $\Psi(L(\lambda)) = L_\lambda$.

Thus, to calculate $m_{\lambda,\mu}$ it suffices to describe the numbers $[W_\lambda : L_\mu]$. To do this, we turn now to the quantum affine enveloping algebra $\mathcal{U}_q(\widehat{\mathfrak{sl}}_r)$ and the Fock space $\mathcal{F}_q$.

3.7. The quantum affine enveloping algebra. Let q be an indeterminate and let I be the set $\{1, \ldots, r\}$. We now we turn our attention to another quantized enveloping algebra, this time of an affine Lie algebra. The *quantum affine enveloping algebra*

---

[6]Note that Rouquier's rational Cherednik algebra is parametrized by $h = -c$.

$\mathcal{U}_q(\widehat{\mathfrak{sl}_r})$ is the $\mathbb{Q}(q)$-algebra generated by E_i, F_i, $K_i^{\pm 1}$ for $i \in I$ and satisfying the relations

$$K_i K_i^{-1} = K_i^{-1} K_i = 1, \qquad K_i K_j = K_j K_i, \qquad \forall 1 \leq i, j \leq r,$$

$$K_i E_j = q^{a_{i,j}} E_j K_i, \qquad K_i F_j = q^{-a_{i,j}} F_j K_i, \quad \forall 1 \leq i, j \leq r, \qquad \text{(3.A)}$$

$$[E_i, F_j] = \delta_{i,j} \frac{K_i - K_i^{-1}}{q - q^{-1}}, \qquad\qquad\qquad \forall 1 \leq i, j \leq r,$$

and the quantum Serre relations

$$E_i^2 E_{i\pm1} - (q + q^{-1}) E_i E_{i\pm1} E_i + E_{i\pm1} E_i^2 = 0, \qquad \text{for all } 1 \leq i \leq r, \qquad \text{(3.B)}$$

$$F_i^2 F_{i\pm1} - (q + q^{-1}) F_i F_{i\pm1} F_i + F_{i\pm1} F_i^2 = 0, \qquad \text{for all } 1 \leq i \leq r, \qquad \text{(3.C)}$$

where the indices in (3.B) and (3.C) are taken modulo r so that $0 = r$ and $r + 1 = 1$. In (3.A), $a_{i,i} = 2$, $a_{i,i\pm1} = -1$ and $a_{i,j}$ is 0 otherwise. In the case $r = 2$, we take $a_{i,j} = -2$ if $i \neq j$.

3.8. The q-deformed Fock space. Let $\mathcal{F}_q$ be the *level one Fock space* for $\mathcal{U}_q(\widehat{\mathfrak{sl}_r})$. It is a $\mathbb{Q}(q)$-vector space with standard basis $\{|\lambda\rangle\}$, labelled by all partitions λ. The action of $\mathcal{U}_q(\widehat{\mathfrak{sl}_r})$ on $\mathcal{F}_q$ is combinatorially defined. Therefore, to describe it we need a little more of the language of partitions.

Let λ be a partition. The *content* of the box $(i, j) \in Y(\lambda)$ is $\mathrm{cont}(i, j) := i - j$. A *removable* box is a box on the boundary of λ which can be removed, leaving a partition of $|\lambda| - 1$. An *indent* box is a concave corner on the rim of λ where a box can be added, giving a partition of $|\lambda| + 1$. For instance, $\lambda = (4, 3, 1)$ has three removable boxes (with content -2, 1 and 3), and four indent boxes (with content -3, -1, 2 and 4). If γ is a box of the Young tableaux corresponding to the partition λ then we say that the *residue* of γ is i, or we say that γ is an i-box of λ, if the content of γ equals i modulo r. Let λ and μ be two partitions such that μ is obtained from λ by adding a box γ with residue i; see Figure 3. We define

$$N_i(\lambda) = |\{\text{indent } i\text{-boxes of } \lambda\}| - |\{\text{removable } i\text{-boxes of } \lambda\}|,$$

$$N_i^l(\lambda, \mu) = |\{\text{indent } i\text{-boxes of } \lambda \text{ situated to the } left \text{ of } \gamma \text{ (not counting } \gamma)\}|$$
$$- |\{\text{removable } i\text{-boxes of } \lambda \text{ situated to the } left \text{ of } \gamma\}|,$$

$$N_i^r(\lambda, \mu) = |\{\text{indent } i\text{-boxes of } \lambda \text{ situated to the } right \text{ of } \gamma \text{ (not counting } \gamma)\}|$$
$$- |\{\text{removable } i\text{-boxes of } \lambda \text{ situated to the } right \text{ of } \gamma\}|,$$

Then,

$$F_i|\lambda\rangle = \sum_\mu q^{N_i^r(\lambda,\mu)}|\mu\rangle, \qquad E_i|\mu\rangle = \sum_\lambda q^{N_i^l(\lambda,\mu)}|\lambda\rangle,$$

where, in each case, the sum is over all partitions such that μ/λ is a i-node, and

$$K_i|\lambda\rangle = q^{N_i(\lambda)}|\lambda\rangle.$$

See [159, Section 4.2] for further details.

EXAMPLE 3.8.1. Let $\lambda = (5, 4, 1, 1)$ and $r = 3$ so that the Young diagram, with residues, of λ is

0				
1				
2	0	1	2	
0	1	2	0	1

Then

$$F_2|\lambda\rangle = |(6, 4, 1, 1)\rangle + |(5, 4, 2, 1)\rangle + q|(5, 4, 1, 1, 1)\rangle,$$
$$E_2|\lambda\rangle = q^2|(5, 3, 1, 1)\rangle,$$
$$K_2|\lambda\rangle = q^2|(5, 3, 1, 1)\rangle.$$

There is an action of a larger algebra, the quantum affine enveloping algebra $\mathcal{U}_q(\widehat{\mathfrak{gl}}_r) \supset \mathcal{U}_q(\widehat{\mathfrak{sl}}_r)$ on the space $\mathcal{F}_q$. The key point for us is that $\mathcal{F}_q$ is an *irreducible* highest weight representation of $\mathcal{U}_q(\widehat{\mathfrak{gl}}_r)$, with highest weight $|\varnothing\rangle$. As a $\mathcal{U}_q(\widehat{\mathfrak{sl}}_r)$-module, the Fock space is actually a direct sum of infinitely many irreducible highest weight modules, see [147].

3.9. The bar involution. The key to showing that the Fock space $\mathcal{F}_q$ admits a canonical basis is to construct a sesquilinear involution on it, that is compatible in a natural way the sesquilinear involution on $\mathcal{U}_q(\widehat{\mathfrak{gl}}_r)$. However, in order to be able to do this we must first write the standard basis of $\mathcal{F}_q$ in terms of infinite q-wedges. The original motivation for doing this came from the boson–fermion correspondence in mathematical physics. Once we have constructed the involution, the existence of a canonical basis comes from general theory developed by Lusztig.

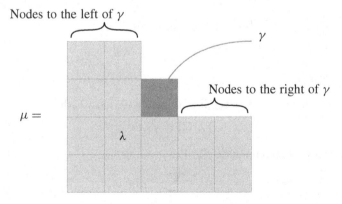

Figure 3. Nodes to the left and right of γ. The total partition is μ, whilst the lightly shaded subpartition is λ.

The $\mathbb{Q}$-linear involution $q \mapsto \bar{q} := q^{-1}$ of $\mathbb{Q}(q)$ extends to an involution $v \mapsto \bar{v}$ of $\mathcal{F}_q$. In order to describe this involution, it suffices to say how to calculate $\overline{|\lambda\rangle}$. We begin by noting that a partition can also be describe as an infinite wedge as follows.

LEMMA 3.9.1. *Let $\mathcal{J}$ be the set of strictly decreasing sequences $i = (i_1, i_2, \dots)$ such that $i_k = -k + 1$ for $k \gg 0$. There is a natural bijection between $\mathcal{J}$ and $\mathcal{P}$, the set of all partitions. This bijection sends*

$$\mathcal{J}_n = \left\{ i \in \mathcal{J} \mid \sum_k (i_k + k - 1) = n \right\}$$

to $\mathcal{P}_n$ the set of all partitions of n.

Thus, to a partition $i \in \mathcal{J} = \mathcal{P}$ we associate the infinite wedge

$$u_i = u_{i_1} \wedge u_{i_2} \wedge u_{i_3} \wedge \cdots,$$

for instance,

$$u_{(3^2,2)} = u_3 \wedge u_2 \wedge u_0 \wedge u_{-3} \wedge u_{-4} \wedge \cdots.$$

An infinite wedge is normally ordered if it equals u_i for some $i \in \mathcal{J}$. Just as for usual wedge products, there is a "normal ordering rule" for the q-deformed wedge product. However, this normal ordering rule depends in a very nontrivial way on the integer r (and more generally, on the level l of the Fock space). In order to describe the normal ordering rule, it suffices to say how to swap two adjacent u_i's. Let $i < j$ be integers with $j - i \equiv m \mod r$ for some $0 \le m < r$. If $m = 0$ then

$$u_i \wedge u_j = -u_j \wedge u_i,$$

and otherwise

$$u_i \wedge u_j = -q^{-1} u_j \wedge u_i + (q^{-2} - 1)[u_{j-m} \wedge u_{i+m} - q^{-1} u_{j-r} \wedge u_{i+r} \\ + q^{-2} u_{j-m-r} \wedge u_{i+m+r} - q^{-3} u_{j-2r} \wedge u_{i+2r} + \cdots],$$

where the sum continues only as long as the terms are normally ordered. Let $i \in \mathcal{J}$ and write $\alpha_{r,k}(i)$ for the number of pairs (a, b) with $1 \le a < b \le k$ and[7] $i_a - i_b \not\equiv 0 \mod r$.

PROPOSITION 3.9.2. *For $k \ge n$, the q-wedge*

$$\overline{u_i} := (-1)^{\binom{k}{2}} q^{\alpha_{r,k}(i)} u_{i_k} \wedge u_{i_{k-1}} \wedge \cdots \wedge u_{i_1} \wedge u_{i_{k+1}} \wedge u_{i_{k+2}} \wedge \cdots$$

is independent of k.

Therefore we can define a semilinear map, $v \mapsto \bar{v}$ on $\mathcal{F}_q$ by

$$\overline{f(q)u_i} = f(q^{-1})\overline{u_i}.$$

This is actually an involution on $\mathcal{F}_q$. For $\mu \vdash n$ define

$$\overline{|\mu\rangle} = \sum_{\lambda \vdash n} a_{\lambda,\mu}(q) |\lambda\rangle.$$

---

[7]There is a typo in definition of $\alpha_{r,k}(i)$ in [161].

3.10. Canonical basis. In order to describe the properties of the polynomials $a_{\lambda,\mu}(q)$ and also define the canonical basis, we need a few more basic properties of partitions. There is a natural partial ordering on $\mathcal{P}_n$, the set of all partitions of n, which is the *dominance ordering* and is defined by $\lambda \trianglelefteq \mu$ if and only if

$$\lambda_1 + \cdots + \lambda_k \leq \mu_1 + \cdots + \mu_k, \quad \text{for all } k.$$

We also require the notion of r-rim-hooks and r-cores. An r-*rim-hook* of λ is a *connected* skew partition $\lambda\backslash\mu$ of length r that does not contain the subpartition $(2, 2)$, i.e., it is a segment of length r of the edge of λ. For example, $\{(2, 2), (3, 2), (3, 1), (4, 1)\}$ is a 4-rim-hook of $(4, 3, 1)$. The r-*core* of λ is the partition $\mu \subset \lambda$ obtained by removing, one after another, all possible r-rim-hooks of λ. It is known that the r-core is independent of the order in which the hooks are removed. For example, the 3-core of $(4, 3, 1)$ is (2).

EXERCISE 3.10.1. Write a program that calculates the r-core of a partition.

Then it is known, [161, Theorem 3.3], that the polynomials $a_{\lambda,\mu}(q)$ have the following properties.

THEOREM 3.10.2. *Let* $\lambda, \mu \vdash n$.

(1) $a_{\lambda,\mu}(q) \in \mathbb{Z}[q, q^{-1}]$.
(2) $a_{\lambda,\mu}(q) = 0$ *unless* $\lambda \triangleleft \mu$ *and* λ, μ *have the same* r-*core*.
(3) $a_{\lambda,\lambda}(q) = 1$.
(4) $a_{\lambda,\mu}(q) = a_{\mu',\lambda'}(q)$.

EXAMPLE 3.10.3. If $r = 3$ then

$$\overline{|(4, 3, 1)\rangle} = |(4, 3, 1)\rangle + (q - q^{-1})|(3, 3, 1, 1)\rangle + (-1 + q^{-2})|(2, 2, 2, 2)\rangle$$
$$+ (q^2 - 1)|(2, 1, 1, 1, 1, 1, 1)\rangle.$$

Leclerc and Thibon showed:

THEOREM 3.10.4 ([161, Theorem 4.1]). *There exist canonical basis* $\{\mathcal{G}^+(\lambda)\}_{\lambda\in\mathcal{P}}$ *and* $\{\mathcal{G}^-(\lambda)\}_{\lambda\in\mathcal{P}}$ *of* $\mathcal{F}_q$ *characterized by*

(1) $\overline{\mathcal{G}^+(\lambda)} = \mathcal{G}^+(\lambda)$, $\overline{\mathcal{G}^-(\lambda)} = \mathcal{G}^-(\lambda)$.
(2) $\mathcal{G}^+(\lambda) \equiv |\lambda\rangle \mod q\mathbb{Z}[q]$ *and* $\mathcal{G}^-(\lambda) \equiv |\lambda\rangle \mod q^{-1}\mathbb{Z}[q^{-1}]$.

It seems that the above result has nothing to do with the fact that $\mathcal{F}_q$ is a $\mathcal{U}_q(\widehat{\mathfrak{sl}}_r)$-module. All that is required is that there is some involution defined on the space. However, for an arbitrary involution, there is no reason to expect a canonical basis to exist (and, indeed, one can check with small examples that it does not). Of course, the involution used by Leclerc and Thibon is not arbitrary. There is a natural involution on the algebra $\mathcal{U}_q(\widehat{\mathfrak{gl}}_r)$. Since $\mathcal{F}_q$ is irreducible as a $\mathcal{U}_q(\widehat{\mathfrak{gl}}_r)$-module, there is a unique involution on $\mathcal{F}_q$ such that $\overline{F|\varnothing\rangle} = \overline{F}|\varnothing\rangle$ for all $F \in \mathcal{U}_q(\widehat{\mathfrak{gl}}_r)$. It is this involution that Leclerc and Thibon use, though, as we have seen, they are able to give an explicit definition of this involution. Then, it follows from a general result by Lusztig, [168,

Section 7.10], that an irreducible, highest weight module equipped with this involution admits a canonical basis. Set

$$\mathcal{G}^+(\mu) = \sum_\lambda d_{\lambda,\mu}(q)|\lambda\rangle, \quad \mathcal{G}^-(\lambda) = \sum_\mu e_{\lambda,\mu}(q)|\mu\rangle.$$

The polynomials $d_{\lambda,\mu}$ and $e_{\lambda,\mu}$ have the following properties:

- they are nonzero only if λ and μ have the same r-core;
- $d_{\lambda,\lambda}(q) = e_{\lambda,\lambda}(q) = 1$;
- $d_{\lambda,\mu}(q) = 0$ unless $\lambda \leq \mu$, and $e_{\lambda,\mu}(q) = 0$ unless $\mu \leq \lambda$.

The key to relating the combinatorics of these canonical basis to the representation theory of the ν-Schur algebra is given by a result of Varagnolo and Vasserot. Assuming that $r > 1$, [237, Theorem 11] says that

$$[W_\lambda : L_\mu] = d_{\lambda',\mu'}(1), \quad [L_\lambda : W_\mu] = e_{\lambda,\mu}(1).$$

Combining the results of [161, 237, 196]:

THEOREM 3.10.5 (Leclerc–Thibon, Vasserot–Varagnolo, Rouquier). *We have*

$$[\Delta(\lambda) : L(\mu)] = d_{\lambda',\mu'}(1) \quad and \quad [L(\lambda) : \Delta(\mu)] = e_{\lambda,\mu}(1). \tag{3.D}$$

3.11. GAP. In order to calculate the polynomial $\mathcal{G}^+(\mu)$ in concrete examples, we use the computer package GAP. The file[8] Canonical.gap contains the functions APolynomial(lambda,mu,r) and DPolynomial(lambda,mu,r) which, when given a pair of partitions and an integer r, polynomials $a_{\lambda,\mu}(q)$ and $d_{\lambda,\mu}(q)$ respectively. For example:

```
gap>Read("Canonical.gap");
gap>APolynomial([2,1,1],[3,1],2);
q^2-1
gap>DPolynomial([1,1,1,1,1],[5],2);
q^2
gap>
```

In order to get a better intuition for these polynomials, we strongly recommend the reader invest some time in playing around with the programs.

3.12. The character of $L(\lambda)$. After our grand tour of combinatorial representation theory, we return to our original problem of calculating the character of the simple $H_c(\mathfrak{S}_n)$-modules $L(\lambda)$. Theorem 3.10.5 gives us a way to express this character in terms of the numbers $e_{\lambda,\mu}(1)$. For the symmetric group, the Euler element in $H_c(\mathfrak{S}_n)$ is[9]

$$\mathbf{eu} = \frac{1}{2}\sum_{i=1}^n x_i y_i + y_i x_i = \sum_{i=1}^n x_i y_i + \frac{n}{2} - \frac{c}{2}\sum_{1 \leq i \neq j \leq n} s_{i,j}.$$

---

[8]Available from maths.gla.ac.uk/~gbellamy/MSRI.html.

[9]The Euler element defined here differs from the one in Section 2.4 by a constant.

Recall that **eu** acts on the space $1 \otimes \lambda \subset \Delta(\lambda)$ by a scalar, denoted c_λ. Associated to the partition $\lambda = (\lambda_1, \ldots, \lambda_k)$ is the *partition statistic*, which is defined to be

$$n(\lambda) := \sum_{i=1}^{k} (i-1)\lambda_i.$$

LEMMA 3.12.1. *For each $\lambda \vdash n$ and $c \in \mathbb{C}$,*

$$c_\lambda = \frac{n}{2} + c(n(\lambda) - n(\lambda')). \tag{3.E}$$

PROOF. The *Jucys–Murphy* elements in $\mathbb{C}[\mathfrak{S}_n]$ are defined to be $\Theta_i = \sum_{j<i} s_{ij}$, for all $i = 2, \ldots, n$ so that

$$\mathbf{eu} = \sum_{i=1}^{n} x_i y_i + \frac{n}{2} - c \sum_{i=2}^{n} \Theta_i.$$

Recall from Example 3.4.1 that the representation λ of $\mathfrak{S}_n$ has a basis v_σ labelled by the standard tableau σ of shape λ. Then it is known that

$$\Theta_i \cdot v_\sigma = \mathrm{ct}_\sigma(i) v_\sigma,$$

where $c_\sigma(i)$ is the column of λ containing i, $r_\sigma(i)$ is the row of λ containing i and $\mathrm{ct}_\sigma(i) := c_\sigma(i) - r_\sigma(i)$ is the content of the node containing i. Note that $\mathrm{ct}_\sigma(1) = 0$ for all standard tableaux σ. Therefore

$$\mathbf{eu} \cdot v_\sigma = \left(\frac{n}{2} - c \sum_{i=2}^{n} \mathrm{ct}_\sigma(i) \right) v_\sigma,$$

and hence

$$c_\lambda = \frac{n}{2} - c \sum_{i=2}^{n} \mathrm{ct}_\sigma(i) = \frac{n}{2} - c \sum_{i=1}^{n} \mathrm{ct}_\sigma(i).$$

Now $\sum_{i=1}^{n} r_\sigma(i) = \sum_{j=1}^{\ell(\lambda)} (j-1)\lambda_j = n(\lambda)$ and similarly $\sum_{i=1}^{n} c_\sigma(i) = n(\lambda')$. This implies Equation (3.E). $\quad\square$

PROPOSITION 3.12.2. *We have*

$$\mathrm{ch}(L(\lambda)) = \frac{1}{(1-t)^n} \cdot \left(\sum_{\mu \leq \lambda} e_{\lambda,\mu}(1) \dim(\mu) t^{c_\mu} \right). \tag{3.F}$$

PROOF. The proposition depends on two key facts about standard modules. Firstly, they form a $\mathbb{Z}$-basis of the Grothendieck ring $K_0(\mathcal{O})$ (exercise 2.8.4) and, secondly, it is easy to calculate the character of $\Delta(\lambda)$. Theorem 3.10.5 implies that we have

$$[L(\lambda)] = \sum_{\mu \leq \lambda} e_{\lambda,\mu}(1)[\Delta(\mu)]$$

in $K_0(\mathcal{O})$. Now the proposition follows from the fact that $\mathrm{ch}(\Delta(\mu)) = \dfrac{\dim(\mu)t^{c_\mu}}{(1-t)^n}$. $\quad\square$

Notice that, though Equation (3.F) looks very simple, it is extremely difficult to extract meaningful information from it. For instance, one cannot tell whether $L(\lambda)$ is finite-dimensional by looking at this character formula. Also, note that the coefficients $e_{\lambda,\mu}(1)$ are often negative so the numerator is a polynomial with some negative coefficients. But expanding the fraction as a power-series around zero gives a series with *only positive integer* coefficients — all negative coefficients magically disappear!

EXAMPLE 3.12.3. Let $n = 4$ and $c = \frac{5}{3}$. In this case, the numbers $e_{\lambda,\mu}(1)$ and c_λ are:

$\mu\backslash\lambda$	(4)	(3,1)	(2,2)	(2,1,1)	(1,1,1,1)
(4)	1	0	0	0	0
(3,1)	0	1	0	0	0
(2,2)	-1	0	1	0	0
(2,1,1)	0	0	0	1	0
(1,1,1,1)	1	0	-1	0	1
c_λ	-8	$-\frac{4}{3}$	2	$\frac{16}{3}$	12

If we define $\mathrm{ch}_\lambda(t) := (1-t)^4 \cdot \mathrm{ch}(L(\lambda))$, then

$$\mathrm{ch}_{(1,1,1,1)}(t) = t^{12}, \qquad\qquad \mathrm{ch}_{(2,1,1)}(t) = 3t^{\frac{16}{3}},$$
$$\mathrm{ch}_{(2,2)}(t) = 2t^2 - t^{12}, \qquad\qquad \mathrm{ch}_{(3,1)}(t) = 3t^{-\frac{4}{3}},$$
$$\mathrm{ch}_{(4)}(t) = t^{-8} - 2t^2 + t^{12}.$$

When $c = \frac{9}{4}$ we get:

$\mu\backslash\lambda$	(4)	(3,1)	(2,2)	(2,1,1)	(1,1,1,1)
(4)	1	0	0	0	0
(3,1)	-1	1	0	0	0
(2,2)	0	0	1	0	0
(2,1,1)	1	-1	0	1	0
(1,1,1,1)	-1	1	0	-1	1
c_λ	$-\frac{23}{2}$	$-\frac{5}{2}$	2	$\frac{13}{2}$	$\frac{31}{2}$

and

$$\mathrm{ch}_{(1,1,1,1)}(t) = t^{\frac{31}{2}}, \qquad \mathrm{ch}_{(2,1,1)}(t) = 3t^{\frac{13}{2}} - t^{\frac{31}{2}},$$
$$\mathrm{ch}_{(2,2)}(t) = 2t^2, \qquad \mathrm{ch}_{(3,1)}(t) = 3t^{-\frac{5}{2}} - 3t^{\frac{13}{2}} + t^{\frac{31}{2}},$$
$$\mathrm{ch}_{(4)}(t) = t^{-\frac{23}{2}} - 3t^{-\frac{5}{2}} + 3t^{\frac{13}{2}} - t^{\frac{31}{2}}.$$

If $\mathfrak{S}_4$ acts on $\mathbb{C}^4$ by permuting the basis elements $\epsilon_1, \ldots, \epsilon_4$, then we let $\mathfrak{h} \subset \mathbb{C}^4$ be the three-dimensional reflection representation, with basis $\{\epsilon_1 - \epsilon_2, \epsilon_2 - \epsilon_3, \epsilon_3 - \epsilon_4\}$. One could also consider representations of $\mathsf{H}_c(\mathfrak{h}, \mathfrak{S}_4)$ instead of $\mathsf{H}_c(\mathbb{C}^4, \mathfrak{S}_4)$. Since $\mathbb{C}^4 = \mathfrak{h} \oplus \mathbb{C}$ as a $\mathfrak{S}_4$-module, the defining relations for $\mathsf{H}_c(\mathbb{C}^4, \mathfrak{S}_4)$ make it clear that we have a decomposition $\mathsf{H}_c(\mathbb{C}^4, \mathfrak{S}_4) = \mathsf{H}_c(\mathfrak{h}, \mathfrak{S}_4) \otimes \mathcal{D}(\mathbb{C})$, which implies that

$\Delta_{\mathbb{C}^4}(\lambda) = \Delta_{\mathfrak{h}}(\lambda) \otimes \mathbb{C}[x]$ for each $\lambda \vdash 4$. Then,

$$\mathrm{ch}(L_{\mathbb{C}^4}(\lambda)) = \frac{1}{1-t} \cdot \mathrm{ch}(L_{\mathfrak{h}}(\lambda)),$$

and hence $\mathrm{ch}_\lambda(t) = (1-t)^3 \cdot \mathrm{ch}(L_{\mathfrak{h}}(\lambda))$. If we consider $\lambda = (4)$ with $c = \frac{9}{4}$ then notice that

$$\frac{t^{-\frac{23}{2}} - 3t^{-\frac{5}{2}} + 3t^{\frac{13}{2}} - t^{\frac{31}{2}}}{(1-t)^3} = t^{-\frac{23}{2}} \left[\frac{1 - 3t^9 + 3t^{18} - t^{27}}{(1-t)^3} \right]$$

$$= t^{-\frac{23}{2}} (t^{24} + 3t^{23} + 6t^{22} + \cdots + 3t + 1).$$

This implies that $L((4))$ is finite-dimensional. Evaluating the above polynomial shows that it actually has dimension 729.

3.13. Yvonne's conjecture. We end the section by explaining how to (conjecturally) generalise the picture for $\mathfrak{S}_n$ to other complex reflection groups (this section is really for those who have a firm grasp of the theory of rational Cherednik algebras, and can be safely ignored if so desired). In many ways, the most interesting, and hence most intensely studied, class of rational Cherednik algebras are those associated to the complex reflection group

$$W = \mathfrak{S}_n \wr \mathbb{Z}_l = \mathfrak{S}_n \ltimes \mathbb{Z}_l^n,$$

the wreath product of the symmetric group with the cyclic group of order l. Fix ζ a primitive l-th root of unity. The reflection representation for $\mathfrak{S}_n \wr \mathbb{Z}_l$ is $\mathbb{C}^n$. If $y_1, \ldots, y_n$ is the standard basis of $\mathbb{C}^n$ then

$$\mathbb{Z}_l^n = \{g_i^j \mid 1 \le i \le n, \ 0 \le j \le l-1\}$$

and

$$g_i^j \cdot y_k = \begin{cases} \zeta^j y_k & \text{if } k = i, \\ y_k & \text{otherwise.} \end{cases}$$

The symmetric group acts as $\sigma \cdot y_i = y_{\sigma(i)}$. This implies that

$$\sigma \cdot g_i = g_{\sigma(i)} \cdot \sigma.$$

The conjugacy classes of reflections in W are

$$R = \{s_{i,j} g_i g_j^{-1} \mid 1 \le i < j \le n\}, \quad S_i = \{g_j^i \mid 1 \le j \le n\}, \ 1 \le i \le l-1.$$

and we define c by $c(s_{i,j}) = c$, $c(g_j^i) = c_i$ so that the rational Cherednik algebra is parametrized by $c, c_1, \ldots, c_{l-1}$. To relate category $\mathcal{O}$ to a certain Fock space, we introduce new parameters $\boldsymbol{h} = (h, h_0, \ldots, h_{l-1})$, where

$$c = 2h, \quad c_i = \sum_{j=0}^{l-1} \zeta^{-ij}(h_j - h_{j+1}), \quad \text{for all } 1 \le i \le l-1.$$

Note that the above equations do not uniquely specify $h_0, \ldots, h_{l-1}$; one can choose $h_0 + \cdots h_{l-1}$ freely. Finally, we define $\boldsymbol{s} = (s_0, \ldots, s_{l-1}) \in \mathbb{Z}^l$ and $e \in \mathbb{N}$ by $h = \frac{1}{e}$ and

$$h_j = \frac{s_j}{e} - \frac{j}{d}.$$

The irreducible representations of $\mathfrak{S}_n \wr \mathbb{Z}_l$ are labelled by *l-multipartitions* of n, where $\lambda = (\lambda^{(1)}, \ldots, \lambda^{(l)})$ is an *l*-multipartition of n if it consists of an *l*-tuple of partitions $\lambda^{(i)}$ such that

$$\sum_{i=1}^{l} |\lambda^{(i)}| = n.$$

The set of all *l*-multipartitions is denoted $\mathcal{P}^l$ and the subset of all *l*-multipartitions of n is denoted $\mathcal{P}_n^l$. Thus, the simple modules in category $\mathcal{O}$ are $L(\lambda)$, for $\lambda \in \mathcal{P}_n^l$. It is actually quite easy to construct the simple W-modules λ once one has constructed all representations of the symmetric group.

Uglov defined, for each $l \geq 1$ and $s \in \mathbb{Z}^l$, a level l Fock space $\mathcal{F}_q^l[s]$ with multicharge s. This is again a representation of the quantum affine algebra $U_q(\widehat{\mathfrak{sl}}_r)$, this time with $\mathbb{Q}(q)$-basis $|\lambda\rangle$ given by *l*-multipartitions. The action of the operators F_i, E_i and K_i have a similar combinatorial flavour as for $l = 1$. The space $\mathcal{F}_q^l[s]$ is also equipped with a $\mathbb{Q}$-linear involution.

THEOREM 3.13.1. *There exists a unique $\mathbb{Q}(q)$-basis $\mathcal{G}(\lambda)$ of $\mathcal{F}_q^l[s]$ such that*

(1) $\overline{\mathcal{G}(\lambda)} = \mathcal{G}(\lambda)$.
(2) $\mathcal{G}(\lambda) - \lambda \in \bigoplus_{\mu \in \mathcal{P}_n^l} q\mathbb{Q}[q]|\mu\rangle$ *if $\lambda \vdash n$.*

Therefore, for $\lambda, \mu \in \mathcal{P}_n^l$, we define $d_{\lambda,\mu}(q)$ by

$$\mathcal{G}(\mu) = \sum_{\lambda \in \mathcal{P}_n^l} d_{\lambda,\mu}(q)|\lambda\rangle.$$

The following conjecture, originally due to Yvonne, but in the generality stated here is due to Rouquier, relates the multiplicities of simple modules inside standard modules for $H_c(\mathfrak{S}_n \wr \mathbb{Z}_l)$ to the polynomials $d_{\lambda,\mu}(q)$.

CONJECTURE 3.13.2. *For all $\lambda, \mu \in \mathcal{P}_n^l$ we have*

$$[\Delta(\lambda) : L(\mu)] = d_{\lambda^\dagger, \mu^\dagger}(1).$$

If $\lambda = (\lambda^{(1)}, \ldots, \lambda^{(m)})$ is an m-multipartition then

$$\lambda^\dagger = ((\lambda^{(m)})', \ldots, (\lambda^{(1)})'),$$

where $(\lambda^{(i)})'$ is the usual transpose of the partition $\lambda^{(i)}$.

3.14. Exercises.

EXERCISE 3.14.1. Let $\lambda = (6, 6, 3, 1, 1)$ and $r = 4$. Calculate the action of F_2, E_4 and K_1 on $|\lambda\rangle$. Hint: draw a picture!

EXERCISE 3.14.2. Write a computer program that calculates the action of the operators K_i, F_i, E_i on $\mathcal{F}_q$.

EXERCISE 3.14.3. Prove Lemma 3.9.1.

EXERCISE 3.14.4. For $r = 2$, describe $\mathcal{G}(\lambda)$ for all $\lambda \vdash 4$ and $\lambda \vdash 5$.

EXERCISE 3.14.5 (harder). Write a program that calculates the polynomials $e_{\lambda,\mu}(q)$.

EXERCISE 3.14.6. (1) For $c = \frac{7}{2}$, compute the character of all the simple modules of category $\mathcal{O}$ for $H_c(\mathfrak{S}_5)$.
(2) For $c = \frac{11}{3}$, compute the character of all the simple modules of category $\mathcal{O}$ for $H_c(\mathfrak{S}_4)$.
(3) What are the blocks of $\mathcal{O}$ for $\mathfrak{S}_5$ at $c = \frac{7}{2}$? at $c = \frac{103}{3}$ or at $c = \frac{29}{5}$? Hint: It is known that two partitions are in the same block of the q-Schur algebra if and only if they have the same r-core.

EXERCISE 3.14.7. Since the action of **eu** on a module $M \in \mathcal{O}$ commutes with the action of W, the multiplicity space of a representation $\lambda \in \mathrm{Irr}(W)$ is a **eu**-module. Therefore, one can refine the character ch to

$$\mathrm{ch}_W(M) = \sum_\lambda \mathrm{ch}(M(\lambda))[\lambda],$$

where $M = \bigoplus_{\lambda \in \mathrm{Irr}(W)} M(\lambda) \otimes \lambda$ as a W-module. Given $\lambda, \mu \in \mathrm{Irr}(W)$, we define the *generalised fake polynomial* $f_{\lambda,\mu}(t)$ by

$$f_{\lambda,\mu}(t) = \sum_{i \in \mathbb{Z}} [\mathbb{C}[\mathfrak{h}]_i^{\mathrm{co}\,W} \otimes \lambda : \mu] t^i,$$

a Laurent polynomial. See Sections 5.5 and 5.7 for an explanation of the terms appearing in the definition of $f_{\lambda,\mu}(t)$. What is the $\mathfrak{S}_4$-graded character of the simple modules, expressed in terms of generalised fake polynomials, in $\mathcal{O}$ when $c = \frac{5}{2}$?

3.15. Additional remarks.

Theorem 3.6.1 is given in [196, Theorem 6.11]. Its proof is based on the uniqueness of quasihereditary covers of the Hecke algebra, as shown in [196].

Theorem 3.13.1, due to Uglov, is Theorem 2.5 of [226].

Yvonne's conjecture was originally stated in [248, Conjecture 2.13]. The refined version, given in Section 3.13, is due to Rouquier, [196, Section 6.5]

4. The KZ functor

Beilinson and Bernstein, in proving their incredible localization theorem for semisimple Lie algebras and hence confirming the Kazhdan–Lusztig conjecture, have shown that it is often very fruitfully to try and understand the representation theory of certain algebras by translating representation theoretic problems into topological problems via $\mathcal{D}$-modules and the Riemann–Hilbert correspondence. This is the motivation behind the Knizhnik–Zamolodchikov (KZ) functor.

In the case of rational Cherednik algebras, the obvious way to try and relate them to $\mathcal{D}$-modules is to use the Dunkl embedding. Recall from Section 1 that this is the embedding $H_c(W) \hookrightarrow \mathcal{D}(\mathfrak{h}_{\mathrm{reg}}) \rtimes W$, which becomes an isomorphism

$$H_c(W)[\delta^{-1}] \xrightarrow{\sim} \mathcal{D}(\mathfrak{h}_{\mathrm{reg}}) \rtimes W$$

after localization. Given a module M in category $\mathcal{O}$, its localization $M[\delta^{-1}]$ becomes a $\mathcal{D}(\mathfrak{h}_{\mathrm{reg}}) \rtimes W$-module. We can understand $\mathcal{D}(\mathfrak{h}_{\mathrm{reg}}) \rtimes W$-modules as $\mathcal{D}$-modules on $\mathfrak{h}_{\mathrm{reg}}$ that are W-equivariant. Since W acts freely on $\mathfrak{h}_{\mathrm{reg}}$, this is the same as considering a $\mathcal{D}$-module on $\mathfrak{h}_{\mathrm{reg}}/W$. Finally, using the powerful Riemann–Hilbert correspondence we can think of such a $\mathcal{D}$-module as a local system on $\mathfrak{h}_{\mathrm{reg}}/W$. That is, it become a representation of the fundamental group $\pi_1(\mathfrak{h}_{\mathrm{reg}}/W)$ of $\mathfrak{h}_{\mathrm{reg}}/W$. This is a purely topological object. The process of passing from a module in category $\mathcal{O}$ to a representation of $\pi_1(\mathfrak{h}_{\mathrm{reg}}/W)$ is exactly what the KZ-functor does. The goal of this section is to try and describe in some geometric way the image of category $\mathcal{O}$ under this procedure. Remarkably, what one gets in the end is a functor

$$\mathsf{KZ} : \mathcal{O} \longrightarrow \mathcal{H}_q(W)\text{-mod},$$

from category $\mathcal{O}$ to the category of finitely generated modules over the *Hecke algebra* $\mathcal{H}_q(W)$ associated to W.

In the case of semisimple Lie algebras, the Beilinson–Bernstein localization result gave us a proof of the Kazhdan–Lusztig conjecture. What does the KZ-functor give us for rational Cherednik algebras? It allows us to play off the representation theory of the rational Cherednik algebra against the representation theory of the Hecke algebra. Via the KZ-functor, the Hecke algebra becomes crucial in proving deep results about the existence of finite-dimensional representations of $H_c(W)$, [32] and [119] (which in turn confirmed a conjecture of Haiman's in algebraic combinatorics), Rouquier's equivalence Theorem 3.6.1, and on the properties of restriction and induction functors for rational Cherednik algebras [199]. Conversely, the rational Cherednik algebra has been used to prove highly nontrivial results about Hecke algebras, e.g., Section 6 of [113] and the article [69].

4.1. Integrable connections. A $\mathcal{D}(\mathfrak{h}_{\mathrm{reg}})$-module which is finitely generated as a $\mathbb{C}[\mathfrak{h}_{\mathrm{reg}}]$-module is called an *integrable connection*. It is a classical result that every integrable connection is actually free as a $\mathbb{C}[\mathfrak{h}_{\mathrm{reg}}]$-module, see [134, Theorem 1.4.10]. Therefore, if N is an integrable connection,

$$N \simeq \bigoplus_{i=1}^{k} \mathbb{C}[\mathfrak{h}_{\mathrm{reg}}]u_i$$

for some $u_i \in N$. If we fix coordinates $x_1, \ldots, x_n$ such that $\mathbb{C}[\mathfrak{h}] = \mathbb{C}[x_1, \ldots, x_n]$ and let $\partial_i = \partial/\partial x_i$, then the action of $\mathcal{D}(\mathfrak{h}_{\mathrm{reg}})$ on N is then completely encoded in the equations

$$\partial_l \cdot u_i = \sum_{j=1}^{k} f_{l,i}^{j} u_j, \tag{4.A}$$

for some polynomials[10] $f_{l,i}^{k} \in \mathbb{C}[\mathfrak{h}_{\mathrm{reg}}]$. The integer k is called the *rank* of N.

A natural approach to studying integrable connections is to look at their space of solutions. Since very few differential equations have polynomial solutions, this approach only makes sense in the analytic topology. So we'll write $\mathfrak{h}_{\mathrm{reg}}^{\mathrm{an}}$ for the same space, but

---

[10]The condition $[\partial_l, \partial_m] = 0$ needs to be satisfied, which implies that one cannot choose arbitrary polynomials $f_{i,i}^{j}$.

now equipped with the analytic topology, and $\mathbb{C}[\mathfrak{h}_{\text{reg}}^{\text{an}}]$ denotes the ring of *holomorphic* functions on $\mathfrak{h}_{\text{reg}}^{\text{an}}$. Since $\mathcal{D}(\mathfrak{h}_{\text{reg}}^{\text{an}}) = \mathbb{C}[\mathfrak{h}_{\text{reg}}^{\text{an}}] \otimes_{\mathbb{C}[\mathfrak{h}_{\text{reg}}]} \mathcal{D}(\mathfrak{h}_{\text{reg}})$, we have a natural functor

$$\mathcal{D}(\mathfrak{h}_{\text{reg}}) \rtimes W\text{-mod} \to \mathcal{D}(\mathfrak{h}_{\text{reg}}^{\text{an}}) \rtimes W\text{-mod}, \quad M \mapsto M^{\text{an}} := \mathbb{C}[\mathfrak{h}_{\text{reg}}^{\text{an}}] \otimes_{\mathbb{C}[\mathfrak{h}_{\text{reg}}]} M.$$

Since $\mathbb{C}[\mathfrak{h}_{\text{reg}}^{\text{an}}]$ is faithfully flat over $\mathbb{C}[\mathfrak{h}_{\text{reg}}]$, this is exact and conservative, i.e., $M^{\text{an}} = 0$ implies that $M = 0$. On any simply connected open subset U of $\mathfrak{h}_{\text{reg}}^{\text{an}}$, the vector space

$$\text{Hom}_{\mathcal{D}(\mathfrak{h}_{\text{reg}}^{\text{an}})}(N^{\text{an}}, \mathbb{C}[U])$$

is k-dimensional because it is the space of solutions of a $k \times k$ matrix of first order linear differential equations. These spaces of local solutions glue together to give a rank k *local system* $\text{Sol}(N)$ on $\mathfrak{h}_{\text{reg}}$, i.e., a locally constant sheaf of $\mathbb{C}$-vector spaces such that each fibre has dimension k; see Figure 5 for an illustration. The reader should have in mind the idea of analytic continuation of a locally defined holomorphic function to a globally defined multivalued function, coming from complex analysis, of which the notion of local system is a generalisation.

4.2. Regular singularities. Assume now that $\dim \mathfrak{h} = 1$, so that $\mathfrak{h} = \mathbb{C}$, $\mathfrak{h}_{\text{reg}} = \mathbb{C}^{\times}$ and $\mathbb{C}[\mathfrak{h}_{\text{reg}}] = \mathbb{C}[x, x^{-1}]$. Then, we say that N is a *regular connection* (or has *regular singularities*) if, with respect to some $\mathbb{C}[\mathfrak{h}_{\text{reg}}^{\text{an}}]$-basis of N^{an}, Equation (4.A) becomes

$$\partial \cdot u_i = \sum_{j=1}^{k} \frac{a_{i,j}}{x} u_j, \quad a_{i,j} \in \mathbb{C}. \tag{4.B}$$

When $\dim \mathfrak{h} > 1$, we say that N has regular singularities if the restriction $N|_C$ of N to any smooth curve $C \subset \mathfrak{h}_{\text{reg}}$ has, after a suitable change of basis, the form (4.B). The category of integrable connections with regular singularities on $\mathfrak{h}_{\text{reg}}$ is denoted $\text{Conn}^{\text{reg}}(\mathfrak{h}_{\text{reg}})$. Similarly, let's write $\text{Loc}(\mathfrak{h}_{\text{reg}})$ for the category of finite-dimensional local systems on $\mathfrak{h}_{\text{reg}}^{\text{an}}$. There are two natural functors from $\text{Conn}^{\text{reg}}(\mathfrak{h}_{\text{reg}})$ to local systems on $\mathfrak{h}_{\text{reg}}^{\text{an}}$. Firstly, there is a *solutions functor*

$$N \mapsto \text{Sol}(N) := \mathcal{H}om_{\mathcal{D}_{\mathfrak{h}_{\text{reg}}^{\text{an}}}}(N^{\text{an}}, \mathcal{O}_{\mathfrak{h}_{\text{reg}}}^{\text{an}}),$$

and secondly the *de Rham functor*

$$N \mapsto \text{DR}(N) := \mathcal{H}om_{\mathcal{D}_{\mathfrak{h}_{\text{reg}}^{\text{an}}}}(\mathcal{O}_{\mathfrak{h}_{\text{reg}}}^{\text{an}}, N^{\text{an}}).$$

There are natural duality (i.e., they are *contravariant* equivalences whose square is the identity) functors

$$\mathbb{D} : \text{Conn}^{\text{reg}}(\mathfrak{h}_{\text{reg}}) \xrightarrow{\sim} \text{Conn}^{\text{reg}}(\mathfrak{h}_{\text{reg}}) \quad \text{and} \quad \mathbb{D}' : \text{Loc}(\mathfrak{h}_{\text{reg}}) \xrightarrow{\sim} \text{Loc}(\mathfrak{h}_{\text{reg}}),$$

on integrable connections (preserving regular singularities) and local systems respectively. These duality functors intertwine the solutions and de Rham functors:

$$\mathbb{D}' \circ \text{DR} = \text{DR} \circ \mathbb{D} = \text{Sol}, \quad \text{and hence} \quad \mathbb{D}' \circ \text{Sol} = \text{Sol} \circ \mathbb{D} = \text{DR}; \tag{4.C}$$

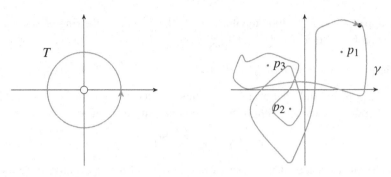

Figure 4. The generator T of the braid group $\pi_1(\mathbb{C}^\times)$, and nontrivial element γ in $\pi_1(\mathbb{C}\setminus\{p_1, p_2, p_2\}, \bullet)$.

see [134, Chapter 7] for details. Sticking with conventions, we'll work with the de Rham functor. There is a natural identification

$$\mathsf{DR}(N) = \{n \in N^{\mathrm{an}} \mid \partial_i \cdot n = 0, \forall i\} =: N^\nabla,$$

where N^∇ is usually referred to as the subsheaf of *horizontal sections* of N^{an}. The fundamental group $\pi_1(\mathfrak{h}_{\mathrm{reg}})$ of $\mathfrak{h}_{\mathrm{reg}}$ is the group of loops in $\mathfrak{h}_{\mathrm{reg}}$ from a fixed point x_0 to itself, considered up to homotopy. Following a local system L along a given loop γ based at x_0 defines an invertible endomorphism $\gamma_* : \mathsf{L}_{x_0} \to \mathsf{L}_{x_0}$ of the stalk of L at x_0. The map γ_* only depends on the homotopy class of γ, i.e., only on the object defined by γ in $\pi_1(\mathfrak{h}_{\mathrm{reg}})$. In this way, the stalk L_{x_0} becomes a representation of the group $\pi_1(\mathfrak{h}_{\mathrm{reg}})$. The following is a well-know result in algebraic topology:

PROPOSITION 4.2.1. *Assume that $\mathfrak{h}_{\mathrm{reg}}$ is connected. The functor*

$$\mathrm{Loc}(\mathfrak{h}_{\mathrm{reg}}) \to \pi_1(\mathfrak{h}_{\mathrm{reg}})\text{-mod}, \quad \mathsf{L} \mapsto \mathsf{L}_{x_0}$$

is an equivalence.

See I, Corollaire 1.4 of [80] for a proof of this equivalence.

4.3. Via the equivalence of Proposition 4.2.1, we may (and will) think of DR and Sol as functors from $\mathrm{Conn}^{\mathrm{reg}}(\mathfrak{h}_{\mathrm{reg}})$ to $\pi_1(\mathfrak{h}_{\mathrm{reg}})$-mod. Deligne's version of the Riemann–Hilbert correspondence, [80], says that:

THEOREM 4.3.1. *The de Rham and solutions functors*

$$\mathsf{DR}, \mathsf{Sol} : \mathrm{Conn}^{\mathrm{reg}}(\mathfrak{h}_{\mathrm{reg}}) \to \pi_1(\mathfrak{h}_{\mathrm{reg}})\text{-mod}$$

are equivalences.

Why is the notion of regular connection crucial in Deligne's version of the Riemann–Hilbert correspondence? A complete answer to this question is beyond the authors understanding. But one important point is that there are, in general, many nonisomorphic integrable connections that will give rise to the same local system. So what Deligne showed is that, for a given local system L, the notion of regular connection gives one a canonical representative in the set of all connections whose solutions are L.

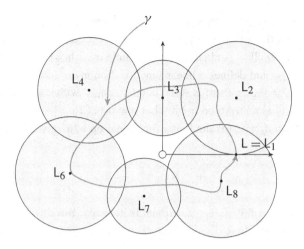

Figure 5. The locally constant sheaf (local system) L on $\mathbb{C}^\times$. On each open set, the local system L_i is constant, and on overlaps we have $L_i \simeq L_{i+1}$. Composing these isomorphisms defines an automorphism $L \xrightarrow{\sim} L$, which is the action of the path γ on the stalk of L at 1.

4.4. Equivariance. Recall that the group W acts freely on the open set $\mathfrak{h}_{\mathrm{reg}}$. This implies that the quotient map $\pi : \mathfrak{h}_{\mathrm{reg}} \to \mathfrak{h}_{\mathrm{reg}}/W$ is a finite covering map. In particular, its differential $d_x\pi : T_x\mathfrak{h}_{\mathrm{reg}} \to T_{\pi(x)}(\mathfrak{h}_{\mathrm{reg}}/W)$ is an isomorphism for all $x \in \mathfrak{h}_{\mathrm{reg}}$.

PROPOSITION 4.4.1. *There is a natural isomorphism* $\mathcal{D}(\mathfrak{h}_{\mathrm{reg}})^W \simeq \mathcal{D}(\mathfrak{h}_{\mathrm{reg}}/W)$.

PROOF. Since $\mathcal{D}(\mathfrak{h}_{\mathrm{reg}})$ acts on $\mathbb{C}[\mathfrak{h}_{\mathrm{reg}}]$, the algebra $\mathcal{D}(\mathfrak{h}_{\mathrm{reg}})^W$ acts on $\mathbb{C}[\mathfrak{h}_{\mathrm{reg}}]^W = \mathbb{C}[\mathfrak{h}_{\mathrm{reg}}/W]$. One can check from the definition of differential operators that this defines a map $\mathcal{D}(\mathfrak{h}_{\mathrm{reg}})^W \to \mathcal{D}(\mathfrak{h}_{\mathrm{reg}}/W)$. Since $\mathcal{D}(\mathfrak{h}_{\mathrm{reg}})^W$ is a simple ring, this map must be injective. Therefore it suffices to show that it is surjective.

Let $x_1, \ldots, x_n$ be a basis of $\mathfrak{h}^*$ so that $\mathbb{C}[\mathfrak{h}] = \mathbb{C}[x_1, \ldots, x_n]$. By the Chevalley–Shephard–Todd theorem, Theorem 1.1.4, $\mathbb{C}[\mathfrak{h}]^W$ is also a polynomial ring with homogeneous, algebraically independent generators $u_1, \ldots, u_n$ say. Moreover, $\mathbb{C}[\mathfrak{h}_{\mathrm{reg}}]^W = \mathbb{C}[u_1, \ldots, u_n][\delta^{-r}]$ for some $r > 0$. Thus,

$$\mathcal{D}(\mathfrak{h}_{\mathrm{reg}}/W) = \mathbb{C}\left\langle u_1, \ldots, u_n, \frac{\partial}{\partial u_1}, \ldots, \frac{\partial}{\partial u_n} \right\rangle [\delta^{-r}]$$

and to show surjectivity it suffices to show that each $\partial/\partial u_i$ belongs to $\mathcal{D}(\mathfrak{h}_{\mathrm{reg}})^W$. That is, we need to find some $f_{i,j} \in \mathbb{C}[\mathfrak{h}_{\mathrm{reg}}]$ such that $v_i := \sum_{j=1}^n f_{i,j} \, \partial/\partial x_j \in \mathcal{D}(\mathfrak{h}_{\mathrm{reg}})^W$ and $v_i(u_k) = \delta_{i,k}$. Let $\Delta(\boldsymbol{u})$ be the n by n matrix with (i, j)-th entry $\partial u_i/\partial x_j$. Then, since $v_i(u_k) = \sum_{j=1}^n f_{i,j} \, \partial u_k/\partial x_j$, we really need to find a matrix $F = (f_{i,j})$ such that $F \cdot \Delta(\boldsymbol{u})$ is the $n \times n$ identity matrix. In other words, F must be the inverse to $\Delta(\boldsymbol{u})$. Clearly, F exists if and only if $\det \Delta(\boldsymbol{u})$ is invertible in $\mathbb{C}[\mathfrak{h}_{\mathrm{reg}}]$. But, for each $t \in \mathfrak{h}_{\mathrm{reg}}$, $\det \Delta(\boldsymbol{u})|_{x=t}$ is just the determinant of the differential $d_t\pi : T_t\mathfrak{h}_{\mathrm{reg}} \to T_{\pi(t)}(\mathfrak{h}_{\mathrm{reg}}/W)$. As noted above, the linear map $d_t\pi$ is always an isomorphism. Thus, since $\mathbb{C}[\mathfrak{h}_{\mathrm{reg}}]$ is a domain, the Nullstellensatz implies that $\det \Delta(\boldsymbol{u})$ is invertible in $\mathbb{C}[\mathfrak{h}_{\mathrm{reg}}]$.

The final thing to check is that if $F = \Delta(\boldsymbol{u})^{-1}$ then each v_i is W-invariant. Let $w \in W$ and consider the derivation $v_i' := w(v_i) - v_i$. It is clear from the definition of v_i that v_i' acts as zero on all u_j and hence on the whole of $\mathbb{C}[\mathfrak{h}_{\text{reg}}]^W$. Geometrically v_i' is a vector field on $\mathfrak{h}_{\text{reg}}$ and defines a linear functional on $\mathfrak{m}/\mathfrak{m}^2$, for each maximal ideal $\mathfrak{m}$ of $\mathbb{C}[\mathfrak{h}_{\text{reg}}]$. If $v_i' \neq 0$ then there is some point $x \in \mathfrak{h}_{\text{reg}}$, with corresponding maximal ideal $\mathfrak{m}$, such that v_i' is nonzero on $\mathfrak{m}/\mathfrak{m}^2$. Let $\mathfrak{n} = \mathfrak{m} \cap \mathbb{C}[\mathfrak{h}_{\text{reg}}]^W$ be the maximal ideal defining $\pi(x)$. Then $d_x \pi(v_i')$ is the linear functional on $\mathfrak{n}/\mathfrak{n}^2$ given by the composite

$$\mathfrak{n}/\mathfrak{n}^2 \to \mathfrak{m}/\mathfrak{m}^2 \xrightarrow{v_i'} \mathbb{C}.$$

Since $d_x \pi$ is an isomorphism, this functional is nonzero. But this contradicts the fact that v_i' acts trivially on $\mathbb{C}[\mathfrak{h}_{\text{reg}}]^W$. Thus, $w(v_i) = v_i$. $\square$

REMARK 4.4.2. The polynomial $\det \Delta(\boldsymbol{u})$ plays an important role in the theory of complex reflection groups, and is closely related to our δ; see [220].

COROLLARY 4.4.3. *The functor $M \mapsto M^W$ defines an equivalence between the category of W-equivariant $\mathcal{D}$-modules on $\mathfrak{h}^{\text{reg}}$ (i.e., the category of $\mathcal{D}(\mathfrak{h}_{\text{reg}}) \rtimes W$-modules) and the category of $\mathcal{D}(\mathfrak{h}_{\text{reg}}/W)$-modules.*

PROOF. Let e be the trivial idempotent in $\mathbb{C}W$. We can identify $\mathcal{D}(\mathfrak{h}_{\text{reg}})^W$ with $e(\mathcal{D}(\mathfrak{h}_{\text{reg}}) \rtimes W)e$. Hence by Proposition 4.4.1, $\mathcal{D}(\mathfrak{h}_{\text{reg}}/W) \simeq e(\mathcal{D}(\mathfrak{h}_{\text{reg}}) \rtimes W)e$. Then the proof of the corollary is identical to the proof of Corollary 1.6.3. We just need to show that if $M^W = eM$ is zero then M is zero. But, since W acts freely on $\mathfrak{h}_{\text{reg}}$, it is already clear for $\mathbb{C}[\mathfrak{h}_{\text{reg}}] \rtimes W$-modules that $M^W = 0$ implies $M = 0$ (see for instance the proof of Theorem 5.1.3). $\square$

One often says that the $\mathcal{D}(\mathfrak{h}_{\text{reg}}) \rtimes W$-module M *descends* to the $\mathcal{D}$-module M^W on $\mathfrak{h}_{\text{reg}}/W$; the terminology coming from descent theory in algebraic geometry.

We are finally in a position to define the KZ-functor. The functor is a composition of four(!) functors, so we'll go through it one step at a time. Recall that we're starting with a module M in category $\mathcal{O}$ and we eventually want a representation of the fundamental group $\pi_1(\mathfrak{h}_{\text{reg}}/W)$. First off, we localize and use the Dunkl embedding: $M \mapsto M[\delta^{-1}]$. This gives us a $\mathcal{D}(\mathfrak{h}_{\text{reg}}) \rtimes W$-module $M[\delta^{-1}]$. By Corollary 4.4.3, $(M[\delta^{-1}])^W$ is a $\mathcal{D}$-module on $\mathfrak{h}_{\text{reg}}/W$. Before going further we need to know what sort of $\mathcal{D}$-module we've ended up with. It's shown in [113] that in fact $(M[\delta^{-1}])^W$ is an integrable connection on $\mathfrak{h}_{\text{reg}}/W$ *with regular singularities*. Thus, we can apply Deligne's equivalence. The deRham functor $\mathrm{DR}(M[\delta^{-1}]^W)$ applied to $(M[\delta^{-1}])^W$ gives us a representation of $\pi_1(\mathfrak{h}_{\text{reg}}/W)$. Thus, we may define the KZ-functor $\mathrm{KZ} : \mathcal{O} \to \pi_1(\mathfrak{h}_{\text{reg}}/W)$-mod by

$$\mathrm{KZ}(M) := \mathrm{DR}(M[\delta^{-1}]^W) = (((M[\delta^{-1}])^W)^{\text{an}})^{\nabla}.$$

The following diagram should help the reader unpack the definition of the KZ-functor.

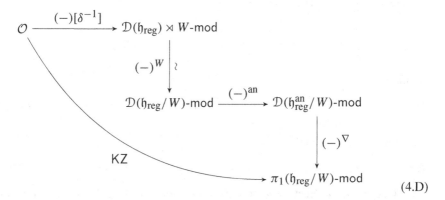

$$\tag{4.D}$$

4.5. A change of parameters. In order to relate, in the next subsection, the rational Cherednik algebra $H_c(W)$, via the KZ-functor, with the cyclotomic Hecke algebra $\mathcal{H}_q(W)$, we need to change the way we parametrize $H_c(W)$ (this is just some technical annoyance and you can skip this section unless you plan some hard core calculations using the KZ-functor). Each complex reflection $s \in S$ defines a reflecting hyperplane $H = \ker \alpha_s \subset \mathfrak{h}$. Let $\mathcal{A}$ denote the set of all hyperplanes arising this way. For a given $H \in \mathcal{A}$, the subgroup $W_H = \{w \in W \mid w(H) \subset H\}$ of W is cyclic. Let $W_H^* = W_H \backslash \{1\}$. Then

$$S = \bigcup_{H \in \mathcal{A}} W_H^*.$$

We may, without loss of generality, assume that

$$\alpha_H := \alpha_s = \alpha_{s'}, \quad \alpha_H^{\vee} := \alpha_s^{\vee} = \alpha_{s'}^{\vee}, \qquad \text{for all } s, s' \in W_H^*.$$

Then the original relations (1.B) in the definition of the rational Cherednik algebra become

$$[y, x] = (y, x) - \sum_{H \in \mathcal{A}} (y, \alpha_H)(\alpha_H^{\vee}, x)\left(\sum_{s \in W_H^*} c(s)s\right), \quad \forall x \in \mathfrak{h}^*, y \in \mathfrak{h}. \tag{4.E}$$

Let $n_H = |W_H|$ and let

$$e_{H,i} = \frac{1}{n_H} \sum_{w \in W_H} (\det w)^i w, \quad 0 \le i \le n_H - 1,$$

be the primitive idempotents in $\mathbb{C}W_H$. For $H \in \mathcal{A}$ and $0 \le i \le n_H - 1$, define $k_{H,i} \in \mathbb{C}$ by

$$\sum_{s \in W_H^*} c(s)s = n_H \sum_{i=0}^{n_H - 1} (k_{H,i+1} - k_{H,i})e_{H,i} \quad \text{and} \quad k_{H,0} = k_{H,n_H} = 0.$$

Note that this forces $k_{H,i} = k_{w(H),i}$ for all $w \in W$ and $H \in \mathcal{A}$. Thus, the parameters $k_{H,i}$ are W-invariant. This implies that

$$c(s) = \sum_{i=0}^{n_H-1} (\det s)^i (k_{H,i+1} - k_{H,i}),$$

and the relation (4.E) becomes

$$[y, x] = (y, x) - \sum_{H \in \mathcal{A}} (y, \alpha_H)(\alpha_H^\vee, x) n_H \sum_{i=0}^{n_H-1} (k_{H,i+1} - k_{H,i}) e_{H,i}, \forall x \in \mathfrak{h}^*, y \in \mathfrak{h}. \quad (4.F)$$

Therefore $\mathsf{H}_c(W) = \mathsf{H}_k(W)$, where

$$k = \{k_{H,i} \mid H \in \mathcal{A}, \ 0 \le i \le n_H - 1, k_{H,0} = k_{H,n_H} = 0 \text{ and } k_{H,i} = k_{w(H),i} \forall w, H, i\}.$$

4.6. The cyclotomic Hecke algebra. The braid group $\pi_1(\mathfrak{h}_{\text{reg}}/W)$ has generators $\{T_s \mid s \in \mathcal{S}\}$, where T_s is an s-generator of the monodromy around H; see [47, Section 4.C] for the precise definition. The T_s satisfy certain "braid relations". Fix $q = \{q_{H,i} \in \mathbb{C}^\times \mid H \in \mathcal{A}, \ 0 \le i \le n_H - 1\}$. The *cyclotomic Hecke algebra* $\mathcal{H}_q(W)$ is the quotient of the group algebra $\mathbb{C}\pi_1(\mathfrak{h}_{\text{reg}}/W)$ by the two-sided ideal generated by

$$\prod_{i=0}^{n_H-1} (T_s - q_{H,i}), \quad \forall s \in \mathcal{S},$$

where H is the hyperplane defined by s.

EXAMPLE 4.6.1. In type A the Hecke algebra is the algebra generated by $T_1, \ldots, T_{n-1}$, satisfying the braid relations

$$T_i T_j = T_j T_i, \qquad \text{if } |i - j| > 1,$$
$$T_i T_{i+1} T_i = T_{i+1} T_i T_{i+1}, \quad \text{if } 1 \le i \le n - 2,$$

and the additional relation

$$(T_i - q)(T_i + 1) = 0, \quad \text{if } 1 \le i \le n - 1.$$

For each $H \in \mathcal{A}$, fix a generator s_H of W_H. Given a parameter k for the rational Cherednik algebra, define q by

$$q_{H,i} = (\det s_H)^{-i} \exp(2\pi\sqrt{-1} k_{H,i}).$$

Based on [47, Theorem 4.12], the following key result was proved in [113, Theorem 5.13].

THEOREM 4.6.2. *The KZ-functor factors through $\mathcal{H}_q(W)$-mod.*

Since each of the functors appearing in diagram (4.D) is exact, the KZ-functor is an exact functor. Therefore, by Watt's theorem, [193, Theorem 5.50], there exists a projective module $P_{\mathsf{KZ}} \in \mathcal{O}$ such that

$$\mathsf{KZ}(-) \simeq \mathrm{Hom}_{\mathsf{H}_c(W)}(P_{\mathsf{KZ}}, -). \quad (4.G)$$

Then, Theorem 4.6.2 implies that P_{KZ} is a $(\mathsf{H}_c(W), \mathcal{H}_q(W))$-bimodule and the action of $\mathcal{H}_q(W)$ on the right of P_{KZ} defines an algebra morphism

$$\phi : \mathcal{H}_q(W) \longrightarrow \mathrm{End}_{\mathsf{H}_c(W)}(P_{KZ})^{op}.$$

LEMMA 4.6.3. *We have a decomposition*

$$P_{KZ} = \bigoplus_{\lambda \in \mathrm{Irr}(W)} (\dim \mathsf{KZ}(L(\lambda))) P(\lambda).$$

PROOF. By definition of projective cover, we have

$$\dim \mathrm{Hom}_{\mathsf{H}_c(W)}(P(\lambda), L(\mu)) = \delta_{\lambda,\mu}.$$

Therefore, if $P_{KZ} = \bigoplus_{\lambda \in \mathrm{Irr}(W)} P(\lambda)^{\oplus n_\lambda}$, then $n_\lambda = \dim \mathrm{Hom}_{\mathsf{H}_c(W)}(P_{KZ}, L(\lambda))$. But, by definition,

$$\mathrm{Hom}_{\mathsf{H}_c(W)}(P_{KZ}, L(\lambda)) = \mathsf{KZ}(L(\lambda)). \qquad \square$$

LEMMA 4.6.4. *Let $\mathcal{A}$ be an abelian, Artinian category and $\mathcal{A}'$ a full subcategory, closed under quotients. Let $F : \mathcal{A}' \to \mathcal{A}$ be the inclusion functor. Define $^\perp F : \mathcal{A} \to \mathcal{A}'$ by setting $^\perp F(M)$ to be the largest quotient of M contained in $\mathcal{A}'$. Then $^\perp F$ is left adjoint to F and the adjunction $\eta : \mathrm{id}_\mathcal{A} \to F \circ (^\perp F)$ is surjective.*

PROOF. We begin by showing that $^\perp F$ is well-defined. We need to show that, for each $M \in \mathcal{A}$, there is a unique maximal quotient N of M contained in $\mathcal{A}'$. Let

$$K = \{N' \subseteq M \mid M/N' \in \mathcal{A}'\}.$$

Note that if N_1' and N_2' belong to K then $N_1' \cap N_2'$ belongs to K. Therefore, if $N_1' \in K$ is not contained in all other $N' \in K$, we choose $N' \in K$ such that $N_1' \not\subset N'$ and set $N_2' = N' \cap N_1' \subsetneq N_1$. Continuing this way we construct a descending chain of submodules $N_1' \supsetneq N_2' \supsetneq \cdots$ of M. Since $\mathcal{A}$ is assumed to be Artinian, this chain must eventually stop. Hence, there is a unique minimal element under inclusion in K. It is clear that $^\perp F$ is left adjoint to F and the adjunction η just sends M to the maximal quotient of M in $\mathcal{A}'$, hence is surjective. $\qquad \square$

THEOREM 4.6.5 (double centralizer theorem). *We have an isomorphism*

$$\phi : \mathcal{H}_q(W) \xrightarrow{\sim} \mathrm{End}_{\mathsf{H}_c(W)}(P_{KZ})^{op}.$$

PROOF. Let $\mathcal{A}$ denote the image of the KZ functor in $\mathcal{H}_q(W)$-mod. It is a full subcategory of $\mathcal{H}_q(W)$-mod. It is also closed under quotients. To see this, notice that it suffices to show that the image of $\mathcal{O}$ under the localization functor $(-)[\delta^{-1}]$ is closed under quotients. If N is a nonzero quotient of $M[\delta^{-1}]$ for some $M \in \mathcal{O}$, then it is easy to check that the preimage N' of M under $M \to M[\delta^{-1}] \twoheadrightarrow N$ is nonzero and generates N. The claim follows. Then, Lemma 4.6.4 implies that ϕ is surjective. Hence to show that it is an isomorphism, it suffices to calculate the dimension of $\mathrm{End}_{\mathsf{H}_c(W)}(P_{KZ})$. By Lemma 4.6.3,

$$P_{KZ} = \bigoplus_{\lambda \in \mathrm{Irr}(W)} \dim \mathsf{KZ}(L(\lambda)) P(\lambda).$$

For any module $M \in \mathcal{O}$, we have $\dim \mathrm{Hom}_{H_c(W)}(P(\lambda), M) = [M : L(\lambda)]$, the multiplicity of $L(\lambda)$ in a composition series for M. This can be proved by induction on the length of M, using the fact that $\dim \mathrm{Hom}_{H_c(W)}(P(\lambda), L(\mu)) = \delta_{\lambda,\mu}$. Hence, using BGG reciprocity, we have

$$\dim \mathrm{End}_{H_c(W)}(P_{KZ}) = \bigoplus_{\lambda,\mu} \dim KZ(L(\lambda)) \dim KZ(L(\mu)) \, \mathrm{Hom}_{H_c(W)}(P(\lambda), P(\mu))$$

$$= \bigoplus_{\lambda,\mu} \dim KZ(L(\lambda)) \dim KZ(L(\mu))[P(\mu) : L(\lambda)]$$

$$= \bigoplus_{\lambda,\mu,\nu} \dim KZ(L(\lambda)) \dim KZ(L(\mu))[P(\mu) : \Delta(\nu)][\Delta(\nu) : L(\lambda)]$$

$$= \bigoplus_{\lambda,\mu,\nu} \dim KZ(L(\lambda)) \dim KZ(L(\mu))[\Delta(\nu) : L(\mu)][\Delta(\nu) : L(\lambda)]$$

$$= \bigoplus_{\nu} (\dim KZ(\Delta(\nu)))^2.$$

Since $\Delta(\nu)$ is a free $\mathbb{C}[\mathfrak{h}]$-module of rank $\dim(\nu)$, its localization $\Delta(\nu)[\delta^{-1}]$ is an integrable connection of rank $\dim(\nu)$. Hence, $\dim KZ(\Delta(\nu)) = \dim(\nu)$ and thus $\dim \mathrm{End}_{H_c(W)}(P_{KZ}) = |W|$.   □

Let $\mathcal{O}_{\mathrm{tor}}$ be the Serre subcategory of $\mathcal{O}$ consisting of all modules that are torsion with respect to the Ore set $\{\delta^N\}_{N \in \mathbb{N}}$. The torsion submodule M_{tor} of $M \in \mathcal{O}$ is the set $\{m \in M \mid \exists \, N \gg 0 \text{ s.t. } \delta^N \cdot m = 0\}$. Then, M is torsion if $M_{\mathrm{tor}} = M$.

COROLLARY 4.6.6. *The* KZ*-functor is a quotient functor with kernel* $\mathcal{O}_{\mathrm{tor}}$*, i.e.,*

$$KZ : \mathcal{O}/\mathcal{O}_{\mathrm{tor}} \xrightarrow{\sim} \mathcal{H}_q(W)\text{-mod.}$$

PROOF. Notice that, of all the functors in diagram (4.D), only the first,

$$M \mapsto M[\delta^{-1}]$$

is not an equivalence. We have $M[\delta^{-1}] = 0$ if and only if M is torsion. Therefore, $KZ(M) = 0$ if and only if $M \in \mathcal{O}_{\mathrm{tor}}$. Thus, we just need to show that KZ is essentially surjective; that is, for each $N \in \mathcal{H}_q(W)$-mod there exists some $M \in \mathcal{O}$ such that $KZ(M) \simeq N$. We fix N to be some finite-dimensional $\mathcal{H}_q(W)$-module. Recall that $P_{KZ} \in \mathcal{O}$ is a $(H_c(W), \mathcal{H}_q(W))$-bimodule. Therefore, $\mathrm{Hom}_{H_c(W)}(P_{KZ}, H_c(W))$ is a $(\mathcal{H}_q(W), H_c(W))$-bimodule and

$$M = \mathrm{Hom}_{\mathcal{H}_q(W)}(\mathrm{Hom}_{H_c(W)}(P_{KZ}, H_c(W)), N)$$

is a module in category $\mathcal{O}$. Applying (4.G) and the double centralizer theorem, Theorem 4.6.5, we have

$$
\begin{aligned}
\mathsf{KZ}(M) &= \mathrm{Hom}_{\mathsf{H}_c(W)}(P_{KZ}, M) \\
&= \mathrm{Hom}_{\mathsf{H}_c(W)}(P_{KZ}, \mathrm{Hom}_{\mathcal{H}_q(W)}(\mathrm{Hom}_{\mathsf{H}_c(W)}(P_{KZ}, \mathsf{H}_c(W)), N)) \\
&\simeq \mathrm{Hom}_{\mathcal{H}_q(W)}(\mathrm{Hom}_{\mathsf{H}_c(W)}(P_{KZ}, \mathsf{H}_c(W)) \otimes_{\mathsf{H}_c(W)} P_{KZ}, N) \\
&\simeq \mathrm{Hom}_{\mathcal{H}_q(W)}(\mathrm{End}_{\mathsf{H}_c(W)}(P_{KZ}), N) \\
&\simeq \mathrm{Hom}_{\mathcal{H}_q(W)}(\mathcal{H}_q(W), N) = N,
\end{aligned}
$$

where we have used [22, Proposition 4.4(b)] in the third line. $\square$

4.7. Example. Let's take $W = \mathbb{Z}_n$. In this case the Hecke algebra $\mathcal{H}_q(\mathbb{Z}_n)$ is generated by a single element $T := T_1$ and satisfies the defining relation

$$
\prod_i (T - q_i^{m_i}) = 0.
$$

Unlike examples of higher rank, the algebra $\mathcal{H}_q(\mathbb{Z}_n)$ is commutative. Let $\zeta \in \mathbb{C}^\times$ be defined by $s(x) = \zeta x$. We fix $\alpha_s = \sqrt{2}x$ and $\alpha_s^\vee = \sqrt{2}y$, which implies that $\lambda_s = \zeta$. Let $\Delta(i) = \mathbb{C}[x] \otimes e_i$ be the standard module associated to the simple $\mathbb{Z}_n$-module e_i, where $s \cdot e_i = \zeta^i e_i$. The module $\Delta(i)$ is free as a $\mathbb{C}[x]$-module and the action of y is uniquely defined by $y \cdot (1 \otimes e_i) = 0$. Since

$$
y = \partial_x - \sum_{i=1}^{m-1} \frac{2c_i}{1 - \zeta^i} \frac{1}{x}(1 - s^i)
$$

under the Dunkl embedding, we have $\Delta(i)[\delta^{-1}] = \mathbb{C}[x, x^{-1}] \otimes e_i$ with connection defined by

$$
\partial_x \cdot e_i = \frac{a_i}{x} e_i,
$$

where

$$
a_i := 2 \sum_{j=1}^{m-1} \frac{c_j(1 - \zeta^{ij})}{1 - \zeta^j}.
$$

It is clear that this connection is regular. The horizontal sections sheaf of $\Delta(i)[\delta^{-1}]$ (i.e., $\mathrm{DR}(\Delta(i)[\delta^{-1}])$) on $\mathfrak{h}_{\mathrm{reg}}^{\mathrm{an}}$ is dual to the sheaf of multivalued solutions $\mathbb{C} \cdot x^{a_i} = \mathrm{Sol}(\Delta(i)[\delta^{-1}])$ of the differential equation $x\partial_x - a_i = 0$. But this is not what we want. We first want to descend the $\mathcal{D}(\mathbb{C}^\times) \rtimes \mathbb{Z}_n$-module to the $\mathcal{D}(\mathfrak{h}_{\mathrm{reg}})^{\mathbb{Z}_n} = \mathcal{D}(\mathfrak{h}_{\mathrm{reg}}/\mathbb{Z}_n)$-module $(\Delta(i)[\delta^{-1}])^{\mathbb{Z}_n}$. Then $\mathsf{KZ}(\Delta(i))$ is defined to be the horizontal sections of $(\Delta(i)[\delta^{-1}])^{\mathbb{Z}_n}$. Let $z = x^n$ so that $\mathbb{C}[\mathfrak{h}_{\mathrm{reg}}/\mathbb{Z}_n] = \mathbb{C}[z, z^{-1}]$. Then, an easy calculation shows that $\partial_z = \frac{1}{nx^{n-1}}\partial_x$ (check this!). Since

$$
(\Delta(i)[\delta^{-1}])^{\mathbb{Z}_n} = \mathbb{C}[z, z^{-1}] \cdot (x^{n-i} \otimes e_i) =: \mathbb{C}[z, z^{-1}] \cdot u_i,
$$

we see that

$$
\partial_z \cdot (x^{n-i} \otimes e_i) = \frac{1}{nx^{n-1}} \partial_x \cdot (x^{n-i} \otimes e_i) = \frac{n - i + a_i}{nz} u_i.
$$

Hence, by Equation (4.C), $\mathsf{KZ}(\Delta(i))$ is the duality functor $\mathbb{D}'$ applied to the local system of solutions $\mathbb{C} \cdot z^{b_i} = \mathrm{Sol}((\Delta(i)[\delta^{-1}])^{\mathbb{Z}_n})$, where

$$b_i = \frac{n - i + a_i}{n}.$$

At this level, the duality functor $\mathbb{D}'$ simply sends the local system $\mathbb{C} \cdot z^{b_i}$ to the local system $\mathbb{C} \cdot z^{-b_i}$. The generator T of $\pi_1(\mathfrak{h}_{\mathrm{reg}}/\mathbb{Z}_n)$ is represented by the loop $t \mapsto \exp(2\pi\sqrt{-1}t)$. Therefore

$$T \cdot z^{-b_i} = \exp(-2\pi\sqrt{-1}b_i)z^{-b_i}.$$

It turns out that, in the rank one case, $L(i)[\delta^{-1}] = 0$ if $L(i) \neq \Delta(i)$. Thus,

$$\mathsf{KZ}(L(i)) = \begin{cases} \mathsf{KZ}(\Delta(i)) & \text{if } L(i) = \Delta(i), \\ 0 & \text{otherwise.} \end{cases}$$

4.8. Application. As an application of the double centralizer theorem, Theorem 4.6.5, we mention the following very useful result due to Vale [227].

THEOREM 4.8.1. *The following are equivalent:*

(1) $\mathsf{H}_k(W)$ *is a simple ring.*
(2) *Category $\mathcal{O}$ is semisimple.*
(3) *The cyclotomic Hecke algebra $\mathcal{H}_q(W)$ is semisimple.*

4.9. The KZ functor for $\mathbb{Z}_2$. In this section we'll try to describe what the KZ functor does to modules in category $\mathcal{O}$ when $W = \mathbb{Z}_2$, our favourite example. The Hecke algebra $\mathcal{H}_q(\mathbb{Z}_2)$ is the algebra generated by $T := T_1$ and satisfying the relation $(T-1)(T-q) = 0$. The defining relation for $\mathsf{H}_c(\mathbb{Z}_2)$ is

$$[y, x] = 1 - 2cs,$$

see Example 1.2.2. We have $q = \det(s)\exp(2\pi\sqrt{-1}c) = -\exp(2\pi\sqrt{-1}c)$. We will calculate $\mathsf{KZ}(P(\rho_1))$, assuming that $c = \frac{1}{2} + m$ for some $m \in \mathbb{Z}_{\geq 0}$. For this calculation, we will use the explicit description of $P(\rho_1)$ given in Section 2.10. Recall that $P(\rho_1)$ is the $\mathsf{H}_c(\mathbb{Z}_2)$-module $\mathbb{C}[x] \otimes \rho_1 \oplus \mathbb{C}[x] \otimes \rho_0$ with

$$x \cdot (1 \otimes \rho_1) = x \otimes \rho_1 + 1 \otimes \rho_0, \quad x \cdot (1 \otimes \rho_0) = x \otimes \rho_0,$$
$$y \cdot (1 \otimes \rho_1) = x^{2m} \otimes \rho_0, \qquad y \cdot (1 \otimes \rho_0) = 0.$$

This implies that

$$\frac{1}{x} \cdot (1 \otimes \rho_1) = \frac{1}{x} \otimes \rho_1 - \frac{1}{x^2} \otimes \rho_0.$$

If we write $P(\rho_1)[\delta^{-1}] = \mathbb{C}[x^{\pm 1}] \cdot a_1 \oplus \mathbb{C}[x^{\pm 1}] \cdot a_0$, where $a_1 = 1 \otimes \rho_1$ and $a_0 = 1 \otimes \rho_0$, then

$$\partial_x \cdot a_1 = \left(y + \frac{c}{x}(1 - s)\right) \cdot a_1 = y \cdot a_1 + \frac{2c}{x} \cdot a_1.$$

Now,

$$y \cdot a_1 = x^{2m} \otimes \rho_0 = x^{2m} \cdot a_0;$$

hence $\partial_x \cdot a_1 = (2c/x) \cdot a_1 + x^{2m} \cdot a_0$. Also, $\partial_x \cdot a_0 = 0$. A free $\mathbb{C}[z^{\pm 1}]$-basis of $P(\rho_0)[\delta^{-1}]^{\mathbb{Z}_2}$ is given by $u_1 = x \cdot a_1$ and $u_0 = a_0$. Therefore

$$\partial_z \cdot u_1 = \frac{1+2c}{2z} u_1 + \frac{1}{2} z^m u_0 = \frac{m+1}{z} u_1 + \frac{1}{2} z^m u_0$$

and $\partial_z \cdot u_0 = 0$, where we used the fact that $c = \frac{1}{2} + m$. Hence $\mathsf{KZ}(P(\rho_1))$ is given by the connection

$$\partial_z + \begin{pmatrix} \frac{m+1}{z} & 0 \\ \frac{1}{2} z^m & 0 \end{pmatrix}.$$

Two linearly independent solutions of this equation are

$$g_1(z) = \begin{pmatrix} z^{-(m+1)} \\ \frac{1}{2} \ln(z) \end{pmatrix}, \quad g_2(z) = \begin{pmatrix} 0 \\ 1 \end{pmatrix}.$$

If, in a small, simply connected neighbourhood of 1, we choose the branch of $\ln(z)$ such that $\ln(1) = 0$, then $\gamma(0) = 0$ and $\gamma(1) = 2\pi\sqrt{-1}$, where

$$\gamma : [0, 1] \to \mathbb{C}, \quad \gamma(t) = \ln(\exp(2\pi\sqrt{-1}t)).$$

Therefore $\mathsf{KZ}(P(\rho_1))$ is the two-dimensional representation of $\mathcal{H}_q(\mathbb{Z}_2)$ given by

$$T \mapsto \begin{pmatrix} 1 & 0 \\ 2\pi\sqrt{-1} & 1 \end{pmatrix}.$$

This is isomorphic to the left regular representation of $\mathcal{H}_q(\mathbb{Z}_2)$.

4.10. Exercises.

EXERCISE 4.10.1. Let $\mathfrak{h}_{\mathrm{reg}} = \mathbb{C}^\times$. For each $\alpha \in \mathbb{C}$, write M_α for the $\mathcal{D}(\mathbb{C}^\times)$-module $\mathcal{D}(\mathbb{C}^\times)/\mathcal{D}(\mathbb{C}^\times)(x\partial - \alpha)$. Show that $M_\alpha \simeq M_{\alpha+1}$ for all α. Describe a multivalued holomorphic function which is a section of the one-dimensional local system $\mathrm{Sol}(M_\alpha)$ (notice that in general, the function is genuinely multivalued and hence not well-defined on the whole of $\mathbb{C}^\times$. This corresponds to the fact that the local system $\mathrm{Sol}(M_\alpha)$ has *no* global sections in general). Finally, the local system $\mathrm{Sol}(M_\alpha)$ defines a one-dimensional representation of the fundamental group $\pi_1(\mathbb{C}^\times) = \mathbb{Z}$. What is this representation?

EXERCISE 4.10.2. By considering the case of $\mathbb{Z}_2$, show that the natural map $\mathcal{D}(\mathfrak{h})^W \to \mathcal{D}(\mathfrak{h}/W)$ is not an isomorphism. Which of injectivity or surjectivity fails? Hint: for complete rigour, consider the associated graded map

$$\mathrm{gr}\, \mathcal{D}(\mathfrak{h})^W \to \mathrm{gr}\, \mathcal{D}(\mathfrak{h}/W).$$

EXERCISE 4.10.3. Assume that $W = \mathbb{Z}_2$, as in Section 4.9. For all c, (1) describe $\mathsf{KZ}(\Delta(\lambda))$ as a $\mathcal{H}_q(\mathbb{Z}_2)$-module, and (2) describe P_{KZ}.

4.11. Additional remarks.

Most of the results of this section first appeared in [113] and our exposition is based mainly on this paper.

Further details on the KZ-functor are also contained in [194].

5. Symplectic reflection algebras at $t = 0$

Recall from Section 1 that we used the Satake isomorphism to show that

- the algebra $Z(H_{0,c}(G))$ is isomorphic to $eH_{0,c}(G)e$ and $H_{0,c}(G)$ is a finite $Z(H_{0,c}(G))$-module, and
- the centre of $H_{1,c}(G)$ equals $\mathbb{C}$.

In this section we'll consider symplectic reflection algebras "at $t = 0$" and, in particular, the geometry of $Z_c(G) := Z(H_{0,c}(G))$.

DEFINITION 5.0.1. The *generalised Calogero–Moser space* $X_c(G)$ is defined to be the affine variety $\operatorname{Spec} Z_c(G)$.

The (classical) Calogero–Moser space was introduced by Kazhdan, Kostant and Sternberg [149] and studied further by Wilson in the wonderful paper [243]. Calogero [60] studied the integrable system describing the motion of n massless particles on the real line with a repulsive force between each pair of particles, proportional to the square of the distance between them. In [149], Kazhdan, Kostant and Sternberg give a description of the corresponding phase space in terms of Hamiltonian reduction. By considering the real line as being the imaginary axis sitting in the complex plane, Wilson interprets the Calogero–Moser phase space as an affine variety

$$\mathcal{C}_n = \big\{ (X, Y; u, v) \in \operatorname{Mat}_n(\mathbb{C}) \times \operatorname{Mat}_n(\mathbb{C}) \times \mathbb{C}^n \times (\mathbb{C}^n)^* \mid [X, Y] + I_n = v \cdot u \big\} /\!\!/ \operatorname{GL}_n(\mathbb{C}).$$
$$(5.A)$$

He showed, [243, Section 1], that $\mathcal{C}_n$ is a smooth, irreducible, symplectic affine variety. For further reading see [95]. The relation to rational Cherednik algebras comes from an isomorphism by Etingof and Ginzburg between the affine variety $X_{c=1}(\mathfrak{S}_n) = \operatorname{Spec} Z_{c=1}(S_n)$, and the Calogero–Moser space $\mathcal{C}_n$:

$$\psi_n : X_{c=1}(\mathfrak{S}_n) \xrightarrow{\sim} \mathcal{C}_n.$$

It is an isomorphism of affine symplectic varieties and implies that $X_c(\mathfrak{S}_n)$ is smooth when $c \neq 0$.

The filtration on $H_{0,c}(G)$ induces, by restriction, a filtration on $Z_c(G)$. Since the associated graded of $Z_c(G)$ is $\mathbb{C}[V]^G$, $X_c(G)$ is reduced and irreducible.

EXAMPLE 5.0.2. When $G = \mathbb{Z}_2$ acts on $\mathbb{C}^2$, the centre of $H_c(\mathbb{Z}_2)$ is generated by $A := x^2$, $B := xy - cs$ and $C = y^2$. Thus,

$$X_c(\mathbb{Z}_2) \simeq \frac{\mathbb{C}[A, B, C]}{(AC - (B + c)(B - c))}$$

is the affine cone over $\mathbb{P}^1 \subset \mathbb{P}^2$ when $c = 0$, but is a smooth affine surface for all $c \neq 0$, see Figure 1.

5.1. Representation theory.

Key point: much of the geometry of the generalised Calogero–Moser space is encoded in the representation theory of the corresponding symplectic reflection algebra (a consequence of the double centralizer property!). In particular, a closed point of X_c is singular if and only if there is a "small" simple module supported at that point — this statement is made precise in Proposition 5.1.4 below.

The fact that $H_{0,c}(G)$ is a finite module over its centre implies that it is an example of a P.I. (*polynomial identity*) ring. This is a very important class of rings in classical ring theory and can be thought of as rings that are "close to being commutative". We won't recall the definition of a P.I. ring here, but refer the reader to Appendix I.13 of the excellent book [50].

LEMMA 5.1.1. *There exists some $N > 0$ such that* $\dim L \leq N$ *for all simple $H_{0,c}(G)$-modules L.*

PROOF. It is a consequence of Kaplansky's theorem, [172, Theorem 13.3.8], that every simple $H_{0,c}(G)$-module is a finite-dimensional vector space over $\mathbb{C}$. More precisely, to every prime P.I. ring is associated its P.I. degree. Then, Kaplansky's theorem implies that if L is a simple $H_{0,c}(G)$-module then

$$\dim L \leq \text{P.I. degree}\,(H_{0,c}(G)) \quad \text{and} \quad H_{0,c}(G)/\operatorname{Ann}_{H_{0,c}(G)} L \simeq \operatorname{Mat}_m(\mathbb{C}). \qquad \square$$

Schur's lemma says that the elements of the centre $Z_c(G)$ of $H_{0,c}(G)$ act as scalars on any simple $H_{0,c}$-module L. Therefore, the simple module L defines a character $\chi_L : Z_c(G) \to \mathbb{C}$ and the kernel of χ_L is a maximal ideal in $Z_c(G)$. Thus, the character χ_L corresponds to a closed point in $X_c(G)$. Without loss of generality, we will refer to this point as χ_L and denote by $Z_c(G)_{\chi_L}$ the localization of $Z_c(G)$ at the maximal ideal Ker χ_L. We denote by $H_{0,c}(G)_\chi$ the central localization $H_{0,c}(G) \otimes_{Z_c(G)} Z_c(G)_\chi$. The Azumaya locus of $H_{0,c}(G)$ over $Z_c(G)$ is defined to be

$$\mathcal{A}_c := \{\chi \in X_c(W) \mid H_{0,c}(W)_\chi \text{ is Azumaya over } Z_c(W)_\chi\}.$$

As shown in [50, Theorem III.1.7], $\mathcal{A}_c$ is a nonempty, open subset of $X_c(W)$.

REMARK 5.1.2. If you are not familiar with the (slightly technical) definition of Azumaya locus, as given in [49, Section 3], then it suffices to note that it is a consequence of the Artin–Procesi theorem [172, Theorem 13.7.14] that the following are equivalent:

(1) $\chi \in \mathcal{A}_c$;
(2) $\dim L = \text{P.I. degree}\,(H_{0,c}(G))$ for all simple modules L such that $\chi_L = \chi$;
(3) there exists a unique simple module L such that $\chi_L = \chi$.

In fact, one can say a great deal more about these simple modules of maximal dimension. The following result strengthens Lemma 5.1.1.

THEOREM 5.1.3. *Let L be a simple $H_{0,c}(G)$-module. Then $\dim L \leq |G|$ and $\dim L = |G|$ implies that $L \simeq \mathbb{C}G$ as a G-module.*

PROOF. We will prove the theorem when $H_{0,c}(G)$ is a rational Cherednik algebra, by using the Dunkl embedding. The proof for arbitrary symplectic reflection algebras is much harder.

By the theory of prime P.I. rings and their Azumaya loci, as described above, it suffices to show that there is some dense open subset U of $X_c(W)$ such that $L \simeq \mathbb{C}G$ for all simple modules L supported on U. Recall that the Dunkl embedding at $t = 0$ gives us an identification $H_{0,c}(G)[\delta^{-r}] \simeq \mathbb{C}[\mathfrak{h}_{reg} \times \mathfrak{h}^*] \rtimes G$, where $r > 0$ such that $\delta^r \in \mathbb{C}[\mathfrak{h}]^G$. Then it suffices to show that every simple $\mathbb{C}[\mathfrak{h}_{reg} \times \mathfrak{h}^*] \rtimes G$-module is isomorphic to $\mathbb{C}G$ as a G-module. The centre of $\mathbb{C}[\mathfrak{h}_{reg} \times \mathfrak{h}^*] \rtimes G$ (which is just the centre of $H_{0,c}(G)$ localized at δ^r) equals $\mathbb{C}[\mathfrak{h}_{reg} \times \mathfrak{h}^*]^G$. For each maximal ideal $\mathfrak{m} \lhd \mathbb{C}[\mathfrak{h}_{reg} \times \mathfrak{h}^*]^G$, we will construct a module $L(\mathfrak{m})$ such that $\mathfrak{m} \cdot L(\mathfrak{m}) = 0$ and $L(\mathfrak{m}) \simeq \mathbb{C}G$ as G-modules. Finally, we show that $L(\mathfrak{m})$ is the unique, up to isomorphism, simple module such that $\mathfrak{m} \cdot L(\mathfrak{m}) = 0$.

So fix a maximal ideal $\mathfrak{m}$ in $\mathbb{C}[\mathfrak{h}_{reg} \times \mathfrak{h}^*]^G$ and let $\mathfrak{n} \lhd \mathbb{C}[\mathfrak{h}_{reg} \times \mathfrak{h}^*]$ be a maximal ideal such that $\mathfrak{n} \cap \mathbb{C}[\mathfrak{h}_{reg} \times \mathfrak{h}^*]^G = \mathfrak{m}$ (geometrically, we have a finite map $\rho : \mathfrak{h}_{reg} \times \mathfrak{h}^* \twoheadrightarrow (\mathfrak{h}_{reg} \times \mathfrak{h}^*)/G$ and we're choosing some point in the preimage of $\mathfrak{m}$). If $\mathbb{C}_{\mathfrak{n}}$ is the one-dimensional $\mathbb{C}[\mathfrak{h}_{reg} \times \mathfrak{h}^*]$-module on which $\mathfrak{n}$ acts trivially, then define

$$L(\mathfrak{m}) = (\mathbb{C}[\mathfrak{h}_{reg} \times \mathfrak{h}^*] \rtimes G) \otimes_{\mathbb{C}[\mathfrak{h}_{reg} \times \mathfrak{h}^*]} \mathbb{C}_{\mathfrak{n}}.$$

The fact that $\rho^{-1}(\mathfrak{m})$ consists of a single free G-orbit implies that $L(\mathfrak{m}) \simeq \mathbb{C}G$ as a G-module. In particular, $\dim L(\mathfrak{m}) = |G|$. To see that it is simple, let N be a nonzero submodule. Then $N = \bigoplus_{p \in G \cdot \mathfrak{n}} N_p$ as a $\mathbb{C}[\mathfrak{h}_{reg} \times \mathfrak{h}^*]$-module, where $N_p = \{ n \in N \mid \mathfrak{n}_p^k \cdot n = 0 \text{ some } k \gg 0 \}$. Since $N \neq 0$, there exists some p such that $N_p \neq 0$. But then multiplication by $g \in G$ defines an isomorphism of vector spaces $N_p \to N_{g(p)}$. Since the G-orbit $G \cdot \mathfrak{n}$ is free, this implies that $\dim N \geq |G|$. Hence $N = L(\mathfrak{m})$.

Finally, let M be a simple module such that $\mathfrak{m} \cdot M = 0$. Arguing as above, this implies that $M_{\mathfrak{n}} \neq 0$. Hence there is a nonzero map $L(\mathfrak{m}) \to M$ induced by the embedding $\mathbb{C}_{\mathfrak{n}} \hookrightarrow M_{\mathfrak{n}}$. Since M is assumed to be simple, $M \simeq L(\mathfrak{m})$. □

THEOREM 5.1.4. *Let L be a simple $H_c(G)$-module then $\dim L = |G|$ if and only if χ_L is a nonsingular point of $X_c(G)$.*

OUTLINE OF PROOF. By Theorem 5.1.3, the dimension of a generic simple module is $|W|$. Since the Azumaya locus $\mathcal{A}_c$ is dense in X_c, it follows that

$$\text{P.I. degree } (H_{0,c}(G)) = |G|.$$

The proposition will then follow from the equality $\mathcal{A}_c = X_c(G)_{sm}$, where $X_c(G)_{sm}$ is the smooth locus of $X_c(G)$. As noted in Corollary 1.3.2, $H_{0,c}(G)$ has finite global dimension. It is known, [50, Lemma III.1.8], that this implies that $\mathcal{A}_c \subseteq (X_c)_{sm}$. The opposite inclusion is an application of a result by Brown and Goodearl [49, Theorem 3.8]. Their theorem says that $(X_c)_{sm} \subseteq \mathcal{A}_c$ (in fact that we have equality) if $H_{0,c}(G)$ has particularly nice homological properties — it must be Auslander-regular and Cohen–Macaulay, and the complement of $\mathcal{A}_c$ has codimension at least two in X_c. The fact that $H_{0,c}(G)$ is Auslander-regular and Cohen–Macaulay can be deduced from the fact that its associated graded, the skew group ring, has these properties (the results that are required to show this are listed in the proof of [48, Theorem 4.4]). The fact that the complement of $\mathcal{A}_c$ has co-dimension at least two in X_c is harder to show. It follows from the fact that X_c is a symplectic variety, Theorem 5.4.5, and that the "representation theory of $H_{0,c}$ is constant along orbits", Theorem 5.4.6. □

Proposition 5.1.4 implies that to answer the question

QUESTION 5.1.5. Is the generalised Calogero–Moser space smooth?

It suffices to compute the dimension of simple $H_c(G)$-modules. Unfortunately, this turns out to be rather difficult to do.

5.2. Poisson algebras. The extra parameter t in $H_{t,c}(G)$ gives us a canonical quantization of the space $X_c(G)$. As a consequence, this implies that $X_c(G)$ is a Poisson variety. Recall:

DEFINITION 5.2.1. A *Poisson algebra* $(A, \{-, -\})$ is a commutative algebra with a bracket $\{-, -\} : A \otimes A \to A$ such that

(1) the pair $(A, \{-, -\})$ is a Lie algebra;
(2) $\{a, -\} : A \to A$ is a derivation for all $a \in A$, i.e.,

$$\{a, bc\} = \{a, b\}c + b\{a, c\}, \quad \forall a, b, c \in A.$$

An ideal I in the Poisson algebra A is called *Poisson* if $\{I, A\} \subseteq I$. As shown in Exercise 5.9.1, if I is a Poisson ideal in the Poisson algebra A then A/I is natural a Poisson algebra.

Hayashi's construction [129]: We may think of t as a variable so that $H_{0,c}(G) = H_{t,c}(G)/t \cdot H_{t,c}(G)$. For $z_1, z_2 \in Z_c(G)$ define

$$\{z_1, z_2\} = \left(\frac{1}{t}[\hat{z}_1, \hat{z}_2]\right) \mod t H_{t,c}(G),$$

where $\hat{z}_1, \hat{z}_2$ are arbitrary lifts of z_1, z_2 in $H_{t,c}(G)$.

PROPOSITION 5.2.2. *Since $H_{0,c}(W)$ is flat over $\mathbb{C}[t]$, $\{-, -\}$ is a well-defined Poisson bracket on $Z_c(G)$.*

PROOF. Write $\rho : H_{t,c}(G) \to H_{0,c}(G)$ for the quotient map. Let us first check that the binary operation is well-defined. Let $\hat{z}_1, \hat{z}_2$ be arbitrary lifts of $z_1, z_2 \in Z_c(G)$. Then $\rho([\hat{z}_1, \hat{z}_2]) = [\rho(\hat{z}_1), \rho(\hat{z}_2)] = 0$. Therefore, there exists some $\hat{z}_3 \in H_{t,c}(G)$ such that $[\hat{z}_1, \hat{z}_2] = t \cdot \hat{z}_3$. Since t is a nonzero divisor, $\hat{z}_3$ is uniquely define. The claim is that $\rho(\hat{z}_3) \in Z_c(G)$: let $h \in H_{0,c}(G)$ and $\hat{h}$ an arbitrary lift of h in $H_{t,c}(G)$. Then

$$[h, \rho(\hat{z}_3)] = \rho([\hat{h}, \hat{z}_3]) = \rho\left(\left[\hat{h}, \frac{1}{t}[\hat{z}_1, \hat{z}_2]\right]\right) = -\rho\left(\frac{1}{t}[\hat{z}_2, [\hat{h}, \hat{z}_1]]\right) - \rho\left(\frac{1}{t}[\hat{z}_2, [\hat{h}, \hat{z}_1]]\right).$$

Since $\hat{z}_1$ and $\hat{z}_1$ are lifts of central elements, the expressions $[\hat{z}_2, [\hat{h}, \hat{z}_1]]$ and $[\hat{z}_2, [\hat{h}, \hat{z}_1]]$ are in $t^2 H_{t,c}(G)$. Hence $[h, \rho(\hat{z}_3)] = 0$. Therefore, the expression $\{z_1, z_2\}$ is well-defined. The fact that $\rho([t \cdot \hat{z}_1, \hat{z}_2]/t) = [\rho(\hat{z}_1), \rho(\hat{z}_2)] = 0$ implies that the bracket is independent of choice of lifts. The fact that the bracket makes $Z_c(G)$ into a Lie algebra and satisfies the derivation property is a consequence of the fact that the commutator bracket of an algebra also has these properties. $\square$

REMARK 5.2.3. The same construction makes $e H_{0,c}(G) e$ into a Poisson algebra such that the Satake isomorphism is an isomorphism of Poisson algebras.

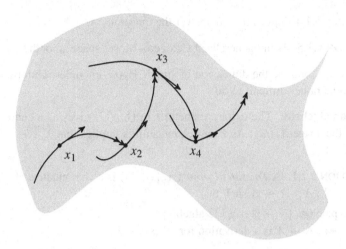

Figure 6. Flow along integral curves in a symplectic leaf. The vectors at x_i indicate a Hamiltonian vector field and the curve through x_i is the corresponding integral curve.

5.3. Symplectic leaves. In the algebraic world there are several different definitions of symplectic leaves, which can be shown to agree in "good" cases. We will define two of them here. First, assume that $X_c(G)$ is smooth. Then $X_c(G)$ may be considered as a complex analytic manifold equipped with the analytic topology. In this case, the *symplectic leaf* through $\mathfrak{m} \in X_c(G)$ is the maximal connected analytic submanifold $\mathcal{L}(\mathfrak{m})$ of $X_c(G)$ which contains $\mathfrak{m}$ and on which $\{-, -\}$ is nondegenerate. An equivalent definition is to say that $\mathcal{L}(\mathfrak{m})$ is the set of all points that can be reached from $\mathfrak{m}$ by travelling along integral curves corresponding to the Hamiltonian vector fields $\{z, -\}$ for $z \in Z_c(G)$.

Let us explain in more detail what is meant by this. Let v be a vector field on a complex analytic manifold X, i.e., v is a holomorphic map $X \to TX$ such that $v(x) \in T_x X$ for all $x \in X$ (v is assigning, in a continuous manner, a tangent vector to each point of x). An integral curve for v through x is a holomorphic function $\Phi_{x,v} : B_\epsilon(0) \to X$, where $B_\epsilon(0)$ is a closed ball of radius ϵ around 0 in $\mathbb{C}$, such that $(d_0 \Phi_{x,v})(1) = v(x)$, i.e., the derivative of $\Phi_{x,v}$ at 0 maps the basis element 1 of $T_0 \mathbb{C} = \mathbb{C}$ to the tangent vector field $v(x)$. The existence and uniqueness of holomorphic solutions to ordinary differential equations implies that $\Phi_{x,v}$ exists, and is unique, for each choice of v and x.

Now assume that $v = \{a, -\}$ is a Hamiltonian vector field, and fix $x \in X$. Then, the image of $\Phi_{x,\{a,-\}}$ is, by definition, contained in the symplectic leaf $\mathcal{L}_x$ through x. Picking another point $y \in \Phi_{x,\{a,-\}}(B_\epsilon(0))$ and another Hamiltonian vector field $\{b, -\}$, we again calculate the integral curve $\Phi_{y,\{b,-\}}$ and its image is again, by definition, contained in $\mathcal{L}_x$. Continuing in this way for as long as possible, $\mathcal{L}_x$ is the set of all points one can reach from x by "flowing along Hamiltonian vector fields".

In particular, this defines a stratification of $X_c(G)$. If, on the other hand, $X_c(G)$ is not smooth, then we first stratify the smooth locus of $X_c(G)$. The singular locus

$X_c(G)_{\text{sing}}$ of $X_c(G)$ is a Poisson subvariety. Therefore, the smooth locus of $X_c(G)_{\text{sing}}$ is again a Poisson manifold and has a stratification by symplectic leaves. We can continue by considering the "the singular locus of the singular locus" of $X_c(G)$ and repeating the argument... This way we get a stratification of the whole of $X_c(G)$ by symplectic leaves.

5.4. Symplectic cores. Let $\mathfrak{p}$ be a prime ideal in $Z_c(G)$. Then there is a (necessarily unique) largest Poisson ideal $\mathcal{P}(\mathfrak{p})$ contained in $\mathfrak{p}$. Define an equivalence relation $\sim$ on $X_c(G)$ by saying

$$\mathfrak{p} \sim \mathfrak{q} \Leftrightarrow \mathcal{P}(\mathfrak{p}) = \mathcal{P}(\mathfrak{q}).$$

The *symplectic cores* of $X_c(G)$ are the equivalence classes defined by $\sim$. We write

$$\mathcal{C}(\mathfrak{p}) = \{\mathfrak{q} \in X_c(G) \mid \mathcal{P}(\mathfrak{p}) = \mathcal{P}(\mathfrak{q})\}.$$

Then, each symplectic core $\mathcal{C}(\mathfrak{p})$ is a locally closed subvariety of $X_c(G)$ and $\overline{\mathcal{C}(\mathfrak{p})} = V(\mathcal{P}(\mathfrak{p}))$. The set of all symplectic cores is a partition of $X_c(G)$ into locally closed subvarieties. As one can see from the examples below, a Poisson variety X will typically have an infinite number of symplectic leaves and an infinite number of symplectic cores.

DEFINITION 5.4.1. We say that the Poisson bracket on X is *algebraic* if X has only finitely many symplectic leaves.

Proposition 3.7 of [51] says:

PROPOSITION 5.4.2. *If the Poisson bracket on X is algebraic then the symplectic leaves are locally closed algebraic sets and that the stratification by symplectic leaves equals the stratification by symplectic cores.*

That is, $\mathcal{L}(\mathfrak{m}) = \mathcal{C}(\mathfrak{m})$ for all maximal ideals $\mathfrak{m} \in X$.

EXAMPLE 5.4.3. We consider the Poisson bracket on $\mathbb{C}^2 = \operatorname{Spec} \mathbb{C}[x, y]$ given by $\{x, y\} = y$, and try to describe the symplectic leaves in $\mathbb{C}^2$. From the definition of a Poisson algebra, it follows that each function $f \in \mathbb{C}[x, y]$ defines a vector field $\{f, -\}$ on $\mathbb{C}^2$. For the generators x, y, these vector fields are $\{x, -\} = y\partial_y$, $\{y, -\} = -y\partial_x$ respectively. In order to calculate the symplectic leaves we need to calculate the integral curve through a point $(p, q) \in \mathbb{C}^2$ for each of these vector fields. Then the leaf through (p, q) will be the submanifold traced out by all these curves. We begin with $y\partial_y$. The corresponding integral curve is $a = (a_1(t), a_2(t)) : B_\epsilon(0) \to \mathbb{C}^2$ such that $a(0) = (p, q)$ and

$$a'(t) = (y\partial_y)_{a(t)}, \quad \forall t \in B_\epsilon(0).$$

Thus, $a_1'(t) = 0$ and $a_2'(t) = a_2(t)$ which means $a = (p, qe^t)$. Similarly, if $b = (b_1, b_2)$ is the integral curve through (p, q) for $-y\partial_x$ then $b = (-qt + p, q)$. Therefore, there are only two symplectic leaves, $\{0\}$ and $\mathbb{C}^2 \backslash \{0\}$.

EXAMPLE 5.4.4. For each finite-dimensional Lie algebra $\mathfrak{g}$, there is a natural Poisson bracket on $\mathbb{C}[\mathfrak{g}^*] = \operatorname{Sym}(\mathfrak{g})$, uniquely defined by $\{X, Y\} = [X, Y]$ for all $X, Y \in \mathfrak{g}$. Recall that $\mathfrak{sl}_2 = \mathbb{C}\{E, F, H\}$ with $[E, F] = H$, $[H, E] = 2E$ and $[H, F] = -2F$. As in the previous example, we will calculate the symplectic leaves of $\mathfrak{sl}_2^*$. The Hamiltonian vector fields of the generators E, F, H of the polynomial ring $\mathbb{C}[\mathfrak{sl}_2]$ are $X_E = H\partial_F - 2E\partial_H$, $X_F = -H\partial_E + 2F\partial_H$ and $X_H = 2E\partial_E - 2F\partial_F$. Let $a = (a_E(t), a_F(t), a_H(t))$ be an integral curve through (p, q, r) for a vector field X.

- For X_E, $a(t) = (p, -pt^2 + rt + q, -2pt + r)$.
- For X_F, $a(t) = (-qt^2 - rt + p, q, 2qt + r)$.
- For X_H, $a(t) = (p\exp(2t), q\exp(-2t), r)$.

Thus, for all $(p, q, r) \neq (0, 0, 0)$, X_E, X_F, X_F span a two-dimensional subspace of $T_{(p,q,r)}\mathfrak{sl}_2^*$. Also, one can check that the expression

$$a_E(t)a_F(t) + \tfrac{1}{4}a_H(t)^2 = pq + \tfrac{1}{4}r^2$$

is independent of t for each of the three integral curves above, e.g., for X_F we have

$$(-qt^2 - rt + p)q + \tfrac{1}{4}(2qt + r)^2 = pq + \tfrac{1}{4}r^2.$$

Therefore, for $s \neq 0$, $V(EF + \tfrac{1}{4}H^2 = s)$ is a smooth, two-dimensional Poisson subvariety on which the symplectic form is everywhere nondegenerate. This implies that it is a symplectic leaf.

The nullcone $\mathcal{N}$ is defined to be $V(EF + \tfrac{1}{4}H^2)$, i.e., we take $s = 0$. If we consider $U = \mathcal{N}\setminus\{(0, 0, 0)\}$, then this is also smooth and the symplectic form is everywhere nondegenerate. Thus, it is certainly contained in a symplectic leaf of $\mathcal{N}$. However, the whole of $\mathcal{N}$ cannot be a leaf because it is singular. Therefore, the symplectic leaves of $\mathcal{N}$ are U and $\{(0, 0, 0)\}$.

In the case of symplectic reflection algebras, we have:

THEOREM 5.4.5. *The symplectic leaves of the Poisson variety $X_c(W)$ are precisely the symplectic cores of $X_c(W)$. In particular, they are finite in number, hence the bracket $\{-, -\}$ is algebraic.*

Remarkably, the representation theory of symplectic reflection algebras is "constant along symplectic leaves", in the following precise sense. For each $\chi \in X_c(G)$, let $\mathsf{H}_{\chi,c}(G)$ be the finite-dimensional quotient $\mathsf{H}_{0,c}(G)/\mathfrak{m}_\chi \mathsf{H}_{0,c}(G)$, where $\mathfrak{m}_\chi$ is the kernel of χ. If $\chi \in \mathcal{A}_c = X_c(G)_{\mathrm{sm}}$ then

$$\mathsf{H}_{\chi,c}(G) \simeq \mathrm{Mat}_{|G|}(\mathbb{C}), \quad \dim \mathsf{H}_{\chi,c}(G) = |G|^2.$$

This is not true if $\chi \in X_c(G)_{\mathrm{sing}}$.

THEOREM 5.4.6. *Let χ_1, χ_2 be two points in $\mathcal{L}$, a symplectic leaf of $X_c(G)$. Then,*

$$\mathsf{H}_{\chi_1,c}(G) \simeq \mathsf{H}_{\chi_2,c}(G).$$

EXAMPLE 5.4.7. Let's consider again our favourite example $W = \mathbb{Z}_2$. When c is nonzero, one can check, as in the right-hand side of Figure 1, that $X_c(\mathbb{Z}_2)$ is smooth. Therefore, it has only one symplectic leaf, i.e., it is a symplectic manifold. Over each closed point of $X_c(\mathbb{Z}_2)$ there is exactly one simple $\mathsf{H}_{0,c}(\mathbb{Z}_2)$-module, which is isomorphic to $\mathbb{C}\mathbb{Z}_2$ as a $\mathbb{Z}_2$-module. If, on the other hand, $c = 0$ so that $\mathsf{H}_{0,0}(\mathbb{Z}_2) =$

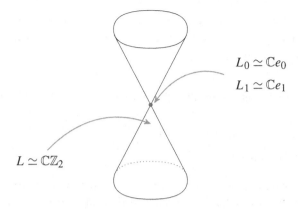

$$L_0 \simeq \mathbb{C}e_0$$
$$L_1 \simeq \mathbb{C}e_1$$

$$L \simeq \mathbb{C}\mathbb{Z}_2$$

Figure 7. The leaves and simple modules of $H_{0,0}(\mathbb{Z}_2)$.

$\mathbb{C}[x, y] \rtimes \mathbb{Z}_2$, then there is one singular point and hence two symplectic leaves — the singular point and its compliment. On each closed point of the smooth locus, there is exactly one simple $\mathbb{C}[x, y] \rtimes \mathbb{Z}_2$-module, which is again isomorphic to $\mathbb{C}\mathbb{Z}_2$ as a $\mathbb{Z}_2$-module. However, above the singular point there are two simple, one-dimensional, modules, isomorphic to $\mathbb{C}e_0$ and $\mathbb{C}e_1$ as $\mathbb{C}\mathbb{Z}_2$-modules. See Figure 7.

5.5. Restricted rational Cherednik algebras. In order to be able to say more about the simple modules for $H_{0,c}(G)$, e.g., to describe their possible dimensions, we restrict ourselves to considering rational Cherednik algebras. Therefore, in this subsection, we let W be a complex reflection group and $H_{0,c}(W)$ the associated rational Cherednik algebra, as defined in Example 1.5. In the case of Coxeter groups, the following was proved in [99, Proposition 4.15], and the general case is due to [118, Proposition 3.6].

PROPOSITION 5.5.1. *Let* $H_{0,c}(W)$ *be a rational Cherednik algebra associated to the complex reflection group* W.

(1) *The subalgebra* $\mathbb{C}[\mathfrak{h}]^W \otimes \mathbb{C}[\mathfrak{h}^*]^W$ *of* $H_{0,c}(W)$ *is contained in* $Z_c(W)$.
(2) *The centre* $Z_c(W)$ *of* $H_{0,c}(W)$ *is a free* $\mathbb{C}[\mathfrak{h}]^W \otimes \mathbb{C}[\mathfrak{h}^*]^W$-*module of rank* $|W|$.

The inclusion of algebras $A := \mathbb{C}[\mathfrak{h}]^W \otimes \mathbb{C}[\mathfrak{h}^*]^W \hookrightarrow Z_c(W)$ allows us to define the *restricted rational Cherednik algebra* $\overline{H}_c(W)$ as

$$\overline{H}_c(W) = \frac{H_c(W)}{A_+ \cdot H_c(W)},$$

where A_+ denotes the ideal in A of elements with zero constant term. This algebra was originally introduced, and extensively studied, in the paper [118]. The PBW theorem implies that

$$\overline{H}_c(W) \cong \mathbb{C}[\mathfrak{h}]^{\mathrm{co}\,W} \otimes \mathbb{C}W \otimes \mathbb{C}[\mathfrak{h}^*]^{\mathrm{co}\,W}$$

as vector spaces. Here

$$\mathbb{C}[\mathfrak{h}]^{\mathrm{co}\,W} = \mathbb{C}[\mathfrak{h}]/\langle \mathbb{C}[\mathfrak{h}]^W_+ \rangle$$

is the *coinvariant algebra*. Since W is a complex reflection group, $\mathbb{C}[\mathfrak{h}]^{\mathrm{co}\,W}$ has dimension $|W|$ and is isomorphic to the regular representation as a W-module. Thus, $\dim \overline{\mathsf{H}}_c(W) = |W|^3$. Denote by $\mathrm{Irr}(W)$ a set of complete, nonisomorphic simple W-modules.

DEFINITION 5.5.2. Let $\lambda \in \mathrm{Irr}(W)$. The *baby Verma module* of $\overline{\mathsf{H}}_c(W)$, associated to λ, is

$$\overline{\Delta}(\lambda) := \overline{\mathsf{H}}_c(W) \otimes_{\mathbb{C}[\mathfrak{h}^*]^{\mathrm{co}\,W} \rtimes W} \lambda,$$

where $\mathbb{C}[\mathfrak{h}^*]_+^{\mathrm{co}\,W}$ acts on λ as zero.

The rational Cherednik algebra is $\mathbb{Z}$-graded (no such grading exists for general symplectic reflection algebras). The grading is defined by $\deg(x) = 1$, $\deg(y) = -1$ and $\deg(w) = 0$ for $x \in \mathfrak{h}^*$, $y \in \mathfrak{h}$ and $w \in W$. At $t = 1$, this is just the natural grading coming from the Euler operator, Section 2.4. Since the restricted rational Cherednik algebra is a quotient of $\mathsf{H}_c(W)$ by an ideal generated by homogeneous elements, it is also a graded algebra. This means that the representation theory of $\overline{\mathsf{H}}_c(W)$ has a rich combinatorial structure and one can use some of the combinatorics to better describe the modules $L(\lambda)$. In particular, since $\mathbb{C}[\mathfrak{h}^*]^{\mathrm{co}\,W} \rtimes W$ is a graded subalgebra of $\overline{\mathsf{H}}_c(W)$, the baby Verma module $\overline{\Delta}(\lambda)$ is a graded $\overline{\mathsf{H}}_c(W)$-module, where $1 \otimes \lambda$ sits in degree zero. By studying quotients of baby Verma modules, it is possible to completely classify the simple $\overline{\mathsf{H}}_c(W)$-module.

PROPOSITION 5.5.3. *Let $\lambda, \mu \in \mathrm{Irr}(W)$.*

(1) *The baby Verma module $\overline{\Delta}(\lambda)$ has a simple head, $L(\lambda)$. Hence $\overline{\Delta}(\lambda)$ is indecomposable.*

(2) *$L(\lambda)$ is isomorphic to $L(\mu)$ if and only if $\lambda \simeq \mu$.*

(3) *The set $\{L(\lambda) \mid \lambda \in \mathrm{Irr}(W)\}$ is a complete set of pairwise nonisomorphic simple $\overline{\mathsf{H}}_c(W)$-modules.*

PROOF. We first recall some elementary facts from the representation theory of finite-dimensional algebras. The *radical* $\mathrm{rad}\,R$ of a ring R is the intersection of all maximal left ideals. When R is a finite-dimensional algebra, $\mathrm{rad}\,R$ is also the union of all nilpotent ideals in R and is itself a nilpotent ideal; see [22, Proposition 3.1].

CLAIM 5.5.4. *If R is $\mathbb{Z}$-graded, finite-dimensional, then $\mathrm{rad}\,R$ is a homogeneous ideal.*

PROOF. By a homogeneous ideal we mean that if $a \in I$ and $a = \sum_{i \in \mathbb{Z}} a_i$ is the decomposition of a into homogeneous pieces, then every a_i belongs to I. If I is an ideal, let $\mathrm{hom}(I)$ denote the ideal in R generated by all homogeneous parts of all elements in I, i.e., $\mathrm{hom}(I)$ is the smallest homogeneous ideal in R containing I. It suffices to show that if I, J are ideals in R such that $IJ = 0$, then $\mathrm{hom}(I)\mathrm{hom}(J) = 0$. In fact, we just need to show that $\mathrm{hom}(I)J = 0$. Let the length of $a \in I$ be the number of integers n such that $a_n \neq 0$. Then, an easy induction on length shows that $a_n J = 0$ for all n. Since I is finite-dimensional, this implies that $\mathrm{hom}(I)J = 0$ as required. $\square$

It is a classical (but difficult) result by G. Bergman that the hypothesis that R is finite-dimensional in the above claim is not necessary.

If M is a finitely generated $\overline{\mathsf{H}}_c(W)$-module, then the radical of M is defined to be the intersection of all maximal submodules of M. It is known, e.g., [22, Section 1.3], that the radical of M equals $(\mathrm{rad}\overline{\mathsf{H}}_c(W))M$ and that the quotient $M/\mathrm{rad}M$ (which is defined to be the *head* of M) is semisimple. Therefore, Claim 5.5.4 implies that if M is graded, then its radical is a graded submodule and equals the intersection of all maximal graded submodules of M. In particular, if M contains a unique maximal graded submodule, then the head of M is a simple, graded module.

Thus, we need to show that $\overline{\Delta}(\lambda)$ has a unique maximal graded submodule. The way we have graded $\overline{\Delta}(\lambda)$, we have $\overline{\Delta}(\lambda)_n = 0$ for $n < 0$ and $\overline{\Delta}(\lambda)_0 = 1 \otimes \lambda$. Let M be the sum of all graded submodules M' of $\overline{\Delta}(\lambda)$ such that $M'_0 = 0$. Clearly, $M \subset \mathrm{rad}\overline{\Delta}(\lambda)$. If this is a proper inclusion then $(\mathrm{rad}\overline{\Delta}(\lambda))_0 \neq 0$. But λ is an irreducible W-module, hence $(\mathrm{rad}\overline{\Delta}(\lambda))_0 = \lambda$. Since $\overline{\Delta}(\lambda)$ is generated by λ this implies that $\mathrm{rad}\overline{\Delta}(\lambda) = \overline{\Delta}(\lambda)$; a contradiction. Thus, $M = \mathrm{rad}\overline{\Delta}(\lambda)$. By the same argument, it is clear that $\overline{\Delta}(\lambda)/M$ is simple. This proves Part (1).

Part (2): For each $\overline{\mathsf{H}}_c(W)$-module L, we define $\mathrm{Sing}(L) = \{l \in L \mid \mathfrak{h} \cdot l = 0\}$. Notice that $\mathrm{Sing}(L)$ is a (graded if L is graded) W-submodule and if $\lambda \subset \mathrm{Sing}(L)$ then there exists a unique morphism $\overline{\Delta}(\lambda) \to L$ extending the inclusion map $\lambda \hookrightarrow \mathrm{Sing}(L)$. Since $L(\lambda)$ is simple, $\mathrm{Sing}(L(\lambda)) = L(\lambda)_0 = \lambda$. Hence $L(\lambda) \simeq L(\mu)$ implies $\lambda \simeq \mu$.

Part (3): It suffices to show that if L is a simple $\overline{\mathsf{H}}_c(W)$-module, then $L \simeq L(\lambda)$ for some $\lambda \in \mathrm{Irr}(W)$. It is well-known that the simple modules of a graded finite-dimensional algebra can be equipped with a (nonunique) grading. So we may assume that L is graded. Since L is finite-dimensional and $\mathfrak{h} \cdot L_n \subset L_{n-1}$, $\mathrm{Sing}(L) \neq 0$. Choose some $\lambda \subset \mathrm{Sing}(L)$. Then there is a nonzero map $\overline{\Delta}(\lambda) \to L$. Hence $L \simeq L(\lambda)$. This concludes the proof of Proposition 5.5.3. $\qquad\square$

5.6. The Calogero–Moser partition. Since the algebra $\overline{\mathsf{H}}_c(W)$ is finite-dimensional, it will decompose into a direct sum of *blocks*:

$$\overline{\mathsf{H}}_c(W) = \bigoplus_{i=1}^{k} B_i,$$

where B_i is a block if it is indecomposable as an algebra. If b_i is the identity element of B_i then the identity element 1 of $\overline{\mathsf{H}}_c(W)$ is the sum $1 = b_1 + \cdots + b_k$ of the b_i. For each simple $\overline{\mathsf{H}}_c(W)$-module L, there exists a unique i such that $b_i \cdot L \neq 0$. In this case we say that L *belongs to the block* B_i. By Proposition 5.5.3, we can (and will) identify $\mathrm{Irr}\,\overline{\mathsf{H}}_c(W)$ with $\mathrm{Irr}\,(W)$. We define the *Calogero–Moser partition* of $\mathrm{Irr}\,(W)$ to be the set of equivalence classes of $\mathrm{Irr}\,(W)$ under the equivalence relation $\lambda \sim \mu$ if and only if $L(\lambda)$ and $L(\mu)$ belong to the same block.

To aid intuition it is a good idea to have a geometric interpretation of the Calogero–Moser partition. The image of the natural map $Z_c/A_+ \cdot Z_c \to \overline{\mathsf{H}}_c(W)$ is clearly contained in the centre of $\overline{\mathsf{H}}_c(W)$. In general it does not equal the centre of $\overline{\mathsf{H}}_c(W)$ (though one can use the Satake isomorphism to show that it is injective). However, it is a consequence of a theorem by Müller, see [52, Corollary 2.7], that the primitive central idempotents of $\overline{H}_c(W)$ (the b_i's) are precisely the images of the primitive idempotents of $Z_c/A_+ \cdot Z_c$. Geometrically, this can be interpreted as follows. The inclusion $A \hookrightarrow Z_c(W)$ defines a finite, surjective morphism

$$\Upsilon : X_c(G) \longrightarrow \mathfrak{h}/W \times \mathfrak{h}^*/W,$$

where $\mathfrak{h}/W \times \mathfrak{h}^*/W = \operatorname{Spec} A$. Müller's theorem is saying that the natural map $\operatorname{Irr}(W) \to \Upsilon^{-1}(0)$, $\lambda \mapsto \operatorname{Supp}(L(\lambda)) = \chi_{L(\lambda)}$, factors through the Calogero–Moser partition (here $\Upsilon^{-1}(0)$ is considered as the set theoretic pull-back):

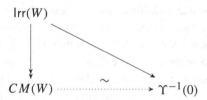

Using this fact, one can show that the geometry of $X_c(W)$ is related to Calogero–Moser partitions in the following way.

THEOREM 5.6.1. *The following are equivalent:*
- *The generalised Calogero–Moser space $X_c(W)$ is smooth.*
- *The Calogero–Moser partition of* $\operatorname{Irr}(W')$ *is trivial for all parabolic subgroup W' of W.*

Here W' is a *parabolic* subgroup of W if there exists some $v \in \mathfrak{h}$ such that $W' = \operatorname{Stab}_W(v)$. It is a remarkable theorem by Steinberg [221, Theorem 1.5] that all parabolic subgroups of W are again complex reflection groups.

5.7. Graded characters. In some cases it is possible to use the $\mathbb{Z}$-grading on $\overline{H}_c(W)$ to calculate the graded character of the simple modules $L(\lambda)$.

LEMMA 5.7.1. *The element $\lambda \in \operatorname{Irr}(W)$ defines a block $\{\lambda\}$ of the restricted rational Cherednik algebra $\overline{H}_c(W)$ if and only if $\dim L(\lambda) = |W|$ if and only if $L(\lambda) \simeq \mathbb{C}W$ as a W-module.*

PROOF. Recall that Müller's theorem from Section 5.6 says that the primitive central idempotents of $\overline{H}_c(W)$ (the b_i's) are the images of the primitive idempotents of $Z_c/A_+ \cdot Z_c$ under the natural map $Z_c/A_+ \cdot Z_c \to \overline{H}_c(W)$. The primitive idempotents of $R := Z_c/A_+ \cdot Z_c$ are in bijection with the maximal ideals in this ring: given a maximal ideal $\mathfrak{m}$ in R there is a unique primitive idempotent b whose image in $R/\mathfrak{m}$ is nonzero. Under this correspondence, the simple module belonging to the block of $\overline{H}_c(W)$ corresponding to b are precisely those simple modules supported at $\mathfrak{m}$. Thus, the block b has just one simple module if and only if there is a unique simple module of $H_c(W)$ supported at $\mathfrak{m}$. But, by the Artin–Procesi Theorem, Remark 5.1.2(3), there is a unique simple module supported at $\mathfrak{m}$ if and only if $\mathfrak{m}$ is in the Azumaya locus of $X_c(W)$. As noted in the proof of Proposition 5.1.4, the smooth locus of $X_c(W)$ equals the Azumaya locus. Thus, to summarise, the module $L(\lambda)$ is on its own in a block if and only if its support is contained in the smooth locus. Then, the conclusions of the lemma follow from Theorem 5.1.3 and Proposition 5.1.4. □

When λ satisfies the conditions of Lemma 5.7.1, we say that λ is in a block on its own. Recall that $\mathbb{C}[\mathfrak{h}]^{\mathrm{co}\,W}$, the coinvariant ring of W, is defined to be the quotient $\mathbb{C}[\mathfrak{h}]/\langle\mathbb{C}[\mathfrak{h}]_+^W\rangle$. It is a graded W-module. Therefore, we can define the *fake degree* of $\lambda \in \mathrm{Irr}(W)$ to be the polynomial

$$f_\lambda(t) = \sum_{i \in Z} [\mathbb{C}[\mathfrak{h}]_i^{\mathrm{co}\,W} : \lambda] t^i,$$

where $\mathbb{C}[\mathfrak{h}]_i^{\mathrm{co}\,W}$ is the part of $\mathbb{C}[\mathfrak{h}]^{\mathrm{co}\,W}$ of degree i and $[\mathbb{C}[\mathfrak{h}]_i^{\mathrm{co}\,W} : \lambda]$ is the multiplicity of λ in $\mathbb{C}[\mathfrak{h}]_i^{\mathrm{co}\,W}$. For each $\lambda \in \mathrm{Irr}(W)$, we define b_λ to be the degree of smallest monomial appearing in $f_\lambda(t)$, e.g., if $f_\lambda(t) = 2t^4 - t^6 +$ higher terms, then $b_\lambda = 4$. Given a finite-dimensional, graded vector space M, the Poincaré polynomial of M is defined to be

$$P(M, t) = \sum_{i \in Z} \dim M_i t^i.$$

LEMMA 5.7.2. *Assume that λ is in a block on its own. Then the Poincaré polynomial of $L(\lambda)$ is given by*

$$P(L(\lambda), t) = \frac{(\dim \lambda) t^{b_\lambda *} P(\mathbb{C}[\mathfrak{h}]^{\mathrm{co}\,W}, t)}{f_{\lambda *}(t)}.$$

PROOF. By Proposition 5.5.3, the baby Verma module $\overline{\Delta}(\lambda)$ is indecomposable. This implies that all its composition factors belong to the same block. We know that $L(\lambda)$ is one of these composition factors. But, by assumption, $L(\lambda)$ is on its own in a block. Therefore every composition factor is isomorphic to $L(\lambda)$. So let's try and calculate the graded multiplicities of $L(\lambda)$ in $\overline{\Delta}(\lambda)$. In the graded Grothendieck group of $\overline{H}_c(W)$, we must have

$$[\overline{\Delta}(\lambda)] = [L(\lambda)][i_1] + \cdots + [L(\lambda)][i_\ell], \tag{5.B}$$

where $[L(\lambda)][k]$ denotes the class of $L(\lambda)$, shifted in degree by k. By Lemma 5.7.1, $\dim L(\lambda) = |W|$ and it is easy to see that $\dim \overline{\Delta}(\lambda) = |W| \dim \lambda$. Thus, $\ell = \dim \lambda$. Since $L(\lambda)$ is a graded quotient of $\overline{\Delta}(\lambda)$ we may also assume that $i_1 = 0$. Recall that $\mathbb{C}[\mathfrak{h}]^{\mathrm{co}\,W}$ is isomorphic to the regular representation as a W-module. Therefore, the fact that $[\mu \otimes \lambda : \mathrm{triv}] \neq 0$ if and only if $\mu \simeq \lambda^*$ (in which case it is one) implies that the multiplicity space of the trivial representation in $\overline{\Delta}(\lambda)$ is $(\dim \lambda)$-dimensional. The Poincaré polynomial of this multiplicity space is precisely the fake polynomial $f_{\lambda *}(t)$. On the other hand, the trivial representation only occurs once in $L(\lambda)$ since it is isomorphic to the regular representation. Therefore, comparing graded multiplicities of the trivial representation on both sides of (5.B) implies that, up to a shift, $t^{i_1} + \cdots + t^{i_\ell} = f_{\lambda *}(t)$. What is the shift? Well, the lowest degree[11] occurring in $t^{i_1} + \cdots + t^{i_\ell}$ is $i_1 = 0$.

---

[11] It is not complete obvious that 0 is the lowest degree in $t^{i_1} + \cdots + t^{i_\ell}$, see [118, Lemma 4.4].

But the lowest degree in $f_\lambda(t)$ is b_{λ^*}. Thus, $t^{i_1} + \cdots + t^{i_\ell} = t^{-b_{\lambda^*}} f_{\lambda^*}(t)$. This implies that

$$P(L(\lambda), t) = \frac{t^{b_{\lambda^*}} P(\overline{\Delta}(\lambda), t)}{f_{\lambda^*}(t)}.$$

Clearly, $P(\overline{\Delta}(\lambda), t) = (\dim \lambda) P(\mathbb{C}[\mathfrak{h}]^{\mathrm{co}\, W}, t)$.   □

The module $L(\lambda)$ is finite-dimensional. Therefore $P(L(\lambda), t)$ is a Laurent polynomial. However, one can often find representations λ for which $f_{\lambda^*}(t)$ does not divide $P(\mathbb{C}[\mathfrak{h}]^{\mathrm{co}\, W}, t)$. In such cases the above calculation show that λ is *never* in a block on its own.

5.8. Symplectic resolutions. Now we return to the original question posed at the start of Section 1: How singular is the space V/G? The usual way of answering this question is to look at resolutions of singularities of V/G.

DEFINITION 5.8.1. A (*projective*) *resolution of singularities* is a birational morphism $\pi : Y \to V/G$ from a smooth variety Y, projective over V/G, such that the restriction of π to $\pi^{-1}((V/G)_{\mathrm{sm}})$ is an isomorphism.

If V_{reg} is the open subset of V on which G acts freely, then $V_{\mathrm{reg}}/G \subset V/G$ is the smooth locus and it inherits a symplectic structure from V, i.e., V_{reg}/G is a symplectic manifold.

DEFINITION 5.8.2. A projective resolution of singularities $\pi : Y \to V/G$ is said to be *symplectic* if Y is a symplectic manifold and the restriction of π to $\pi^{-1}((V/G)_{\mathrm{sm}})$ is an isomorphism of symplectic manifolds.

The existence of a symplectic resolution for V/G is a very strong condition and implies that the map π has some very good properties, e.g., π is *semismall*. Therefore, as one might expect, symplectic resolutions exist only for very special groups.

THEOREM 5.8.3. *Let (V, ω, G) be an irreducible symplectic reflection group.*

- *The quotient singularity V/G admits a symplectic resolution if and only it admits a smooth Poisson deformation.*
- *The quotient singularity V/G admits a smooth Poisson deformation if and only if $X_c(G)$ is smooth for generic parameters c.*

The irreducible symplectic reflection groups have been classified by Cohen, [71]. Using the above theorem, work of several people (Verbitsky [238], Ginzburg and Kaledin [114], Gordon [118], Bellamy [27], Bellamy and Schedler [29]) means that the classification of quotient singularities admitting symplectic resolutions is (almost) complete.

EXAMPLE 5.8.4. Let $G \subset \mathrm{SL}_2(\mathbb{C})$ be a finite group. Since $\dim \mathbb{C}^2/G = 2$, there is a *minimal resolution* $\widetilde{\mathbb{C}^2}/G$ of $\mathbb{C}^2/G$ through which all other resolutions factor. This resolution can be explicitly constructed as a series of blowups. Moreover, $\widetilde{\mathbb{C}^2}/G$ is a symplectic manifold and hence provides a symplectic resolution of $\mathbb{C}^2/G$. The corresponding generalised Calogero–Moser space $X_c(G)$ is smooth for generic parameters

Figure 8. A representation of the resolution of the D_4 Kleinian singularity.

c. The corresponding symplectic reflection algebras are closely related to deformed preprojective algebras [75].

5.9. Exercises.

EXERCISE 5.9.1. Let I be a Poisson ideal in the Poisson algebra A. Show that A/I naturally inherits a Poisson bracket from A, making it a Poisson algebra.

EXERCISE 5.9.2. The character table of the Weyl group G_2 is given by

Class	1	2	3	4	5	6
Size	1	1	3	3	2	2
Order	1	2	2	2	3	6
T	1	1	1	1	1	1
S	1	1	−1	−1	1	1
V_1	1	−1	1	−1	1	−1
V_2	1	−1	−1	1	1	−1
$\mathfrak{h}_1$	2	2	0	0	−1	−1
$\mathfrak{h}_2$	2	2	0	0	−1	1

The fake polynomials are

$$f_T(t) = 1, \quad f_{V_1}(t) = t^3, \quad f_{\mathfrak{h}_1}(t) = t^2 + t^4,$$
$$f_S(t) = t^6, \quad f_{V_2}(t) = t^3, \quad f_{\mathfrak{h}_2}(t) = t + t^5.$$

Is $X_c(G_2)$ ever smooth?

EXERCISE 5.9.3 (harder). For this exercise you'll need to have GAP 3, together with the package "CHEVIE" installed. Using the code fake.gap,[12] show that there is (at most one) exceptional complex reflection group W for which the space $X_c(W)$ can ever hope to be smooth. Which exceptional group is this? For help with this exercise, read [27].

---

[12]Available from maths.gla.ac.uk/~gbellamy/MSRI.html.

5.10. Additional remarks.

Theorem 5.1.3 is proven in [99, Theorem 1.7].

The fact that $X_c(W)$ has finitely many symplectic leaves, Theorem 5.4.5, is [51, Theorem 7.8].

The beautiful result that the representation theory of symplectic reflection algebras is constant along leaves, Theorem 5.4.6, is due to Brown and Gordon, [51, Theorem 4.2].

Proposition 5.5.3 is Proposition 4.3 of [118]. It is based on the general results of [133], applied to the restricted rational Cherednik algebra.

The Calogero–Moser partition was first defined in [122].

Theorem 5.6.1 is stated for rational Cherednik algebras at $t = 1$ in positive characteristic in [28, Theorem 1.3]. However, the proof given there applies word for word to rational Cherednik algebras at $t = 0$ in characteristic zero.

Theorem 5.8.3 follows from the results in [114] and [176].

CHAPTER IV

Noncommutative resolutions

Michael Wemyss

Introduction

The notion of a noncommutative crepant resolution (NCCR) was introduced by Van den Bergh [231], following his interpretation [230] of work of Bridgeland [42] and Bridgeland, King and Reid [44]. Since then, NCCRs have appeared prominently in both the mathematics and physics literature as a general homological structure that underpins many topics currently of interest, for example moduli spaces, dimer models, curve counting Donaldson–Thomas invariants, spherical–type twists, the minimal model program and mirror symmetry.

My purpose in writing these notes is to give an example based approach to some of the ideas and constructions for NCCRs, with latter sections focussing more on the explicit geometry and restricting mainly to dimensions two and three, rather than simply presenting results in full generality. The participants at the MSRI Summer School had a wonderful mix of diverse backgrounds, so the content and presentation of these notes reflect this. There are exercises scattered throughout the text, at various levels of sophistication, and also computer exercises that hopefully add to the intuition.

The following is a brief outline of the content of the notes. In Section 1 we begin by outlining some of the motivation and natural questions for NCCRs through the simple example of the $\mathbb{Z}_3$ surface singularity. We then progress to the setting of two-dimensional Gorenstein quotient singularities for simplicity, although most things work much more generally. We introduce the notion of Auslander algebras, and link to the idea of finite CM type. We then introduce skew group rings and use this to show that a certain endomorphism ring in the running example has finite global dimension.

Section 2 begins with the formal definitions of Gorenstein and CM rings, depth and CM modules, before giving the definition of a noncommutative crepant resolution (NCCR). This comes in two parts, and the second part is motivated using some classical commutative algebra. We then deal with uniqueness issues, showing that in dimension two NCCRs are unique up to Morita equivalence, whereas in dimension three they are unique up to derived equivalence. We give examples to show that these results are the best possible. Along the way, the three key technical results of the depth lemma, reflexive equivalence and the Auslander–Buchsbaum formula are formulated.

Section 3 breaks free of the algebraic shackles imposed in the previous two sections, by giving a brief overview of quiver GIT. This allows us to extract geometry from NCCRs, and we illustrate this in examples of increasing complexity.

In Section 4 we go homological so as to give a language in which to compare the geometry to NCCRs. We sketch some aspects of derived categories, and give an outline of tilting theory. We then illustrate tilting explicitly in the examples from Section 3. In a purely homological section we then relate the crepancy of birational morphisms to the condition $\mathrm{End}_R(M) \in \mathrm{CM}\, R$. The section then considers CY categories and algebras, and we prove that NCCRs are d-CY. We formulate singular derived categories as a mechanism to relate the constructions involving CM modules to the CY property, and also as an excuse to introduce AR duality, which links questions from Section 1 to AR sequences, which appear in Section 5.

Section 5 begins by overviewing the McKay correspondence in dimension two. It starts with the classical combinatorial version, before giving the Auslander version using AR sequences and the category of CM modules. We then upgrade this and give the derived version, which homologically relates minimal resolutions to NCCRs for ADE surface singularities. The last, and main, section gives the three-dimensional version, which is the original motivation for introducing NCCRs. We sketch the proof.

A short appendix (see p. 297) gives very basic background on quiver representations, and sets the notation that is used in the examples and exercises.

Acknowledgments

I would like to thank all the participants and co-organizers of the summer school for their questions, and for making the course such fun to teach. In addition, thanks go to Pieter Belmans, Kosmas Diveris, Will Donovan, Martin Kalck, Joe Karmazyn, Boris Lerner and Alice Rizzardo for their many comments on previous drafts of these notes.

1. Motivation and first examples

1.1. The basic idea. The classical method for resolving singularities is to somehow associate to the singularity $X = \mathrm{Spec}\, R$ an ideal I. This becomes the centre of the blowup and we hope to resolve the singularity via the picture

Although this is entirely inside the world of commutative algebra, we hope to obtain a better understanding of this process by introducing noncommutative methods. Instead of finding an ideal I, we want to, without referring to a resolution (i.e., the answer), produce a noncommutative ring A from which we can extract resolution(s) of X.

$$
\begin{array}{c}
\mathcal{M} \\
\downarrow \qquad \rightsquigarrow \quad A \\
\text{Spec } R
\end{array}
\tag{1.A}
$$

Just as there are many different ideal sheaves which give the same blowup, there are in general many different noncommutative rings that can be used to resolve the singularity and so the subtlety comes through asking for the "best" one — for a noncommutative ring to be called a noncommutative resolution it needs to satisfy some extra conditions.

The purpose of these lectures is to explain how to go about constructing such an A, and they will also outline some of the methods that are used to extract the geometry. There are both algebraic and geometric consequences. The main benefit of this noncommutative approach is that we equip the geometry with extra structure in the form of tautological bundles, which can then be used in various homological (and explicit) constructions.

1.2. Motivation and questions. Here we input a finite subgroup G of SL$(2, \mathbb{C})$. Then G acts on $\mathbb{C}^2$, so it acts on $\mathbb{C}[\![x, y]\!]$ via inverse transpose. We define $R := \mathbb{C}[\![x, y]\!]^G$ and consider the quotient germ $\mathbb{C}^2 / G = \text{Spec } R$.

EXAMPLE 1.2.1. The running example will be

$$
G = \tfrac{1}{3}(1, 2) := \left\langle g := \begin{pmatrix} \varepsilon_3 & 0 \\ 0 & \varepsilon_3^2 \end{pmatrix} \right\rangle,
$$

where ε_3 is a primitive third root of unity. Here g sends x to $\varepsilon_3^2 x$ and y to $\varepsilon_3 y$, so it is clear that x^3, xy, y^3 are all invariants. In fact they generate the invariant ring, so

$$
R = \mathbb{C}[\![x, y]\!]^{\frac{1}{3}(1,2)} = \mathbb{C}[\![x^3, y^3, xy]\!] \cong \mathbb{C}[\![a, b, c]\!]/(ab - c^3).
$$

SETTING 1.2.2. Throughout the remainder of this section, R will always denote $\mathbb{C}[\![x, y]\!]^G$ for some $G \leq \text{SL}(2, \mathbb{C})$. We remark that experts can instead take their favourite complete local Gorenstein ring R with $\dim R = 2$, as all results remain true. Indeed most results still hold when R is a complete local CM normal ring of dimension two, provided that R has a canonical module. Note that dropping the "complete local" is possible, but at the expense of making the language a bit more technical, and the proofs much more so.

Recall that if M is a finitely generated R-module (written $M \in \text{mod } R$), there exists a surjection $R^n \twoheadrightarrow M$ for some $n \in \mathbb{N}$, and the kernel is denoted ΩM. This is called the syzygy of M.

TEMPORARY DEFINITION 1.2.3. $M \in \text{mod } R$ is called a Cohen–Macaulay (CM) module if $M \cong \Omega(\Omega X)$ for some $X \in \text{mod } R$. We denote the category of CM modules by CM R.

We remark that the definition should be treated with caution, as assumptions are needed (which are satisfied in setting 1.2.2 above) for it to be equivalent to the more standard definition that will be explained later (Definition 2.0.6). Note that Definition 1.2.3 ensures that $R \in \mathrm{CM}\, R$.

EXAMPLE 1.2.4. Let

$$G = \tfrac{1}{3}(1, 2) := \left\langle \begin{pmatrix} \varepsilon_3 & 0 \\ 0 & \varepsilon_3^2 \end{pmatrix} \right\rangle,$$

where ε_3 is a primitive third root of unity. In this situation $R = \mathbb{C}[\![x^3, y^3, xy]\!] \cong \mathbb{C}[\![a, b, c]\!]/(ab - c^3)$. We claim that R, together with the ideals $M_1 := (a, c)$ and $M_2 := (a, c^2)$, are all CM R-modules. In fact, the situation is particularly nice since the calculation below shows that $\Omega M_1 \cong M_2$ and $\Omega M_2 \cong M_1$, hence $\Omega^2 M_1 \cong M_1$ and $\Omega^2 M_2 \cong M_2$.

For the calculation, just note that we have a short exact sequence

$$0 \to (a, c^2) \xrightarrow{\left(-\frac{c}{a} \ \mathrm{inc}\right)} R^2 \xrightarrow{\binom{a}{c}} (a, c) \to 0, \tag{1.B}$$

where the second map sends $(r_1, r_2) \mapsto r_1 a + r_2 c$, and the first map sends $ra + sc^2 \mapsto ((ra + sc^2)(-\frac{c}{a}), ra + sc^2) = (-rc - sb, ra + sc^2)$. In a similar way, we have an exact sequence

$$0 \to (a, c) \xrightarrow{\left(-\frac{c^2}{a} \ \mathrm{inc}\right)} R^2 \xrightarrow{\binom{a}{c^2}} (a, c^2) \to 0. \tag{1.C}$$

Notice that with our Definition 1.2.3, the above example shows that CM modules are in fact quite easy to produce — just find a module, then syzygy twice. Note at this stage it is not clear how many other CM modules there are in Example 1.2.4, never mind what this has to do with Section 1.1 and the relationship to the geometry.

EXAMPLE 1.2.5. Continuing the above example, in the spirit of discovery let's compute $\mathrm{End}_R(R \oplus (a, c) \oplus (a, c^2))$. Why we do this will only become clear afterwards. We write each indecomposable module as a vertex

$$(a, c)$$

$$R \qquad\qquad (a, c^2)$$

Clearly we have the inclusions $(a, c^2) \subseteq (a, c) \subseteq R$ and so we have morphisms

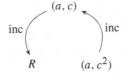

If we multiply an element of R by c we get an element of (a, c), and similarly if we multiply an element of (a, c) by c we get an element of (a, c^2). Thus we add in

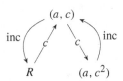

We can multiply an element of R by a to get an element of (a, c^2) and so obtain

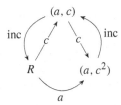

It is possible to multiply an element of R by a and get an element of (a, c), but we don't draw it since this map is just the composition of the arrow a with the arrow inc, so it is already taken care of. The only morphism which is not so obvious is the map $(a, c^2) \to R$ given by $\frac{c}{a}$ (as in (1.B)), thus this means that we have guessed the following morphisms between the modules:

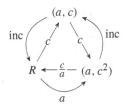

It turns out that these are in fact all necessary morphisms, in that any other must be a linear combination of compositions of these. This can be shown directly, but at this stage it is not entirely clear. Is this algebra familiar?

The above simple example already illustrates some interesting phenomenon that later we will put on a more firm theoretical basis. Indeed, the following five questions emerge naturally, and motivate much of the content of these notes.

Q1. Is there a systematic way of computing the quiver? We just guessed.

Q2. Are the three CM R-modules we guessed in Example 1.2.4 all the indecomposable CM R-modules up to isomorphism?

Q3. Is it a coincidence that $\Omega^2 = \mathrm{Id}$ on the CM modules?

Q4. Why are we only considering noncommutative rings that look like $\mathrm{End}_R(M)$? It seems that Section 1.1 allows for almost arbitrary rings.

Q5. How do we extract the geometry from these noncommutative rings, as in (1.A)?

In short, the answers are:

A1. Yes. This is one of the things that Auslander–Reiten (AR) theory does.

A2. Yes. The proof, using Auslander algebras, is remarkably simple. This will lead to our main definition of noncommutative crepant resolutions (NCCRs).

A3. No. This is a special case of matrix factorizations, which appear for any hypersurface. Amongst other things, this leads to connections with Mirror Symmetry.

A4. Mainly for derived category reasons. See Section 4 and remarks there.

A5. We use quiver GIT. See Section 3.

The remainder of Section 1 will focus on Q2.

1.3. Auslander algebras and finite type. This section is purely algebraic, and show-cases Auslander's philosophy that endomorphism rings of finite global dimension are important from a representation–theoretic viewpoint. In algebraic geometry commutative rings with finite global dimension correspond precisely to nonsingular varieties (see Section 2.2), so Auslander's philosophy will guide us forward.

THEOREM 1.3.1 (Auslander). *Let R be as in setting 1.2.2, and let $\mathcal{C}$ be a finite set of indecomposable CM R-modules, such that $R \in \mathcal{C}$. Then the following are equivalent:*

(1) $\mathrm{gl.dim}\,\mathrm{End}_R\left(\bigoplus_{C \in \mathcal{C}} C\right) \leq 2$.

(2) $\mathcal{C}$ *contains all indecomposable CM R-modules (up to isomorphism).*

In fact, as the proof below shows, $(1) \Rightarrow (2)$ holds in arbitrary dimension, whereas $(2) \Rightarrow (1)$ needs $\dim R = 2$. To prove Theorem 1.3.1 will require two facts. The first is quite easy to prove, the second requires a little more technology. Recall if $M \in \mathrm{mod}\,R$ we denote add M to be the collection of all direct summands of all finite direct sums of M. If Λ is a ring, then proj $\Lambda := \mathrm{add}\,\Lambda$, the category of projective Λ-modules.

FACTS 1.3.2. Let the notation be as above.

(1) If $M \in \mathrm{mod}\,R$ contains R as a summand, then the functor

$$\mathrm{Hom}_R(M, -) : \mathrm{mod}\,R \to \mathrm{mod}\,\mathrm{End}_R(M)$$

is fully faithful, restricting to an equivalence add $M \xrightarrow{\simeq} \mathrm{proj}\,\mathrm{End}_R(M)$.

(2) Since R is Gorenstein (or normal CM), CM R-modules are always reflexive, i.e., the natural map $X \to X^{**} = \mathrm{Hom}_R(\mathrm{Hom}_R(X, R), R)$ is an isomorphism.

With these facts, the proof of $(1) \Rightarrow (2)$ is quite straightforward. The proof $(2) \Rightarrow (1)$ uses the depth lemma, which will be explained in Section 2.

PROOF OF THEOREM 1.3.1. Denote $M := \bigoplus_{C \in \mathcal{C}} C$ and $\Lambda := \mathrm{End}_R(M)$.

$(1) \Rightarrow (2)$ Suppose that gl.dim $\Lambda \leq 2$ and let $X \in \mathrm{CM}\,R$. Consider a projective resolution $R^b \to R^a \to X^* \to 0$, then dualizing via $(-)^* = \mathrm{Hom}_R(-, R)$ and using Fact 1.3.2(2) gives an exact sequence $0 \to X \to R^a \to R^b$. Applying $\mathrm{Hom}_R(M, -)$ then gives an exact sequence

$$0 \to \mathrm{Hom}_R(M, X) \to \mathrm{Hom}_R(M, R^a) \to \mathrm{Hom}_R(M, R^b).$$

Both $\text{Hom}_R(M, R^a)$ and $\text{Hom}_R(M, R^b)$ are projective Λ-modules by Fact 1.3.2(1), so since by assumption gl.dim $\Lambda \leq 2$ it follows that $\text{Hom}_R(M, X)$ is a projective Λ-module. One last application of Fact 1.3.2(1) shows that $X \in \text{add } M$.

$(2) \Rightarrow (1)$ Suppose that dim $R = 2$, and that C contains all indecomposable CM R-modules up to isomorphism. Let $Y \in \text{mod } \Lambda$, and consider the initial terms in a projective resolution $P_1 \overset{f}{\to} P_0 \to Y \to 0$. By Fact 1.3.2(1) there exists a morphism $M_1 \overset{g}{\to} M_0$ in add M such that

$$(P_1 \overset{f}{\to} P_0) = (\text{Hom}_R(M, M_1) \overset{\cdot g}{\to} \text{Hom}_R(M, M_0)).$$

Put $X := \text{Ker } g$. Since dim $R = 2$, by the depth lemma (see Lemma 2.0.7 in the next section) we have $X \in \text{CM } R$, so by assumption $X \in C$. Hence we have an exact sequence

$$0 \to X \to M_1 \to M_0,$$

with each term in add M. Simply applying $\text{Hom}_R(M, -)$ gives an exact sequence

$$0 \to \text{Hom}_R(M, X) \to \text{Hom}_R(M, M_1) \to \text{Hom}_R(M, M_0) \to Y \to 0.$$

Thus we have proj.dim $_\Lambda Y \leq 2$. Since this holds for all Y, gl.dim $\text{End}_R(M) \leq 2$. In fact, since $\text{End}_R(M)$ has finite length modules, a global dimension of less than 2 would contradict the depth lemma. $\square$

Thus to answer Q2, by Theorem 1.3.1 we show that

$$\text{gl.dim End}_R(R \oplus (a, c) \oplus (a, c^2)) \leq 2.$$

It is possible just to do this directly, using the calculation in Example 1.2.5, but the next subsection gives a non-explicit proof.

1.4. Skew group rings. Recall our setting $G \leq \text{SL}(2, \mathbb{C})$ and $R = \mathbb{C}[[x, y]]^G$. Since R is defined as a quotient of the smooth space $\text{Spec } \mathbb{C}[[x, y]]$ by G, the basic idea is that we should use the module theory of $\mathbb{C}G$, together with the module theory of $\mathbb{C}[[x, y]]$, to encode some of the geometry of the quotient. The (false in general) slogan is that 'G-equivariant sheaves on $\mathbb{C}[[V]]$ encode the geometry of the resolution of V/G'.

EXAMPLE 1.4.1. Let $G = \frac{1}{3}(1, 2) = \langle g \rangle$ as in Example 1.2.4. Consider the one-dimensional representations ρ_0, ρ_1 and ρ_2 of G, and denote their bases by e_0, e_1 and e_2. Our convention is that g acts on e_i with weight ε_3^i. Recall g acts on the polynomial ring via $x \mapsto \varepsilon_3^{-1} x$ and $y \mapsto \varepsilon_3 y$, hence G acts on both side of the tensor $\mathbb{C}[[x, y]] \otimes_\mathbb{C} \rho_i$ and so we can consider the invariants $(\mathbb{C}[[x, y]] \otimes_\mathbb{C} \rho_i)^G$. Note that since G acts trivially on ρ_0, we have that $R = \mathbb{C}[[x, y]]^G = (\mathbb{C}[[x, y]] \otimes_\mathbb{C} \rho_0)^G$. Denote $N_1 := (\mathbb{C}[[x, y]] \otimes_\mathbb{C} \rho_1)^G$ and $N_2 := (\mathbb{C}[[x, y]] \otimes_\mathbb{C} \rho_2)^G$.

Note that $x \otimes e_1$ belongs to N_1 since under the action of G,

$$x \otimes e_1 \mapsto \varepsilon_3^{-1} x \otimes \varepsilon_3 e_1 = x \otimes e_1.$$

Similarly $y^2 \otimes e_1 \in N_1$. In fact $x \otimes e_1$ and $y^2 \otimes e_1$ generate N_1 as an R-module. Similarly $x^2 \otimes e_2$ and $y \otimes e_2$ generate N_2 as an R-module. In fact, $N_2 \cong M_1$ and $N_1 \cong M_2$ where the M_i are as in Example 1.2.4, and in these new coordinates, dropping tensors we have

$$\operatorname{End}_R(R \oplus M_1 \oplus M_2) \cong \operatorname{End}_R(R \oplus N_2 \oplus N_1) \cong$$

DEFINITION 1.4.2. For a $\mathbb{C}$-algebra A and a finite group G together with a group homomorphism $G \to \operatorname{Aut}_{\mathbb{C}-\mathrm{alg}}(A)$, we define the skew group ring $A\#G$ as follows: as a vector space it is $A \otimes_{\mathbb{C}} \mathbb{C}G$, with multiplication defined as

$$(f_1 \otimes g_1)(f_2 \otimes g_2) := (f_1 \cdot g_1(f_2)) \otimes g_1 g_2,$$

for any $f_1, f_2 \in A$ and $g_1, g_2 \in G$, extended by linearity.

In these notes we will always use Definition 1.4.2 in the setting where A is a power series (or polynomial) ring in finitely many variables, and G is a finite subgroup of $\mathrm{GL}(n, \mathbb{C})$. The following theorem is due to Auslander.

THEOREM 1.4.3 (Auslander). *Let* $G \leq \mathrm{SL}(n, \mathbb{C})$ *be a finite subgroup and denote* $S := \mathbb{C}[\![x_1, \ldots, x_n]\!]$ *and* $R := \mathbb{C}[\![x_1, \ldots, x_n]\!]^G$. *Then*

$$S\#G \cong \operatorname{End}_R\left(\bigoplus_{\rho \in \operatorname{Irr} G} ((S \otimes \rho)^G)^{\oplus \dim_{\mathbb{C}} \rho} \right).$$

We remark that the theorem also holds if G is a subgroup of $\mathrm{GL}(n, \mathbb{C})$ which contains no complex reflections (in the sense of Definition 1.1.3 in Chapter III) except the identity.

By Theorem 1.4.3, in the running Example 1.4.1 we have an isomorphism

$$\operatorname{End}_R(R \oplus (a, c) \oplus (a, c^2)) \cong \mathbb{C}[\![x, y]\!]\#\tfrac{1}{3}(1, 2).$$

Thus, via Theorem 1.3.1, to show that $\{R, (a, c), (a, c^2)\}$ are all the indecomposable CM R-modules, it suffices to prove that $\mathrm{gl.dim}\,\mathbb{C}[\![x, y]\!]\#\tfrac{1}{3}(1, 2) \leq 2$.

Now if $M, N \in \operatorname{mod} S\#G$ then G acts on $\operatorname{Hom}_S(M, N)$ by $(gf)(m) := g \cdot f(g^{-1}m)$ for all $g \in G$, $f \in \operatorname{Hom}_S(M, N)$ and $m \in M$. It is easy to check that

$$\operatorname{Hom}_{S\#G}(M, N) = \operatorname{Hom}_S(M, N)^G.$$

Further, since taking G-invariants is exact (since G is finite, and we are working over $\mathbb{C}$), this induces a functorial isomorphism

$$\operatorname{Ext}^i_{S\#G}(M, N) = \operatorname{Ext}^i_S(M, N)^G$$

for all $i \geq 0$. In particular, $\mathrm{gl.dim}\,S\#G \leq \mathrm{gl.dim}\,S$ holds, and so in our setting we have $\mathrm{gl.dim}\,\mathbb{C}[\![x, y]\!]\#\tfrac{1}{3}(1, 2) \leq \mathrm{gl.dim}\,\mathbb{C}[\![x, y]\!] = 2$, as required.

REMARK 1.4.4. The above can be strengthened to show that

$$\mathrm{gl.dim}\,S\#G = \mathrm{gl.dim}\,S.$$

Credits: The material in this section is now quite classical. The ideas around Theorem 1.3.1 were originally developed for representation dimension of Artin algebras [16], but they came across to CM modules following Auslander's version of the McKay correspondence [19]. These ideas were pursued later by Iyama in his higher-dimensional AR theory [138], who first observed the link to NCCRs. See also the paper by Leuschke [163]. Fact 1.3.2(1) is known as "projectivization" (see [22, II.2.1], for example) and Fact 1.3.2(2) can be found in most commutative algebra textbooks. Skew group rings are also a classical topic. There are now many proofs of Theorem 1.4.3; see for example [99, 170, 140]. Auslander's original proof is outlined in [247].

1.5. Exercises.

EXERCISE 1.5.1. Let R be a commutative ring and let M be an R-module. Define

$$\text{End}_R(M) := \{f : M \to M \mid f \text{ is an } R\text{-module homomorphism}\}.$$

(1) Verify that $\text{End}_R(M)$ is indeed a ring, and has the structure of an R-module.

(2) Give an example of R and M for which $\text{End}_R(M)$ is a commutative ring, and give an example for which $\text{End}_R(M)$ is noncommutative. Roughly speaking, given an R and M how often is the resulting endomorphism ring $\text{End}_R(M)$ commutative?

(3) We say that M is a *simple* R-module if the only submodules of M are $\{0\}$ and M. Prove that if R is any ring and M is a simple R-module then $\text{End}_R(M)$ is a division ring.

EXERCISE 1.5.2. As in Section 1, consider the group

$$\frac{1}{r}(1,a) := \left\langle g := \begin{pmatrix} \varepsilon_r & 0 \\ 0 & \varepsilon_r^a \end{pmatrix} \right\rangle,$$

where ε_r is a primitive r-th root of unity. We assume that r and a are coprime, and denote the representations of G by $\rho_0, \ldots, \rho_{r-1}$.

(1) Show that $S_i := (\mathbb{C}[\![x, y]\!] \otimes \rho_i)^G \cong \{f \in \mathbb{C}[\![x, y]\!] \mid g \cdot f = \varepsilon_r^i f\}$. (The exact superscript on ε will depend on conventions).

(2) Suppose $a = r - 1$ (i.e., the group G is inside $\text{SL}(2, \mathbb{C})$).

    (a) Determine $R = S_0$, and find generators for the R-modules S_i.

    (b) Hence or otherwise, determine the quiver of $\text{End}_R\left(\bigoplus_{i=0}^{r-1} S_i\right)$.

(3) (This will be helpful for counterexamples later.) Consider in turn $G = \frac{1}{3}(1, 1)$ and $\frac{1}{5}(1, 2)$.

    (a) For each of these cases, determine $R = S_0$, and find generators for the R-modules S_i. This should be quite different from (2)(a).

    (b) Hence or otherwise, determine the quiver of $\text{End}_R\left(\bigoplus_{i=0}^{r-1} S_i\right)$.

    (c) Consider only the modules in (3)(a) that have two generators. Sum them together, along with R. Determine the quiver of the resulting endomorphism ring.

Computer exercises.

EXERCISE 1.5.3 (When is a ring CM?). Consider the ring $R := k[\![a, b, c]\!]/(ab - c^3)$ from Section 1, where k has characteristic zero. We code this into Singular as

> LIBhomolog.lib";
> LIBsing.lib";
> ring   S = 0, (a, b, c), ds;

The first two commands loads libraries that we will use. The last command defines the power series rings $S := k[\![a, b, c]\!]$ (this is the ds; use dp for the polynomial ring) in the variables a, b, c. Now

> ideal   i = ab − c3;
> dim_slocus(std(i));

The first command specifies the ideal i (for more than one generator, separate with commas; e.g., > ideal   i = ab − c3, a4 − b2;). The second command asks for the dimension of the singular locus of the variety cut out by the ideal i. Here the answer given is zero, which means it is an isolated singularity.

> qring   R = std(i);

This specifies our ring R to be the factor S/I. We now define the free rank one R-module $F := R_R$

> module   F = [0];

and ask whether it is CM via

> depth(F);
> dim(F);

If these two numbers agree, then the ring is CM. Using a similar procedure, calculate whether the following are CM rings.

(1) (Whitney umbrella.) $\mathbb{C}[\![u, v, x]\!]/(uv^2 - x^2)$.

(2) The ring of invariants of $\frac{1}{4}(1, 1)$, i.e., $\mathbb{C}[\![x^4, x^3y, x^2y^2, xy^3, y^4]\!]$. This is isomorphic to $\mathbb{C}[\![a, b, c, d, e]\!]$ factored by the 2×2 minors of

$$\begin{pmatrix} a & b & c & d \\ b & c & d & e \end{pmatrix}.$$

(3) $\mathbb{C}[\![x^4, x^3y, xy^3, y^4]\!]$. This is isomorphic to $\mathbb{C}[\![a, b, d, e]\!]$ factored by the 2×2 minors of

$$\begin{pmatrix} a & b^2 & be & d \\ b & ad & d^2 & e \end{pmatrix}.$$

(4) $\mathbb{C}[\![u, v, x, y]\!]/(uv - f(x, y))$ where $f(x, y) \in \mathbb{C}[\![x, y]\!]$.

(5) Try experimenting with other commutative rings. Roughly, how often are they CM?

EXERCISE 1.5.4 (When is a module CM?). The procedure to determine whether a module is CM is similar to the above. Singular encodes modules as factors of free modules, so for example the ideal $M := (a, c)$ in the ring $R = k[\![a, b, c]\!]/(ab - c^3)$, i.e.,

$$R^2 \xrightarrow{\begin{pmatrix} c & -b \\ -a & c^2 \end{pmatrix}} R^2 \xrightarrow{(a\ c)} (a, c) \to 0,$$

is coded using the columns of the matrix as

> module M = [c, −a], [−b, c2];

Alternatively, to automatically work out the relations between a and c, code

> module Na = [a], [c];

> module N = syz(Na);

since the first line codes the factor $R/(a, b)$, and the second takes the kernel of the natural map $R \to R/(a, b)$, hence giving (a, b). Now endomorphism rings are also easy to code, for example

> module E = Hom(N, N);

The procedure for checking the depth and dimension of a module is exactly the same as in the previous example, namely,

> depth(N);

> dim(N);

> depth(E);

> dim(E);

(1) E_7 surface singularity $\mathbb{C}[\![x, y, z]\!]/(x^3 + xy^3 + z^2)$. Determine whether the following are CM modules, and whether their endomorphism rings are CM.

(a) The quotient field $k = R/\mathfrak{m}$, i.e.,

$$R^3 \xrightarrow{(x\ y\ z)} R \to k \to 0.$$

(b) The module Ωk.

(c) The module $\Omega^2 k$.

(d) The module given by

$$R^4 \xrightarrow{\begin{pmatrix} -z & y^2 & 0 & x \\ xy & z & -x^2 & 0 \\ 0 & -x & -z & y \\ x^2 & 0 & xy^2 & z \end{pmatrix}} R^4 \to M \to 0.$$

(2) The ring $\mathbb{C}[\![u, v, x, y]\!]/(uv - xy)$. Determine whether the following are CM modules, and whether their endomorphism rings are CM.

(a) The quotient field $k = R/\mathfrak{m}$, i.e.,

$$R \xrightarrow{(u\ v\ x\ y)} R \to k \to 0.$$

(b) The module Ωk.

(c) The module $\Omega^2 k$.

(d) The module $\Omega^3 k$.

(e) The modules (u, x), (u, y), and (u^2, ux, x^2).

(3) As (2), but with $\mathbb{C}[\![u, v, x, y]\!]/(uv - x^2 y)$.

Note that provided R is CM of dimension d, $\Omega^d X$ is CM for all $X \in \text{mod } R$, so Singular can be used to produce many CM modules. If $d \geq 3$, if $Y \in \text{CM } R$ then it is quite rare that $\text{End}_R(Y) \in \text{CM } R$.

EXERCISE 1.5.5 (When is a ring Gorenstein?). If $(R, \mathfrak{m})$ is local of dimension d and $k = R/\mathfrak{m}$, then $\text{Ext}^{d+1}(k, R) = 0$ implies that R is Gorenstein. This can be coded using the techniques from above. For each ring in Exercise 1.5.3 and Exercise 1.5.4 above, check whether it is Gorenstein. Try also some other commutative rings. Roughly, how often are they Gorenstein?

2. NCCRs and uniqueness issues

In the last section we started with a ring $R := \mathbb{C}[\![a, b, c]\!]/(ab - c^3)$, and guessed a CM module $M := R \oplus (a, c) \oplus (a, c^2)$ such that $\text{gl.dim End}_R(M) = 2$. To go further requires more technology. The following is the homological definition of depth, and the usual definition of a CM module.

DEFINITION 2.0.6. If $(R, \mathfrak{m})$ is a local ring and $M \in \text{mod } R$, we define the *depth* of M to be

$$\text{depth}_R M := \min\{i \geq 0 \mid \text{Ext}^i_R(R/\mathfrak{m}, M) \neq 0\}.$$

For $M \in \text{mod } R$ it is always true that

$$\text{depth } M \leq \dim M \leq \dim R \leq \dim_{R/\mathfrak{m}} \mathfrak{m}/\mathfrak{m}^2 < \infty.$$

We say that M is a (maximal) CM module if $\text{depth } M = \dim R$, and in this case we write $M \in \text{CM } R$. We say that R is a CM ring if $R_R \in \text{CM } R$, and we say that R is Gorenstein if it is CM and further $\text{inj.dim } R < \infty$.

The definition is stated to make it clear that Gorenstein rings are a special class of CM rings. It turns out that in fact $\text{inj.dim } R < \infty$ implies that R is CM, so the above definition can be simplified. When R is not necessarily local, we define $M \in \text{mod } R$ to be CM by reducing to the local case, namely M is defined to be CM if $M_\mathfrak{m} \in \text{CM } R_\mathfrak{m}$ for all $\mathfrak{m} \in \text{Max } R$.

To show that Definition 2.0.6 is equivalent to the temporary definition from the last section (at least in the setting there) will require the following easy lemma, which will turn out to be one of our main tools.

LEMMA 2.0.7 (the depth lemma). *Suppose that $(R, \mathfrak{m})$ is a local ring and let $0 \rightarrow A \rightarrow B \rightarrow C \rightarrow 0$ be a short exact sequence of finitely generated R-modules. Then:*

(1) *If $\text{depth } B > \text{depth } C$ then $\text{depth } A = \text{depth } C + 1$.*

(2) *$\text{depth } A \geq \min\{\text{depth } B, \text{depth } C\}$.*

PROOF. This just follows by applying $\text{Hom}_R(R/\mathfrak{m}, -)$ and applying the definition of depth to the resulting long exact sequence. $\square$

LEMMA 2.0.8. *Suppose that R is a local Gorenstein (or local normal CM) ring of dimension 2. Then* $\text{CM } R = \{\Omega^2 X \mid X \in \text{mod } R\}$.

PROOF. ($\subseteq$) Let $X \in \text{CM } R$. By Fact 1.3.2(2), X is reflexive. Take a projective resolution $R^b \to R^a \to X^* \to 0$, then dualizing via $(-)^* = \text{Hom}_R(-, R)$ and using the fact that X is reflexive gives an exact sequence $0 \to X \to R^a \to R^b$. This shows that X is a second syzygy.

($\supseteq$) If X is a second syzygy, we have short exact sequences

$$0 \to X \to R^a \to C \to 0, \tag{2.A}$$

$$0 \to C \to R^b \to D \to 0. \tag{2.B}$$

We know that depth $R = 2$, and $0 \leq \text{depth } D \leq \dim R = 2$. We go through each of the three cases:

- If depth $D = 0$, then the depth lemma applied to (2.B) shows that depth $C = 1$. The depth lemma applied to (2.A) then shows that depth $X = 2$, so $X \in \text{CM } R$.
- If depth $D = 1$, then the depth lemma applied to (2.B) shows that depth $C = 2$. The depth lemma applied to (2.A) then shows that depth $X = 2$, so $X \in \text{CM } R$.
- If depth $D = 2$, then the depth lemma applied to (2.B) shows that depth $C = 2$. The depth lemma applied to (2.A) then shows that depth $X = 2$, so $X \in \text{CM } R$.

In all cases, we deduce that $X \in \text{CM } R$.   □

LEMMA 2.0.9. *If $R = \mathbb{C}[\![a, b, c]\!]/(ab - c^3)$ and $M := R \oplus (a, c) \oplus (a, c^2)$ (our running example, 1.2.5), we have* $\text{End}_R(M) \in \text{CM } R$.

PROOF. Just take a projective resolution $R^s \to R^t \to M \to 0$, and apply $\text{Hom}_R(-, M)$ to obtain an exact sequence

$$0 \to \text{End}_R(M) \to \text{Hom}_R(R, M)^s \to \text{Hom}_R(R, M)^t.$$

This is just $0 \to \text{End}_R(M) \to M^s \to M^t$. Since $M \in \text{CM } R$, both M^s and M^t have depth 2. Repeating the argument in the proof of Lemma 2.0.8, using the depth lemma, shows that depth $\text{End}_R(M) = 2 = \dim R$.   □

2.1. Definition of NCCRs. The upshot so far is that in our running example $R = \mathbb{C}[\![a, b, c]\!]/(ab - c^3)$ and $M := R \oplus (a, c) \oplus (a, c^2)$, we have discovered that $\text{End}_R(M) \in \text{CM } R$ and further gl.dim $\text{End}_R(M) = \dim R$.

DEFINITION 2.1.1. Let R be a (equicodimensional normal) CM ring. A noncommutative crepant resolution (NCCR) of R is by definition a ring of the form $\text{End}_R(M)$ for some $M \in \text{ref } R$, such that

(1) $\text{End}_R(M) \in \text{CM } R$,
(2) gl.dim $\text{End}_R(M) = \dim R$.

The first important remark is that although the definition is made in the CM setting, to get any relationship with the geometry it turns out to be necessary to require that R is Gorenstein. So, although we can always do algebra in the CM setting, when we turn to geometry we will restrict to Gorenstein rings.

In the definition of NCCR, the first condition $\operatorname{End}_R(M) \in \operatorname{CM} R$ turns out to correspond to the geometric property of crepancy (a map $f \colon X \to Y$ is called crepant if $f^*\omega_X = \omega_Y$), but it is hard to explain this without the derived category, so we postpone explanation until Section 4. The second condition $\operatorname{gl.dim} \operatorname{End}_R(M) = \dim R$ is explained below.

2.2. Global dimension, krull dimension and smoothness.

As motivation, suppose that V is an irreducible variety and that $\mathbb{C}[V]$ denotes its coordinate ring. By the work of Auslander–Buchsbaum and Serre in the 1950s, it is known that for $(R, \mathfrak{m})$ a commutative noetherian local ring, R is a regular local ring if and only if $\operatorname{gl.dim} R < \infty$. In fact,

$$V \text{ is nonsingular} \iff \operatorname{gl.dim} \mathbb{C}[V] < \infty \iff \operatorname{gl.dim} \mathbb{C}[V] = \dim \mathbb{C}[V].$$

Thus as soon as the global dimension is finite, necessarily it is equal to $\dim \mathbb{C}[V]$. When asking for the noncommutative analogue of smoothness, it is natural to hope that something similar happens. However, as the exercises should demonstrate, the noncommutative world is not so well behaved.

REMARK 2.2.1. Suppose that R is a CM ring, and $M \in \operatorname{CM} R$. Then it is possible that $\operatorname{gl.dim} \operatorname{End}_R(M) < \infty$ without $\operatorname{gl.dim} \operatorname{End}_R(M) = \dim R$ (see Exercise 2.5.3).

Thus in the noncommutative situation we have to make a choice, either we use $\operatorname{gl.dim} \operatorname{End}_R(M) < \infty$ or $\operatorname{gl.dim} \operatorname{End}_R(M) = \dim R$. Which to choose? To motivate, consider the resolution of the cone singularity

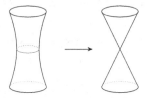

Imagine an ant standing at some point on the cooling tower. It wouldn't know precisely which point it is at, since each point is indistinguishable from every other point. Since points correspond (locally) to simple modules, every simple module should thus be expected to behave in the same way.

Now if we have $\Lambda := \operatorname{End}_R(M) \in \operatorname{CM} R$, by the depth lemma necessarily any simple Λ-module S has $\operatorname{proj.dim}_\Lambda S \geq \dim R$. We don't want erratic behaviour like the projective dimension jumping (as in Exercise 2.5.3), so we choose the $\operatorname{gl.dim} \operatorname{End}_R(M) = \dim R$ definition to ensure homogeneity.

REMARK 2.2.2. When R is Gorenstein, $M \in \operatorname{ref} R$ such that $\operatorname{End}_R(M) \in \operatorname{CM} R$ (which is satisfied in the geometric setting in Section 4), by Lemma 2.3.3 below $\operatorname{gl.dim} \operatorname{End}_R(M) < \infty \iff \operatorname{gl.dim} \operatorname{End}_R(M) = \dim R$. Thus in the main geometric setting of interest, homogeneity of the projective dimension of the simples is not an extra condition.

2.3. NCCRs are Morita equivalent in dimension 2. In the algebraic geometric theory of surfaces, there exists a *minimal* resolution through which all others factor. This is unique up to isomorphism. We can naively ask whether the same is true for NCCRs. It is not, for stupid reasons:

EXAMPLE 2.3.1. Consider the ring $R = \mathbb{C}[x, y]$. Then both $\operatorname{End}_R(R) \cong R$ and $M_2(R) \cong \operatorname{End}_R(R \oplus R)$ are NCCRs of R. They are clearly not isomorphic.

It is well known (see also the exercises) that R and $M_2(R)$ are Morita equivalent, meaning that $\operatorname{mod} R \simeq \operatorname{mod} M_2(R)$ as categories. Thus, even in dimension two, the best we can hope for is that NCCRs are unique up to Morita equivalence.

Recall that if R is a domain with field of fractions F, then $a \in F$ is called *integral over* R if it is the root of a monic polynomial in $R[X]$. Clearly we have $R \subseteq \{a \in F \mid a$ is integral over $R\}$. We say that R is *normal* if equality holds. The key reason we are going to assume that R is normal is Fact 2.3.2(1) below, which will act as the replacement for our previous Fact 1.3.2(1) (note that Fact 1.3.2(1) required that $R \in \operatorname{add} M$).

In the following there is a condition on the existence of a canonical module, which is needed for various technical commutative algebra reasons. In all the geometric situations we will be interested in (or when R is Gorenstein) a canonical module does exist.

FACTS 2.3.2. Suppose that $(R, \mathfrak{m})$ is a local CM normal domain of dimension d with a canonical module, and let $M \in \operatorname{ref} R$.

(1) (Reflexive equivalence) M induces equivalences of categories

$$
\begin{array}{ccc}
\operatorname{ref} R & \xrightarrow[\sim]{\operatorname{Hom}_R(M,-)} & \operatorname{ref}_R \operatorname{End}_R(M) \\
\Big\uparrow & & \Big\uparrow \\
\operatorname{add} M & \xrightarrow[\sim]{\operatorname{Hom}_R(M,-)} & \operatorname{proj} \operatorname{End}_R(M)
\end{array}
$$

where $\operatorname{ref}_R \operatorname{End}_R(M)$ denotes the category of those $\operatorname{End}_R(M)$ modules which are reflexive when considered as R-modules.

(2) (The Auslander–Buchsbaum formula)

    (a) If $\Lambda := \operatorname{End}_R(M)$ is a NCCR, then for all $X \in \operatorname{mod} \Lambda$ we have

$$\operatorname{depth}_R X + \operatorname{proj.dim}_\Lambda X = \dim R.$$

    (b) If R is Gorenstein and $\Lambda := \operatorname{End}_R(M) \in \operatorname{CM} R$, then for all $X \in \operatorname{mod} \Lambda$ with $\operatorname{proj.dim}_\Lambda M < \infty$ we have

$$\operatorname{depth}_R X + \operatorname{proj.dim}_\Lambda X = \dim R.$$

The special case $M = R$ in (2)(b), namely $\Lambda := \operatorname{End}_R(R) \cong R$, gives the classical Auslander–Buchsbaum formula. As a first application, we have:

LEMMA 2.3.3. *Suppose that R is a local Gorenstein normal domain, $M \in CM\ R$ with $\mathrm{End}_R(M) \in CM\ R$. Then $\mathrm{gl.dim\ End}_R(M) < \infty$ if and only if $\mathrm{gl.dim\ End}_R(M) = \dim R$.*

PROOF. Global dimension can be computed as the supremum of the $\mathrm{End}_R(M)$-modules that have finite length as R-modules.

($\Rightarrow$) Suppose that $\mathrm{gl.dim\ End}_R(M) < \infty$. Each finite length $\mathrm{End}_R(M)$-module has depth zero, and by assumption has finite projective dimension. Hence by Auslander–Buchsbaum, each finite length module has projective dimension equal to $\dim R$, so $\mathrm{gl.dim\ End}_R(M) = \dim R$. □

The second application of Auslander–Buchsbaum is our first uniqueness theorem.

THEOREM 2.3.4. *Let $(R, \mathfrak{m})$ be a local CM normal domain of dimension 2 with a canonical module. If R has a NCCR, then all NCCRs of R are Morita equivalent.*

PROOF. Let $\mathrm{End}_R(M)$ and $\mathrm{End}_R(N)$ be NCCRs. Consider $X \in \mathrm{ref}_R \mathrm{End}_R(M)$. We know that depth $X \geq 2$ by the depth lemma (exactly as in the proof of Lemma 2.0.8). By Auslander–Buchsbaum (Fact 2.3.2(2)) we conclude that X is a projective $\mathrm{End}_R(M)$-module. This shows that $\mathrm{ref}_R \mathrm{End}_R(M) = \mathrm{proj\ End}_R(M)$. By Fact 2.3.2(1), this in turn implies that $\mathrm{ref}\ R = \mathrm{add}\ M$.

Repeating the argument with $\mathrm{End}_R(N)$ shows that $\mathrm{ref}\ R = \mathrm{add}\ N$, so combining we see that $\mathrm{add}\ M = \mathrm{add}\ N$. From here it is standard that $\mathrm{End}_R(M)$ and $\mathrm{End}_R(N)$ are Morita equivalent, via the progenerator $\mathrm{Hom}_R(M, N)$. □

We remark that all these theorems hold in the nonlocal setting, provided that we additionally assume that R is equicodimensional (i.e., $\dim R_{\mathfrak{m}} = \dim R$ for all $\mathfrak{m} \in \mathrm{Max}\ R$). This assumption allows us to reduce to the local case without the dimension dropping, and so the global–local arguments work nicely.

REMARK 2.3.5. Theorem 2.3.4 only gives uniqueness, it does not give existence. Indeed, NCCRs do not exist for all local CM normal domain of dimension 2, since as a consequence of Theorem 1.3.1 if such an R admits an NCCR, necessarily it must have finite CM type. If we work over $\mathbb{C}$, another theorem of Auslander (see [247, Section 11]) says that the only such R are the two-dimensional quotient singularities.

2.4. NCCRs are derived equivalent in dimension 3. In dimension three, the situation is more complicated, but can still be controlled. In algebraic geometry, when passing from surfaces to 3-folds we (often) replace the idea of a minimal resolution by a crepant resolution, and these are definitely not unique up to isomorphism. However, by a result of Bridgeland, all crepant resolutions of a given Spec R are unique up to derived equivalence. Using this as motivation, we thus ask whether all NCCRs for a given R are derived equivalent.

As a first remark, this is the best that we can hope for. In contrast to the previous subsection, NCCRs are definitely not unique up to Morita equivalence in dimension three, as the next example demonstrates.

EXAMPLE 2.4.1. Consider the ring $R = \mathbb{C}[[u, v, x, y]]/(uv - x^2 y^2)$. In this example, although it might not be immediately obvious, $\mathrm{End}_R(R \oplus (u, x) \oplus (u, xy) \oplus (u, xy^2))$ and $\mathrm{End}_R(R \oplus (u, x) \oplus (u, xy) \oplus (u, x^2 y))$ are both NCCRs. This can be proved in a variety of ways, for example using dimers and toric geometry [175], arguing directly with tilting bundles [143, Section 5], or by commutative algebra and Calabi–Yau reduction [144]. Regardless, the two NCCRs above can be presented as

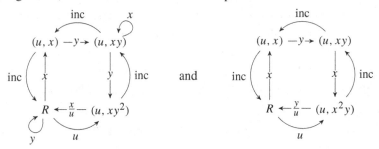

respectively, and they are not Morita equivalent (for example, examine the Ext groups of the vertex simples).

As in the previous subsection, the following theorem holds in the more general setting where R is equicodimensional, but the proof is given in the local case for simplicity.

THEOREM 2.4.2. *Suppose that $(R, \mathfrak{m})$ is a normal CM domain with a canonical module, such that $\dim R = 3$. Then all NCCRs of R are derived equivalent.*

The proof has very little to do with derived categories, and can be understood without even knowing the definition, given the knowledge that classical tilting modules induce derived equivalences.

DEFINITION 2.4.3. Let Λ be a ring. Then $T \in \mathrm{mod}\, \Lambda$ is called a *classical partial tilting module* if $\mathrm{proj.dim}_\Lambda T \le 1$ and $\mathrm{Ext}^1_\Lambda(T, T) = 0$. If further there exists an exact sequence

$$0 \to \Lambda \to T_0 \to T_1 \to 0$$

with each $T_i \in \mathrm{add}\, T$, we say that T is a *classical tilting module*.

If T is a classical tilting Λ-module, it is standard that there is a derived equivalence between Λ and $\mathrm{End}_\Lambda(T)$. Using only this and the facts we have already, we can now prove Theorem 2.4.2.

PROOF. Suppose that $\mathrm{End}_R(M)$ and $\mathrm{End}_R(N)$ are NCCRs of R. Our strategy is to prove that $T := \mathrm{Hom}_R(M, N)$ is a tilting $\Lambda := \mathrm{End}_R(M)$-module. By the remark above, this then shows that $\mathrm{End}_R(M)$ and $\mathrm{End}_{\mathrm{End}_R(M)}(T)$ are derived equivalent. Since $\mathrm{End}_{\mathrm{End}_R(M)}(T) \cong \mathrm{End}_R(N)$ by Fact 2.3.2(1), this will then show that $\mathrm{End}_R(M)$ and $\mathrm{End}_R(N)$ are derived equivalent.

(1) We first show $\mathrm{proj.dim}_\Lambda T \le 1$. This is really just Auslander–Buchsbaum. Take $R^a \to R^b \to M \to 0$ and apply $\mathrm{Hom}_R(-, N)$ to obtain

$$0 \to T \to \mathrm{Hom}_R(R, N)^b \to \mathrm{Hom}_R(R, N)^a. \tag{2.C}$$

Since depth $N \geq 2$ (since N is reflexive), by the depth lemma applied to (2.C) we have depth $T \geq 2$. Hence, by Auslander–Buchsbaum (Fact 2.3.2(2)), proj.dim$_\Lambda T \leq 1$.

(2) We next show that $\text{Ext}^1_\Lambda(T, T) = 0$. This is really just the depth lemma, and using localization to induct. For all primes $\mathfrak{p}$ with ht $\mathfrak{p} = 2$, $T_\mathfrak{p} \in \text{CM } R_\mathfrak{p}$ (since $T_\mathfrak{p} \in \text{ref } R_\mathfrak{p}$, but for Gorenstein surfaces ref $R_\mathfrak{p} = \text{CM } R_\mathfrak{p}$, as in Section 1). Hence by Auslander–Buchsbaum applied to $\Lambda_\mathfrak{p}$, it follows that $T_\mathfrak{p}$ is a projective $\Lambda_\mathfrak{p}$-module for all such primes. This in turn shows that R-module $\text{Ext}^1_\Lambda(T, T)$ is supported only on the maximal ideal, hence has finite length. In particular, provided that $\text{Ext}^1_\Lambda(T, T)$ is nonzero, there is an injection $R/\mathfrak{m} \hookrightarrow \text{Ext}^1_\Lambda(T, T)$ and so by the definition of depth, necessarily depth $\text{Ext}^1_\Lambda(T, T) = 0$. But on the other hand $\text{End}_\Lambda(T) \cong \text{End}_R(N) \in \text{CM } R$ (the isomorphism is as above, by Fact 2.3.2(1)) and so the depth lemma applied to

$$0 \to \text{Hom}_\Lambda(T, T) \cong \text{End}_R(N)$$
$$\to \text{Hom}_\Lambda(\Lambda^a, T) \to \text{Hom}_\Lambda(\Omega T, T) \to \text{Ext}^1_\Lambda(T, T) \to 0$$

forces depth $\text{Ext}^1_\Lambda(T, T) > 0$. This is a contradiction, unless $\text{Ext}^1_\Lambda(T, T) = 0$.

(3) We lastly show that there is an exact sequence $0 \to \Lambda \to T_0 \to T_1 \to 0$ with each $T_i \in \text{add } T$. This involves a duality trick. Denote $(-)^* := \text{Hom}_R(-, R)$, then certainly $\Gamma := \text{End}_R(N^*)$ is also a NCCR.

Consider a projective Γ-module P surjecting as $P \xrightarrow{\psi} \mathbb{F}M^*$, where $\mathbb{F} = \text{Hom}_R(N^*, -)$. By reflexive equivalence, we know that $P = \mathbb{F}N_0^*$ for some $N_0^* \in \text{add } N^*$, and further $\psi = \mathbb{F}f$ for some $f : N_0^* \to M^*$. Taking the kernel of f, this all means that we have an exact sequence

$$0 \to K \to N_0^* \xrightarrow{f} M^*,$$

such that

$$0 \to \mathbb{F}K \to \mathbb{F}N_0^* \to \mathbb{F}M^* \to 0 \tag{2.D}$$

is exact. By the depth lemma $\mathbb{F}K \in \text{CM } R$ and so by Auslander–Buchsbaum $\mathbb{F}K$ is a projective Γ-module. Thus $K \in \text{add } N^*$ by reflexive equivalence; say $K = N_1^*$.

Now $\text{Ext}^1_\Gamma(\mathbb{F}M^*, \mathbb{F}M^*) = 0$ by applying the argument in (2) to Γ. Thus applying $\text{Hom}_\Gamma(-, \mathbb{F}M^*)$ to (2.D) gives us the commutative diagram

$$0 \to \text{Hom}_\Gamma(\mathbb{F}M^*, \mathbb{F}M^*) \longrightarrow \text{Hom}_\Gamma(\mathbb{F}N_0^*, \mathbb{F}M^*) \longrightarrow \text{Hom}_\Gamma(\mathbb{F}N_1^*, \mathbb{F}M^*) \to 0$$
$$\qquad\qquad \| \qquad\qquad\qquad\qquad \| \qquad\qquad\qquad\qquad \|$$
$$0 \longrightarrow \text{Hom}_R(M^*, M^*) \longrightarrow \text{Hom}_R(N_0^*, M^*) \longrightarrow \text{Hom}_R(N_1^*, M^*) \longrightarrow 0$$

where the top row is exact. Hence the bottom row is exact. Since

$$(-)^* : \text{ref } R \to \text{ref } R$$

is a duality, this means that

$$0 \to \text{Hom}_R(M, M) \to \text{Hom}_R(M, N_0) \to \text{Hom}_R(M, N_1) \to 0$$

is exact. But this is simply

$$0 \to \Lambda \to T_0 \to T_1 \to 0$$

with each $T_i \in \mathrm{add}\, T$. Hence T is a tilting Λ-module. $\square$

Credits: The material on depth and maximal CM modules is well-known and can be found in most commutative algebra books. The definition of an NCCR is due to Van den Bergh [231], modelled both on the skew group ring and also on his interpretation [230] of the flops paper of Bridgeland [42]. The idea that finite global dimension is not enough and we need homogeneity appears in the "homologically homogeneous" rings of Brown–Hajarnavis [53], and also as "nonsingular orders" in the language of Auslander [17, 18].

Reflexive equivalence is also well-known, it appears in Reiten–Van den Bergh [182]. The Auslander–Buchsbaum formula in the commutative setting is much more general than the version presented here, and first appeared in [20]. The proof can be found in most commutative algebra or homological algebra books. The noncommutative version of the Auslander–Buchsbaum formula for NCCRs in Fact 2.3.2(2)(a) was first established by Iyama and Reiten [139], whereas the version presented in Fact 2.3.2(2)(b), which is valid in the infinite global dimension case, is taken from [141]. The "correct" setting is that of a Gorenstein R-order.

Lemma 2.3.3 appears in [231], but it is also explained by Iyama and Reiten [139] and Dao and Huneke [76]. The fact that NCCRs are unique in dimension two was well-known to experts, but is only written down in [142]. The fact that NCCRs are all derived equivalent in dimension three when the base ring R is Gorenstein is due to Iyama and Reiten [139], but presented here is the simplified proof from [142] since it holds in the more general CM setting.

2.5. Exercises.

EXERCISE 2.5.1 (common examples and counterexamples for surfaces). Consider the following quivers with relations (Q, R):

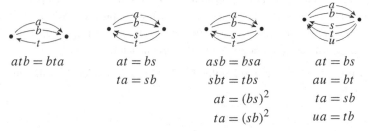

$atb = bta$	$at = bs$	$asb = bsa$	$at = bs$
	$ta = sb$	$sbt = tbs$	$au = bt$
		$at = (bs)^2$	$ta = sb$
		$ta = (sb)^2$	$ua = tb$

For each $\Lambda := kQ/R$:
  (1) Determine the centre $Z(\Lambda)$. Is it CM? Is it Gorenstein? Is it smooth?
  (2) Is $\Lambda \cong \mathrm{End}_{Z(\Lambda)}(M)$ for some $M \in Z(\Lambda)$? If so, is $M \in \mathrm{CM}\, Z(\Lambda)$?
  (3) Is $\Lambda \in \mathrm{CM}\, Z(\Lambda)$?
  (4) What are the projective dimension of the vertex simples?
  (5) In the situation when the vertex simples have infinite projective dimension, is there anything remarkable about their projective resolutions?
  (6) Using (2) and (4), compute gl.dim Λ.

(7) Using (2) and (6), which Λ are NCCRs over $Z(\Lambda)$?

(8) (harder) To which spaces are the Λ derived equivalent? (Aside: does this explain (5)?)

EXERCISE 2.5.2 (example of a nonlocal NCCR in dimension two). Consider (Q, R) given by

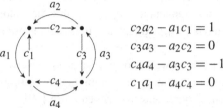

$$c_2 a_2 - a_1 c_1 = 1$$
$$c_3 a_3 - a_2 c_2 = 0$$
$$c_4 a_4 - a_3 c_3 = -1$$
$$c_1 a_1 - a_4 c_4 = 0$$

Set $\Lambda := kQ/R$ and $Z := \mathbb{C}[u, v, x]/(uv - x^2(x - 1)^2)$. Note that Z has only two singular points, which locally are just the $\frac{1}{2}(1, 1)$ surface singularity.

(1) Show that $kQ/R \cong \mathrm{End}_Z(Z \oplus (u, x - 1) \oplus (u, x(x - 1)) \oplus (u, x^2(x - 1)))$.

(2) Using (1), deduce that Λ is a NCCR.

(3) Find some algebras that are Morita equivalent to Λ.

EXERCISE 2.5.3 (Left and right modules can matter). Consider the algebra Λ given by

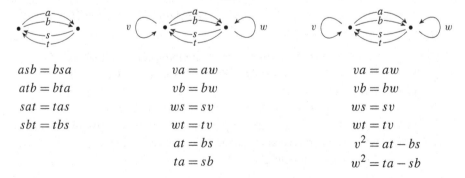

$$c_1 a_1 = a_3 k_1 \quad c_1 c_2 k_1 = a_3 c_3$$
$$c_2 a_2 = a_1 c_1 \quad k_1 c_1 c_2 = c_3 a_3$$
$$k_1 a_3 = a_2 c_2$$

This should be familiar from Exercise 1.5.2. Denote the vertex simples by S_1, S_2, S_3.

(1) Show that as left Λ-modules proj.dim $S_1 =$ proj.dim $S_2 = 2$ whilst proj.dim $S_3 = 3$.

(2) Show that as right Λ-modules proj.dim $S_1 = 3$ whilst proj.dim $S_2 =$ proj.dim $S_3 = 2$.

(3) In this example $\Lambda \cong \mathrm{End}_{Z(\Lambda)}(M)$, where $Z(\Lambda)$ is a CM ring, and $M \in \mathrm{CM}\, Z(\Lambda)$. Why does this not contradict Auslander–Buchsbaum?

EXERCISE 2.5.4 (common examples for 3-folds). Consider the following quivers with relations (Q, R):

$asb = bsa$	$va = aw$	$va = aw$
$atb = bta$	$vb = bw$	$vb = bw$
$sat = tas$	$ws = sv$	$ws = sv$
$sbt = tbs$	$wt = tv$	$wt = tv$
	$at = bs$	$v^2 = at - bs$
	$ta = sb$	$w^2 = ta - sb$

$$x_i y_{i+1} = y_i x_{i+1}$$
$$x_i z_{i+1} = z_i x_{i+1}$$
$$y_i z_{i+1} = z_i y_{i+1}$$

where in the last example the relations are taken over all $1 \le i \le 3$, with the subscripts taken mod 3 if necessary. For each $\Lambda = kQ/R$:

(1) Check that the relations can be packaged as a superpotential.
(2) Determine the centre $Z(\Lambda)$. It should be a three-dimensional Gorenstein ring (if necessary, check using Singular).
(3) (harder) Show that Λ is an NCCR over $Z(\Lambda)$.
(4) (much harder) To which space(s) are the Λ derived equivalent? (aside: the first three examples capture a certain geometric phenomenon regarding curves in 3-folds. Which phenomenon?)

Computer exercises.

EXERCISE 2.5.5 (CM via Ext groups). Let $M, N \in \mathrm{CM}\, R$, where R is a CM ring. This exercise will explore whether the property $\mathrm{Hom}_R(M, N) \in \mathrm{CM}\, R$ can be characterised in terms of Ext groups. To calculate the depth of $\mathrm{Ext}^1_R(M, N)$ and $\mathrm{Ext}^2_R(M, N)$, use

> depth(Ext(1, M, M));

> depth(Ext(2, M, N));

(1) We restrict to $\dim R = 3$ with the assumption that R is CM, with an isolated singularity.

(a) Let $R = \mathbb{C}[\![u, v, x, y]\!]/(uv - xy)$. Consider the modules R, (u, x), (u, y), $R \oplus (u, x)$, $R \oplus (u, y)$ and $R \oplus (u, x) \oplus (u, y)$. Check they are all CM. For each M, compute whether $\mathrm{End}_R(M) \in \mathrm{CM}\, R$, and compute $\mathrm{Ext}^1_R(M, M)$.

(b) Let $R = \mathbb{C}[\![u, v, x, y]\!]/(uv - (x + y)(x + 2y)(x + 3y))$. To ease notation set $f_a = x + ay$. Consider the modules R, (u, f_1), (u, f_2), (u, f_3), $(u, f_1 f_2)$, $(u, f_1 f_3)$, $(u, f_2 f_3)$ and all direct sum combinations. All are CM. For each M, compute whether $\mathrm{End}_R(M) \in \mathrm{CM}\, R$, and compute $\mathrm{Ext}^1_R(M, M)$.

(2) Now $\dim R = 3$ with R CM, but the singular locus is no longer isolated (for example check using Singular).

(a) Let $R = \mathbb{C}[\![u, v, x, y]\!]/(uv - x^2 y)$. Consider the modules R, (u, x), (u, y), (u, x^2), (u, xy) and all direct sum combinations. All are CM. For each M, compute whether $\mathrm{End}_R(M) \in \mathrm{CM}\, R$, and compute $\mathrm{Ext}^1_R(M, M)$.

(3) Is there a pattern from (1) and (2)? Prove this relationship — it should really only involve the depth lemma.

EXERCISE 2.5.6 (how to determine invariant rings). Computing invariant rings is generally quite a grim task. This exercise will show how to get a computer to (i) compute the generators of the invariant ring (ii) compute all the relations between them.

(1) Consider $G = \langle g := \left(\begin{smallmatrix} -1 & 0 \\ 0 & -1 \end{smallmatrix} \right) \rangle$. The invariants are calculated using the code

```
> LIBfinvar.lib";
> ring   S = complex, (x, y), dp;
> matrixA[2][2] = −1, 0, 0, −1;
> list   L = group_reynolds(A);
> matrix   T = invariant_algebra_reynolds(L[1], 1);
> print(T);
```

The output should be y^2, xy, x^2, which we know generate. To calculate this in terms of generators and relations, we code

```
> string   newring = E";
> orbit_variety(T, newring);
> print(G);
> basering;
```

The output is y(2)^2 − y(1) * y(3), which is the equation we already know. Using this presentation, we can now plug it into Exercises 1.5.3 and 1.5.5 and ask whether the invariant ring is CM, or even Gorenstein.

(2) The next group is

$$BD_8 := \langle \left(\begin{smallmatrix} i & 0 \\ 0 & -i \end{smallmatrix} \right), \left(\begin{smallmatrix} 0 & 1 \\ -1 & 0 \end{smallmatrix} \right) \rangle.$$

To code two generators requires

```
> LIBfinvar.lib";
> ring   S = complex, (x, y), dp;
> matrix   A[2][2] = i, 0, 0, −i;
> matrix   B[2][2] = 0, 1, 1, 0;
> list   L = group_reynolds(A, B);
> matrix   T = invariant_algebra_reynolds(L[1], 1);
> print(T);
```

By using the same method as in (1), find the generators and relations. Try to prove this without using the computer.

(3) Try the same question with the following groups.

(a) $\frac{1}{4}(1, 1) := \langle \left(\begin{smallmatrix} i & 0 \\ 0 & i \end{smallmatrix} \right) \rangle$.

(b) $\frac{1}{4}(1, 1, 2) := \langle \left(\begin{smallmatrix} i & 0 & 0 \\ 0 & i & 0 \\ 0 & 0 & -1 \end{smallmatrix} \right) \rangle$.

(c) $\mathbb{Z}_2 \times \mathbb{Z}_2 := \langle \left(\begin{smallmatrix} -1 & 0 & 0 \\ 0 & -1 & 0 \\ 0 & 0 & 1 \end{smallmatrix} \right), \left(\begin{smallmatrix} 1 & 0 & 0 \\ 0 & -1 & 0 \\ 0 & 0 & -1 \end{smallmatrix} \right) \rangle$.

Is there a pattern as to when the invariant ring is Gorenstein, and when it is CM?

3. From algebra to geometry: quiver GIT

Sections 1 and 2 contain only algebra, and use geometry only to motivate some of the results. In this section the process begins to reverse, and we will begin extracting geometry (and obtain geometric theorems) starting from NCCRs.

The setup is that Z is a Gorenstein normal domain, and we continue our original motivation of trying to resolve Spec Z via the picture

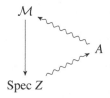

The first arrow is the process of associating to Z an NCCR $A := \mathrm{End}_Z(M)$. We have changed notation to Z (from R), since it is always the centre of A. We remark that NCCRs do not exist in general, even for easy examples like $\mathbb{C}[[u, v, x, y]]/(uv - x(x^2 + y^7))$. The content of this section is to explain the second arrow in the above diagram, i.e., how to extract the geometry from the noncommutative ring A.

To simplify the exposition, although it is not strictly necessary, we will assume that we have written A as a quiver with relations $A = kQ/R$ (see the Appendix on page 297 for a brief overview). We are going to define various moduli spaces of finite-dimensional representations, and to do this requires geometric invariant theory (GIT).

For a fixed dimension vector α we may consider all representations of $A = kQ/R$ with dimension vector α, namely

$$\mathcal{R} := \mathrm{Rep}(A, \alpha) = \{\text{representations of } A \text{ of dimension } \alpha\}.$$

This is an affine variety, so denote the coordinate ring by $k[\mathcal{R}]$. The variety, and hence the coordinate ring, carries a natural action of $G := \prod_{i \in Q_0} \mathrm{GL}(\alpha_i, k)$, where Q_0 denotes the set of vertices and $\mathrm{GL}(\alpha_i, k)$ denotes the group of invertible $\alpha_i \times \alpha_i$ matrices with entries in k. The action is via conjugation; g acts on an arrow a as $g \cdot a = g_{t(a)}^{-1} a g_{h(a)}$. It is actually an action by PGL, since the diagonal one-parameter subgroup $\Delta = \{(\lambda 1, \cdots, \lambda 1) : \lambda \in k^*\}$ acts trivially. By linear algebra the *isomorphism classes* of representations of $A = kQ/R$ are in natural one-to-one correspondence with the orbits of this action.

To understand the space of isomorphism classes is normally an impossible problem (for example, the algebra might have wild representation type), so we want to throw away some representations and take what is known as a GIT quotient. The key point is that to make a GIT quotient requires an addition piece of data in the form of a character χ of G.

The characters χ of $G = \prod_{i \in Q_0} \mathrm{GL}(\alpha_i, k)$ are known to be just the powers of the determinants

$$\chi(g) = \prod_{i \in Q_0} \det(g_i)^{\theta_i},$$

for some collection of integers $\theta_i \in \mathbb{Z}^{Q_0}$. Since such a χ determines and is determined by the θ_i, we usually denote χ by χ_θ. We now consider the map

$$\theta : \texttt{fdmod}A \to \mathbb{Z}, \quad M \mapsto \sum_{i \in Q_0} \theta_i \dim M_i,$$

which is additive on short exact sequences and so induces a map $K_0(\texttt{fdmod}A) \to \mathbb{Z}$.

We assume that our character satisfies $\chi_\theta(\Delta) = \{1\}$ (for experts — this is needed to use Mumford's numerical criterion in [153, Proposition 2.5]). It not too hard to see that this condition translates into $\sum_{i \in Q_0} \theta_i \alpha_i = 0$ and so for these χ_θ, $\theta(M) = 0$ whenever M has dimension vector α.

We arrive at a key definition introduced by King [153, Definition 1.1].

DEFINITION 3.0.7. Let $\mathcal{A}$ be an abelian category, and $\theta : K_0(\mathcal{A}) \to \mathbb{Z}$ an additive function. We call θ a character of $\mathcal{A}$. An object $M \in \mathcal{A}$ is called θ-semistable if $\theta(M) = 0$ and every subobject $M' \subseteq M$ satisfies $\theta(M') \geq 0$. Such an object M is called θ-stable if the only subobjects M' with $\theta(M') = 0$ are M and 0. We call θ generic if every M which is θ-semistable is actually θ-stable.

For $A = kQ/R$ as before, we are interested in this definition for the case $\mathcal{A} = \texttt{fdmod}A$. We shall see how this works in practice in the next section. The reason King gave the above definition is that it is equivalent to the other notion of stability from GIT, which we now describe.

As stated above, $\mathcal{R}$ is an affine variety with an action of a linearly reductive group $G = \prod_{i \in Q_0} \text{GL}(\alpha_i, k)$. Since G is reductive, we have a quotient

$$\mathcal{R} \to \mathcal{R}/\!\!/G = \text{Spec } k[\mathcal{R}]^G,$$

which is dual to the inclusion $k[\mathcal{R}]^G \to k[\mathcal{R}]$. The reductiveness of the group ensures that $k[\mathcal{R}]^G$ is a finitely generated k-algebra, and so $\text{Spec } k[\mathcal{R}]^G$ is a variety, not just a scheme.

To make a GIT quotient we have to add to this picture the extra data of χ, some character of G.

DEFINITION 3.0.8. $f \in k[\mathcal{R}]$ is a semi-invariant of weight χ if $f(g \cdot x) = \chi(g)f(x)$ for all $g \in G$ and all $x \in \mathcal{R}$. We write the set of such f as $k[\mathcal{R}]^{G,\chi}$. We define

$$\mathcal{R}/\!\!/_\chi G := \text{Proj}\left(\bigoplus_{n \geq 0} k[\mathcal{R}]^{G,\chi^n}\right).$$

DEFINITION 3.0.9.   (1) $x \in \mathcal{R}$ is called χ-semistable (in the sense of GIT) if there exists some semi-invariant f of weight χ^n with $n > 0$ such that $f(x) \neq 0$, otherwise $x \in \mathcal{R}$ is called unstable.

(2) $x \in \mathcal{R}$ is called χ-stable (in the sense of GIT) if it is χ-semistable, the G orbit containing x is closed in $\mathcal{R}^{ss}$ and further the stabilizer of x is finite.

The set of semistable points $\mathcal{R}^{ss}$ forms an open subset of $\mathcal{R}$; in fact we have a morphism

$$q : \mathcal{R}^{ss} \to \mathcal{R}/\!/_\chi G,$$

which is a good quotient. In fact q is a geometric quotient on the stable locus $\mathcal{R}^s$, meaning that $\mathcal{R}^s /\!/_\chi G$ really is an orbit space.

The point in the above discussion is the following result, which says that the two notions coincide.

THEOREM 3.0.10 (King). *Let* $M \in Rep(A, \alpha) = \mathcal{R}$, *choose* θ *as in Definition 3.0.7. Then* M *is* θ-*semistable (in the categorical sense of Definition 3.0.7) if and only if* M *is* χ_θ-*semistable (in the sense of GIT). The same holds replacing semistability with stability. If* θ *is generic, then* $\mathcal{R}/\!/_\chi G$ *parametrises the* θ-*stable modules up to isomorphism.*

Thus we use the machinery from the GIT side to define for quivers the following:

DEFINITION 3.0.11. For $A = kQ/R$ choose dimension vector α and character θ satisfying $\sum_{i \in Q_0} \alpha_i \theta_i = 0$. Denote $Rep(A, \alpha) = \mathcal{R}$ and $G = \prod_{i \in Q_0} GL(\alpha_i, k)$. We define

$$\mathcal{M}_\theta^{ss}(A, \alpha) := \mathcal{R}/\!/_{\chi_\theta} G := \mathrm{Proj}\left(\bigoplus_{n \geq 0} k[\mathcal{R}]^{G, \chi^n} \right),$$

and call it the moduli space of θ-semistable representations of dimension vector α.

This is by definition projective over the ordinary quotient $\mathcal{R}/\!/ G = \mathrm{Spec}\, k[\mathcal{R}]^G$. Hence for example if $k[\mathcal{R}]^G = k$ then $\mathcal{M}_\theta^{ss}(A, \alpha)$ is a projective variety, but in our setting this will not be the case.

REMARK 3.0.12. If Spec Z is a singularity that we would like to resolve, ideally we would like the zeroth piece $k[\mathcal{R}]^G$ to be Z, since then the moduli space is projective over Spec Z. However, even in cases where we use NCCRs to resolve singularities, $k[\mathcal{R}]^G$ might not be Z (see Exercise 3.2.4).

Note that $\mathcal{M}_\theta^{ss}(A, \alpha)$ may be empty, and in fact it is often very difficult to determine whether this is true or not. In our NCCR setting, this won't be a problem since by choice of dimension vector later it will always contain the Azumaya locus. Even when $\mathcal{M}_\theta^{ss}(A, \alpha)$ is not empty, computing it explicitly can often be hard.

We remark that $\mathcal{M}_\theta^{ss}(A, \alpha)$ is a moduli space in the strict sense that it represents a functor. This functorial viewpoint is very important, but in these notes we gloss over it, and the other related technical issues.

3.1. Examples. We want to input an NCCR, and by studying the moduli we hope to output some crepant resolution. The problem is that so far we have no evidence that this is going to work (!), so in this section we will explicitly compute some moduli spaces to check that this strategy is not entirely unreasonable.

We assume that we have already presented our NCCR as $A = kQ/R$. We follow the exposition from King:

"To specify such a moduli space we must give a dimension vector α and a weight vector (or "character") θ satisfying $\sum_{i \in Q_0} \theta_i \alpha_i = 0$. The moduli space of θ-stable A-modules of dimension vector α is then the parameter space for those A-modules which have no proper submodules with any dimension vector β for which $\sum_{i \in Q_0} \theta_i \beta_i \leq 0$."

Even though two choices are needed to create a moduli space, namely α and θ, for NCCRs the α is given naturally by the ranks of the reflexive modules that we have summed together to create the NCCR (see Section 5.4).

Before computing examples with NCCRs, we begin with something easier.

EXAMPLE 3.1.1. Consider the quiver

with no relations. Choose $\alpha = (1, 1)$ and $\theta = (-1, 1)$. With these choices, since $\sum \theta_i \alpha_i = 0$ we can form the moduli space. Now a representation of dimension vector $\alpha = (1, 1)$ is θ-semistable by definition if $\theta(M') \geq 0$ for all subobjects M'. But the only possible subobjects in this example are of dimension vector $(0, 0)$, $(0, 1)$ and $(1, 0)$, and θ is ≥ 0 on all but the last (in fact its easy to see that θ is generic in this example). Thus a representation of dimension vector $(1, 1)$ is θ-semistable if and only if it has no submodules of dimension vector $(1, 0)$. Now take an arbitrary representation M of dimension vector $(1, 1)$

$$M = \mathbb{C} \underset{b}{\overset{a}{\rightrightarrows}} \mathbb{C} \ .$$

Notice that M has a submodule of dimension vector $(1, 0)$ if and only if $a = b = 0$, since the diagram

$$\begin{array}{ccc} \mathbb{C} \underset{b}{\overset{a}{\rightrightarrows}} \mathbb{C} \\ \cong \uparrow \qquad \uparrow 0 \\ \mathbb{C} \underset{0}{\overset{0}{\rightrightarrows}} 0 \end{array}$$

must commute. Thus by our choice of stability θ,

M is θ-semistable $\iff$ M has no submodule of dim vector $(1, 0)$ $\iff$ $a \neq 0$ or $b \neq 0$,

and so we see that the semistable objects parametrise $\mathbb{P}^1$ via the ratio $(a : b)$, so the moduli space is just $\mathbb{P}^1$. Another way to see this: we have two open sets, one corresponding to $a \neq 0$ and the other to $b \neq 0$. After changing basis we can set them to be the identity, and so we have

$$U_0 = \{\ \mathbb{C} \underset{b}{\overset{1}{\rightrightarrows}} \mathbb{C} \ | b \in \mathbb{C}\}, \qquad U_1 = \{\ \mathbb{C} \underset{1}{\overset{a}{\rightrightarrows}} \mathbb{C} \ | a \in \mathbb{C}\}.$$

Now the gluing is given by, whenever $U_0 \ni b \neq 0$

$$U_0 \ni b = \mathbb{C} \underset{b}{\overset{1}{\rightrightarrows}} \mathbb{C} \cong \mathbb{C} \underset{1}{\overset{b^{-1}}{\rightrightarrows}} \mathbb{C} = b^{-1} \in U_1,$$

which is evidently just $\mathbb{P}^1$.

Although the following is not an NCCR, it has already appeared on the example sheets (Exercise 2.5.1). Whenever resolving singularities, it is traditional to begin by blowing up the origin of $\mathbb{C}^2$.

EXAMPLE 3.1.2. Consider the quiver with relations

$$\bullet \underset{t}{\overset{a}{\underset{b}{\rightrightarrows}}} \bullet \quad atb = bta,$$

and again choose the dimension vector to be $(1,1)$ and the stability $\theta_0 = (-1,1)$. Exactly as above, if

$$M = \quad \mathbb{C} \underset{t}{\overset{a}{\underset{b}{\rightrightarrows}}} \mathbb{C}$$

then

M is θ-semistable $\iff$ M has no submodule of dim vector $(1,0)$ $\iff$ $a \neq 0$ or $b \neq 0$.

For the first open set in the moduli U_0 (when $a \neq 0$): after changing basis so that $a = 1$ we see that the open set is parametrized by the two scalars b and t subject to the single relation (substituting $a = 1$ into the quiver relations) $tb = bt$. But this always holds, thus the open set U_0 is just $\mathbb{C}^2$ with coordinates b, t. We write this as $\mathbb{C}^2_{b,t}$. Similarly for the other open set:

$$\mathbb{C} \underset{t}{\overset{1}{\underset{b}{\rightrightarrows}}} \mathbb{C} \qquad \mathbb{C} \underset{t}{\overset{a}{\underset{1}{\rightrightarrows}}} \mathbb{C}$$

$$U_0 = \mathbb{C}^2_{b,t} \qquad U_1 = \mathbb{C}^2_{a,t}$$

Now the gluing is given by, whenever $b \neq 0$,

$$U_0 \ni (b,t) = \quad \mathbb{C} \underset{t}{\overset{1}{\underset{b}{\rightrightarrows}}} \mathbb{C} \quad \cong \quad \mathbb{C} \underset{bt}{\overset{b^{-1}}{\underset{1}{\rightrightarrows}}} \mathbb{C} \quad = (b^{-1}, bt) \in U_1,$$

and so we see that this is just the blowup of the origin of $\mathbb{C}^2$.

With the two above examples in hand, we can now compute an example where an NCCR gives a crepant resolution. There are many more examples in the exercises. We begin in dimension two.

EXAMPLE 3.1.3. We return to the running example from Sections 1 and 2, namely $R := \mathbb{C}[a, b, c]/(ab - c^3)$. We now know that $\operatorname{End}_R(R \oplus (a, c) \oplus (a, c^2))$ is a NCCR, and we can believe (from Example 1.2.5) that it is presented as

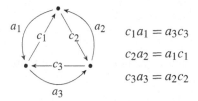

$$c_1 a_1 = a_3 c_3$$
$$c_2 a_2 = a_1 c_1$$
$$c_3 a_3 = a_2 c_2$$

We consider the dimension vector $(1, 1, 1)$ (corresponding to the ranks of the CM modules that we have summed), and stability $(-2, 1, 1)$, where the -2 sits at the bottom left vertex. As above, for a module to be θ-stable corresponds to there being, for every vertex i, a nonzero path from the bottom left vertex to i. Thus we have three open sets, corresponding to the following pictures

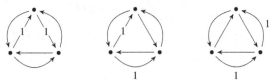

Accounting for the relations, these open sets are parametrized by

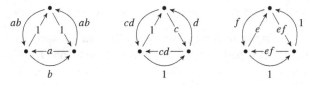

The first, U_1, is just affine space $\mathbb{A}^2$ with coordinates a, b (written $\mathbb{A}^2_{a,b}$), whilst the second and third are $U_2 = \mathbb{A}^2_{c,d}$ and $U_3 = \mathbb{A}^2_{e,f}$ respectively. We immediately see that our moduli space is smooth, since it is covered by three affine opens, each of which is smooth.

Now we ask how these open sets glue. Visually, it is clear that U_1 and U_2 glue if and only if the arrow b in U_1 is not equal to zero (in which case we can base change to make it the identity, and hence land in U_2). Similarly U_2 and U_3 glue if and only if the arrow d in U_2 is nonzero. In principle there could also be a glue between U_1 and U_3, but for these to glue certainly the arrows b and ab in U_1 must be nonzero, hence the glue between U_1 and U_3 is already covered by the previous two glues. Explicitly, the two glues are

$$\mathbb{A}^2_{a,b} \ni (a,b) = \quad\cong\quad = (b^{-1}, ab^2) \in \mathbb{A}^2_{c,d}$$

and

$$\mathbb{A}^2_{c,d} \ni (c,d) = \quad\cong\quad = (d^{-1}, cd^2) \in \mathbb{A}^2_{e,f}.$$

The upshot is that the moduli space looks (very roughly) like the following:

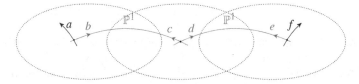

where the black dots correspond to the origins of the coordinate charts, and the two curved lines joining them give two $\mathbb{P}^1$s.

It is instructive to see explicitly why the two $\mathbb{P}^1$s form the exceptional divisor. By the nature of the Proj construction, there is a natural map from the moduli space to $\operatorname{Spec} \mathbb{C}[\mathcal{R}]^G$. What is this map? First notice that $\mathbb{C}[\mathcal{R}]$ is, by definition, the commutative ring $\mathbb{C}[c_1, c_2, c_3, a_1, a_2, a_3]/(c_1 a_1 = c_2 a_2 = c_3 a_3)$ where we have just taken the arrows and relations and made everything commute. Now since the dimension vector is $\alpha = (1, 1, 1)$, the group G is precisely $\operatorname{GL}(1, \mathbb{C}) \times \operatorname{GL}(1, \mathbb{C}) \times \operatorname{GL}(1, \mathbb{C}) = \mathbb{C}^* \times \mathbb{C}^* \times \mathbb{C}^*$. The action is by base change $g \cdot a = g_{t(a)}^{-1} a g_{h(a)}$, which means

$$(\lambda_1, \lambda_2, \lambda_3) \cdot c_1 := \lambda_1^{-1} c_1 \lambda_2,$$

$$(\lambda_1, \lambda_2, \lambda_3) \cdot c_2 := \lambda_2^{-1} c_2 \lambda_3,$$

$$(\lambda_1, \lambda_2, \lambda_3) \cdot c_3 := \lambda_3^{-1} c_3 \lambda_1,$$

$$(\lambda_1, \lambda_2, \lambda_3) \cdot a_1 := \lambda_2^{-1} a_1 \lambda_1,$$

$$(\lambda_1, \lambda_2, \lambda_3) \cdot a_2 := \lambda_3^{-1} a_2 \lambda_2,$$

$$(\lambda_1, \lambda_2, \lambda_3) \cdot a_3 := \lambda_1^{-1} a_3 \lambda_3.$$

For a product to be invariant under this action the λ's must cancel, and visually these correspond to cycles in the quiver. Here, the invariants are generated by $A := c_1 c_2 c_3$, $B := a_1 a_2 a_3$ and $C := c_1 a_1 = c_2 a_2 = c_3 a_3$. Note that $AB = C^3$, so $\mathbb{C}[\mathcal{R}]^G \cong \mathbb{C}[A, B, C]/(AB - C^3)$, which is our base singularity.

Using this information, the natural map takes a stable representation to the point $(c_1 c_2 c_3, a_1 a_2 a_3, c_1 a_1)$ of $\operatorname{Spec} \mathbb{C}[\mathcal{R}]^G$, which in the case of the three open sets gives

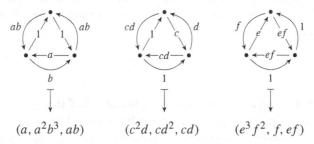

Now we look above the singular point $(0, 0, 0)$, and in U_1 we see $\{(0, b) \mid b \in \mathbb{C}\}$, in U_2 we see $\{(c, d) \mid cd = 0\}$, and in U_3 we see $\{(e, 0) \mid e \in \mathbb{C}\}$. Thus, the curved lines in the rough picture at the bottom of the previous page constitute the exceptional locus, and they are the union of two $\mathbb{P}^1$s. The space is smooth, and is in fact the minimal resolution of our original motivating singularity.

REMARK 3.1.4. There is a pattern evident in the previous examples. If we consider dimension vector $(1, 1, \ldots, 1)$ and stability condition $(-n, 1, \ldots, 1)$, where the $-n$ corresponds to a vertex $\star$, then a module M of dimension vector $(1, 1, \ldots, 1)$ is θ-stable if and only if for every vertex in the quiver representation of M, there is a nonzero path from $\star$ to that vertex. This is quite a pleasant combinatorial exercise (see Exercise 3.2.5).

EXAMPLE 3.1.5 (the suspended pinch point). This example illustrates the phenomenon of nonisomorphic crepant resolutions in dimension three. The example to consider is $R := \mathbb{C}[u, v, x, y]/(uv - x^2 y)$. This has a 1-dimensional singular locus, namely the y-axis $u = v = x = 0$.

Here $\mathrm{End}_R(R \oplus (u, x) \oplus (u, x^2))$ is a NCCR, and in fact it can be presented as

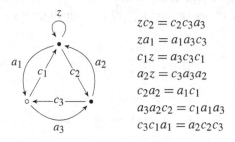

$$zc_2 = c_2 c_3 a_3$$
$$za_1 = a_1 a_3 c_3$$
$$c_1 z = a_3 c_3 c_1$$
$$a_2 z = c_3 a_3 a_2$$
$$c_2 a_2 = a_1 c_1$$
$$a_3 a_2 c_2 = c_1 a_1 a_3$$
$$c_3 c_1 a_1 = a_2 c_2 c_3$$

Since this is a NCCR, we pick the dimension vector corresponding to the ranks of the CM modules, which in this case is $(1, 1, 1)$. By a similar calculation as in Example 3.1.3 (but bearing in mind Remark 3.1.4), computing the moduli space for the stability $(-2, 1, 1)$ then looking above the y-axis ($=$ the singular locus) we see

where the curves indicate the two $\mathbb{P}^1$s that are above the origin. However, if taking the stability $(-1, 2, -1)$ (where the 2 is on the top vertex) and looking above the y-axis gives the picture

This is recommended as an instructive exercise in quiver GIT (Exercise 3.2.3 is similar). The left hand curve in our original picture has been flopped into the exceptional gray surface. These two moduli spaces are both smooth (in fact, they are crepant resolutions of Spec R), but they are not isomorphic.

In the language of toric geometry, the two crepant resolutions above correspond to the following pictures

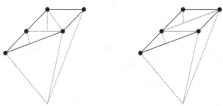

Since in toric geometry crepant resolutions correspond to subdividing the cone, there is one further crepant resolution. This too can be obtained using quiver GIT (in a similar way to Exercise 3.2.3).

Credits: All the material on quiver GIT is based on the original paper of King [153], and also on the lecture course he gave in Bath in 2006. Examples 3.1.1 and 3.1.2 were two of the motivating examples for the theory. The example of the $\mathbb{Z}_3$ singularity has been studied by so many people that it is hard to properly credit each; certainly it appears in the work of Kronheimer [158] and Cassens–Slodowy [61], and is used (and expanded) in the work of Craw–Ishii [73] and many others. The fact that $\mathbb{C}[\mathcal{R}]^G = \mathbb{C}[a, b, c]/(ab - c^3)$ in Example 3.1.3 is a general theorem for Kleinian singularities, but also appears in this special case in the lectures of Le Bruyn [160]. The suspended pinch point is the name given by physicists to the singularity $uv = x^2 y$. The geometry is toric and so well-known. The NCCRs in this case can be found in either Van den Bergh [231, Section 8], Nagao [175, Section 1.4], or [143], but also in many papers by various physicists.

3.2. Exercises.

EXERCISE 3.2.1. Consider the examples in Exercise 2.5.1. For each, consider the dimension vector $(1, 1)$. There are essentially only two generic stability conditions, namely $(-1, 1)$ and $(1, -1)$. For each of the above examples, compute the spaces given by these two stabilities. Each of the examples should illustrate a different phenomenon.

EXERCISE 3.2.2. Consider some of the examples in Exercise 2.5.4, namely,

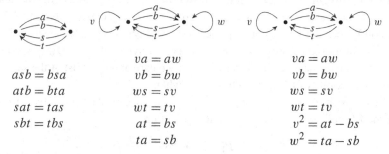

$$asb = bsa$$
$$atb = bta$$
$$sat = tas$$
$$sbt = tbs$$

$$va = aw$$
$$vb = bw$$
$$ws = sv$$
$$wt = tv$$
$$at = bs$$
$$ta = sb$$

$$va = aw$$
$$vb = bw$$
$$ws = sv$$
$$wt = tv$$
$$v^2 = at - bs$$
$$w^2 = ta - sb$$

In each of the these, compute the spaces given by the two generic stabilities for the dimension vector $(1, 1)$. Are the spaces isomorphic? Are you sure?

EXERCISE 3.2.3. Consider the ring $R := \mathbb{C}[u, v, x, y]/(uv - x^2 y)$. In this example $\mathrm{End}_R(R \oplus (u, x) \oplus (u, xy))$ is a NCCR, and in fact it can be presented as

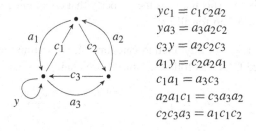

$$yc_1 = c_1 c_2 a_2$$
$$ya_3 = a_3 a_2 c_2$$
$$c_3 y = a_2 c_2 c_3$$
$$a_1 y = c_2 a_2 a_1$$
$$c_1 a_1 = a_3 c_3$$
$$a_2 a_1 c_1 = c_3 a_3 a_2$$
$$c_2 c_3 a_3 = a_1 c_1 c_2$$

(the relations can in fact be packaged as a superpotential). Consider the dimension vector $(1, 1, 1)$. For this example there are essentially six generic stability conditions. Compute each. How many nonisomorphic crepant resolutions are obtained?

EXERCISE 3.2.4. (Shows that Rep $/\!\!/$ GL might not be what you want) Consider (Q, R) given by

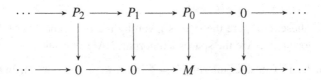

$$va = aw$$
$$vb = bw$$
$$ws = sv$$
$$wt = tv$$
$$at = bs$$
$$ta = sb$$

and set $\Lambda := kQ/R$. We know from Exercise 2.5.4 that this is a NCCR over $Z(\Lambda)$. Show that $\mathrm{Rep}(kQ, (1, 1))/\!\!/ \mathrm{GL}$ is not isomorphic to $Z(\Lambda)$.

EXERCISE 3.2.5. Prove Remark 3.1.4.

4. Into derived categories

We retain the setup that R is a commutative noetherian Gorenstein normal domain, and assume that $\Lambda = \mathrm{End}_R(M)$ is a NCCR. In this section, we begin to relate this homologically to geometric crepant resolutions $Y \to \mathrm{Spec}\, R$.

4.1. Derived categories: motivation and definition. Let $\mathcal{A}$ and $\mathcal{B}$ denote abelian categories. In our case $\mathcal{A} = \mathrm{mod}\,\Lambda$ and $\mathcal{B} = \mathrm{coh}\, Y$. It is very unlikely that $\mathcal{A} \simeq \mathcal{B}$ since usually $\mathrm{coh}\, Y$ does not have enough projectives, whereas $\mathrm{mod}\, A$ always does. But we still want to homologically relate Y and Λ. The derived category $\mathrm{D}(\mathcal{A})$ solves this issue since it carries many of the invariants that we care about, whilst at the same time allowing the flexibility of $\mathrm{D}(\mathcal{A}) \simeq \mathrm{D}(\mathcal{B})$ even when $\mathcal{A} \not\simeq \mathcal{B}$.

Now to create the derived category, we observe that we can take a projective resolution of $M \in \mathrm{mod}\,\Lambda$ and view it as a commutative diagram

$$
\begin{array}{ccccccccc}
\cdots & \longrightarrow & P_2 & \longrightarrow & P_1 & \longrightarrow & P_0 & \longrightarrow & 0 & \longrightarrow & \cdots \\
& & \downarrow & & \downarrow & & \downarrow & & \downarrow & & \\
\cdots & \longrightarrow & 0 & \longrightarrow & 0 & \longrightarrow & M & \longrightarrow & 0 & \longrightarrow & \cdots
\end{array}
$$

We write this $P_\bullet \xrightarrow{f} M$. This map has the property that cohomology $\mathrm{H}^i(f) : \mathrm{H}^i(P_\bullet) \to \mathrm{H}^i(M)$ is an isomorphism for all $i \in \mathbb{Z}$.

DEFINITION 4.1.1. For any abelian category $\mathcal{A}$, we define the category of chain complexes, denoted $\mathrm{C}(\mathcal{A})$, as follows. Objects are chain complexes, i.e.,

$$
\cdots \xrightarrow{d_{-1}} C_{-1} \xrightarrow{d_0} C_0 \xrightarrow{d_1} C_1 \xrightarrow{d_2} \cdots
$$

with each $C_i \in \mathcal{A}$, such that $d_i d_{i+1} = 0$ for all $i \in \mathbb{Z}$, and the morphisms $C_\bullet \to D_\bullet$ are collections of morphisms in $\mathcal{A}$ such that

$$
\begin{array}{ccccccc}
\cdots \longrightarrow & C_{-1} & \longrightarrow & C_0 & \longrightarrow & C_1 & \longrightarrow \cdots \\
& \downarrow & & \downarrow & & \downarrow & \\
\cdots \longrightarrow & D_{-1} & \longrightarrow & D_0 & \longrightarrow & D_1 & \longrightarrow \cdots
\end{array}
$$

commutes. A map of chain complexes $f : C_\bullet \to D_\bullet$ is called a quasi-isomorphism (qis) if cohomology $\mathrm{H}^i(f) : \mathrm{H}^i(C_\bullet) \to \mathrm{H}^i(D_\bullet)$ is an isomorphism for all $i \in \mathbb{Z}$. The derived category $\mathrm{D}(\mathcal{A})$, is defined to be $\mathrm{C}(\mathcal{A})[\{qis\}^{-1}]$, where we just formally invert all quasi-isomorphisms. The bounded derived category $\mathrm{D}^b(\mathcal{A})$ is defined to be the full subcategory of $\mathrm{D}(\mathcal{A})$ consisting of complexes isomorphic (in the derived category) to bounded complexes

$$
\cdots \to 0 \to C_i \to C_{i+1} \to \cdots \to C_j \to 0 \to \cdots .
$$

Thus in the derived category we just formally identify M and its projective resolution. Now much of what we do on the abelian category level is very formal — the building blocks of homological algebra are short exact sequences, and we have constructions like kernels and cokernels. Often for many proofs (e.g., in Section 1 and Section 2) we just need the fact that these constructions exist, rather than precise knowledge of the form they take.

When passing from abelian categories to derived categories, the building blocks are no longer short exact sequences, instead these are replaced by a weaker notion of *triangles*. As in the abelian setting, many constructions and proofs follow formally from the properties of triangles. The derived category is an example of a triangulated category, which is defined as follows.

DEFINITION 4.1.2. A triangulated category is an additive category $\mathcal{C}$ together with an additive autoequivalence $[1] : \mathcal{C} \to \mathcal{C}$ and a class of sequences

$$
X \to Y \to Z \to X[1]
$$

called triangles, satisfying the following:

T1(a). Every sequence $X' \to Y' \to Z' \to X'[1]$ isomorphic to a triangle is itself a triangle.

T1(b). For every object $X \in \mathcal{C}$, $0 \to X \xrightarrow{\mathrm{id}} X \to 0[1]$ is a triangle.

T1(c). Every map $f : X \to Y$ can completed to a triangle

$$
X \xrightarrow{f} Y \to Z \to X[1].
$$

T2 (Rotation). We have

$$
X \xrightarrow{f} Y \xrightarrow{g} Z \xrightarrow{h} X[1] \text{ is a triangle} \iff Y \xrightarrow{g} Z \xrightarrow{h} X[1] \xrightarrow{-f[1]} Y[1] \text{ is a triangle}.
$$

T3. Given a commutative diagram

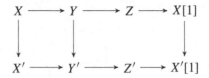

where the two rows are triangles, then there exists $Z \to Z'$ such that the whole diagram commutes.

T4 (Octahedral axiom). Given

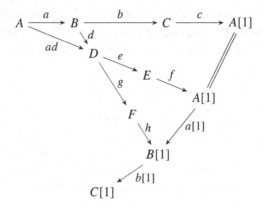

where (a, b, c), (d, g, h) and (ad, e, f) are triangles, there exists morphisms such that $C \to E \to F \to C[1]$ is a triangle, and the whole diagram commutes.

The only fact needed for now is that short exact sequences of complexes give triangles in the derived category.

4.2. Tilting. We now return to our setup. We are interested in possible equivalences between $D^b(\operatorname{coh} Y)$ and $D^b(\operatorname{mod} \Lambda)$. How to achieve this? We first note that there are two nice subcategories of $D^b(\operatorname{coh} Y)$ and $D^b(\operatorname{mod} \Lambda)$.

DEFINITION 4.2.1. We define $\mathfrak{Perf}(Y) \subseteq D^b(\operatorname{coh} Y)$ to be all those complexes that are (locally) quasiisomorphic to bounded complexes consisting of vector bundles of finite rank. We denote $K^b(\operatorname{proj} \Lambda) \subseteq D^b(\operatorname{mod} \Lambda)$ to be all those complexes isomorphic to bounded complexes of finitely generated projective Λ-modules.

From now on, to simplify matters we will always assume that our schemes are quasiprojective over a commutative noetherian ring of finite type over $\mathbb{C}$, since in our NCCR quiver GIT setup, this will always be true. We could get by with less, but the details become more technical.

Under these assumptions, $\mathfrak{Perf}(Y)$ can be described as all those complexes that are isomorphic (in the derived category) to bounded complexes consisting of vector bundles of finite rank [179, Lemma 1.6 and Remark 1.7]. Furthermore, any equivalence between $D^b(\operatorname{coh} Y)$ and $D^b(\operatorname{mod} \Lambda)$ must restrict to an equivalence between $\mathfrak{Perf}(Y)$

and $K^b(\text{proj } \Lambda)$, since both can be characterized intrinsically as the homologically finite complexes.

Now the point is that $K^b(\text{proj } \Lambda)$ has a very special object $_\Lambda\Lambda$, considered as a complex in degree zero. For $D^b(\text{coh } Y) \simeq D^b(\text{mod } \Lambda)$ we need $\mathfrak{Perf}(Y) \simeq K^b(\text{proj } \Lambda)$, so we need $\mathfrak{Perf}(Y)$ to contain an object that behaves in the same way as $_\Lambda\Lambda$ does. But what properties does $_\Lambda\Lambda$ have?

The first property is Hom-vanishing in the derived category.

FACT 4.2.2. *If M and N are Λ-modules, thought of as complexes in degree zero, we have*

$$\text{Hom}_{D^b(\text{mod } \Lambda)}(M, N[i]) \cong \text{Ext}^i_\Lambda(M, N)$$

for all $i \in \mathbb{Z}$. In particular $\text{Hom}_{D^b(\text{mod } \Lambda)}(_\Lambda\Lambda, {}_\Lambda\Lambda[i]) = 0$ *for all* $i \neq 0$.

Next, we have to develop some language to say that $K^b(\text{proj } \Lambda)$ is "built" from $_\Lambda\Lambda$.

DEFINITION 4.2.3. Let $\mathcal{C}$ be a triangulated category. A full subcategory $\mathcal{D}$ is called a triangulated subcategory if (a) $0 \in \mathcal{D}$ (b) $\mathcal{D}$ is closed under finite sums (c) $\mathcal{D}$ is closed under shifts (d) (2 out of 3 property) If $X \to Y \to Z \to X[1]$ is a triangle in $\mathcal{C}$, then if any two of $\{X, Y, Z\}$ is in $\mathcal{D}$, then so is the third. If further $\mathcal{D}$ is closed under direct summands (i.e., $X \oplus Y \in \mathcal{D}$ implies that $X, Y \in \mathcal{D}$), then we say that $\mathcal{D}$ is thick.

NOTATION 4.2.4. Let $\mathcal{C}$ be a triangulated category, $M \in \mathcal{C}$. We denote by $\text{thick}(M)$ the smallest full thick triangulated subcategory containing M.

Using this, the second property that $_\Lambda\Lambda$ possesses is *generation*.

EXAMPLE 4.2.5. Consider $_\Lambda\Lambda \in D^b(\text{mod } \Lambda)$, considered as a complex in degree zero. We claim that $\text{thick}(_\Lambda\Lambda) = K^b(\text{proj } \Lambda)$. Since $\text{thick}(_\Lambda\Lambda)$ is closed under finite sums, it contains all finitely generated free Λ-modules, and further since it is closed under summands it contains all projective Λ-modules. It is closed under shifts, so it contains $P[i]$ for all finitely generated projectives P and all $i \in \mathbb{Z}$. Now consider a 2-term complex

$$\cdots \longrightarrow 0 \longrightarrow P_1 \longrightarrow P_0 \longrightarrow 0 \longrightarrow \cdots$$

with $P_0, P_1 \in \text{proj } \Lambda$. We have a commutative diagram

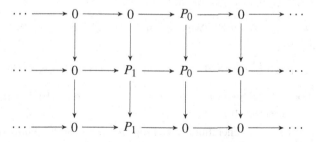

which is a short exact sequence of complexes. But short exact sequences of complexes give triangles in the derived category, so since the outer two terms belong to $\text{thick}(_\Lambda\Lambda)$, so does the middle (using the 2 out of 3 property). This shows that all 2-term complexes of finitely generated projectives belong to thick($_\Lambda \Lambda$). By induction, we have that all bounded complexes of finitely generated projectives belong to thick($_\Lambda \Lambda$), i.e., $K^b(\text{proj } \Lambda) \subseteq \text{thick}(_\Lambda \Lambda)$. But $K^b(\text{proj } \Lambda)$ is a full thick triangulated subcategory containing $_\Lambda \Lambda$, so since thick($_\Lambda \Lambda$) is the smallest such, we conclude that $K^b(\text{proj } \Lambda) = \text{thick}(_\Lambda \Lambda)$.

Thus, combining Fact 4.2.2 and Example 4.2.5, a necessary condition for $D^b(\text{coh } Y) \simeq D^b(\text{mod } \Lambda)$ is that there exists a complex $\mathcal{V} \in \mathfrak{Perf}(Y)$ for which $\text{Hom}_{D^b(\text{coh } Y)}(\mathcal{V}, \mathcal{V}[i])$ vanishes for all $i \neq 0$, such that thick($\mathcal{V}$) = $\mathfrak{Perf}(Y)$. Tilting theory tells us that these properties are in fact sufficient.

DEFINITION 4.2.6. We say that $\mathcal{V} \in \mathfrak{Perf}(Y)$ is a tilting complex if

$$\text{Hom}_{D^b(\text{coh } Y)}(\mathcal{V}, \mathcal{V}[i]) = 0$$

for all $i \neq 0$, and further thick($\mathcal{V}$) = $\mathfrak{Perf}(Y)$. If further $\mathcal{V}$ is a vector bundle (not just a complex), then we say that $\mathcal{V}$ is a tilting bundle.

The following is stated for the case when $\mathcal{V}$ is a vector bundle (not a complex), since in these notes this is all that is needed.

THEOREM 4.2.7. *With our running hypothesis on Y (namely it is quasiprojective over a commutative noetherian ring of finite type over $\mathbb{C}$), assume that $\mathcal{V}$ is a tilting bundle.*
(1) $\mathbf{R}\text{Hom}_Y(\mathcal{V}, -)$ *induces an equivalence between $D^b(\text{coh } Y)$ and $D^b(\text{mod End}_Y(\mathcal{V}))$.*
(2) *Y is smooth if and only if* gl.dim $\text{End}_Y(\mathcal{V}) < \infty$.

In practice, to check the Ext vanishing in the definition of a tilting bundle can be quite mechanical, whereas establishing generation is more of an art. Below, we will often use the following trick to simplify calculations.

PROPOSITION 4.2.8 (Neeman's generation trick). *Say Y has an ample line bundle $\mathcal{L}$. Pick $\mathcal{V} \in \mathfrak{Perf}(Y)$. If $(\mathcal{L}^{-1})^{\otimes n} \in \text{thick}(\mathcal{V})$ for all $n \geq 1$, then* thick($\mathcal{V}$) = $\mathfrak{Perf}(Y)$.

4.3. Tilting examples. We now illustrate tilting in the three examples from the previous section on quiver GIT, namely $\mathbb{P}^1$, the blowup of $\mathbb{A}^2$ at the origin, then our running $\mathbb{Z}_3$ example. This will explain where the algebras used in Section 3 arose.

EXAMPLE 4.3.1. Consider $\mathbb{P}^1$. We claim that $\mathcal{V} := \mathcal{O}_{\mathbb{P}^1} \oplus \mathcal{O}_{\mathbb{P}^1}(1)$ is a tilting bundle. First, we have

$$\text{Ext}^i_{\mathbb{P}^1}(\mathcal{V}, \mathcal{V}) \cong H^i(\mathcal{V}^{-1} \otimes \mathcal{V}) = H^i(\mathcal{O}_{\mathbb{P}^1}) \oplus H^i(\mathcal{O}_{\mathbb{P}^1}(1)) \oplus H^i(\mathcal{O}_{\mathbb{P}^1}(-1)) \oplus H^i(\mathcal{O}_{\mathbb{P}^1}),$$

which is zero for all $i > 0$ by a Čech cohomology calculation in Hartshorne [128, III.5]. Thus $\text{Ext}^i_{\mathbb{P}^1}(\mathcal{V}, \mathcal{V}) = 0$ for all $i > 0$.

Now we use Neeman's generation trick (Proposition 4.2.8). We know that $\mathcal{O}_{\mathbb{P}^1}(1)$ is an ample line bundle on $\mathbb{P}^1$. Further, we have the Euler short exact sequence

$$0 \to \mathcal{O}_{\mathbb{P}^1}(-1) \to \mathcal{O}_{\mathbb{P}^1}^{\oplus 2} \to \mathcal{O}_{\mathbb{P}^1}(1) \to 0, \tag{4.A}$$

which gives a triangle in the derived category. Since the rightmost two terms both belong to thick($\mathcal{V}$), by the 2 out of 3 property we deduce that $\mathcal{O}_{\mathbb{P}^1}(-1) \in$ thick($\mathcal{V}$). Now twisting (4.A) we obtain another short exact sequence

$$0 \to \mathcal{O}_{\mathbb{P}^1}(-2) \to \mathcal{O}_{\mathbb{P}^1}(-1)^{\oplus 2} \to \mathcal{O}_{\mathbb{P}^1} \to 0. \tag{4.B}$$

Again this gives a triangle in the derived category, and since the rightmost two terms both belong to thick($\mathcal{V}$), by the 2 out of 3 property we deduce that $\mathcal{O}_{\mathbb{P}^1}(-2) \in$ thick($\mathcal{V}$). Continuing like this we deduce that $\mathcal{O}_{\mathbb{P}^1}(-n) \in$ thick($\mathcal{V}$) for all $n \geq 1$, and so thick($\mathcal{V}$) = $\mathfrak{Perf}(\mathbb{P}^1)$ by Proposition 4.2.8.

Thus $\mathcal{V}$ is a tilting bundle, so by Theorem 4.2.7 we deduce that $D^b(\text{coh}\,\mathbb{P}^1) \simeq D^b(\text{mod}\,\text{End}_{\mathbb{P}^1}(\mathcal{V}))$. We now identify the endomorphism ring with an algebra with which we are more familiar. We have

$$\text{End}_{\mathbb{P}^1}(\mathcal{V}) = \text{End}_{\mathbb{P}^1}(\mathcal{O} \oplus \mathcal{O}(1)) \cong \begin{pmatrix} \text{Hom}_{\mathbb{P}^1}(\mathcal{O}, \mathcal{O}) & \text{Hom}_{\mathbb{P}^1}(\mathcal{O}, \mathcal{O}(1)) \\ \text{Hom}_{\mathbb{P}^1}(\mathcal{O}(1), \mathcal{O}) & \text{Hom}_{\mathbb{P}^1}(\mathcal{O}(1), \mathcal{O}(1)) \end{pmatrix},$$

which, again by the Čech cohomology calculation in Hartshorne, is isomorphic to

$$\begin{pmatrix} H^0(\mathcal{O}) & H^0(\mathcal{O}(1)) \\ H^0(\mathcal{O}(-1)) & H^0(\mathcal{O}) \end{pmatrix} \cong \begin{pmatrix} \mathbb{C} & \mathbb{C}^2 \\ 0 & \mathbb{C} \end{pmatrix} \cong \bullet \rightrightarrows \bullet$$

Thus $D^b(\text{coh}\,\mathbb{P}^1) \simeq D^b(\text{mod}\ \bullet \rightrightarrows \bullet)$.

EXAMPLE 4.3.2. Consider now the blowup of $\mathbb{A}^2$ at the origin.

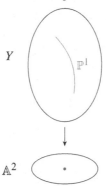

We constructed Y explicitly in Example 3.1.2, where we remarked that $Y = \mathcal{O}_{\mathbb{P}^1}(-1)$. Being the total space of a line bundle over $\mathbb{P}^1$, in this example we have extra information in the form of the diagram

$$Y = \mathcal{O}_{\mathbb{P}^1}(-1) \xrightarrow{\pi} \mathbb{P}^1$$
$$f \downarrow \qquad\qquad$$
$$\mathbb{A}^2 \qquad\qquad$$

Let $\mathcal{V} := \mathcal{O}_{\mathbb{P}^1} \oplus \mathcal{O}_{\mathbb{P}^1}(1)$, as in the previous example, and set $\mathcal{W} := \pi^*(\mathcal{V})$. We claim that $\mathcal{W}$ is a tilting bundle on Y. To visualise this (we will need to in the next example), we denote $\mathcal{W} = \mathcal{O}_Y \oplus \mathcal{L}_1$ and draw

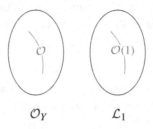

$$\mathcal{O}_Y \qquad\qquad \mathcal{L}_1$$

Since $\pi_*(\mathcal{O}_Y) \cong \bigoplus_{p \leq 0} \mathcal{O}_{\mathbb{P}^1}(-1)^{\otimes p} = \bigoplus_{k \geq 0} \mathcal{O}_{\mathbb{P}^1}(k)$, the projection formula yields $\pi_*\pi^*(\mathcal{V}) \cong \bigoplus_{k \geq 0} \mathcal{V} \otimes_{\mathbb{P}^1} \mathcal{O}_{\mathbb{P}^1}(k)$. Thus the Ext vanishing condition on $\mathcal{W}$ follows from properties of adjoint functors, namely

$$\begin{aligned}
\mathrm{Ext}^i_Y(\mathcal{W}, \mathcal{W}) &= \mathrm{Ext}^i_Y(\pi^*(\mathcal{V}), \pi^*(\mathcal{V})) \\
&\cong \mathrm{Ext}^i_{\mathbb{P}^1}(\mathcal{V}, \pi_*\pi^*(\mathcal{V})) \\
&\cong \bigoplus_{k \geq 0} \mathrm{Ext}^i_{\mathbb{P}^1}(\mathcal{V}, \mathcal{V} \otimes_{\mathbb{P}^1} \mathcal{O}_{\mathbb{P}^1}(k)),
\end{aligned}$$

which is zero for all $i > 0$ again by the Čech cohomology calculation in Hartshorne [128, III.5]. Generation also follows immediately from our previous example, since $\mathcal{L}_1$ is ample and π^* is exact on short exact sequences of vector bundles. Thus $\mathcal{W} = \mathcal{O}_Y \oplus \mathcal{L}_1$ is a tilting bundle, so by Theorem 4.2.7 Y and $\mathrm{End}_Y(\mathcal{O}_Y \oplus \mathcal{L}_1)$ are derived equivalent. It is an instructive exercise (see Exercise 4.9.5) to show that

$$\mathrm{End}_Y(\mathcal{O}_Y \oplus \mathcal{L}_1) \cong \bullet \underset{t}{\overset{\overset{a}{\underset{b}{\rightrightarrows}}}{\leftarrow}} \bullet \qquad atb = bta.$$

EXAMPLE 4.3.3. Consider the running example $R := \mathbb{C}[a, b, c]/(ab - c^3)$. We have, by Example 3.1.3, the picture

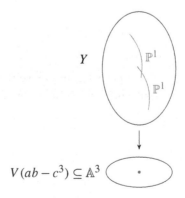

$$V(ab - c^3) \subseteq \mathbb{A}^3$$

where the dot downstairs represents the singular point. We would like Y to be derived equivalent to our original algebra $\mathrm{End}_R(R \oplus (a, c) \oplus (a, c^2))$, so we need to find a tilting bundle on Y with three summands. Which to choose? We have to construct

bundles, and the natural candidates are

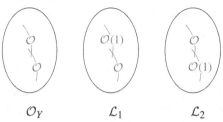

$$\mathcal{O}_Y \qquad \mathcal{L}_1 \qquad \mathcal{L}_2$$

It turns out, but more technology is needed to prove it, that $\mathcal{W} := \mathcal{O}_Y \oplus \mathcal{L}_1 \oplus \mathcal{L}_2$ is a tilting bundle on Y with $\mathrm{End}_Y(\mathcal{O}_Y \oplus \mathcal{L}_1 \oplus \mathcal{L}_2) \cong \mathrm{End}_R(R \oplus (a, c) \oplus (a, c^2))$.

4.4. Derived categories and crepant resolutions. In this section we explain the geometric origin of the condition $\mathrm{End}_R(M) \in \mathrm{CM}\, R$, and also why we only consider rings of the form $\mathrm{End}_R(M)$, answering Q4 from Section 1.

The key theorem is the following.

THEOREM 4.4.1. *Suppose that* $f : Y \to \mathrm{Spec}\, R$ *is a projective birational map, where* Y *and* R *are both normal Gorenstein of dimension* d. *If* Y *is derived equivalent to a ring* Λ, *then the following are equivalent.*

(1) f *is crepant (i.e.,* $f^*\omega_R = \omega_Y$).

(2) $\Lambda \in \mathrm{CM}\, R$.

In this case $\Lambda \cong \mathrm{End}_R(M)$ *for some* $M \in \mathrm{ref}\, R$.

It follows that the only possible algebras derived equivalent to crepant partial resolutions of d-dimensional Gorenstein singularities have the form $\mathrm{End}_R(M)$, and ultimately this is the reason why we restrict to studying rings of this form. We outline the main ingredients of the proof below.

REMARK 4.4.2. By Theorem 4.4.1 the condition $\mathrm{End}_R(M) \in \mathrm{CM}\, R$ corresponds precisely to the geometric notion of crepancy, provided that we can actually find some Y which is derived equivalent to $\mathrm{End}_R(M)$. This explains the first condition in the definition of a NCCR.

REMARK 4.4.3. Suppose that $\mathrm{End}_R(M)$ is a NCCR. We remark that when $d = \dim R \geq 4$ there may be no scheme Y projective birational over $\mathrm{Spec}\, R$ for which Y is derived equivalent to $\mathrm{End}_R(M)$. This is because NCCRs can exist even when commutative crepant resolutions do not. A concrete example is invariants by the group $\frac{1}{2}(1, 1, 1, 1)$. Also, when $d \geq 4$ it is possible that a crepant resolution Y exists but there is no algebra that is derived equivalent to Y. Thus the correspondence between NCCRs and crepant resolutions breaks down completely, even when $d = 4$. Thus Theorem 4.4.1 is usually only used for analogies (or very specific situations) in high dimension.

Theorem 4.4.1 has the following important corollary.

COROLLARY 4.4.4. *Suppose that* $Y \to \mathrm{Spec}\, R$ *is a projective birational map between* d-*dimensional Gorenstein normal varieties. If* Y *is derived equivalent to a ring* Λ, *then the following are equivalent.*

(1) Λ *is an NCCR.*

(2) *Y is a crepant resolution of* Spec *R.*

PROOF. $(1) \Rightarrow (2)$ If Λ is an NCCR then Λ has finite global dimension. This means that $\mathrm{K}^{\mathrm{b}}(\mathrm{proj}\,\Lambda) = \mathrm{D}^{\mathrm{b}}(\mathrm{mod}\,\Lambda)$ and so via the derived equivalence $\mathfrak{Perf}(Y) \simeq \mathrm{D}^{\mathrm{b}}(\mathrm{coh}\,Y)$. Hence Y is smooth. Further f is crepant by Theorem 4.4.1.

$(2) \Rightarrow (1)$ Since f is crepant, $\Lambda \cong \mathrm{End}_R(M) \in \mathrm{CM}\,R$ by Theorem 4.4.1. Since Y is smooth $\mathfrak{Perf}(Y) \simeq \mathrm{D}^{\mathrm{b}}(\mathrm{coh}\,Y)$ induces $\mathrm{K}^{\mathrm{b}}(\mathrm{proj}\,\Lambda) = \mathrm{D}^{\mathrm{b}}(\mathrm{mod}\,\Lambda)$ and so every $N \in \mathrm{mod}\,\Lambda$ has $\mathrm{proj.dim}_{\Lambda}\,N < \infty$. By Auslander–Buchsbaum, necessarily $\mathrm{proj.dim}_{\Lambda}\,N \le \dim R$ and so $\Lambda \cong \mathrm{End}_R(M)$ is a NCCR. $\qquad\square$

EXAMPLE 4.4.5. To make this a little more concrete, consider our running example $R := \mathbb{C}[a, b, c]/(ab - c^3)$. We know from Section 1 and Section 2 that $\Lambda := \mathrm{End}_R(R \oplus (a, c) \oplus (a, c^3)) \in \mathrm{CM}\,R$ is a NCCR, and we constructed, using quiver GIT (3.1.3), a space Y which we then showed (4.3.2) is derived equivalent to Λ. Since $\Lambda \in \mathrm{CM}\,R$, we can deduce from Theorem 4.4.1 that Y is a crepant resolution. Alternatively, knowing a bit of geometry, the two exceptional curves calculated in Example 3.1.3 are both (-2)-curves, so it follows that $Y \to \mathrm{Spec}\,R$ is crepant. Hence we could alternatively use Corollary 4.4.4 to give another proof that Λ is a NCCR.

The main ingredient in the proof of Theorem 4.4.1 is Grothendieck duality and relative Serre functors, which we now review.

4.5. Relative Serre functors. The following is based on [112, Definition 7.2.6].

DEFINITION 4.5.1. Suppose that $Z \to \mathrm{Spec}\,T$ is a morphism where T is a CM ring with a canonical module C_T. We say that a functor $\mathbb{S}\colon \mathfrak{Perf}(Z) \to \mathfrak{Perf}(Z)$ is a *Serre functor relative to* C_T if there are functorial isomorphisms

$$\mathbf{RHom}_T(\mathbf{RHom}_Z(\mathcal{F}, \mathcal{G}), C_T) \cong \mathbf{RHom}_Z(\mathcal{G}, \mathbb{S}(\mathcal{F}))$$

in $\mathrm{D}(\mathrm{Mod}\,T)$ for all $\mathcal{F} \in \mathfrak{Perf}(Z)$, $\mathcal{G} \in \mathrm{D}^{\mathrm{b}}(\mathrm{coh}\,Z)$. If Λ is a module-finite T-algebra, we define a Serre functor $\mathbb{S}\colon \mathrm{K}^{\mathrm{b}}(\mathrm{proj}\,\Lambda) \to \mathrm{K}^{\mathrm{b}}(\mathrm{proj}\,\Lambda)$ relative to C_T in a similar way.

The ability to consider different canonicals in the above definition is convenient when comparing the geometry to the algebra. For example, when T is Gorenstein, there is a geometrically–defined canonical module ω_T, but T itself is also a canonical module. It turns out that from the NCCR perspective T is the most natural (see Theorem 4.6.3(1)), whereas the notion of crepancy is defined with respect to ω_T.

LEMMA 4.5.2. *Suppose that* $f\colon Z \to \mathrm{Spec}\,T$ *is a projective morphism, where Z and T are both Gorenstein varieties.*

(1) $f^{!}(\mathcal{O}_T) \cong \mathcal{L}[\dim Z - \dim T]$, *where* $\mathcal{L}$ *is some line bundle on Z.*

(2) $\mathbb{S}_Z := - \otimes_Z f^{!}\mathcal{O}_T\colon \mathfrak{Perf}(Z) \to \mathfrak{Perf}(Z)$ *is a Serre functor relative to the canonical* T.

PROOF. (1) Since T is Gorenstein, ω_T is a line bundle and thus is a compact object in $\mathrm{D}(\mathrm{Mod}\,T)$. Hence by [178, pp. 227–228] we have $f^{!}\omega_T = \mathbf{L}f^{*}\omega_T \otimes_Z^{\mathbf{L}} f^{!}\mathcal{O}_T =$

$f^*\omega_T \otimes_Z f^!\mathcal{O}_T$, and so

$$\omega_Z \cong f^!\omega_T[-\dim Z + \dim T] \cong f^*\omega_T \otimes_Z f^!\mathcal{O}_T[-\dim Z + \dim T].$$

Since both ω_Y and $f^*\omega_T$ are line bundles,

$$f^!\mathcal{O}_T = (f^*\omega_T)^{-1} \otimes_Z \omega_Z[\dim Z - \dim T].$$

(2) Since $f^!\mathcal{O}_T$ is a shift of a bundle by (1), it follows that tensoring gives a functor $-\otimes_Z f^!\mathcal{O}_T \colon \mathfrak{Perf}(Z) \to \mathfrak{Perf}(Z)$. The result then follows since

$$\mathbf{RHom}_Z(\mathcal{G}, \mathcal{F} \otimes_Z f^!\mathcal{O}_T) \cong \mathbf{RHom}_Z(\mathbf{R}\mathcal{H}om_Z(\mathcal{F}, \mathcal{G}), f^!\mathcal{O}_T)$$
$$\cong \mathbf{RHom}_T(\mathbf{RHom}_Z(\mathcal{F}, \mathcal{G}), \mathcal{O}_T)$$

for all $\mathcal{F} \in \mathfrak{Perf}(Z)$, $\mathcal{G} \in D^b(\text{coh } Z)$, where the last isomorphism is sheafified Grothendieck duality. $\qquad\square$

We consider the Serre functor for NCCRs in the next section.

4.6. Calabi–Yau categories. Related to Serre functors are CY categories and algebras. We retain our setup, namely R denotes an (equicodimensional) Gorenstein normal domain of dimension d. To keep the technicalities to a minimum, in this section we will assume that R is of finite type over an algebraically closed field k, but we could get by with much less. In this section we show that NCCRs are d-CY.

DEFINITION 4.6.1. Suppose that $\mathcal{C}$ is a triangulated category in which the Hom spaces are all k-vector spaces. We say that $\mathcal{C}$ is d-CY if there exists a functorial isomorphism

$$\text{Hom}_{\mathcal{C}}(x, y[d]) \simeq \text{Hom}_{\mathcal{C}}(y, x)^* \tag{4.C}$$

for all $x, y \in \mathcal{C}$, where $(-)^*$ denotes the k-dual.

For a k-algebra Λ, we naively ask whether $D^b(\text{mod } \Lambda)$ is d-CY. It is (almost) never in the strict sense above, since then

$$\Lambda^* \cong \text{Hom}_\Lambda(\Lambda, \Lambda)^* \cong \text{Hom}_{D^b}(\Lambda, \Lambda)^* \cong \text{Hom}_{D^b}(\Lambda, \Lambda[d]) \cong \text{Ext}^d_\Lambda(\Lambda, \Lambda) = 0,$$

whenever $d \neq 0$. Also, note that for k-duality to work well requires the Hom spaces to be finite-dimensional. Hence, for an algebra to be CY, we must ask for (4.C) to be true for only certain classes of objects x and y. This is done as follows.

DEFINITION 4.6.2. Let Λ be a module finite R-algebra, then for $d \in \mathbb{Z}$ we call Λ d-Calabi–Yau (d-CY) if there is a functorial isomorphism

$$\text{Hom}_{D^b(\text{mod } \Lambda)}(x, y[d]) \cong \text{Hom}_{D^b(\text{mod } \Lambda)}(y, x)^* \tag{4.D}$$

for all $x \in D^b(\text{fl } \Lambda)$, $y \in D^b(\text{mod } \Lambda)$, where $D^b(\text{fl } \Lambda)$ denotes all complexes x for which $\dim_k \bigoplus_{i \in \mathbb{Z}} H^i(x) < \infty$. Similarly we call Λ singular d-Calabi–Yau (d-sCY) if (4.D) holds for all $x \in D^b(\text{fl } \Lambda)$ and $y \in K^b(\text{proj } \Lambda)$.

Since R is Gorenstein, and Λ is a module-finite R-algebra, there is a functor

$$\mathbb{S}_\Lambda := \mathbf{R}\mathrm{Hom}_R(\Lambda, R) \otimes_\Lambda^{\mathbf{L}} -: \ \mathrm{D}^-(\mathrm{mod}\,\Lambda) \to \mathrm{D}^-(\mathrm{mod}\,\Lambda).$$

By [139, Proposition 3.5(2),(3)], there exists a functorial isomorphism

$$\mathbf{R}\mathrm{Hom}_\Lambda(a, \mathbb{S}(b)) \cong \mathbf{R}\mathrm{Hom}_R(\mathbf{R}\mathrm{Hom}_\Lambda(b, a), R) \tag{4.E}$$

in $\mathrm{D}(R)$ for all $a \in \mathrm{D}^b(\mathrm{mod}\,\Lambda)$ and all $b \in \mathrm{K}^b(\mathrm{proj}\,\Lambda)$. This is not quite a Serre functor, since we don't yet know whether $\mathbb{S}_\Lambda$ preserves $\mathrm{K}^b(\mathrm{proj}\,\Lambda)$.

THEOREM 4.6.3 (Iyama–Reiten). *Let R an be equicodimensional Gorenstein normal domain over an algebraically closed field k, and let Λ be an NCCR.*

(1) $\mathbb{S}_\Lambda = \mathrm{Id}$, *and so* Id *is a Serre functor on Λ relative to the canonical R.*

(2) Λ *is d-CY, that is*

$$\mathrm{Hom}_{\mathrm{D}^b(\mathrm{mod}\,\Lambda)}(x, y[d]) \cong \mathrm{Hom}_{\mathrm{D}^b(\mathrm{mod}\,\Lambda)}(y, x)^*$$

for all $x \in \mathrm{D}^b(\mathrm{fl}\,\Lambda)$ *and* $y \in \mathrm{K}^b(\mathrm{proj}\,\Lambda) = \mathrm{D}^b(\mathrm{mod}\,\Lambda)$.

PROOF. (1) By definition of $\mathbb{S}_\Lambda$, we just need to establish that $\mathbf{R}\mathrm{Hom}_R(\Lambda, R) \cong \Lambda$ as Λ-Λ bimodules. But by the definition of an NCCR, $\Lambda \in \mathrm{CM}\,R$ and so $\mathrm{Ext}_R^i(\Lambda, R) = 0$ for all $i > 0$. This shows that $\mathbf{R}\mathrm{Hom}_R(\Lambda, R) \cong \mathrm{Hom}_R(\Lambda, R)$. The fact that $\mathrm{Hom}_R(\Lambda, R) \cong \Lambda$ as Λ-Λ bimodules follows from the fact that symmetric R-algebras are closed under reflexive equivalences (this uses the fact that R is normal — see Fact 2.3.2(1)). We conclude that $\mathbb{S}_\Lambda = \mathrm{Id}$, and this clearly preserves $\mathrm{K}^b(\mathrm{proj}\,\Lambda)$. Thus (4.E) shows that Id is a Serre functor relative to the canonical R.

(2) This is in fact a consequence of (1), using local duality. See [139, Lemma 3.6 and Theorem 3.7]. □

4.7. Singular derived categories. From before recall that $\mathrm{K}^b(\mathrm{proj}\,R) \subseteq \mathrm{D}^b(\mathrm{mod}\,R)$.

DEFINITION 4.7.1. We define the singular derived category (sometimes called the triangulated category of singularities, or the singularity category) $\mathrm{D}_{\mathrm{sg}}(R)$ to be the quotient category $\mathrm{D}^b(\mathrm{mod}\,R)/\,\mathrm{K}^b(\mathrm{proj}\,R)$.

Since $\mathrm{K}^b(\mathrm{proj}\,R)$ is a full thick triangulated subcategory, for general abstract reasons the quotient $\mathrm{D}_{\mathrm{sg}}(R)$ is also triangulated. Also, being a localization, morphisms in $\mathrm{D}_{\mathrm{sg}}(R)$ are equivalence classes of morphisms in $\mathrm{D}^b(\mathrm{mod}\,R)$. But the derived category is itself a localization, so the morphisms in $\mathrm{D}_{\mathrm{sg}}(R)$ are equivalence classes of equivalence classes. From this perspective, there is no reason to expect that this category should behave well.

If R is a Gorenstein ring, it is thus remarkable that $\mathrm{D}_{\mathrm{sg}}(R)$ can be described easily by using CM R-modules. There is a natural functor

$$\mathrm{CM}\,R \longrightarrow \mathrm{D}^b(\mathrm{mod}\,R) \longrightarrow \mathrm{D}_{\mathrm{sg}}(R) = \mathrm{D}^b(\mathrm{mod}\,R)/\,\mathrm{K}^b(\mathrm{proj}\,R).$$

This is not an equivalence, since projective modules P are CM, and these get sent to complexes which by definition are zero in $\mathrm{D}_{\mathrm{sg}}(R)$. Hence we must "remove" the projectives in CM R.

How to do this is standard, and is known as taking the stable category $\underline{\text{CM}}R$. By definition the objects in $\underline{\text{CM}}R$ are just the same objects as in CM R, but morphism spaces are defined as $\underline{\text{Hom}}_R(X, Y) := \text{Hom}_R(X, Y)/\mathcal{P}(X, Y)$ where $\mathcal{P}(X, Y)$ is the subspace of morphisms factoring through proj R. If $P \in \text{proj } R$ then id_P clearly factors through a projective and so $id_P = 0_P$ in $\underline{\text{CM}}R$. This shows that $P \simeq 0$ in $\underline{\text{CM}}R$ for all $P \in \text{proj } R$, and consequently the above functor induces a functor

$$\underline{\text{CM}}R \xrightarrow{\quad F \quad} D_{sg}(R) = D^b(\text{mod } R)/ K^b(\text{proj } R).$$

When R is Gorenstein, the category $\underline{\text{CM}}R$ is actually triangulated, being the stable category of a Frobenius category. The shift functor $[1]$ is given by the inverse of the syzygy functor.

THEOREM 4.7.2 (Buchweitz). *Let R be a Gorenstein ring. Then the natural functor F above is a triangle equivalence, so $\underline{\text{CM}}R \simeq D_{sg}(R)$ as triangulated categories.*

This shows, at least when R is Gorenstein, why the CM modules encode much of the singular behaviour of R.

REMARK 4.7.3. When R is Gorenstein with only isolated singularities, the triangulated category $\underline{\text{CM}}R$ is in fact (dim $R - 1$)-CY. This follows from Auslander–Reiten (AR) duality in Theorem 4.8.1 below. The existence of AR duality links to NCCRs via the CY property, it can be used to prove the existence of AR sequences (which in turn answers the motivating Q1 from Section 1), and will appear again in the McKay correspondence in Section 5.

REMARK 4.7.4. If R is a hypersurface, we have already seen in examples (and it is true generally via matrix factorizations) that $\Omega^2 = \text{Id}$, where Ω is the syzygy functor introduced in Section 1. Consequently $[1]^2 = \text{Id}$, so $\underline{\text{CM}}R$ can be 2-CY, 4-CY, etc. This shows that the precise value of the CY property is not unique.

4.8. Auslander–Reiten duality.

THEOREM 4.8.1 (AR duality). *Let R be a d-dimensional equicodimensional Gorenstein ring, with only isolated singularities. Then there exists a functorial isomorphism*

$$\text{Hom}_{\underline{\text{CM}}R}(X, Y) \cong D(\text{Hom}_{\underline{\text{CM}}R}(Y, X[d-1]))$$

for all $X, Y \in$ CM R.

The proof is actually quite straightforward, using only fairly standard homological constructions, given the following two commutative algebra facts. For a finitely generated R-module M, we denote $E_R(M)$ to be the injective hull of M.

FACTS 4.8.2. Let R be a d-dimensional equi-codimensional Gorenstein ring.

(1) The minimal R-injective resolution of R has the form

$$0 \to R \to I_0 := \bigoplus_{\mathfrak{p}:\text{ht } \mathfrak{p}=0} E(R/\mathfrak{p}) \to \cdots \to I_d := \bigoplus_{\mathfrak{p}:\text{ht } \mathfrak{p}=d} E(R/\mathfrak{p}) \to 0. \qquad (4.\text{F})$$

In particular the Matlis dual is $D = \text{Hom}_R(-, I_d)$.

(2) If $W \in \text{mod } R$ with $\dim W = 0$, then $\text{Hom}_R(W, I_i) = 0$ for all $0 \leq i \leq d - 1$.

PROOF. (1) is one of Bass's original equivalent characterizations of Gorenstein rings [23]. For the more general CM version, see [54, Proposition 3.2.9, Theorem 3.3.10(b)]. (2) follows from (1), together with the knowledge that (i) Ass $E_R(R/\mathfrak{p}) = \{\mathfrak{p}\}$, and (ii) if $X \in \text{mod } R$ and $Y \in \text{Mod } R$ satisfies Supp $X \cap \text{Ass } Y = \varnothing$, then $\text{Hom}_R(X, Y) = 0$. $\square$

Set $(-)^* := \text{Hom}_R(-, R)$. For any $X \in \text{mod } R$, consider the start of a projective resolution

$$P_1 \xrightarrow{f} P_0 \to X \to 0.$$

We define $\text{Tr } X \in \text{mod } R$ to be the cokernel of f^*, that is

$$0 \to X^* \to P_0^* \xrightarrow{f^*} P_1^* \to \text{Tr } X \to 0.$$

This gives a duality

$$\text{Tr}: \underline{\text{mod}}\, R \xrightarrow{\sim} \underline{\text{mod}}\, R$$

called the Auslander–Bridger transpose. We denote $\Omega: \underline{\text{mod}}\, R \to \underline{\text{mod}}\, R$ to be the syzygy functor. Combining these we define the Auslander–Reiten translation τ to be

$$\tau(-) := \text{Hom}_R(\Omega^d \text{Tr}(-), R): \underline{\text{CM}}R \to \underline{\text{CM}}R.$$

We let $D := \text{Hom}(-, I_d)$ denote Matlis duality.

LEMMA 4.8.3. *Suppose that R is Gorenstein with $d := \dim R$. Then $\tau \cong \Omega^{2-d}$.*

PROOF. We have $\Omega^2 \text{Tr}(-) \cong \text{Hom}_R(-, R)$. Thus

$$\begin{aligned}
\tau = \text{Hom}_R(\Omega^d \text{Tr}(-), R) &\cong \text{Hom}_R(\Omega^{d-2} \text{Hom}_R(-, R), R) \\
&\cong \Omega^{2-d} \text{Hom}_R(\text{Hom}_R(-, R), R) \\
&\cong \Omega^{2-d}.
\end{aligned}$$

$\square$

PROOF OF THEOREM 4.8.1. Set $T := \text{Tr } X$. Since $Y \in \text{CM } R$, $\text{Ext}^i_R(Y, R) = 0$ for all $i > 0$ and so applying $\text{Hom}_R(Y, -)$ to (4.F) gives an exact sequence

$$0 \to {}_R(Y, R) \to {}_R(Y, I_0) \to {}_R(Y, I_1) \to \cdots \to {}_R(Y, I_{d-1}) \to {}_R(Y, I_d) \to 0$$

of R-modules, which we split into short exact sequences as

Applying $\text{Hom}_R(T, -)$ gives exact sequences

$$\text{Ext}^1_R(T, {}_R(Y, I_{d-1})) \longrightarrow \text{Ext}^1_R(T, {}_R(Y, I_d)) \longrightarrow \text{Ext}^2_R(T, C_{d-1}) \longrightarrow \text{Ext}^2_R(T, {}_R(Y, I_{d-1}))$$
$$\text{Ext}^2_R(T, {}_R(Y, I_{d-2})) \longrightarrow \text{Ext}^2_R(T, C_{d-1}) \longrightarrow \text{Ext}^3_R(T, C_{d-2}) \longrightarrow \text{Ext}^3_R(T, {}_R(Y, I_{d-2}))$$
$$\vdots$$
$$\text{Ext}^{d-1}_R(T, {}_R(Y, I_1)) \longrightarrow \text{Ext}^{d-1}_R(T, C_2) \longrightarrow \text{Ext}^d_R(T, C_1) \longrightarrow \text{Ext}^d_R(T, {}_R(Y, I_1))$$
$$\text{Ext}^d_R(T, {}_R(Y, I_0)) \longrightarrow \text{Ext}^d_R(T, C_1) \longrightarrow \text{Ext}^{d+1}_R(T, {}_R(Y, R)) \longrightarrow \text{Ext}^{d+1}_R(T, {}_R(Y, I_0))$$

But whenever I is an injective R-module, we have a functorial isomorphism

$$\text{Ext}^j_R(A, {}_R(B, I)) \cong \text{Hom}_R(\text{Tor}^R_j(A, B), I)$$

and so we may rewrite the above as

$$\boxed{{}_R(\text{Tor}^R_1(T, Y), I_{d-1})} \longrightarrow {}_R(\text{Tor}^R_1(T, Y), I_d) \longrightarrow \text{Ext}^2_R(T, C_{d-1}) \longrightarrow \boxed{{}_R(\text{Tor}^R_2(T, Y), I_{d-1})}$$
$$\boxed{{}_R(\text{Tor}^R_2(T, Y), I_{d-2})} \longrightarrow \text{Ext}^2_R(T, C_{d-1}) \longrightarrow \text{Ext}^3_R(T, C_{d-2}) \longrightarrow \boxed{{}_R(\text{Tor}^R_3(T, Y), I_{d-2})}$$
$$\vdots$$
$$\boxed{{}_R(\text{Tor}^R_{d-1}(T, Y), I_1)} \longrightarrow \text{Ext}^{d-1}_R(T, C_2) \longrightarrow \text{Ext}^d_R(T, C_1) \longrightarrow \boxed{{}_R(\text{Tor}^R_d(T, Y), I_1)}$$
$$\boxed{{}_R(\text{Tor}^R_d(T, Y), I_0)} \longrightarrow \text{Ext}^d_R(T, C_1) \longrightarrow \text{Ext}^{d+1}_R(T, {}_R(Y, R)) \longrightarrow \boxed{{}_R(\text{Tor}^R_{d+1}(T, Y), I_0)}$$

Since R is an isolated singularity, and CM modules are free on regular local rings, $X_{\mathfrak{p}} \in \text{proj}\, R_{\mathfrak{p}}$ and so $T_{\mathfrak{p}} \in \text{proj}\, R_{\mathfrak{p}}$ for all non-maximals primes $\mathfrak{p}$. It follows that $\text{Tor}^R_i(T, Y)$ all have finite length, so by Fact 4.8.2(2) all the terms in the dotted boxes are zero. Thus the above reduces to

$$D(\text{Tor}^R_1(T, Y)) \cong \text{Ext}^2_R(T, C_{d-1}) \cong \cdots \cong \text{Ext}^{d+1}_R(T, {}_R(Y, R)). \tag{4.G}$$

But now

$$\text{Ext}^{d+1}_R(T, {}_R(Y, R)) \cong \text{Ext}^1_R(\Omega^d T, {}_R(Y, R)) \cong \text{Ext}^1_R(Y, {}_R(\Omega^d T, R)) = \text{Ext}^1_R(Y, \tau X), \tag{4.H}$$

and so combining (4.G) and (4.H) gives

$$D(\text{Tor}^R_1(T, Y)) \cong \text{Ext}^1_R(Y, \tau X).$$

Thus the standard functorial isomorphism $\text{Tor}^R_1(\text{Tr}\, X, Y) \cong \text{Hom}_{\underline{\text{CM}}R}(X, Y)$ (see e.g., [247]) yields

$$D(\text{Hom}_{\underline{\text{CM}}R}(X, Y)) \cong \text{Ext}^1_R(Y, \tau X) \cong \text{Hom}_{\underline{\text{CM}}R}(\Omega Y, \tau X) = \text{Hom}_{\underline{\text{CM}}R}(Y[-1], \tau X).$$

But now by Lemma 4.8.3 we have $\tau X = \Omega^{2-d} X = X[d-2]$, so

$$D(\text{Hom}_{\underline{\text{CM}}R}(X, Y)) \cong \text{Hom}_{\underline{\text{CM}}R}(Y[-1], X[d-2]) = \text{Hom}_{\underline{\text{CM}}R}(Y, X[d-1]),$$

as required. □

Credits: The derived category section is a very brief summary of a course given by Jeremy Rickard in Bristol in 2005. Further information on derived categories can now be found in many places, for example the notes of Keller [152] or Milicic [174]. Perfect complexes originate from SGA6, but came to prominence first through the work of

Thomason–Trobaugh [225], then by viewing them as compact objects via Neeman [178]. Many of the technical results presented here can be found in Orlov [179, 180] and Rouquier [195].

Tilting modules existed in the 1980s, but tilting complexes first appeared in Rickard's Morita theory for derived categories [185]. Tilting bundles at first were required to have endomorphism rings with finite global dimension but this is not necessary; the precise statement of Theorem 4.2.7 is taken from Hille–Van ben Bergh [131], based on the tricks and ideas of Neeman [178].

Example 4.3.1 is originally due to Beilinson [24], who proved it by resolving the diagonal. The example of the blowup of the origin and the $\mathbb{Z}_3$ singularity both follow from Artin–Verdier [12], Esnault [94], and Wunram [244], but those papers do not contain the modern derived category language. A much more general setting was provided by Van den Bergh [230], which subsumes all these results.

Crepancy and the CM property in the smooth setting is used in [231], being somewhat implicit in [44]. The proof here (in the singular setting) can be found in [143], based on the ideas of relative Serre functors due to Ginzburg [112]. The fact that all such algebras have the form $\mathrm{End}_R(M)$ is really just a result of Auslander–Goldman [21]. All the CY algebra section can be found in Iyama–Reiten [139, Section 3].

The singular derived category section is based on the very influential preprint of Buchweitz [56]. The proof of Auslander–Reiten duality can be found in [19], and the technical background on injective hulls can be found in commutative algebra textbooks, e.g., [54].

4.9. Exercises.

EXERCISE 4.9.1 (torsion-free modules are not needed). Suppose that R is a normal domain, and that M is a torsion-free R-module.

(1) Is $\mathrm{End}_R(M) \cong \mathrm{End}_R(M^{**})$ true in general?
(2) Suppose further $\mathrm{End}_R(M) \in \mathrm{ref}\, R$ (e.g., if R is CM, this happens when $\mathrm{End}_R(M) \in \mathrm{CM}\, R$). In this case, show that $\mathrm{End}_R(M) \cong \mathrm{End}_R(M^{**})$. Since $M^{**} \in \mathrm{ref}\, R$, this is why in the definition of NCCR we don't consider torsion-free modules.

EXERCISE 4.9.2 (much harder, puts restrictions on the rings that can admit NCCRs). Suppose that R is a d-dimensional CM domain admitting an NCCR $\Lambda := \mathrm{End}_R(R \oplus M)$. Show:

(1) The Azumaya locus of Λ is equal to the nonsingular locus of R.
(2) The singular locus $\mathrm{Sing}(R)$ must be closed (!).
(3) If further $(R, \mathfrak{m})$ is local, the class group of R must be finitely generated.

EXERCISE 4.9.3. Give an example of a CM ring that is not equicodimensional.

EXERCISE 4.9.4 (NCCR characterization on the base, without finite global dimension mentioned). Let R be a d-dimensional local Gorenstein ring. Then for any $M \in \mathrm{CM}\, R$ with $R \in \mathrm{add}\, M$, show that the following are equivalent:

(1) $\operatorname{add} M = \{X \in \operatorname{CM} R \mid \operatorname{Hom}_R(M, X) \in \operatorname{CM} R\}$.
(2) $\operatorname{add} M = \{X \in \operatorname{CM} R \mid \operatorname{Hom}_R(X, M) \in \operatorname{CM} R\}$.
(3) $\operatorname{End}_R(M)$ is a NCCR.
(4) $\operatorname{End}_R(M)^{\operatorname{op}} \cong \operatorname{End}_R(M^*)$ is a NCCR.

(Hint: the proof should be similar to the Auslander gl.dim $- 2$ proof in Section 1. The key, as always, is the Auslander–Buchsbaum formula and the depth lemma.)

EXERCISE 4.9.5. With the setting as in Example 4.3.2, prove that

$$\operatorname{End}_Y(\mathcal{O}_Y \oplus \mathcal{L}_1) \cong \quad \bullet \overset{a}{\underset{t}{\overset{b}{\rightleftarrows}}} \bullet \quad atb = bta.$$

5. McKay and beyond

This section gives an overview of the McKay correspondence in dimension two, in some of its various different forms. This then leads into dimension three, where we sketch some results of Bridgeland–King–Reid [44] and Van den Bergh [231]. The underlying message is that there is a very tight relationship between crepant resolutions and NCCRs in low dimension.

5.1. McKay correspondence (surfaces). Let $G \leq \operatorname{SL}(2, \mathbb{C})$ be a finite subgroup, and consider $R := \mathbb{C}[x, y]^G$. So far we know that R has a NCCR, since $\mathbb{C}[x, y]\#G \cong \operatorname{End}_R\left(\bigoplus_{\rho \in \operatorname{Irr} G}((\mathbb{C}[x, y] \otimes \rho)^G)^{\oplus \dim_{\mathbb{C}} \rho}\right)$ by Auslander (Theorem 1.4.3), and we have already observed that as a quiver algebra this NCCR has the form

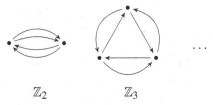

$$\mathbb{Z}_2 \qquad\qquad \mathbb{Z}_3$$

The first is in the exercises (Exercise 2.5.1), the second is our running example. There is a pattern, and to describe it requires the concept of the dual graph, which can be viewed as a simplified picture of the minimal resolution of $\operatorname{Spec} R$.

DEFINITION 5.1.1. Denote by $\{E_i\}$ the exceptional set of $\mathbb{P}^1$s in the minimal resolution $Y \to \operatorname{Spec} R$ (i.e., those $\mathbb{P}^1$s above the origin, as in Example 3.1.3). Define the dual graph as follows: for every E_i draw a dot, and join two dots if the corresponding $\mathbb{P}^1$'s intersect.

EXAMPLE 5.1.2. For the $\mathbb{Z}_4$ singularity we obtain

and so the dual graph is simply $\bullet\!\!-\!\!-\!\!\bullet\!\!-\!\!-\!\!\bullet$

The finite subgroups of $\operatorname{SL}(2, \mathbb{C})$ are classified.

THEOREM 5.1.3. *Let G be a finite subgroup of* $\mathrm{SL}(2, \mathbb{C})$. *Then G is isomorphic to one of the following groups:*

Type	Definition
$\mathbb{A}_n$	$\frac{1}{n}(1, n-1) := \langle \psi_n \rangle$
$\mathbb{D}_n$	$BD_{4n} := \langle \psi_{2n}, \tau \rangle$
$\mathbb{E}_6$	$\langle \psi_4, \tau, \eta \rangle$
$\mathbb{E}_7$	$\langle \psi_8, \tau, \eta \rangle$
$\mathbb{E}_8$	$\langle \kappa, \omega, \iota \rangle$

Here

$$\psi_k = \begin{pmatrix} \varepsilon_k & 0 \\ 0 & \varepsilon_k^{-1} \end{pmatrix}, \quad \tau = \begin{pmatrix} 0 & \varepsilon_4 \\ \varepsilon_4 & 0 \end{pmatrix}, \quad \eta = \frac{1}{\sqrt{2}} \begin{pmatrix} \varepsilon_8 & \varepsilon_8^3 \\ \varepsilon_8 & \varepsilon_8^7 \end{pmatrix}, \quad \kappa = \begin{pmatrix} 0 & -1 \\ 1 & 0 \end{pmatrix},$$

$$\omega = \begin{pmatrix} \varepsilon_5^3 & 0 \\ 0 & \varepsilon_5^2 \end{pmatrix}, \quad \iota = \frac{1}{\sqrt{5}} \begin{pmatrix} \varepsilon_5^4 - \varepsilon_5 & \varepsilon_5^2 - \varepsilon_5^3 \\ \varepsilon_5^2 - \varepsilon_5^3 & \varepsilon_5 - \varepsilon_5^4 \end{pmatrix},$$

where ε_t denotes a primitive t-th root of unity.

Now $G \leq \mathrm{SL}(2, \mathbb{C})$ so G acts on $V := \mathbb{C}^2$. In general, if $G \leq \mathrm{GL}(n, \mathbb{C})$ then the resulting geometry of $\mathbb{C}^n / G$ depends on two parameters, namely the group G and the natural representation $V = \mathbb{C}^n$. For example, if we fix the group $\mathbb{Z}_2$, then we could consider $\mathbb{Z}_2$ as a subgroup in many different ways, for example,

$$\left\langle \begin{pmatrix} -1 & 0 \\ 0 & 1 \end{pmatrix} \right\rangle \quad \text{and} \quad \left\langle \begin{pmatrix} -1 & 0 \\ 0 & -1 \end{pmatrix} \right\rangle.$$

The first gives invariants $\mathbb{C}[x^2, y]$, which is smooth, whilst the latter gives invariants $\mathbb{C}[x^2, xy, y^2]$, which is singular.

Consequently the representation theory by itself will tell us nothing about the geometry (since it only depends on one variable, namely the group G), so we must enrich the representation theory with the action of G on V. In the following definition, this is why we tensor with V.

DEFINITION 5.1.4. For a given finite group G acting on $\mathbb{C}^2 = V$, the McKay quiver is defined to be the quiver with vertices corresponding to the isomorphism classes of indecomposable representations of G, and the number of arrows from ρ_1 to ρ_2 is defined to be

$$\dim_{\mathbb{C}} \mathrm{Hom}_{\mathbb{C}G}(\rho_1, \rho_2 \otimes V),$$

i.e., the number of times ρ_1 appears in the decomposition of $\rho_2 \otimes V$ into irreducibles.

EXAMPLE 5.1.5. We return to our running Example 1.2.1 of the $\mathbb{Z}_3$ singularity. Over $\mathbb{C}$, the group $\mathbb{Z}_3$ has only three irreducible representations ρ_0, ρ_1, ρ_2. The natural representation is, by definition,

$$\left\langle \begin{pmatrix} \varepsilon_3 & 0 \\ 0 & \varepsilon_3^2 \end{pmatrix} \right\rangle,$$

which splits as $\rho_1 \oplus \rho_2$. Thus $\rho_0 \otimes V \cong \rho_1 \oplus \rho_2$, and so we conclude that there is precisely one arrow from ρ_1 to ρ_0, one arrow from ρ_2 to ρ_0, and no arrows from ρ_0 to ρ_0. Thus, so far we have

Continuing the calculation, decomposing $\rho_1 \otimes V$ and $\rho_2 \otimes V$ gives the McKay quiver

which coincides with the quiver from Example 1.2.5.

EXAMPLE 5.1.6. As a second example, consider the group BD_8. Being non-abelian of order 8, necessarily it must have four one-dimensional representations and one irreducible two-dimensional representation. The natural representation V is the irreducible two-dimensional representation. A calculation, using only character theory (Exercise 5.5.2), shows that the McKay quiver is

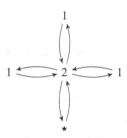

where $\star$ is the trivial representation. The numbers on the vertices correspond to the dimension of the corresponding irreducible representations.

Given the character table, the McKay quiver is a combinatorial object which is easy to construct.

THEOREM 5.1.7 (SL(2, $\mathbb{C}$) McKay correspondence, combinatorial version). *Let $G \le SL(2, \mathbb{C})$ be a finite subgroup and let $Y \to \mathbb{C}^2/G = \operatorname{Spec} R$ denote the minimal resolution.*

(1) *There exist one-to-one correspondences*

$$\{exceptional\ curves\} \longleftrightarrow \{nontrivial\ irreducible\ representations\}$$
$$\longleftrightarrow \{nonfree\ indecomposable\ CM\ R\text{-modules}\},$$

where the right hand sides are taken up to isomorphism.

(2) *The dual graph is an ADE Dynkin diagram, of type corresponding to* $\mathbb{A}$, $\mathbb{D}$ *and* $\mathbb{E}$ *in the classification of the possible groups.*

(3) *(McKay) There is a correspondence*

$$\{dual\ graph\} \quad \rightleftarrows \quad \{McKay\ quiver\}$$

To go from the McKay quiver to the dual graph, simply kill the trivial representation, then merge every pair of arrows in opposite directions (to get an undirected graph). To go from the dual graph to the McKay quiver, simply add a vertex to make the corresponding extended Dynkin diagram, then double the resulting graph to get a quiver.

The last part of the theorem is illustrated by the picture

EXAMPLE 5.1.8. The correspondence for our running $\mathbb{Z}_3$ example is precisely

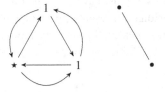

EXAMPLE 5.1.9. The correspondence for the previous Example 5.1.6 is precisely

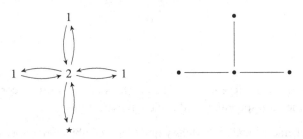

This all takes place at a combinatorial level, to gain more we must add structure.

5.2. Auslander's McKay correspondence (surfaces). In its simplest form, Auslander's version of the McKay correspondence states that the geometry of the minimal resolution can be reconstructed using homological algebra on the category CM $\mathbb{C}[[x, y]]^G$. Reinterpreted, the CM modules on the singularity Spec $\mathbb{C}[[x, y]]^G$ encode information about the resolution.

For what follows we must use the completion $\mathbb{C}[[x, y]]^G$ (not $\mathbb{C}[x, y]^G$) so that the relevant categories are Krull–Schmidt. This is mainly just for technical simplification;

with more work it is possible to phrase the results in the noncomplete setting. Throughout this subsection, to simplify notation we set

$$R := \mathbb{C}[\![x, y]\!]^G.$$

DEFINITION 5.2.1. We say that a short exact sequence

$$0 \to A \to B \overset{f}{\to} C \to 0$$

in the category CM R is an Auslander–Reiten sequence if every $D \to C$ in CM R which is not a split epimorphism factors through f.

Since R is an isolated singularity (normal surface singularities are always isolated), it is a consequence of AR duality (Section 4.8) that AR sequences exist in the category CM R, see for example [247, Section 3]. Now given the existence of AR sequences, we attach to the category CM R the AR quiver as follows.

DEFINITION 5.2.2. The AR quiver of the category CM R has as vertices the isomorphism classes of the indecomposable CM R-modules. As arrows, for each indecomposable M we consider the AR sequence ending at M

$$0 \to \tau(M) \to E \overset{f}{\to} M \to 0,$$

decompose E into indecomposable modules as $E \cong \bigoplus_{i=1}^{n} M_i^{\oplus a_i}$ and for each i draw a_i arrows from M_i to M.

As a consequence of Section 1 (precisely Theorem 1.3.1 and Theorem 1.4.3) we know that there are only finitely many CM R-modules up to isomorphism. Further, by projectivization Fact 1.3.2(1) there is an equivalence of categories

$$\text{CM } R \simeq \text{proj } \mathbb{C}[\![x, y]\!] \# G.$$

The quiver of the skew group ring is well-known.

LEMMA 5.2.3. Let $G \leq \text{GL}(V)$ be a finite subgroup. Then the skew group ring $\mathbb{C}[V] \# G$ is always Morita equivalent to the McKay quiver, modulo some relations.

The relations in Lemma 5.2.3 can be described, but are not needed below. Some further work involving the Koszul complex gives the following.

THEOREM 5.2.4 (Auslander McKay correspondence). Let G be a finite subgroup of $\text{SL}(2, \mathbb{C})$. Then the AR quiver of CM $\mathbb{C}[\![x, y]\!]^G$ equals the McKay quiver of G, and so there is a correspondence

$$\{dual\ graph\} \quad \rightleftarrows \quad \{AR\ quiver\ of\ \text{CM } \mathbb{C}[\![x, y]\!]^G\}.$$

5.3. Derived McKay correspondence (surfaces). The Auslander McKay correspondence upgraded the combinatorial version by considering CM modules over the commutative ring R (to give more structure), and applied methods in homological algebra to obtain the dual graph. The derived version of the McKay correspondence improves this further, since it deals with not just the dual graph, which is a simplified version of the minimal resolution, but with the minimal resolution itself. The upshot is that by considering the NCCR given by the skew group ring (the Auslander algebra of Section 1) seriously as an object in its own right, we are in fact able to obtain the minimal resolution as a quiver GIT, and thus obtain the geometry from the noncommutative resolution without assuming that the geometry exists.

THEOREM 5.3.1 (derived $SL(2, \mathbb{C})$ McKay correspondence). *Let* $G \leq SL(2, \mathbb{C})$ *be a finite subgroup and let* $Y \to \mathbb{C}^2/G = \operatorname{Spec} R$ *denote the minimal resolution. Then:*

(1) $D^b(\operatorname{mod} \mathbb{C}[x, y]\#G) \simeq D^b(\operatorname{coh} Y)$.

(2) *Considering* $\mathbb{C}[x, y]\#G$ *as the McKay quiver subject to the preprojective relations, choose the dimension vector* α *corresponding to the dimensions of the irreducible representations. Then for any generic stability condition* θ, *we have*

    (a) $D^b(\operatorname{coh} \mathcal{M}_\theta^s) \simeq D^b(\operatorname{mod} \mathbb{C}[x, y]\#G)$;

    (b) $\mathcal{M}_\theta^s$ *is the minimal resolution of* $\operatorname{Spec} R$.

There are now many different proofs of both (1) and (2), and some are mentioned in the credits below. A sketch proof is given in the next section, which also covers the dimension three situation.

REMARK 5.3.2. With regards to the choice of stability, once Theorem 5.3.1 has been established for one generic θ, it actually follows that the theorem holds for all generic stability conditions simply by tracking θ through derived equivalences induced by spherical twists and their inverses. This observation also has applications in higher dimensions.

EXAMPLE 5.3.3. Returning to the running $\mathbb{Z}_3$ example, we computed in Example 3.1.3 the moduli space for a very specific generic parameter, remarked it was the minimal resolution and showed in Example 4.3.2 that the NCCR and the minimal resolution were derived equivalent. We know that NCCRs are unique up to Morita equivalence in dimension two (Theorem 2.3.4), so this establishes the above theorem in the special case $G = \mathbb{Z}_3$ and $\theta = (-2, 1, 1)$. The theorem extends our result by saying we could have taken *any* generic parameter, and in fact any group G.

5.4. From NCCRs to crepant resolutions. The main theorem in the notes, which encapsulates both Gorenstein quotient singularities in dimension two and three, and also many other situations, is the following. Everything in this section is finite type over $\mathbb{C}$, and by a point we mean a *closed* point.

THEOREM 5.4.1 (NCCRs give crepant resolutions in dimension ≤ 3). *Let R be a Gorenstein normal domain with* $\dim R = n$, *and suppose that there exists an NCCR*

$\Lambda := \mathrm{End}_R(M)$. *Choose the dimension vector corresponding to the ranks of the inde-composable summands of M (in some decomposition of M into indecomposables), and for any generic stability θ consider the moduli space $\mathcal{M}_\theta^s$. Let X_1 be the locus of $\mathrm{Spec}\, R$ where M is locally free, and define $\mathcal{U}_\theta$ to be the unique irreducible component of $\mathcal{M}_\theta^s$ that maps onto X_1. Assume moreover that $\dim(\mathcal{U}_\theta \times_R \mathcal{U}_\theta) \leq n + 1$.*

(1) *There is an equivalence of triangulated categories*

$$\mathrm{D}^b(\mathrm{coh}\,\mathcal{U}_\theta) \simeq \mathrm{D}^b(\mathrm{mod}\,\Lambda).$$

(2) $\mathcal{U}_\theta$ *is a crepant resolution of* $\mathrm{Spec}\, R$.

Part (2) is an immediate consequence of part (1) by Corollary 4.4.4, and note that the assumption on the dimension of the product is automatically satisfied if $n \leq 3$. We sketch the proof of Theorem 5.4.1 below. One of the key innovations in [44], adapted from [45], is that the New Intersection Theorem from commutative algebra can be used to establish regularity.

Recall that if $E \in \mathrm{D}^b(\mathrm{coh}\, Y)$, then the support of E, denoted $\mathrm{Supp}\, E$, is the closed subset of Y obtained as the union of the support of its cohomology sheaves. On the other hand, the homological dimension of E, denoted $\mathrm{homdim}\, E$, is the smallest integer i such that E is quasi-isomorphic to a complex of locally free sheaves of length i. Note that by convention, a locally free sheaf considered as a complex in degree zero has length 0. We need the following lemma due to Bridgeland–Maciocia [45, Lemma 5.3 and Proposition 5.4].

LEMMA 5.4.2. *Let Z be a scheme of finite type over $\mathbb{C}$, and take $E \in \mathrm{D}^b(\mathrm{coh}\, Z)$.*

(1) *Fix a point $z \in Z$. Then*

$$z \in \mathrm{Supp}\, E \iff \mathrm{Hom}_{\mathrm{D}(Z)}(E, \mathcal{O}_z[i]) \neq 0 \text{ for some } i.$$

(2) *Suppose further that Z is quasi-projective. If there exists $j \in \mathbb{Z}$ and $s \geq 0$ such that for all points $z \in Z$*

$$\mathrm{Hom}_{\mathrm{D}(Z)}(E, \mathcal{O}_z[i]) = 0 \text{ unless } j \leq i \leq j + s,$$

then $\mathrm{homdim}\, E \leq s$.

The next result is much deeper, and follows from the New Intersection Theorem in commutative algebra [43, Corollary 1.2].

THEOREM 5.4.3. *Let Z be an irreducible scheme of dimension n over $\mathbb{C}$, and let E be a nonzero object of $\mathrm{D}^b(\mathrm{coh}\, Z)$. Then:*

(1) $\dim(\mathrm{Supp}\, E) \geq \dim Z - \mathrm{homdim}\, E$.

(2) *Suppose there is a point $z' \in Z$ such that the skyscraper $\mathcal{O}_{z'}$ is a direct summand of $H^0(E)$, and further*

$$\mathrm{Hom}_{\mathrm{D}(Z)}(E, \mathcal{O}_z[i]) = 0 \quad \text{unless } z = z' \text{ and } 0 \leq i \leq n.$$

Then Z is nonsingular at z' and $E \cong H^0(E)$ in $\mathrm{D}(Z)$.

Recall that a set $\Omega \subseteq D^b(\mathrm{coh}\, Y)$ is called a *spanning class* if (a) $\mathbf{RHom}_Y(a, c) = 0$ for all $c \in \Omega$ implies $a = 0$, and (b) $\mathbf{RHom}_Y(c, a) = 0$ for all $c \in \Omega$ implies $a = 0$.

LEMMA 5.4.4. *Suppose that Z is a smooth variety of finite type over $\mathbb{C}$, projective over $\mathrm{Spec}\, T$ where T is a Gorenstein ring. Then $\Omega := \{\mathcal{O}_z \mid z \in Z\}$ is a spanning class.*

PROOF. By Lemma 5.4.2(1) we have $z \in \mathrm{Supp}\, E \iff \mathbf{RHom}_Z(E, \mathcal{O}_z) \neq 0$; hence $\mathbf{RHom}_Z(E, \mathcal{O}_z) = 0$ for all $\mathcal{O}_z \in \Omega$ implies that $\mathrm{Supp}\, E$ is empty, thus $E = 0$. This shows (a). To check the condition (b), suppose that $\mathbf{RHom}_Z(\mathcal{O}_z, E) = 0$ for all $\mathcal{O}_z \in \Omega$. Then

$$0 = \mathbf{RHom}_T(\mathbf{RHom}_Z(\mathcal{O}_z, E), T) \overset{4.5.2}{\cong} \mathbf{RHom}_Z(E, \mathbb{S}_Z \mathcal{O}_z).$$

But since Z and T are Gorenstein, again by Lemma 4.5.2 $\mathbb{S}_Z$ is simply tensoring by a line bundle and (possibly) shifting. Tensoring a skyscraper by a line bundle gives back the same skyscraper, so the above implies that $\mathbf{RHom}_Z(E, \mathcal{O}_z) = 0$ for all $\mathcal{O}_z \in \Omega$. As above, this gives $E = 0$. $\qquad\square$

As one final piece of notation, if Y is a $\mathbb{C}$-scheme and Γ is a $\mathbb{C}$-algebra, we define

$$\mathcal{O}_Y^\Gamma := \mathcal{O}_Y \otimes_{\underline{\mathbb{C}}} \underline{\Gamma},$$

where $\underline{\mathbb{C}}$ and $\underline{\Gamma}$ are the constant sheaves associated to $\mathbb{C}$ and Γ.

We can now sketch the proof of Theorem 5.4.1.

PROOF. For notational simplicity in the proof, denote $\mathcal{U}_\theta$ simply as Y. The tautological bundle from the GIT construction restricted to Y gives us a bundle $\mathcal{M}$ on Y, which is a sheaf of Λ-modules. There is a projective birational morphism $f : Y \twoheadrightarrow \mathrm{Spec}\, R$, see Remark 5.4.6.

Consider the commutative diagram

$$\begin{array}{ccc} Y \times Y & \xrightarrow{\ p_1\ } & Y \\ {\scriptstyle p_2}\Big\downarrow & & \Big\downarrow{\scriptstyle \pi} \\ Y & \xrightarrow{\ \pi\ } & \mathrm{Spec}\,\mathbb{C} \end{array} \qquad\qquad (5.\mathrm{A})$$

There is a natural functor

$$\Phi := \mathbf{R}\pi_*(- \otimes^{\mathbf{L}}_{\mathcal{O}_Y} \mathcal{M}) : D(\mathrm{Qcoh}\, Y) \to D(\mathrm{Mod}\,\Lambda),$$

which has left adjoint $\Psi := \pi^*(-) \otimes^{\mathbf{L}}_{\mathcal{O}_Y^\Lambda} \mathcal{M}^\vee$. There are natural isomorphisms

$$\Psi\Phi(-) = \pi^*(\mathbf{R}\pi_*(- \otimes^{\mathbf{L}}_{\mathcal{O}_Y} \mathcal{M})) \otimes^{\mathbf{L}}_{\mathcal{O}_Y^\Lambda} \mathcal{M}^\vee$$

$$\cong \mathbf{R}p_{2*} p_1^*(- \otimes^{\mathbf{L}}_{\mathcal{O}_Y} \mathcal{M}) \otimes^{\mathbf{L}}_{\mathcal{O}_Y^\Lambda} \mathcal{M}^\vee \qquad \text{(flat base change)}$$

$$\cong \mathbf{R}p_{2*}\big(p_1^*(- \otimes^{\mathbf{L}}_{\mathcal{O}_Y} \mathcal{M}) \otimes^{\mathbf{L}}_{\mathcal{O}_{Y\times Y}^\Lambda} p_2^*\mathcal{M}^\vee\big) \qquad \text{(projection formula)}$$

$$\cong \mathbf{R}p_{2*}\big(p_1^*(-) \otimes^{\mathbf{L}}_{\mathcal{O}_{Y\times Y}} p_1^*\mathcal{M} \otimes^{\mathbf{L}}_{\mathcal{O}_{Y\times Y}^\Lambda} p_2^*\mathcal{M}^\vee\big).$$

It follows that $\Psi\Phi$ is the Fourier–Mukai functor with respect to

$$\mathcal{Q} := \mathcal{M} \boxtimes^{\mathbf{L}}_{\Lambda} \mathcal{M}^{\vee} := p_1^* \mathcal{M} \otimes^{\mathbf{L}}_{\mathcal{O}^{\Lambda}_{Y \times Y}} p_2^* \mathcal{M}^{\vee}.$$

It is known that any Fourier–Mukai functor applied to a skyscraper $\mathcal{O}_y$ is the derived restriction of the corresponding kernel via the morphism $i_y \colon \{y\} \times Y \to Y \times Y$, so since $\Psi\Phi$ is the Fourier–Mukai functor with kernel $\mathcal{Q}$,

$$\Psi\Phi\mathcal{O}_y \cong \mathbf{L}i_y^* \mathcal{Q}. \tag{5.B}$$

We claim that $\mathcal{Q}$ is supported on the diagonal. For this, we let $\mathcal{M}_y$ be the fibre of $\mathcal{M}$ at y, and first remark that

$$\mathrm{Ext}^i_{\Lambda}(\mathcal{M}_y, \mathcal{M}_{y'}) = 0 \quad \text{for all } y \neq y' \text{ if } i \notin [1, n-1]. \tag{5.C}$$

The case $i = 0$ is clear, the case $i = n$ follows from CY duality Theorem 4.6.3(2) on Λ, and the remaining cases $i < 0$ and $i > n$ follow since $\mathcal{M}_y$ and $\mathcal{M}_{y'}$ are modules, and gl.dim $\Lambda = n$.

Now the chain of isomorphisms

$$\begin{aligned}
\mathrm{Hom}_{D(Y \times Y)}(\mathcal{Q}, \mathcal{O}_{y,y'}[i]) &\cong \mathrm{Hom}_{D(Y \times Y)}(\mathcal{Q}, i_{y_*}\mathcal{O}_{y'}[i]) \\
&\cong \mathrm{Hom}_{D(Y)}(\mathbf{L}i_y^* \mathcal{Q}, \mathcal{O}_{y'}[i]) && \text{(adjunction)} \\
&\cong \mathrm{Hom}_{D(Y)}(\Psi\Phi\mathcal{O}_y, \mathcal{O}_{y'}[i]) && \text{(by (5.B))} \\
&\cong \mathrm{Hom}_{D(\Lambda)}(\Phi\mathcal{O}_y, \Phi\mathcal{O}_{y'}[i]) && \text{(adjunction)} \\
&\cong \mathrm{Hom}_{D(\Lambda)}(\mathcal{M}_y, \mathcal{M}_{y'}[i]) \\
&\cong \mathrm{Ext}^i_{\Lambda}(\mathcal{M}_y, \mathcal{M}_{y'})
\end{aligned}$$

shows two things.

(1) First, if $(y, y') \notin Y \times_R Y$ then $f(y) \neq f(y')$ and so $\mathcal{M}_y$ and $\mathcal{M}_{y'}$ are finite length Λ-modules which, when viewed as R-modules, are supported at different points of R. Since

$$\mathrm{Ext}^i_{\Lambda}(\mathcal{M}_y, \mathcal{M}_{y'})_{\mathfrak{m}} \cong \mathrm{Ext}^i_{\Lambda_{\mathfrak{m}}}(\mathcal{M}_{y_{\mathfrak{m}}}, \mathcal{M}_{y'_{\mathfrak{m}}})$$

for all maximal ideals $\mathfrak{m}$ of R, we deduce that there are no Homs or Exts between $\mathcal{M}_y$ and $\mathcal{M}_{y'}$, so the above chain of isomorphisms gives $\mathrm{Hom}_{Y \times Y}(\mathcal{Q}, \mathcal{O}_{y,y'}[i]) = 0$ for all i. Hence, by Lemma 5.4.2(1), we see that $\mathcal{Q}$ is supported on $Y \times_R Y$, and so

$$\dim(\mathrm{Supp}\,\mathcal{Q}|_{(Y \times Y) \setminus \Delta}) \leq \dim(Y \times_R Y) \leq n + 1,$$

where the last inequality holds by assumption.

(2) Second, if $(y, y') \in (Y \times Y) \setminus \Delta$ (i.e., $y \neq y'$), then by (5.C)

$$\mathrm{Hom}_{D(Y \times Y)}(\mathcal{Q}, \mathcal{O}_{y,y'}[i]) = 0 \quad \text{unless } 1 \leq i \leq n - 1.$$

Applying Lemma 5.4.2(2) with $j = 1$ gives homdim $\mathcal{Q}|_{(Y \times Y) \setminus \Delta} \leq n - 2$, thus by Theorem 5.4.3(1)

$$\begin{aligned}
\dim(\mathrm{Supp}\,\mathcal{Q}|_{(Y \times Y) \setminus \Delta}) &\geq \dim((Y \times Y) \setminus \Delta) - \text{homdim}\,\mathcal{Q}|_{(Y \times Y) \setminus \Delta} \\
&\geq 2n - (n - 2) = n + 2.
\end{aligned}$$

Combining (1) and (2) gives a contradiction unless $\mathcal{Q}|_{(Y \times Y) \setminus \Delta} = 0$, so indeed $\mathcal{Q}$ is supported on the diagonal. In particular, by Lemma 5.4.2(1) it follows that $\text{Hom}_{D(Y \times Y)}(\mathcal{Q}, , \mathcal{O}_{y,y'}[i]) = 0$ for all $i \in \mathbb{Z}$ (provided that $y \neq y'$), so the above chain of isomorphisms shows that

$$\text{Ext}^i_\Lambda (\mathcal{M}_y, \mathcal{M}_{y'}) = 0 \text{ for all } i \in \mathbb{Z}, \quad \text{provided that } y \neq y'. \tag{5.D}$$

Pick a point $y \in Y$ and next consider $\Psi \Phi \mathcal{O}_y$. The counit of the adjunction gives us a natural map $\Psi \Phi \mathcal{O}_y \to \mathcal{O}_y$ and thus a triangle

$$c_y \to \Psi \Phi \mathcal{O}_y \to \mathcal{O}_y \to . \tag{5.E}$$

Applying $\text{Hom}_Y(-, \mathcal{O}_y)$ and using the adjunction gives an exact sequence

$$0 \to \text{Hom}_Y(c_y, \mathcal{O}_y[-1]) \to \text{Hom}_Y(\mathcal{O}_y, \mathcal{O}_y) \xrightarrow{\alpha} \text{Hom}_\Lambda (\Phi \mathcal{O}_y, \Phi \mathcal{O}_y)$$

$$\to \text{Hom}_Y(c_y, \mathcal{O}_y) \to \text{Hom}_Y(\mathcal{O}_y, \mathcal{O}_y[1]) \xrightarrow{\beta} \text{Hom}_\Lambda (\Phi \mathcal{O}_y, \Phi \mathcal{O}_y[1])$$

$$\to \text{Hom}_Y(c_y, \mathcal{O}_y[1]) \to \cdots .$$

The domain and the codomain of α are isomorphic to $\mathbb{C}$, and Φ (being a functor) takes the identity to the identity, thus α is an isomorphism. By a result of Bridgeland [41, Lemma 4.4], β can be identified with the Kodaira–Spencer map, and is injective. We conclude that $\text{Hom}_Y(c_y, \mathcal{O}_y[i]) = 0$ for all $i \leq 0$, from which the spectral sequence

$$E_2^{p,q} = \text{Ext}^p (H^{-q}(c_y), \mathcal{O}_y) \Rightarrow \text{Hom}(c_y, \mathcal{O}_y[p+q])$$

(see e.g., [137, Example 2.70(ii)]) yields $H^i(c_y) = 0$ for all $i \geq 0$. Taking cohomology of (5.E) then gives $H^0(\Psi \Phi \mathcal{O}_y) \cong \mathcal{O}_y$. Set $E := \Psi \Phi \mathcal{O}_y$, then since further

$$\text{Hom}_{D(Y)}(E, \mathcal{O}_{y'}[i]) \cong \text{Ext}^i_\Lambda (\mathcal{M}_y, \mathcal{M}_{y'}) = 0$$

for all $y \neq y'$ by (5.D), and also for all $i < 0$ and all $i > n$ in the case $y = y'$ (since Λ has global dimension n), it follows from the New Intersection Theorem 5.4.3(2) that Y is nonsingular at y, and $E \cong H^0(E)$, thus also $c_y = 0$.

Since this holds for all $y \in Y$ we see that Y is smooth. Thus $\mathfrak{Perf}(Y) = D^b(\text{coh } Y)$, and further since Λ has finite global dimension it is also true that $K^b(\text{proj } \Lambda) = D^b(\text{mod } \Lambda)$. Consequently it is easy to see that

$$\Phi = \mathbf{R}\pi_*(- \otimes^{\mathbf{L}}_Y \mathcal{M}) : \ D^b(\text{coh } Y) \to D^b(\text{mod } \Lambda)$$

and its left adjoint $\Psi = - \otimes^{\mathbf{L}}_\Lambda \mathcal{M}^\vee$ also preserves boundedness and coherence. Further, by Lemma 4.5.2(2), $\mathbb{S}_Y = - \otimes f^! \mathcal{O}_R$ is a Serre functor on Y relative to the canonical R, and $\mathbb{S}_\Lambda = \text{Id}$ is a Serre functor on Λ relative to the canonical R by Theorem 4.6.3(1). Using

$$\mathbf{R}\text{Hom}_\Lambda (\Phi(a), b) \cong \mathbf{R}\text{Hom}_R (\mathbf{R}\text{Hom}_\Lambda (\mathbb{S}^{-1}_\Lambda b, \Phi(a)), R)$$

$$\cong \mathbf{R}\text{Hom}_R (\mathbf{R}\text{Hom}_Y (\Psi \mathbb{S}^{-1}_\Lambda b, a), R) \quad \text{(adjunction)}$$

$$\cong \mathbf{R}\text{Hom}_Y (a, \mathbb{S}_Y \Psi \mathbb{S}^{-1}_\Lambda b),$$

taking degree zero cohomology we see that $\mathbb{S}_Y \circ \Psi \circ \mathbb{S}^{-1}_\Lambda$ is right adjoint to Φ.

Now, since Y is smooth, $\{\mathcal{O}_y \mid y \in Y\}$ is a spanning class by Lemma 5.4.4. Hence since Φ has both right and left adjoints, and further $\Phi\mathcal{O}_y = \mathcal{M}_y$, to show that Φ is fully faithful we just have to check that the natural maps

$$\mathrm{Hom}_{\mathrm{D^b(coh}\,Y)}(\mathcal{O}_y, \mathcal{O}_{y'}[i]) \to \mathrm{Hom}_{\mathrm{D^b(mod}\,\Lambda)}(\mathcal{M}_y, \mathcal{M}_{y'}[i]) \qquad (5.F)$$

are bijections for all $y, y' \in Y$ and $i \in \mathbb{Z}$ [137, Proposition 1.49].

But by (5.E) (and the fact we now know that $c_y = 0$), the counit $\Psi\Phi \to 1$ is an isomorphism on all the skyscrapers $\mathcal{O}_y$, and hence also on their shifts. Thus all the skyscrapers and all their shifts belong to the full subcategory on which Φ is fully faithful, and so in particular (5.F) must be bijective for all $y, y' \in Y$ and all $i \in \mathbb{Z}$. We deduce that Φ is fully faithful.

Finally, since $\mathrm{D^b(coh}\,Y)$ is nontrivial and $\mathrm{D^b(mod}\,\Lambda)$ is indecomposable, to show that Φ is an equivalence we just need that $\mathbb{S}_\Lambda \Phi\mathcal{O}_y \cong \Phi\mathbb{S}_Y\mathcal{O}_y$ for all $y \in Y$ (the proof of this fact follows in a similar way to [137, Corollary 1.56], with Serre functors replaced by the RHom versions). But, as above $\mathbb{S}_\Lambda = \mathrm{Id}$ and $\mathbb{S}_Y$ is tensoring by a line bundle, so

$$\Phi\mathbb{S}_Y\mathcal{O}_y = \Phi(\mathcal{L} \otimes \mathcal{O}_y) \cong \Phi\mathcal{O}_y \cong \mathrm{Id}\Phi\mathcal{O}_y \cong \mathbb{S}_\Lambda \Phi\mathcal{O}_y.$$

It follows that Φ is an equivalence. $\square$

REMARK 5.4.5. The irreducible component appears in the statement of the theorem since (as in Exercise 3.2.1) often components arise in quiver GIT. However, with the hypotheses as in Theorem 5.4.1, there is still no known example of when $\mathcal{U}_\theta \neq \mathcal{M}_\theta^s$. In the case when R is complete local and $\dim R = 3$, $\mathcal{U}_\theta = \mathcal{M}_\theta^s$ by [231, Remark 6.6.1].

REMARK 5.4.6. The above proof of Theorem 5.4.1 skips over the existence of a projective birational map $Y \to \mathrm{Spec}\,R$. Recall (from Remark 3.0.12) that the quiver GIT only gives a projective map $Y \to \mathrm{Spec}\,\mathbb{C}[\mathcal{R}]^G$, but $\mathbb{C}[\mathcal{R}]^G$ need not equal R by Exercise 3.2.4. For details on how to overcome this problem, see [231, Section 6.2] or [73, Proposition 2.2, Remark 2.3].

Credits: The geometry of the minimal resolutions of the quotient singularities arising from finite subgroups of $\mathrm{SL}(2, \mathbb{C})$ has a long history stretching back to at least du Val. The relationship with the representation theory was discovered by McKay in [173], which in turned produced many geometric interpretations and generalisations, e.g., [12, 94, 244]. Auslander's version of the McKay correspondence was proved in [19], and is summarised in [247, Section 10].

The fact that the skew group ring is always Morita equivalent to the McKay quiver was written down in [39], but it was well-known to Reiten–Van den Bergh [182], Crawley-Boevey and many others well before then.

It is possible to use [12, 94, 244] to establish the derived equivalence in Theorem 5.3.1, but there are many other proofs of this. The first was Kapranov–Vasserot [146], but it also follows from [230], [44] or [231]. The main result, Theorem 5.4.1, is Van den Bergh's [231] interpretation of Bridgeland–King–Reid [44]. See also [72, Section 7].

5.5. Exercises.

EXERCISE 5.5.1. We can use weighted $\mathbb{C}^*$-actions on polynomial rings as an easy way to produce NCCRs. Consider the polynomial ring $S = \mathbb{C}[x_1, \ldots, x_n, y_1, \ldots, y_m]$, with $n, m \geq 2$, and non-negative integers $a_1, \ldots, a_n, b_1, \ldots, b_m \in \mathbb{N}$. We define a $\mathbb{C}^*$-action on S by

$$\lambda \cdot x_i := \lambda^{a_i} x_i \quad \text{and} \quad \lambda \cdot y_i := \lambda^{-b_i} y_i.$$

As shorthand, we denote this action by $(a_1, \ldots, a_n, -b_1, \ldots, -b_m)$. We consider

$$S_i := \{f \in S \mid \lambda \cdot f = \lambda^i f\}$$

for all $i \in \mathbb{Z}$. It should be fairly clear that $S = \bigoplus_{i \in \mathbb{Z}} S_i$. We will consider the invariant ring S_0, which is known to be CM, of dimension $n + m - 1$. It is Gorenstein if the sum of the weights is zero, i.e., $a_1 + \cdots + a_n - b_1 - \cdots b_m = 0$. Note that the S_i are modules over S_0, so this gives a cheap supply of S_0-modules.

(1) (The easiest case.) Consider the weights $(1, 1, -1, -1)$. Here

$$S_0 = \mathbb{C}[x_1 y_1, x_1 y_2, x_2 y_1, x_2 y_2] \cong \mathbb{C}[a, b, c, d]/(ac - bd)$$

We have seen this example before. Which of the $S_i \in \mathrm{CM}\, S_0$? (The result is combinatorial, but quite hard to prove). Show that $\mathrm{End}_{S_0}(S_1 \oplus S_2) \cong \mathrm{End}_{S_0}(S_0 \oplus S_1)$. This gives an example of an NCCR given by a reflexive module that is not CM.

(2) Find S_0 for the action $(2, 1, -2, -1)$. We have also seen this before. How to build an NCCR for this example?

(3) Experiment with other $\mathbb{C}^*$ actions. Is there any structure?

EXERCISE 5.5.2. Compute the character table for the groups $\mathbb{A}_n$ and $\mathbb{D}_n$ and hence determine their McKay quivers.

EXERCISE 5.5.3. For $R = \mathbb{C}[\![x, y]\!]^G$ with G a finite subgroup of $\mathrm{SL}(2, \mathbb{C})$, prove that for each nonfree CM R-module, the minimal number of generators is precisely twice its rank.

Computer exercises.

EXERCISE 5.5.4 (how to show infinite global dimension without actually computing it). By Exercise 4.9.4, when R is local Gorenstein, $M \in \mathrm{CM}\, R$ with $R \in \mathrm{add}\, M$, and $\mathrm{End}_R(M) \in \mathrm{CM}\, R$, then

$$\mathrm{gl.dim}\, \mathrm{End}_R(M) < \infty \iff \mathrm{add}\, M = \{X \in \mathrm{CM}\, R \mid \mathrm{Hom}_R(M, X) \in \mathrm{CM}\, R\}.$$

Thus to show $\mathrm{End}_R(M)$ has infinite global dimension, we just need to find a CM module X, with $X \notin \mathrm{add}\, M$, such that $\mathrm{Hom}_R(M, X) \in \mathrm{CM}\, R$. Once we have guessed an X, we can check the rest on Singular. Consider the ring $R := \mathbb{C}[\![u, v, x, y]\!]/(uv - x(x^2 + y^7))$.

(1) Let $M := R \oplus (u, x)$. Show that $\mathrm{End}_R(M) \in \mathrm{CM}\, R$.

(2) Consider the module X given as

$$R^4 \xrightarrow{\begin{pmatrix} -x & -y & -u & 0 \\ -y^6 & x & 0 & -u \\ v & 0 & x^2 & xy \\ 0 & v & xy^6 & -x^2 \end{pmatrix}} R^4 \to X \to 0.$$

Show that $\operatorname{Hom}_R(M, X) \in \operatorname{CM} R$, and so $\operatorname{End}_R(M)$ has infinite global dimension. (aside: how we produce such a counterexample X is partially explained in Exercise 5.5.5)

EXERCISE 5.5.5 (Knörrer periodicity). Consider a hypersurface $R:=\mathbb{C}[\![x, y,]\!]/(f)$ where $f \in \mathfrak{m} := (x, y)$. We could take $f = x(x^2 + y^7)$, as in the last example. Consider a CM R-module X with given projective presentation

$$R^a \xrightarrow{\varphi} R^a \xrightarrow{\psi} R^a \to X \to 0$$

(the free modules all having the same rank is actually forced). The Knörrer functor takes X to a module $K(X)$ for the ring $R' := \mathbb{C}[\![u, v, x, y]\!]/(uv - f)$, defined as the cokernel

$$(R')^{2a} \xrightarrow{\begin{pmatrix} -\varphi & -u\mathbb{I} \\ v\mathbb{I} & \psi \end{pmatrix}} (R')^{2a} \to K(X) \to 0.$$

(1) Experimenting with different f, show that $K(X) \in \operatorname{CM} R'$. For example, if we write f into irreducibles $f = f_1 \ldots f_n$, then

$$R^a \xrightarrow{f_1} R^a \xrightarrow{f_2 \cdots f_n} R^a \to X \to 0$$

are examples of CM R-modules on which to test the hypothesis.
(2) When $X, Y \in \operatorname{CM} R$, compute both $\operatorname{Ext}^1_R(X, Y)$ and $\operatorname{Ext}^1_{R'}(K(X), K(Y))$. Is there a pattern?

6. Appendix: Quiver representations

Quivers provide a method to visualise modules (= representations) and are very useful tool to explicitly write down examples of modules. They have many uses throughout mathematics.

DEFINITION 6.0.6. A (finite) quiver Q is a directed graph with finitely many vertices and finitely many arrows.

We often label the vertices with numbers.

EXAMPLE 6.0.7. For example

$$Q_1 = \underset{1}{\bullet} \longrightarrow \underset{2}{\bullet} \longleftarrow \underset{3}{\bullet}\!\!\circlearrowright \qquad \text{and} \qquad Q_2 = \underset{1}{\bullet} \underset{\longleftarrow}{\longrightarrow} \underset{2}{\bullet} \underset{\longleftarrow}{\longrightarrow} \underset{3}{\bullet} \qquad \underset{4}{\bullet}\!\!\circlearrowright$$

are both quivers.

Important technical point: for every vertex i in a quiver we should actually add in a loop at that vertex (called the trivial loop, or trivial path) and denote it by e_i, but we do not draw these loops. Thus really in Example 6.0.7 the quivers are

$$Q_1 = \underset{e_1 \quad e_2 \quad e_3}{\bullet \longrightarrow \bullet \longleftarrow \bullet \circlearrowright} \qquad \text{and} \qquad Q_2 = \underset{e_1 \quad e_2 \quad e_3 \quad e_4}{\bullet \rightrightarrows \bullet \leftleftarrows \bullet \quad \bullet \circlearrowright}$$

but we do not usually draw the dotted loops.

Given any quiver we can produce a k-algebra as follows:

DEFINITION 6.0.8. (1) For a quiver Q denote the set of vertices by Q_0 and the set of arrows by Q_1. For every arrow $a \in Q_1$ we define the head of a (denoted $h(a)$) to be the vertex that a points to, and we define the tail of a to be the vertex that a starts from. For example if

$$Q = \underset{1}{\bullet} \overset{a}{\longrightarrow} \underset{2}{\bullet}$$

then $h(a) = 2$ and $t(a) = 1$.

(2) A nontrivial path in Q is just a formal expression $a_1 \cdot a_2 \cdot \ldots \cdot a_n$ where $a_1, \ldots, a_n$ are nontrivial arrows in Q satisfying $h(a_i) = t(a_{i+1})$ for all i such that $1 \le i \le n-1$. Pictorially this means we have a sequence of arrows

$$\bullet \overset{a_1}{\longrightarrow} \bullet \overset{a_2}{\longrightarrow} \bullet \ \cdots \ \bullet \overset{a_n}{\longrightarrow} \bullet$$

in Q and we just write down the formal expression $a_1 \cdot a_2 \cdot \ldots \cdot a_n$. We define the head and the tail of a path in the obvious way, namely $h(a_1 \cdot a_2 \cdot \ldots \cdot a_n) = h(a_n)$ and $t(a_1 \cdot a_2 \cdot \ldots \cdot a_n) = t(a_1)$.

(3) For a quiver Q we define the path algebra kQ as follows. kQ has a k-basis given by all nontrivial paths in Q together with the trivial loops. Multiplication is defined by

$$pq := \begin{cases} p \cdot q & \text{if } h(p) = t(q), \\ 0 & \text{else,} \end{cases} \qquad e_i p := \begin{cases} p & \text{if } t(p) = i, \\ 0 & \text{else,} \end{cases}$$

$$pe_i := \begin{cases} p & \text{if } h(p) = i, \\ 0 & \text{else,} \end{cases}$$

for all paths p and q and then extend by linearity.

REMARK 6.0.9. (1) Although kQ may be infinite-dimensional, by definition every element of kQ is a *finite* sum $\sum \lambda_p p$ over some paths in Q.

(2) Pictorially multiplication is like composition, i.e., if

$$p = \bullet \overset{a_1}{\longrightarrow} \bullet \overset{a_2}{\longrightarrow} \bullet \cdots \bullet \overset{a_n}{\longrightarrow} \bullet \qquad \text{and} \qquad q = \bullet \overset{b_1}{\longrightarrow} \bullet \overset{b_2}{\longrightarrow} \bullet \cdots \bullet \overset{b_m}{\longrightarrow} \bullet$$

then

$$pq = \begin{cases} \bullet \overset{a_1}{\longrightarrow} \bullet \overset{a_2}{\longrightarrow} \bullet \cdots \bullet \overset{a_n}{\longrightarrow} \bullet \overset{b_1}{\longrightarrow} \bullet \overset{b_2}{\longrightarrow} \bullet \cdots \bullet \overset{b_m}{\longrightarrow} \bullet & \text{if } h(a_n) = t(b_1), \text{ i.e., } h(p) = t(q), \\ 0 & \text{else.} \end{cases}$$

This means that pq is equal to the formal expression $a_1 \cdot \ldots \cot a_n \cdot b_1 \cdot \ldots \cot b_m$ if p and q can be composed, and is equal to zero if p and q cannot be composed. In practice this means that kQ is often noncommutative.

(3) kQ is an algebra, with identity $1_{kQ} = \sum_{i \in Q_0} e_i$.

EXAMPLES 6.0.10. (1) Consider $Q = \bullet$ (recall we never draw the trivial loop, but it is there). Then the basis of kQ is given by e_1, the only path. Hence every element in kQ looks like λe_1 for some $\lambda \in k$. Multiplication is given by $e_1 e_1 = e_1$ extended by linearity, which means $(\lambda e_1)(\mu e_1) = (\lambda \mu) e_1 e_1 = (\lambda \mu) e_1$. This implies that kQ is just the field k.

(2) Consider $Q = \underset{1}{\bullet} \overset{a}{\longrightarrow} \underset{2}{\bullet}$. The basis for kQ is given by e_1, e_2, a. An element of kQ is by definition $\lambda_1 e_1 + \lambda_2 e_2 + \lambda_3 a$ for some $\lambda_1, \lambda_2, \lambda_3 \in k$. Multiplication is given by

$$(\lambda_1 e_1 + \lambda_2 e_2 + \lambda_3 a)(\mu_1 e_1 + \mu_2 e_2 + \mu_3 a) = \lambda_1 \mu_1 e_1 e_1 + \lambda_1 \mu_2 e_1 e_2 + \lambda_1 \mu_3 e_1 a$$
$$+ \lambda_2 \mu_1 e_2 e_1 + \lambda_2 \mu_2 e_2 e_2 + \lambda_2 \mu_3 e_2 a$$
$$+ \lambda_3 \mu_1 a e_1 + \lambda_3 \mu_2 a e_2 + \lambda_3 \mu_3 a a,$$

which is equal to $\lambda_1 \mu_1 e_1 + 0 + \lambda_1 \mu_3 a + 0 + \lambda_2 \mu_2 e_2 + 0 + 0 + \lambda_3 \mu_2 a + 0$. That is,

$$(\lambda_1 e_1 + \lambda_2 e_2 + \lambda_3 a)(\mu_1 e_1 + \mu_2 e_2 + \mu_3 a) = \lambda_1 \mu_1 e_1 + \lambda_2 \mu_2 e_2 + (\lambda_1 \mu_3 + \lambda_3 \mu_2) a.$$

This should be familiar. If we write $\lambda_1 e_1 + \lambda_2 e_2 + \lambda_3 a$ as

$$\begin{pmatrix} \lambda_1 & \lambda_3 \\ 0 & \lambda_2 \end{pmatrix},$$

then the above multiplication is simply

$$\begin{pmatrix} \lambda_1 & \lambda_3 \\ 0 & \lambda_2 \end{pmatrix} \begin{pmatrix} \mu_1 & \mu_3 \\ 0 & \mu_2 \end{pmatrix} = \begin{pmatrix} \lambda_1 \mu_1 & \lambda_1 \mu_3 + \lambda_3 \mu_2 \\ 0 & \lambda_2 \mu_2 \end{pmatrix},$$

which shows that $kQ \cong U_2(k)$, upper triangular matrices. More formally define $\psi :$ $kQ \to U_2(k)$ by $e_1 \mapsto E_{11}$, $e_2 \mapsto E_{22}$ and $a \mapsto E_{12}$ and extend by linearity. By above this is a k-algebra homomorphism which is clearly surjective. Since both sides have dimension three, ψ is also injective.

(3) Consider

$$Q = \overset{\alpha}{\underset{\bullet}{\circlearrowright}} .$$

The basis of kQ is given by $e_1, \alpha, \alpha \cdot \alpha, \alpha \cdot \alpha \cdot \alpha, \ldots$, and so kQ is infinite-dimensional. If we agree to write

$$\underbrace{\alpha \cdot \ldots \cdot \alpha}_{n} := \alpha^n,$$

then every element of kQ is by definition a *finite* sum of paths in Q, i.e., a polynomial in α. Since all paths can be composed the multiplication in kQ is

$$\alpha^i \alpha^j = (\underbrace{\alpha \cdot \ldots \cdot \alpha}_{i})(\underbrace{\alpha \cdot \ldots \cdot \alpha}_{j}) = \underbrace{\alpha \cdot \ldots \cdot \alpha}_{i+j} = \alpha^{i+j}$$

extended by linearity, i.e., polynomial multiplication. This shows that $kQ \cong k[X]$.

(4) Consider $Q = \overset{a}{\underset{1}{\bullet} \longrightarrow} \overset{b}{\underset{2}{\bullet} \longrightarrow} \underset{3}{\bullet}$. The basis for kQ is given by e_1, e_2, e_3, a, b and $a \cdot b$. and so kQ is six-dimensional. In kQ the product ba equals zero whereas the product ab equals the path $a \cdot b$. Some other products:

$$aa = 0, \quad e_1 a = a, \quad e_2 a = 0, \quad e_1 e_2 = 0, \quad (a \cdot b)e_3 = a \cdot b.$$

In fact, $kQ \cong U_3(k)$ in a similar way to (2). See Exercise 6.1.2.

Thus by studying quivers we have recovered many of the algebras that we already know. In fact if we now study *quivers with relations* we can obtain even more:

DEFINITION 6.0.11. For a given quiver Q a relation is just a k-linear combination of paths in Q, each with the same head and tail. Given a finite number of specified relations $R_1, \ldots, R_n$ we can form the two-sided ideal

$$R := kQR_1kQ + \cdots + kQR_nkQ$$

of kQ. We call (Q, R) a quiver with relations and we call kQ/R the path algebra of a quiver with relations.

REMARK 6.0.12. Informally think of a relation $p - q$ as saying "going along path p is the same as going along path q" since $p = q$ in kQ/R. Because of this we sometimes say "subject to the relation $p = q$" when we really mean "subject to the relation $p - q$".

EXAMPLES 6.0.13. (1) Consider

$$Q = \overset{\alpha}{\underset{\bullet}{\bigcirc}}$$

subject to the relation $\alpha \cdot \alpha$. Then $kQ \cong k[X]$ and under this isomorphism the two-sided ideal generated by $\alpha \cdot \alpha$ corresponds to the ideal generated by X^2 in $k[X]$. Thus $kQ/R \cong k[X]/(X^2)$.

(2) Consider

$$Q = \underset{1}{\bullet} \overset{a}{\underset{b}{\rightleftarrows}} \underset{2}{\bullet}$$

subject to the relations $a \cdot b - e_1$ and $b \cdot a - e_2$. Then $kQ/R \cong M_2(k)$. To see this notice (in a very similar way to Example 6.0.10(2)) that there is a k-algebra homomorphism $\psi : kQ \to M_2(k)$ by sending $e_1 \mapsto E_{11}, e_2 \mapsto E_{22}, a \mapsto E_{12}, b \mapsto E_{21}$. Now

$$\psi(a \cdot b - e_1) = E_{12}E_{21} - E_{11} = E_{11} - E_{11} = 0,$$
$$\psi(b \cdot a - e_2) = E_{21}E_{12} - E_{22} = E_{22} - E_{22} = 0,$$

and so ψ induces a well-defined algebra homomorphism $kQ/R \to M_2(k)$. It is clearly surjective. But the dimension of kQ/R is four which is the same as the dimension of $M_2(k)$. Hence the map is also injective, so $kQ/R \cong M_2(k)$.

DEFINITION 6.0.14. (1) Let kQ be the path algebra of a quiver Q. A finite-dimensional quiver representation of Q is the assignment to every vertex $i \in Q_0$ a finite-dimensional vector space V_i and to every arrow a a linear map $f_a : V_{t(a)} \to V_{h(a)}$. We sometimes denote this data by (V_i, f_a). Note that by convention we always assign to the trivial loops e_i the identity linear map.

(2) If (Q, R) is a quiver with relations, we define a finite-dimensional quiver representation of (Q, R) to be a finite-dimensional quiver representation of Q such that for all relations R_i, if $R_i = \sum \lambda_p p$ then $\sum \lambda_p f_p = 0_{\text{map}}$.

Note that if there are no relations (i.e., $R = 0$) then trivially a finite-dimensional quiver representation of Q is the same thing as a finite-dimensional quiver representation of (Q, R).

EXAMPLE 6.0.15. Consider

$$Q = \bullet \underset{b}{\overset{a}{\rightrightarrows}} \bullet \overset{c}{\longrightarrow} \bullet$$

subject to the relation $a \cdot c - b \cdot c$. If we set

$$M := \quad k \underset{f_b=id}{\overset{f_a=id}{\rightrightarrows}} k \overset{f_c=id}{\longrightarrow} k \quad \text{and} \quad N := \quad k \underset{f_b=0}{\overset{f_a=id}{\rightrightarrows}} k \overset{f_c=id}{\longrightarrow} k$$

then M is a quiver representation of (Q, R) since the relation $f_a \cdot f_c - f_b \cdot f_c = 0$ holds. However N is not a quiver representation of (Q, R) since the relation does not hold.

To make quiver representations into a category, we must define morphisms:

DEFINITION 6.0.16. Suppose (Q, R) is a quiver with relations, and $V = (V_i, f_a)$ and $W = (W_i, g_a)$ are quiver representations for (Q, R). A morphism of quiver representations ψ from V to W is given by specifying, for every vertex i, a linear map $\psi_i : V_i \to W_i$ such that for every arrow $a \in Q_1$

$$
\begin{array}{ccc}
V_{t(a)} & \overset{f_a}{\longrightarrow} & V_{h(a)} \\
\psi_{t(a)} \downarrow & & \downarrow \psi_{h(a)} \\
W_{t(a)} & \overset{g_a}{\longrightarrow} & W_{h(a)}
\end{array}
$$

we have $f_a \cdot \psi_{h(a)} = \psi_{t(a)} \cdot g_a$.

EXAMPLES 6.0.17. (1) Consider $Q = \bullet \overset{a}{\longrightarrow} \bullet$ (no relations). Consider M and N defined in Example 6.0.21(2). To specify a morphism of quiver representations from M to N we must find linear maps ψ_1 and ψ_2 such that

$$
\begin{array}{ccc}
k & \overset{id}{\longrightarrow} & k \\
\psi_1 \downarrow & & \downarrow \psi_2 \\
k & \overset{0}{\longrightarrow} & 0
\end{array}
$$

commutes. Note ψ_2 is the zero map, whereas ψ_1 can be an arbitrary scalar. Now to specify a morphism of quiver representations from N to M we must find linear maps

ϕ_1 and ϕ_2 such that

$$
\begin{array}{ccc}
k & \xrightarrow{\ \mathrm{id}\ } & k \\
\phi_1 \uparrow & & \uparrow \phi_2 \\
k & \xrightarrow{\ 0\ } & 0
\end{array}
$$

commutes. Note ϕ_2 is the zero map, and the fact that the diagram commutes forces ϕ_1 to be the zero map too. This shows that the only morphism of quiver representations from N to M is the zero morphism.

(2) Consider

$$
Q = \bullet \underset{b}{\overset{a}{\rightleftarrows}} \bullet
$$

subject to the relations $a \cdot b - e_1$ and $b \cdot a - e_2$. By Example 6.0.13(2)

$$
kQ/R \cong M_2(k).
$$

Suppose that

$$
M := k^n \underset{f_b}{\overset{f_a}{\rightleftarrows}} k^n
$$

is a quiver representation of (Q, R). Note that since $f_a \cdot f_b = id$ and $f_b \cdot f_a = id$, both f_a and f_b must be linear isomorphisms. Now it is clear that

$$
k^n \underset{\mathrm{id}}{\overset{\mathrm{id}}{\rightleftarrows}} k^n
$$

is a quiver representation of (Q, R) and further

$$
\begin{array}{ccc}
k^n & \underset{f_b}{\overset{f_a}{\rightleftarrows}} & k^n \\
\mathrm{id} \uparrow & & \uparrow f_a \\
k^n & \underset{\mathrm{id}}{\overset{\mathrm{id}}{\rightleftarrows}} & k^n
\end{array}
$$

is a morphism of quiver representations since both

$$
\begin{array}{ccc}
k^n & \xrightarrow{\ f_a\ } & k^n \\
\mathrm{id} \uparrow & & \uparrow f_a \\
k^n & \xrightarrow{\ \mathrm{id}\ } & k^n
\end{array}
\qquad \text{and} \qquad
\begin{array}{ccc}
k^n & \xleftarrow{\ \ } & k^n \\
\mathrm{id} \uparrow & {\scriptstyle f_b} & \uparrow f_a \\
k^n & \xleftarrow{\ \mathrm{id}\ } & k^n
\end{array}
$$

commute.

DEFINITION 6.0.18. For a quiver with relations (Q, R) we define $\mathtt{fRep}(kQ, R)$ to be the category of all finite-dimensional quiver representations, where the morphisms are defined to be all morphisms of quiver representations. We denote by $\mathrm{fdmod}\, kQ/R$ the category of finite-dimensional right kQ/R-modules

The category $\mathtt{fRep}(kQ, R)$ is visual and computable; in contrast, the category $\mathrm{fdmod}\, kQ/R$ is more abstract. The following result is one of the main motivations for studying quivers.

THEOREM 6.0.19. *Suppose (Q, R) is a quiver with relations. Then finite-dimensional quiver representations of (Q, R) are the same as finite-dimensional right kQ/R-modules. More specifically there is an equivalence of categories*

$$\mathtt{fRep}(kQ, R) \simeq \mathrm{fdmod}\, kQ/R.$$

SKETCH PROOF. If M is a finite-dimensional right kQ/R-module, define a finite-dimensional quiver representation of (Q, R) by setting $V_i = Me_i$ and $f_a : Me_{t(a)} \to Me_{h(a)}$ by $f_a(x) = xa$. Conversely, given a quiver representation (V_i, f_a) of (Q, R), define $M = \bigoplus_{i \in Q_0} V_i$. Denote by $V_i \overset{\iota_i}{\to} M \overset{\pi_i}{\to} V_i$ inclusion and projection; then M is a right kQ/R-module via

$$x \cdot (a_1 \cdot \ldots \cdot a_m) := \iota_{h(a_m)} f_{a_m} \circ \ldots \circ f_{a_1} \pi_{t(a_1)}(x),$$

$$x \cdot e_i := \iota_i \circ \pi_i(x).$$

It is fairly straightforward to show that these are inverses.   □

REMARK 6.0.20. (1) Suppose we want to understand the modules of some algebra A. Theorem 6.0.19 says that, provided we can find a quiver Q with relations R such that $A \cong kQ/R$, then A-modules are precisely the same as quiver representations of (Q, R). This means that A-modules are very easy to write down! Hence Theorem 6.0.19 gives us a method to visualise modules.

(2) Note that in the proof of Theorem 6.0.19 if (V_i, f_a) is a quiver representation of (Q, R) then the corresponding kQ/R module has dimension $\sum_{i \in Q_0} \dim_k V_i$.

EXAMPLES 6.0.21. (1) Consider $Q = \bullet$ (no relations). Then by Example 6.0.10(1) $kQ \cong k$ and so k-modules are the same as quiver representations for Q. But here to specify a quiver representation we just need to assign a vector space to the only vertex, and so quiver representations are precisely the same as vector spaces. This just says that k-modules are the same as vector spaces.

(2) Consider $Q = \bullet \overset{a}{\longrightarrow} \bullet$ (no relations). By Example 6.0.10(2) $kQ \cong U_2(k)$ and so $U_2(k)$-modules are the same as quiver representations of Q. Hence examples of $U_2(k)$-modules include

$$M := \left(k \overset{\mathrm{id}}{\longrightarrow} k \right) \quad \text{and} \quad N := \left(k \overset{0}{\longrightarrow} 0 \right).$$

(3) Consider $Q = \overset{\alpha}{\underset{\bullet}{\circlearrowleft}}$ (no relations). By Example 6.0.10(3) $kQ \cong k[X]$ and so $k[X]$-modules are the same as quiver representations of Q. For example,

$$\overset{\text{id}}{\underset{k}{\circlearrowleft}}$$

is a quiver representation of Q and so is a $k[X]$-module. Now here a quiver representation of Q is given by specifying a vector space V together with a linear map from V to itself and so (since $kQ \cong k[X]$) $k[X]$-modules are given by specifying a vector space V and a linear map $V \xrightarrow{\alpha} V$.

(4) Consider

$$Q = \overset{\alpha}{\underset{\bullet}{\circlearrowleft}}$$

subject to the relation $\alpha \cdot \alpha$. By Example 6.0.13(1) $kQ/R \cong k[X]/(X^2)$. Now

$$\overset{f_\alpha = \text{id}}{\underset{k}{\circlearrowleft}}$$

is not a quiver representation for (Q, R) since the relation $f_\alpha \cdot f_\alpha = 0$ does not hold $(\text{id} \cdot \text{id} \neq 0)$, hence it is not a module for $k[X]/(X^2)$.

REMARK 6.0.22. If (Q, R) is a quiver with relations, then under Theorem 6.0.19 morphisms of quiver representations correspond to kQ/R-module homomorphisms. Further,

$$\psi = (\psi_i) \text{ corresponds to a } \begin{cases} \text{monomorphism} \\ \text{epimorphism} \\ \text{isomorphism} \end{cases} \iff \text{ each } \psi_i \text{ is an } \begin{cases} \text{injective} \\ \text{surjective} \\ \text{bijective} \end{cases}$$

linear map. Thus if we have a morphism of quiver representations $\psi = (\psi_i) : N \to M$ in which each ψ_i is an injective linear map, we call N a subrepresentation of M (since under the above correspondence N embeds into the module of M, so we can view N as a submodule of M).

EXAMPLE 6.0.23. In Examples 6.0.17(2) the morphism of quiver representations is an isomorphism since both the connector maps id and f_a are linear isomorphisms. This shows that as $kQ/R \cong M_2(k)$-modules,

$$M = \left(\ k^n \underset{f_b}{\overset{f_a}{\rightleftarrows}} k^n \ \right) \cong \left(\ k^n \underset{\text{id}}{\overset{\text{id}}{\rightleftarrows}} k^n \ \right).$$

Taking the direct sum of modules can also be visualised easily in the language of quivers:

DEFINITION 6.0.24. Suppose (Q, R) is a quiver with relations, and $V = (V_i, f_a)$, $W = (W_i, g_a)$ are quiver representations of (Q, R). Then we define the direct sum $V \oplus W$ to be the quiver representation of (Q, R) given by

$$\left(V_i \oplus W_i, \begin{pmatrix} f_a & 0 \\ 0 & g_a \end{pmatrix} \right).$$

EXAMPLE 6.0.25. (1) Consider $Q = \bullet \overset{a}{\longrightarrow} \bullet$ (no relations) as in Example 6.0.21(2). Then

$$\left(\ k \overset{\text{id}}{\longrightarrow} k \ \right) \bigoplus \left(\ k \overset{0}{\longrightarrow} 0 \ \right) = k \oplus k \overset{\left(\begin{smallmatrix} \text{id} & 0 \\ 0 & 0 \end{smallmatrix}\right)}{\longrightarrow} k \oplus 0 = k^2 \overset{\left(\begin{smallmatrix} \text{id} \\ 0 \end{smallmatrix}\right)}{\longrightarrow} k \ .$$

(2) In Example 6.0.23

$$M \cong \left(\ k^n \underset{\text{id}}{\overset{\text{id}}{\rightleftarrows}} k^n \ \right) = \overset{n}{\underset{i=1}{\bigoplus}} \left(\ k \underset{\text{id}}{\overset{\text{id}}{\rightleftarrows}} k \ \right).$$

6.1. Exercises.

EXERCISE 6.1.1. Write down the dimension (if it is finite) of the following quiver algebras, where there are no relations.

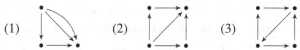

(1)   (2)   (3)

What is the general result?

EXERCISE 6.1.2. (1) Show that the algebra $U_n(k)$ of upper triangular matrices is algebra-isomorphic to the path algebra of the quiver

$$\underset{1}{\bullet} \longrightarrow \underset{2}{\bullet} \longrightarrow \underset{3}{\bullet} \cdots \underset{n-1}{\bullet} \longrightarrow \underset{n}{\bullet}$$

subject to no relations. How do we view the algebra $D_n(k)$ of diagonal matrices in the above picture?

(2) (This question shows that $U_2(k)$ is not a semisimple algebra.) Consider the case $n = 2$ (i.e., $U_2(k) \cong \bullet \longrightarrow \bullet$) and let M be the quiver representation ($U_2(k)$-module)

$$k \overset{\text{id}}{\longrightarrow} k \ .$$

Show that M has a subrepresentation ($U_2(k)$ submodule) $N := 0 \xrightarrow{0} k$ and that there does not exist a submodule N' such that $M = N \oplus N'$.

EXERCISE 6.1.3 (from quivers to algebras). Consider the following quivers with relations

(1) $\quad \overset{\alpha}{\underset{\bullet}{\circlearrowright}}$
$\alpha^3 = e_1$

(2) $\quad \alpha \overset{\curvearrowright}{\underset{\bullet}{}} \beta$
$\alpha\beta = \beta\alpha$

(3) $\quad \bullet \quad \overset{\alpha}{\underset{\bullet}{\circlearrowright}}$
(no relations)

Identify each with an algebra you are already familiar with. If $k = \mathbb{C}$ is there a quiver with no relations which is isomorphic to the quiver with relations in (1)?

EXERCISE 6.1.4. (From algebras to quivers). Write the following algebras as quivers with relations (there is not a unique way of answering these — to solve them in as many ways as possible):

(1) The free algebra in n variables.
(2) The polynomial ring in n variables.
(3) $R \times S$, given knowledge of R and S as quivers with relations.
(4) The group algebra $\mathbb{C}G$, where G is any finite group.
(5) Any k-algebra given by a finite number of generators and a finite number of relations.

EXERCISE 6.1.5. (1) Show that $M_n(k)$ is algebra-isomorphic to the quiver

$$\underset{1}{\bullet} \overset{f_1}{\underset{g_1}{\rightleftarrows}} \underset{2}{\bullet} \overset{f_2}{\underset{g_2}{\rightleftarrows}} \underset{3}{\bullet} \quad \cdots \quad \underset{n-1}{\bullet} \overset{f_{n-1}}{\underset{g_{n-1}}{\rightleftarrows}} \underset{n}{\bullet}$$

subject to the relations $f_i g_i = e_i$ and $g_i f_i = e_{i+1}$ for all i such that $1 \le i \le n-1$.

(2) (The quiver proof that $M_n(k)$ is semisimple.) By (1) representations of the above quiver with relations are the same thing as finite-dimensional $M_n(k)$-modules. Using the quiver with relations, show that there is precisely one simple $M_n(k)$-module, and it has dimension n. Further show directly that every finite-dimensional $M_n(k)$-module is the finite direct sum of this simple module.

(Direct proof of Morita equivalence.) Using (1) and (2), show that there is an equivalence of categories between $\operatorname{mod} k$ and $\operatorname{mod} M_n(k)$.

EXERCISE 6.1.6. Show, using quivers, that $k[x]/x^n$ has precisely one simple module (up to isomorphism), and it has dimension one.

EXERCISE 6.1.7. Let k be a field of characteristic zero. Let A be the first Weyl algebra, that is the path algebra of the quiver

$$X \overset{\curvearrowright}{\underset{\bullet}{}} Y$$

subject to the relation $XY - YX = 1$. Show that $\{0\}$ is the only finite-dimensional A-module.

Solutions to the exercises

This chapter provides partial solutions to the exercises at the end of each section of the book.

I. Noncommutative projective geometry

SOLUTION TO EXERCISE 1.6.1. (1) Write $A = k[x_1, \ldots, x_n] = B[x_n]$, where $B = k[x_1, \ldots, x_{n-1}]$. Since $A = B \oplus Bx_n \oplus Bx_n^2 \oplus \cdots$ we have

$$h_A(t) = h_B(t)[1 + t^{d_n} + t^{2d_n} + \cdots] = h_B(t)/(1 - t^{d_n}),$$

and we are done by induction on n.

(2) Since F_m is spanned by words of degree m, considering the last letter in a word we have the direct sum decomposition $F_m = \bigoplus_{i=1}^n F_{m-d_i} x_i$. Writing $h_F(t) = \sum a_i t^i$ we get $a_m = \sum_{i=1}^n a_{m-d_i}$ (where $a_i = 0$ for $i < 0$). This leads to the Hilbert series equation $h_F(t) = \sum_{i=1}^n h_F(t) t^{d_i}$; now solve for $h_F(t)$. □

SOLUTION TO EXERCISE 1.6.2. Suppose that x is a nonzerodivisor and a normal homogeneous element in A, such that A/xA is a domain. If A is not a domain, it must have homogeneous nonzero elements y, z with $yz = 0$. Choose such nonzero $y \in A_m$, $z \in A_n$ with $m + n$ minimal. Then $\overline{yz} = 0$ in A/xA, where $\overline{y} = y + xA$ is the image of y in the factor ring. So $\overline{y} = 0$ or $\overline{z} = 0$; without loss of generality, assume the former. Then $y \in xA$, so $y = xy'$ with $0 \neq y'$ of smaller degree than y. Then $xy'z = 0$, and x is a nonzerodivisor so $y'z = 0$, contradicting minimality.

In the example $A = k\langle x, y \rangle/(yx^2 - x^2 y, y^2 x - xy^2)$, the element $xy - yx$ is normal since $x(xy - yx) = -(xy - yx)x$ and similarly for y. Clearly $A/(xy - yx) \cong k[x, y]$ is a noetherian domain. Since the Hilbert series of A is known to be $1/(1-t)^2(1-t^2)$, a similar argument as in the proof of Lemma 1.3.3 shows that $(xy - yx)$ must be a nonzerodivisor, so Lemma 1.3.2 applies. □

SOLUTION TO EXERCISE 1.6.3. Let v and w be reduction-unique elements. Let s be any composition of reductions such that $s(v + \lambda w)$ is a linear combination of reduced words, where $\lambda \in k$. Let t be some composition of reductions such that $ts(v)$ is a linear combination of reduced words, and let u be some composition of reductions such that $uts(w)$ is a linear combination of reduced words. Then $s(v + \lambda w) = uts(v + \lambda w)$ (since $s(v + \lambda w)$ is already a linear combination of reduced words) and $uts(v + \lambda w) = uts(v) + \lambda uts(w) = \text{red}(v) + \lambda \text{red}(w)$. Thus $s(v + \lambda w)$ is independent of s, and the set of reduction-unique elements is a subspace. Also, $s(v + \lambda w) = \text{red}(v + \lambda w) = \text{red}(v) + \lambda \text{red}(w)$ and so red$(-)$ is linear on this subspace.

For (3) $\implies$ (2), the hypothesis is equivalent to $F = I \oplus V$ as k-spaces, where V is the k-span of the set of reduced words. Given any element h, let s and t be compositions of reductions such that $s(h)$ and $t(h)$ are both linear combinations of reduced words. Then $s(h) - t(h) \in I$ (since a reduction changes an element to one equivalent modulo I) and $s(h) - t(h) \in V$. So $s(h) - t(h) \in I \cap V = 0$.

For (2) $\implies$ (1), if a word w contains a non-resolving ambiguity, then there are reductions r_1, r_2 such that $r_1(w)$ and $r_2(w)$ cannot be made equal by performing further reductions to each. In particular, if s and t are compositions of reductions such that $sr_1(w)$ and $tr_2(w)$ are both linear combinations of reduced words, they are not equal, so w is not reduction-unique. $\qquad\square$

SOLUTION TO EXERCISE 1.6.4. The word z^2x may first be resolved using g_1 to give $(xy+yx)x = xyx+yx^2$, which is a linear combination of reduced words, or using g_2 to give $z(xz)$, which may be further reduced giving xz^2 and then $x(xy+yx) = x^2y+xyx$. These are distinct so their difference is added as the new relation $g_4 = yx^2 - x^2y$. Similarly, $g_5 = y^2x - xy^2$ is added from resolving z^2y two ways. It is straightforward to check that all ambiguities now resolve.

The basis of reduced words is all words not containing any of z^2, zx, zy, y^2x, yx^2, which is

$$\{x^{i_1}(yx)^{i_2}y^{i_3}z^\epsilon \,|\, i_1, i_2, i_3 \geq 0, \epsilon \in \{0, 1\}\}.$$

The set of such words with $\epsilon = 0$ has the same Hilbert series as a polynomial ring in variables of weights $1, 1, 2$, namely $1/(1-t)^2(1-t^2)$ using Exercise 1.6.1. Thus $h_A(t) = (1+t)/(1-t)^2(1-t^2) = 1/(1-t)^3$. $\qquad\square$

SOLUTION TO EXERCISE 1.6.5. (1) The usual isomorphism $\underline{\text{Hom}}_A(A, A) \cong A$ given by $\phi \mapsto \phi(1)$ is easily adjusted to prove that $\underline{\text{Hom}}_A(A(-s_i), A) \cong A(s_i)$ in the graded setting. The indicated formula follows since finite direct sums pull out of either coordinate of $\underline{\text{Hom}}$.

(2) If $e_1, \ldots, e_m$ is the standard basis for the free right module P and $f_1, \ldots, f_n$ is the standard basis for the free right module Q, then the matrix $M = (m_{ij})$ is determined by $\phi(e_j) = \sum_i f_i m_{ij}$. We can take as a basis for $\underline{\text{Hom}}(P, A)$ the dual basis $e_1^*, \ldots, e_m^*$ such that $e_i^*(e_i) = 1$ and $e_i^*(e_j) = 0$ for $i \neq j$. Identifying $\underline{\text{Hom}}(P, A)$ and $\bigoplus_{i=1}^m A(s_i)$ using part (a), then $\{e_i^*\}$ is just the standard basis of $\bigoplus_{i=1}^m A(s_i)$. Similarly, the basis $f_1^*, \ldots, f_n^*$ of $\underline{\text{Hom}}(Q, A)$ is identified with the standard basis of $\bigoplus_{j=1}^n A(t_j)$.

Now the matrix $N = (n_{ij})$ of ϕ^* should satisfy $\phi^*(f_i^*) = \sum_j n_{ij}e_j^*$ since these are left modules. By definition we have $\phi^*(f_i^*) = f_i^* \circ \phi$ and $f_i^* \circ \phi(e_k) = f_i^*(\sum_l f_l m_{lk}) = \sum_l f_i^*(f_l)m_{lk} = m_{ik}$. We also have $(\sum_j n_{ij}e_j^*)(e_k) = \sum_j n_{ij}(e_j^*(e_k)) = n_{ik}$, using the definition of the left A-module structure on $\underline{\text{Hom}}(P, A)$. This proves that $M = N$, as required. $\qquad\square$

SOLUTION TO EXERCISE 1.6.6. (1) This varies the argument of Lemma 1.5.2.

(2) Using the minimal graded free resolution $P_\bullet$ of M to calculate Tor, we see that $\operatorname{Tor}_i^A(M_A, {}_Ak)$ is the i-th homology of the complex $P_\bullet \otimes_A k$, where $P_i \otimes_A k \cong P_i / P_i A_{\geq 1}$. Since the resolution is minimal, the maps in this complex are 0 by Lemma 1.5.2, and thus $\operatorname{Tor}_i^A(M, k) \cong P_i / P_i A_{\geq 1}$ is a k-space of dimension equal to the minimal number of homogeneous generators of P_i. In particular, the minimal free resolution has length $\max\{i \mid \operatorname{Tor}_i(M_A, {}_Ak) \neq 0\}$, so the statement follows from (1).

(3) Clearly if ${}_Ak$ has projective dimension d, using a projective resolution in the second coordinate to calculate Tor gives $\max\{i \mid \operatorname{Tor}_i(M_A, {}_Ak) \neq 0\} \leq d$. Combined with part (1) we get the first statement. Obviously we can prove all the same results on the other side to obtain proj. $\dim({}_AN) \leq$ proj. $\dim(k_A)$ for any left bounded graded left module N.

(4) By part (b) we have both proj. $\dim(k_A) \leq$ proj. $\dim({}_Ak)$ and proj. $\dim({}_Ak) \leq$ proj. $\dim(k_A)$, so proj. $\dim({}_Ak) =$ proj. $\dim(k_A)$. Finitely generated graded modules are left bounded, and we only need to consider finitely generated modules by the result of Auslander. Any finitely generated graded right module M satisfies proj. $\dim(M) \leq$ proj. $\dim({}_Ak)$ by parts (1) and (2), so r. gl. $\dim(A) =$ proj. $\dim({}_Ak)$. Similarly we obtain l. gl. $\dim(A) =$ proj. $\dim(k_A)$. $\square$

SOLUTION TO EXERCISE 2.4.1. The required isomorphism follows from an application of Lemma 1.5.5(1) followed by its left-sided analog. If A is weakly AS-regular, that is if (1) and (3) of Definition 2.0.8 hold, then if $P_\bullet$ is the minimal graded free resolution of k_A, we know that the P_i have finite rank ([218]) and that $Q_\bullet = \underline{\operatorname{Hom}}_A(P_\bullet, A)$ is a minimal graded free resolution of ${}_Ak$. Applying $\underline{\operatorname{Hom}}(-, {}_AA)$ to $Q_\bullet$ yields a complex of free right modules isomorphic to the original $P_\bullet$, by the isomorphism above. This implies that $\underline{\operatorname{Ext}}_A^i({}_Ak, {}_AA)$ is isomorphic to $k_A(\ell)$ if $i = d$, and is 0 otherwise. $\square$

SOLUTION TO EXERCISE 2.4.2. By the Diamond Lemma, A has k-basis

$$\{x^i(yx)^j y^k \mid i, j, k \geq 0\}$$

since the overlap y^2x^2 resolves. Thus $h_A(t) = 1/(1-t)^2(1-t^2)$. Now using Lemma 2.1.3 and guessing at the final map we write down the potential free resolution of k_A as follows:

$$0 \to A(-4) \xrightarrow{\binom{y}{x}} A(-3)^{\oplus 2} \xrightarrow{\begin{pmatrix} xy-2yx & y^2 \\ x^2 & yx-2xy \end{pmatrix}} A(-1)^{\oplus 2} \xrightarrow{(x\ y)} A \to 0.$$

The proof that this is indeed a free resolution of k_A, and the verification of the AS-Gorenstein condition, is similar to the proof in Example 2.1.5. $\square$

SOLUTION TO EXERCISE 2.4.3. (1) It is easy to see since the algebra has one relation that the existence of a graded isomorphism $A(\tau) \to A(\tau')$ is equivalent to the existence of a change of variable $x' = c_{11}x + c_{12}y$, $y' = c_{21}x + c_{22}y$, such that $x'\tau'(x') + y'\tau'(y')$ and $x\tau(x) + y\tau(y)$ generate the same ideal of the free algebra, that is, they are nonzero scalar multiples. By adjusting the change of variable by a scalar, this is equivalent to finding such a change of variable with $x'\tau'(x') + y'\tau'(y') = x\tau(x) + y\tau(y)$, since our base field is algebraically closed. We have

$$\begin{pmatrix} \tau(x) \\ \tau(y) \end{pmatrix} = B\begin{pmatrix} x \\ y \end{pmatrix}, \qquad \begin{pmatrix} \tau'(x') \\ \tau'(y') \end{pmatrix} = B'\begin{pmatrix} x' \\ y' \end{pmatrix} \quad \text{for some matrix } B', \qquad \begin{pmatrix} x' \\ y' \end{pmatrix} = C\begin{pmatrix} x \\ y \end{pmatrix}.$$

So

$$(x' \; y')\begin{pmatrix} \tau'(x') \\ \tau'(y') \end{pmatrix} = (x \; y)C^t B'C\begin{pmatrix} x \\ y \end{pmatrix}, \quad \text{while} \quad (x \; y)\begin{pmatrix} \tau(x) \\ \tau(y) \end{pmatrix} = (x \; y)B\begin{pmatrix} x \\ y \end{pmatrix}.$$

Thus $A(\tau) \cong A(\tau')$ if and only if $B = C^t B'C$ for some invertible matrix C, that is, if B and B' are congruent.

(2) This is a tedious but elementary computation.

(3) Let $A = k\langle x, y\rangle/(x^2)$. To calculate the Hilbert series of A, note that the overlap of x^2 with itself trivially resolves, so the set of words not containing x^2 is a k-basis for A. let W_n be the set of words of degree n not containing x^2, and note that $W_n = W_{n-2}yx \cup W_{n-1}y$. Thus $\dim_k A_n = \dim_k A_{n-1} + \dim_k A_{n-2}$ for $n \geq 2$, while $\dim_k A_0 = 1$ and $\dim_k A_1 = 2$. In terms of Hilbert series we have $h_A(t) = th_A(t) + t^2 h_A(t) + 1 + t$ and thus $h_A(t) = (1+t)/(1-t-t^2)$. Since $\dim_k A_n$ is part of the Fibonacci sequence, these numbers grow exponentially and A has exponential growth; in particular $\mathrm{GKdim}(A) = \infty$. The minimal free resolution of k_A is easily calculated to begin

$$\cdots \longrightarrow A(-4) \xrightarrow{(x)} A(-3) \xrightarrow{(x)} A(-2) \xrightarrow{\binom{x}{0}} A(-1)^{\oplus 2} \xrightarrow{(x \; y)} A \longrightarrow 0,$$

after which it simply repeats, since $xA = \{a \in A | xa = 0\}$. In particular, A has infinite global dimension.

To see that $k\langle x, y\rangle/(yx)$ is not right noetherian, show that the right ideal generated by $y, xy, x^2y, \ldots$ is not finitely generated. Similarly, for $k\langle x, y\rangle/(x^2)$ show that the right ideal generated by $x, yx, y^2x, \ldots$ is not finitely generated. $\qquad\square$

SOLUTION TO EXERCISE 2.4.4. We claim that the minimal free resolution of k is

$$0 \to A(-2) \xrightarrow{\begin{pmatrix} \tau(x_1) \\ \cdots \\ \tau(x_n) \end{pmatrix}} \bigoplus_{i=1}^{n} A(-1) \xrightarrow{(x_1 \; \cdots \; x_n)} A \to 0.$$

This complex is exact by Lemma 2.1.3, except maybe in the final spot, in other words, the last map may not be injective. Thus it is exact if and only if it has the Hilbert series predicted by (2.E), that is $h_A(t) = 1/(1 - nt + t^2)$. In this case, the AS-Gorenstein condition follows in the same way and A is weakly AS-regular.

Assume now that the leading term of $f = \sum x_i \tau(x_i)$ is $x_n x_i$ for some $i < n$, under the degree lex order with $x_1 < \cdots < x_n$. Then clearly there are no overlaps to check and a k-basis of A consists of words not containg $x_n x_i$. If W_m is the set of such words in degree m, then $W_m = (\bigcup_{1 \leq j \leq n} W_{m-1}x_j) \setminus W_{m-2}x_n x_1$. This leads to the Hilbert series equation $h_A(t) = nh_A(t)t - h_A(t)t^2$ and hence $h_A(t) = 1/(1 - nt + t^2)$. Thus A is weakly AS-regular by the argument above.

If instead f has leading term x_n^2, the same argument as in Exercise 2.4.3 shows that a change of variable leads to a new relation $f = \sum x_i' \tau'(x_i')$, where τ' corresponds to a matrix congruent to the matrix representing τ. But every matrix is congruent to a matrix which is 0 in the (n, n)-spot: this amounts to finding a nontrivial zero of some homogeneous degree 2 polynomial, which is always possible since $n \geq 2$ and k is algebraically closed. $\qquad\square$

SOLUTION TO EXERCISE 2.4.5. The argument in the proof of Theorem 2.2.1 shows that if A is AS-regular of global dimension 2, minimally generated by elements of degrees $d_1 \leq d_2 \leq \cdots \leq d_n$, then the free resolution of k_A must have the form

$$0 \to A(-\ell) \xrightarrow{\begin{pmatrix} \tau(x_1) \\ \cdots \\ \tau(x_n) \end{pmatrix}} \bigoplus_{i=1}^{n} A(-d_i) \xrightarrow{(x_1 \cdots x_n)} A \to 0,$$

where $\ell = d_i + d_{n-i}$ for all i, and where the $\tau(x_i)$ are another minimal generating set for the algebra. Then $A = k\langle x_1, \ldots, x_n \rangle / (\sum x_i \tau(x_i))$ and A has Hilbert series $1/(1 - \sum_{i=1}^{n} t^{d_i} + t^\ell)$. It is easy to check that whenever $n \geq 3$, necessarily the denominator of this Hilbert series has a real root bigger than 1 and hence A has exponential growth. Thus $n = 1$ or $n = 2$, and $n = 1$ is easily ruled out as in the proof of Theorem 2.2.1. Thus the main case left is $n = 2$ and $\ell = d_1 + d_2$. Write $x_1 = x$ and $x_2 = y$.

If $d_1 = d_2$, then we can reduce to the degree 1 generated case simply by reassigning degrees to the elements; so we know from Theorem 2.2.1 that up to isomorphism we have one of the relations $yx - qxy$ or $yx - xy - x^2$. If $d_2 > d_1$ but $d_1 i = d_2$ for some i, then $A_{d_2} = kx^i + ky$, $A_{d_1} = kx$, and so clearly the relation has the form $x(ax^i + by) + cyx = 0$ with b and c nonzero. A change of variables sends this to $yx - xy - x^{i+1}$ or else $yx - qxy$. Finally, if d_1 does not divide d_2, then $A_{d_2} = ky$, $A_{d_1} = kx$, and so the relation is necessarily of the form $yx - qxy$ (after scaling).

This limits the possible regular algebras to those on the given list. That these algebras really are regular is proved in the same way as for the Jordan and quantum planes: the Diamond Lemma easily gives their Hilbert series, which is used to prove the obvious potential resolution of k_A is exact. $\qquad\square$

SOLUTION TO EXERCISE 2.4.6. (1) Recall that $v = (x_1, \ldots, x_n)$. We want to express $\pi' = \sum \alpha_{i_1, \ldots, i_d} \tau(x_{i_d}) x_{i_1} \ldots x_{i_{d-1}}$ as a matrix product. We have

$$\pi = \sum_{i_d} \left(\sum_{i_1, \ldots, i_{d-1}} \alpha_{i_1, \ldots, i_d} x_{i_1} \ldots x_{i_{d-1}} \right) x_{i_d}$$

and thus vM is the row vector with i_d-th coordinate

$$(vM)_{i_d} = \sum_{i_1, \ldots, i_{d-1}} \alpha_{i_1, \ldots, i_d} x_{i_1} \ldots x_{i_{d-1}}.$$

We see from this that π' is equal to $(\tau(x_1), \ldots, \tau(x_n))(vM)^t = vQ^{-1}(vM)^t$. Now using that $QMv^t = (vM)^t$, we get $\pi' = vQ^{-1}QMv^t = vMv^t = \pi$ as claimed. Iterating this result d times gives $\pi = \tau(\pi)$.

(2) Writing $\pi = \sum r_i x_i$ for some uniquely determined r_i, we see that the r_i are the coordinates of the row vector vM, and hence they are a k-basis for the minimal set of the relations of the algebra A by the construction of the free resolution of k_A (Lemma 2.1.3). If we have some other k-basis $\{y_i\}$ of $kx_1 + \cdots + kx_n$, writing $\pi = \sum s_i y_i$, the s_i are a basis for the same vector space as the r_i, so they are also a minimal set of relations for A. In particular, taking $y_{i_d} = \tau(x_{i_d})$ and applying this to $\tau(\pi) = \pi$ shows that

$$\left\{ \sum_{i_1,\ldots,i_{d-1}} \alpha_{i_1,\ldots,i_d} \tau(x_{i_1}) \ldots \tau(x_{i_{d-1}}) \;\middle|\; 1 \leq i_d \leq n \right\}$$

is a minimal set of relations for A. Thus τ preserves the ideal of relations and induces an automorphism of A.   □

SOLUTION TO EXERCISE 3.3.1. The multilinearized relations can be written in the matrix form

$$\begin{pmatrix} 0 & z_0 & -ry_0 \\ -qz_0 & 0 & x_0 \\ y_0 & -px_0 & 0 \end{pmatrix} \begin{pmatrix} x_1 \\ y_1 \\ z_1 \end{pmatrix} = 0.$$

The solutions $\{(x_0 : y_0 : z_0), (x_1 : y_1 : z_1)\} \subseteq \mathbb{P}^2 \times \mathbb{P}^2$ to this equation give X_2. Thus the first projection E of X_2 is equal to those $(x_0 : y_0 : z_0)$ for which the 3×3 matrix M above is singular.

Taking the determinant gives $\det M = (1 - pqr)x_0 y_0 z_0$. Then either $pqr = 1$ and $\det M = 0$ identically so $E = \mathbb{P}^2$, or $pqr \neq 1$ and $\det M = 0$ when $x_0 y_0 z_0 = 0$, that is when E is the union of the three coordinate lines in $\mathbb{P}^2$. The equations for X_2 can also be written in the matrix form

$$\begin{pmatrix} x_0 & y_0 & z_0 \end{pmatrix} \begin{pmatrix} 0 & -rz_1 & y_1 \\ z_1 & 0 & -qx_1 \\ -py_1 & x_1 & 0 \end{pmatrix},$$

and a similar calculation shows that the second projection of X_2 is also equal to E.

Now it is easy to check in each case that for $(x_0 : y_0 : z_0)$ such that M is singular, it has rank exactly 2. In fact, in either case for E, the first two rows of M are independent when $z_0 \neq 0$, the first and third when $y_0 \neq 0$, and the last two when $x_0 \neq 0$. This implies that for each $p \in E$, there is a unique $q \in E$ such that $(p, q) \in X_2$. A similar argument using the second projection shows that for each $q \in E$ there is a unique $p \in E$ such that $(p, q) \in X_2$. Thus $X_2 = \{(p, \sigma(p)) | p \in E\}$ for some bijection σ, and we can find a formula for σ by taking the cross product of the first two rows when $z_0 \neq 0$ and similarly in the other two cases.

Thus when $z_0 \neq 0$ we get $\sigma(x_0 : y_0 : z_0) = (x_0 : rqy_0 : qz_0)$, when $y_0 \neq 0$ the formula is $\sigma(x_0 : y_0 : z_0) = (prx_0 : ry_0 : z_0)$, and when $x_0 \neq 0$ the formula is $\sigma(x_0 : y_0 : z_0) = (px_0 : y_0 : pqz_0)$. These formulas are correct in both cases for E. (When $pqr = 1$ and $E = \mathbb{P}^2$, one can easily see that the three formulas match up to give a single formula.)   □

SOLUTION TO EXERCISE 3.3.2. This is a similar calculation as in Exercise 3.3.1. The multilinearized relations can be written in two matrix forms

$$\begin{pmatrix} x_0 & y_0 \end{pmatrix} \begin{pmatrix} -y_1 y_2 & y_1 x_2 \\ -x_1 y_2 & x_1 x_2 \end{pmatrix} = \begin{pmatrix} y_0 y_1 & -x_0 y_1 \\ y_0 x_1 & -x_0 x_1 \end{pmatrix} \begin{pmatrix} x_2 \\ y_2 \end{pmatrix}.$$

The determinants of both 2×2 matrices appearing are identically 0. Thus $X_3 \subseteq \mathbb{P}^1 \times \mathbb{P}^1 \times \mathbb{P}^1$ has projections π_{12} and π_{23} which are onto. On the other hand, the matrices have rank exactly 1 for each point in $\mathbb{P}^1 \times \mathbb{P}^1$, so that for each $(p_0, p_1) \in \mathbb{P}^1 \times \mathbb{P}^1$ there is a unique $p_2 \in \mathbb{P}^1$ with $(p_0, p_1, p_2) \in X_3$; explicitly, it is easy to see that $p_2 = p_0$. Thus $X_3 = \{(p_0, p_1, p_0) | p_0, p_1 \in \mathbb{P}^1\}$ and clearly the full set of point modules is in bijection already with $X_3 \cong \mathbb{P}^1 \times \mathbb{P}^1$. Moreover, for the automorphism σ of $\mathbb{P}^1 \times \mathbb{P}^1$ given by $(p_0, p_1) \mapsto (p_1, p_0)$, we have $X_3 = \{(p, q, r) \in (\mathbb{P}^1)^{\times 3} | (q, r) = \sigma(p, q)\}$. $\square$

SOLUTION TO EXERCISE 3.3.3. The multilinearization of the relation is $y_0 x_1$. Its vanishing set in $\mathbb{P}^1 \times \mathbb{P}^1$, with coordinates $((x_0 : y_0), (x_1 : y_1))$, is the set

$$X_2 = ((1 : 0) \times \mathbb{P}^1) \bigcup (\mathbb{P}^1 \times (0 : 1)).$$

Then by construction, $X_n \subseteq (\mathbb{P}^1)^{\times n}$ consists of sequences of points $(p_0, \ldots p_{n-1})$ such that $(p_i, p_{i+1}) \in E$ for all $0 \leq i \leq n - 2$. Thus

$$X_n = \left\{ (p_0, \ldots, p_{n-1}) \, \middle| \, \exists i \in \{0, \ldots, n-1\} \text{ s.t. } p_j = (1:0) \text{ for } j < i, \; p_j = (0:1) \text{ for } j > i \right\}.$$

In particular, the projection map $X_{n+1} \twoheadrightarrow X_n$ collapses the set

$$\left\{ (\overbrace{(1:0), (1:0), \ldots, (1:0)}^{n}, q) \mid q \in \mathbb{P}^1 \right\}$$

to a single point, and thus the inverse limit of Proposition 3.0.10 does not stabilize for any n. $\square$

SOLUTION TO EXERCISE 3.3.4. (1) Assume A is quadratic regular. This exercise simply formalizes the general pattern seen in all the examples so far. The entries of the matrix M are of degree 1. Write $M = M(x, y, z)$. Taking the three relations to be the entries of vM in the free algebra, then the multilinearized relations can be put into either of the two forms

$$[QM](x_0, y_0, z_0) \begin{pmatrix} x_1 \\ y_1 \\ z_1 \end{pmatrix} = 0, \qquad \begin{pmatrix} x_0 & y_0 & z_0 \end{pmatrix} M(x_1, y_1, z_1) = 0.$$

Thus if E is the vanishing of $\det M$ in $\mathbb{P}^2$, then $\det QM = \det Q \det M$ differs only by a scalar, so also has vanishing set E. Thus E is equal to both the first and second projections of $X_2 \subseteq \mathbb{P}^2 \times \mathbb{P}^2$, by a similar argument as we have seen in the examples. Either $\det M$ vanishes identically and so $E = \mathbb{P}^2$, or else $\det M$ is a cubic polynomial, and so E is a degree 3 hypersurface in $\mathbb{P}^2$.

(2) Now let A be cubic regular. In this case M has entries of degree 2. We let $N = N(x_0, y_0; x_1, y_1)$ be the matrix M with its entries multilinearized. Then the

multilinearized relations of A can be put into either of the two forms

$$[QN](x_0, y_0; x_1, y_1) \begin{pmatrix} x_2 \\ y_2 \end{pmatrix} = 0, \qquad (x_0 \quad y_0) \, N(x_1, y_1; x_2, y_2) = 0.$$

The determinant of N is a degree $(2, 2)$ multilinear polynomial and so its vanishing is all of $\mathbb{P}^1 \times \mathbb{P}^1$ if $\det N$ is identically 0, or a degree $(2, 2)$-hypersurface in $\mathbb{P}^1 \times \mathbb{P}^1$ otherwise. Again $\det N$ and $\det QN$ have the same vanishing set E and so the projections $p_{12}(X_3)$ and $p_{23}(X_3)$ are both equal to E, where $X_3 \subseteq \mathbb{P}^1 \times \mathbb{P}^1 \times \mathbb{P}^1$.   □

SOLUTION TO EXERCISE 4.5.1. We do the quantum plane case; the Jordan plane can be analyzed with a similar idea. Let $A = k\langle x, y\rangle/(yx - qxy)$ where q is not a root of 1. Recall that $\{x^i y^j \,|\, i, j \geq 0\}$ is a k-basis for A. Let I be any nonzero ideal of A and choose a nonzero element $f \in I$. Write $f = \sum_{i=0}^n p_i(x)y^i$ for some $p_i(x) \in k[x]$ with $p_n \neq 0$. Choose such an element with n as small as possible. Then look at $fx - q^i xf \in I$, which is of smaller degree in y and so must be zero. But this can happen only if $f = p_n(x)y^n$, since the powers of q are distinct. A similar argument in the other variable forces $x^m y^n \in I$ for some m, n. But x and y are normal, and so $x^m y^n \in I$ implies $(x)^m (y)^n \subseteq I$. Now if I is also prime, then either $(x) \subseteq I$ or else $(y) \subseteq I$. Thus every nonzero prime ideal I of A contains either x or y.

Now $A/(x) \cong k[y]$ and $k[y]$ has only 0 and (y) as graded prime ideals. Similarly, $A/(y) \cong k[x]$ which has only 0 and (x) as graded prime ideals. It follows that $0, (x), (y), (x, y)$ are the only graded primes of A.   □

SOLUTION TO EXERCISE 4.5.2. If $\theta \in \mathrm{Hom}_{\mathrm{qgr}\text{-}A}(\pi(M), \pi(N))$ is an isomorphism, with inverse $\psi \in \mathrm{Hom}_{\mathrm{qgr}\text{-}A}(\pi(N), \pi(M))$, find module maps

$$\tilde{\theta} : M_{\geq n} \to N \quad \text{and} \quad \tilde{\psi} : N_{\geq m} \to M$$

representing these morphisms in the respective direct limits $\lim_{n \to \infty} \mathrm{Hom}_{\mathrm{gr}\text{-}A}(M_{\geq n}, N)$ and $\lim_{n \to \infty} \mathrm{Hom}_{\mathrm{gr}\text{-}A}(N_{\geq n}, M)$.

Then for any $q \geq \max(m, n)$, $\tilde{\psi}|_{N_{\geq q}} \circ \tilde{\theta}|_{M_{\geq q}} : M_{\geq q} \to M$ represents $\psi \circ \theta = 1$. Thus is equal in the direct limit $\lim_{n \to \infty} \mathrm{Hom}_{\mathrm{gr}\text{-}A}(M_{\geq n}, M)$ to the identity map, so for some (possibly larger) q, the map $\tilde{\psi}|_{N_{\geq q}} \circ \tilde{\theta}|_{M_{\geq q}}$ is the identity map from $M_{\geq q}$ onto $M_{\geq q}$. Similarly, there must be r such that $\tilde{\theta}|_{M_{\geq r}} \circ \tilde{\psi}|_{N_{\geq r}}$ gives an isomorphism from $N_{\geq r}$ onto $N_{\geq r}$. This forces $M_{\geq s} \cong N_{\geq s}$ for $s = \max(q, r)$. The converse is similar.   □

SOLUTION TO EXERCISE 4.5.3. Consider a Zhang twist $S = R^\sigma$ for some graded automorphism σ. Thus σ acts as a bijection of $R_1 = kx + ky + xz$. We claim that $w \in S_1 = R_1$ is normal in S if and only if w is an eigenvector for σ. Since $w * S_1 = wR_1$ and $S_1 * w = R_1\sigma(w)$, we have w normal in S if and only if $wR_1 = \sigma(w)R_1$, which happens if and only if $\sigma(w) = \lambda w$ for some λ (for example, by unique factorization), proving the claim.

Now the quantum polynomial ring A of Example 4.1.4 has normal elements x, y, z which are a basis for A_1. By the previous paragraph, if $\phi : A \to R^\sigma$ is an isomorphism, we can take the images of x, y, z to be a basis of eigenvectors for σ; after a change of variables, we can take these to be the elements with those names already in R_1, and thus σ is now diagonalized with $\sigma(x) = ax$, $\sigma(y) = by$, $\sigma(z) = cz$ for nonzero a, b, c. But then the relations of R^σ are

$$y * x - ab^{-1}x * y, \quad z * y - bc^{-1}y * z, \quad x * z - ca^{-1}z * x,$$

where $(ab^{-1})(bc^{-1})(ca^{-1}) = 1$. Conversely, if $pqr = 1$ then taking $a = p$, $b = 1$, $c = q^{-1}$ we have $ab^{-1} = p$, $bc^{-1} = q$, $ca^{-1} = r$ so that the twist R^σ by the σ above will have the relations of the quantum polynomial ring in Example 4.1.4.   □

SOLUTION TO EXERCISE 4.5.4. (1) We have the following portion of the long exact sequence in Ext:

$$\cdots \longrightarrow \underline{\mathrm{Ext}}^i(A/A_{\geq n}, A) \overset{\phi}{\longrightarrow} \underline{\mathrm{Ext}}^i(A/A_{\geq n+1}, A) \longrightarrow \underline{\mathrm{Ext}}^i\left(\bigoplus k(-n), A\right) \longrightarrow \cdots$$

as indicated in the hint. If $\underline{\mathrm{Ext}}^i(k, A)$ is finite-dimensional, say it is contained in degrees between m_1 and m_2, then $\underline{\mathrm{Ext}}^i\left(\bigoplus k(-n), A\right)$ is finite-dimensional and contained in degrees between $m_1 - n$ and $m_2 - n$. In particular, $\underline{\mathrm{Ext}}^i(A/A_{\geq n+1}, A)/(\mathrm{im}\,\phi)$ is finite-dimensional, and is contained in negative degrees for $n \gg 0$. This shows by induction on n that the nonnegative part of the direct limit $\lim_{n \to \infty} \underline{\mathrm{Ext}}^i(A/A_{\geq n}, A)_{\geq 0}$ is also finite-dimensional. If $\underline{\mathrm{Ext}}^i(k, A) = 0$, then this same exact sequence implies by induction that $\underline{\mathrm{Ext}}^i(A/A_{\geq n}, A) = 0$ for all n, and so $\lim_{n \to \infty} \underline{\mathrm{Ext}}^i(A/A_{\geq n}, A) = 0$ in this case.

(2) Calculating $\underline{\mathrm{Ext}}^i(k, A)$ with a minimal projective resolution of k, since A is noetherian each term in the resolution is noetherian, and so the homology groups $\underline{\mathrm{Ext}}^i(k, A)$ will also be noetherian A-modules. Calculating with an injective resolution of A instead, after applying $\underline{\mathrm{Hom}}(k_A, -)$ each term is an A-module killed by $A_{\geq 1}$. Thus the homology groups $\underline{\mathrm{Ext}}^i(k, A)$ will also be killed by $A_{\geq 1}$. Thus $\underline{\mathrm{Ext}}^i(k, A)$ is both finitely generated and killed by $A_{\geq 1}$, so it is a finite-dimensional $A/A_{\geq 1} = k$-module. (The reason this argument fails in the noncommutative case is that calculating with the projective resolution gives a right A-module structure to the Ext groups, while the calculation with the injective resolution gives a left A-module structure to the Ext groups. The Ext groups are then (A, A)-bimodules which are finitely generated on one side and killed by $A_{\geq 1}$ on the other, which does not force them to be finite-dimensional. In the commutative case the left and right module structures must coincide.)   □

SOLUTION TO EXERCISE 4.5.5. (1) We have seen that $\{x^i y^j | i, j \geq 0\}$ is a k-basis of B, by the Diamond Lemma. Thus By has k-basis $\{x^i y^j | i \geq 0, j \geq 1\}$ and B/By has $\{1, x, x^2, \ldots\}$ as k-basis. The idealizer A' of By is certainly a graded subring, so if it is larger than A, then $x^n \in A'$ for some $n \geq 1$. But then $yx^n = x^n y + nx^{n+1} \in By$, which is clearly false since char $k = 0$. Obviously $k \subseteq A'$ and thus $A' = A$.

(2) Since $x^n By \in A$ for all $n \geq 1$, we see that each $x^n \in B$ is annihilated by $A_{\geq 1} = By$ as a right A-module. Thus the right A-module structure of B/A is the same as its $A/A_{\geq 1} = k$-vector space structure. In particular, as a graded module it is isomorphic to $\bigoplus_{n \geq 1} k(-n)$.

(3) Left multiplication by x^n for any $n \geq 1$ gives a homomorphism of degree n in $\underline{\mathrm{Hom}}_A(A_{\geq 1}, A)$, because $x^n A_{\geq 1} \subseteq A$. Any homomorphism in $\underline{\mathrm{Hom}}_A(A, A)$ is equal to left multiplication by some $a \in A$, and if its restriction to $\underline{\mathrm{Hom}}_A(A_{\geq 1}, A)$ is the same

as left multiplication by x^n we get $ab = x^n b$ for all $b \in A_{\geq 1}$ and thus $a = x^n$ since B is a domain. This contradicts $x^n \notin A$. Thus x^n is an element of $\underline{\mathrm{Hom}}_A(A_{\geq 1}, A)$ not in the image of $\underline{\mathrm{Hom}}_A(A, A)$ for each $n \geq 1$. The map $\underline{\mathrm{Hom}}_A(A, A) \to \underline{\mathrm{Hom}}_A(A_{\geq 1}, A)$ is also clearly injective, and so from the long exact sequence we get that $\underline{\mathrm{Ext}}^1_A(k_A, A)$ is infinite-dimensional.

More generally, all of the maps in the direct limit $\lim_{n \to \infty} \underline{\mathrm{Hom}}_A(A_{\geq n}, A)$ are injective, since A is a domain; hence the cokernel $\lim_{n \to \infty} \underline{\mathrm{Ext}}^1_A(A/A_{\geq n}, A)$ of the map $A \to \mathrm{Hom}_{\mathrm{qgr}\text{-}A}(\pi(A), \pi(A))$ is also infinite-dimensional. (In fact one may show that $\mathrm{Hom}_{\mathrm{qgr}\text{-}A}(\pi(A), \pi(A)) \cong B$.) $\qquad\square$

II. Deformations of algebras in noncommutative geometry

SOLUTION TO EXERCISE 1.2.1. (a) Write $\mathrm{Weyl}_n := A/(R)$ where A is the free algebra on $x_1, \ldots, x_n$ and $y_1, \ldots, y_n$, and R is the span of the defining relations $x_i y_j - y_j x_i = \delta_{ij}$, $x_i x_j - x_j x_i$, and $y_i y_j - y_j y_i$. We then have a unique homomorphism $\phi : A \to \mathcal{D}(\mathbb{A}^n)$ by the assignment $\phi(x_i) = x_i$ and $\phi(y_i) = -\partial_i$. We show that $\phi(R) = 0$, and hence ϕ factors through a homomorphism $\bar{\phi} : \mathrm{Weyl}_n \to \mathcal{D}(\mathbb{A}^n)$. This follows from a direct computation: to verify $x_i(-\partial_j) - (-\partial_j)x_i = \delta_{ij}$, we check that $x_i(-\partial_j)(f) - (-\partial_j)(x_i f) = -(-\partial_j)(x_i) \cdot f = \delta_{ij} f$; then note that multiplication operators by x_i and x_j commute, as do partial derivative operators ∂_i, ∂_j. It is clear that $\bar{\phi}$ is surjective, since it sends generators to generators.

(b) We need to show that $\bar{\phi}$ is injective. We first claim that Weyl_n is spanned by the operators $f \cdot P$ where $f \in k[x_1, \ldots, x_n]$ and $P \in k[y_1, \ldots, y_n]$ are monomials. This is easy to see, because given any monomial in x_i and y_j, we can apply the relations to push the y_j to the right and the x_i to the left, resulting in a linear combination of elements of the desired form. Next, we claim that the resulting elements $\bar{\phi}(fP)$ are linearly independent (again assuming f and P are monomials). This implies not only that $\bar{\phi}$ is an isomorphism, but in fact that these elements fP form a basis of Weyl_n (note that, in Exercise 1.2.8 below, we will actually see that this is a basis for arbitrary commutative rings k, and not merely for characteristic zero fields; but note that $\bar{\phi}$ will no longer be an isomorphism in this generality).

It remains to prove the final claim. For a contradiction, suppose that a nonzero linear combination $F := \sum_{f,P} \lambda_{f,P} \bar{\phi}(fP)$ of such monomials is zero, with $\lambda_{f,P} \in k$. Let P be of maximal degree such that $\lambda_{f,P} \neq 0$. Let $g \in k[x_1, \ldots, x_n]$ be the monomial corresponding to P, i.e., such that, viewing g as a function in n variables, $P = g(\partial_1, \ldots, \partial_n)$. Then, applying the operator F to g, we get $F(g) = cf$ where $c \geq 1$ is a positive integer (specifically, if $g = x_1^{r_1} \cdots x_n^{r_n}$, then $c = r_1! r_2! \cdots r_n!$). Since we assumed that k had characteristic zero, we have $cf \neq 0$, which is a contradiction, since F was assumed to be zero. (Note that this proof works when k is any ring of characteristic zero, not necessarily a field.) $\qquad\square$

SOLUTION TO EXERCISE 1.2.3. It can easily be seen from the definition that the map $\phi : \mathrm{Weyl}_n \to \mathrm{Weyl}(V)$ such that $\phi(x_i) = x_i$ and $\phi(y_i) = y_i$ is a homomorphism. It is surjective since V is spanned by the x_i and y_i. We only need to show it is injective.

The kernel is (R) where R is the span of the relations $xy - yx - (x, y)$, for $x, y \in V$. We only need to show these are zero in Weyl_n. To do so, write x and y as a linear combination of the x_i and y_i. Then $xy - yx - (x, y)$ decomposes as a linear combination of the defining relations of Weyl_n (where x and y are replaced by basis elements). $\square$

SOLUTION TO EXERCISE 1.2.7. Following the hint, we showed above in the solution to Exercise 1.2.1.(b) that a basis for $\mathcal{D}(\mathbb{A}^n)$ is of the desired form. Since the associated graded elements of this basis are clearly a basis for $\mathsf{Sym}\, V$, it remains only to see that the degree of these elements matches the description: under the additive filtration, x_i and y_i are both in degree one, whereas under the geometric filtration, the x_i are in degree zero and the y_i are in degree one. $\square$

SOLUTION TO EXERCISE 1.2.8. Setting $\Bbbk = \mathbb{Z}$, the result actually follows from the solution given for Exercise 1.2.1.(b) above. But then, applying $\otimes_{\mathbb{Z}} \Bbbk$ gives the result for $\mathsf{Weyl}_n(\Bbbk)$ defined over an arbitrary rings $\Bbbk$, again by the same relations (since $\mathsf{Weyl}_n(\Bbbk) = \mathsf{Weyl}_n(\mathbb{Z}) \otimes_{\mathbb{Z}} \Bbbk$). In the case that $\Bbbk$ is a field and V is a vector space, we then know that $\mathsf{Weyl}_n = \mathsf{Weyl}(V)$ by Exercise 1.2.3. $\square$

SOLUTION TO EXERCISE 1.3.4. Let $\phi : TV = \mathsf{gr}(TV) \to \mathsf{gr}\, A$ be the tautological surjection. We show that $\phi(R) = 0$, so that it descends to a homomorphism $\bar{\phi} : B \to \mathsf{gr}\, A$. Indeed, $R = \mathsf{gr}(R) = \mathsf{gr}(E) \subseteq \mathsf{gr}(TV) = TV$, which implies that $\phi(R) = 0$. $\square$

SOLUTION TO EXERCISE 1.4.2. The skew-symmetry and Jacobi identities for $\{-,-\}$ follow immediately from those identities for the commutator $[a, b] = ab - ba$, by taking associated graded, using the identities $\{\mathsf{gr}_m a, \mathsf{gr}_n b\} = \mathsf{gr}_{m+n-d}[a, b]$ and $\{\mathsf{gr}_m a, \{\mathsf{gr}_n b, \mathsf{gr}_p c\}\} = \mathsf{gr}_{m+n+p-2d}[a, [b, c]]$, for $a \in A_{\leq m}, b \in A_{\leq n}$, and $c \in A_{\leq p}$ (which follow immediately). The Leibniz rule is then equivalent to the statement that $\{\mathsf{gr}_m a, -\}$ is a derivation for all $a \in A_{\leq m}$ (and all $m \geq 0$). This follows because $[a, -]$ is a derivation already in A: $[a, bc] = [a, b]c + b[a, c]$; we then apply $\mathsf{gr}_{m+n+p-d}$, assuming $a \in A_{\leq m}, b \in A_{\leq n}$, and $c \in A_{\leq p}$. $\square$

SOLUTION TO EXERCISE 1.4.3. If d is not maximal such that $\mathsf{gr}(A)$ is commutative, then $[A_{\leq m}, A_{\leq n}] \subseteq A_{\leq(m+n-d-1)}$ for all $m, n \geq 0$. Thus $\mathsf{gr}_{m+n-d}[A_{\leq m}, A_{\leq n}] = 0$, and hence $\{\mathsf{gr}_m A, \mathsf{gr}_n A\} = 0$. $\square$

SOLUTION TO EXERCISE 1.4.4. For $x, y \in \mathfrak{g}$, we have

$$\{\mathsf{gr}_1 x, \mathsf{gr}_1 y\} = \mathsf{gr}_1(xy - yx) = \mathsf{gr}_1\{x, y\}$$

(denoting the Lie bracket on $\mathfrak{g}$ also by $\{-, -\}$), which is the bracket of (1.D). Then, (1.D) follows from the Leibniz identity (indeed, it is clear that the Leibniz identity uniquely determines the Poisson bracket from the Lie bracket on generators, so there is a unique formula for the LHS of (1.D), and it is easy to check this formula is the RHS). $\square$

SOLUTION TO EXERCISE 1.4.5. By Exercise 1.2.7 (solved above), we only need to identify the Poisson bracket on $\mathsf{gr}\, \mathsf{Weyl}(V)$ with the bracket on $\mathsf{Sym}\, V$. By the Leibniz rule, it suffices to check this on homogeneous generators. In the case of the additive filtration, gr V is homogeneous of degree one, and we have $uv - vw = (u, v)$ in Weyl(V), matching $\{u, v\} = (u, v)$ in Sym V, as desired. In the case of the geometric filtration, the same identity holds, requiring u and v to be homogeneous (of degree zero or one) in V. $\qquad\square$

SOLUTION TO EXERCISE 1.5.2. By definition, Diff is nonnegatively filtered. We only need to show that its associated graded algebra is commutative, i.e., that

$$[\mathrm{Diff}_{\le m}(B), \mathrm{Diff}_{\le n}(B)] \subseteq \mathrm{Diff}_{\le m+n-1}(B).$$

We prove this inductively on the sum $m + n$. Applying ad(b) for $b \in B$, we obtain by the Leibniz identity for commutators,

$\mathrm{ad}(b)([\mathrm{Diff}_{\le m}(B), \mathrm{Diff}_{\le n}(B)])$

$$\subseteq [\mathrm{ad}(b)(\mathrm{Diff}_{\le m}(B)), \mathrm{Diff}_{\le n}(B)] + [\mathrm{Diff}_{\le m}(B), \mathrm{ad}(b)\,\mathrm{Diff}_{\le n}(B)]$$

$$\subseteq [\mathrm{Diff}_{\le m-1}(B), \mathrm{Diff}_{\le n}(B)] + [\mathrm{Diff}_{\le m}(B), \mathrm{Diff}_{\le n-1}(B)],$$

and both terms on the RHS lie in $\mathrm{Diff}_{m+n-1}(B)$ by the inductive hypothesis. $\qquad\square$

SOLUTION TO EXERCISE 1.9.5. (i) This is a matter of checking the definition. Given a representation of Q, we let every path in $\Bbbk Q$ act by the corresponding composite of linear transformations. Given a representation of $\Bbbk Q$, we restrict the representation to the subset $Q \subseteq \Bbbk Q$ to obtain a representation of Q.

(ii) More generally, if we fix (V_i), then every representation of the form $(\rho, (V_i))$ is clearly given by the choices of linear operators $\rho(a) : V_{a_t} \to V_{a_h}$ for all $a \in Q_1$. $\qquad\square$

SOLUTION TO EXERCISE 1.9.6. If there is only one vertex, then paths in the quiver are the same as words in the arrows, which form a basis for the tensor algebra over $\Bbbk Q_1$. $\qquad\square$

SOLUTION TO EXERCISE 1.9.8. For any vector space, $T^*V \cong V \oplus V^*$, canonically. Thus the assertion follows from Exercise 1.9.5.(ii). $\qquad\square$

SOLUTION TO EXERCISE 1.9.10. (i) This follows because the relations are quadratic (hence homogeneous).

(ii) The associated graded space of the relations for $\Pi_\lambda(Q)$ identifies with the span of the relations for $\Pi_0(Q)$, so by Exercise 1.3.4 we have a surjection gr $\Pi_\lambda(Q) \twoheadrightarrow \Pi_0(Q)$.

(iii) For the quiver Q with two vertices and one arrow, from one vertex to the other, we can consider $\lambda = (0, 1)$. Let a, a^* be the two arrows in $\bar{Q}$, with $\lambda_{a_t} = 0$ and $\lambda_{a_h} = 1$. Then $\Pi_\lambda(Q) = \Bbbk \bar{Q}/(a^*a, aa^* - a_h)$. Thus $a^*(aa^*) = a^*a_h = a^*$ in $\Pi_\lambda(Q)$, but also $(a^*a)a^* = 0$. So $a^* = 0$ in $\Pi_\lambda(Q)$, hence also in gr $\Pi_\lambda(Q)$, but this is not true in $\Pi_0(Q)$. $\qquad\square$

SOLUTION TO EXERCISE 1.9.13. By definition, $\mathrm{Rep}_d(\Pi_\lambda(Q))$ is the subspace of the space of all representations of $\bar{Q}$ satisfying the relation $\mu(\rho) = \lambda \cdot \mathrm{Id}$. $\qquad\square$

SOLUTION TO EXERCISE 1.10.1. The first statement follows because the monomials given are those in which xy and yx (which are zero) do not appear. The next statement, that $A_{a,b} = \{0\}$ for $a \neq b$, follows because in this case $xyx = ax = bx$ implies $x = 0$, and then $0 = xy = a$ and $0 = yx = b$ shows that, since one of a and b is nonzero, we obtain the relation $1 = 0$, i.e., $A_{a,b} = \{0\}$.

The final statement, that $A_{a,a} \cong \Bbbk[x, x^{-1}]$, follows because in this case the relations are equivalent to $y = ax^{-1}$. It is clear that a basis of $\Bbbk[x, x^{-1}]$ is given by monomials in either x or $x^{-1} = a^{-1}y$, but not both, proving the next statement. Since the basis is the same as for the case $a = 0$, we obtain that $A_{a,b}$ is flat along the diagonal, as desired. □

SOLUTION TO EXERCISE 1.10.2. (a) One way to correct this, which we will use below, is $e = -\frac{1}{2}D^2$, $f = \frac{1}{2}x^2$, and $h = [e, f] = -\frac{1}{2}(xD + Dx)$.

(b) For this, we first compute that, in a highest weight representation of highest weight m, the element C acts by $\frac{1}{2}m^2 + m$. Indeed, since C is central, it suffices to compute this on the highest weight vector v, where $(ef + fe + \frac{1}{2}h^2)(v) = (h + \frac{1}{2}h^2)(v) = (\frac{1}{2}m^2 + m)v$. Thus we only have to solve $\frac{1}{2}m^2 + m = -\frac{3}{8}$. We obtain the solutions $-\frac{1}{2}, -\frac{3}{2}$, as desired.

(c) Clearly $\{1, x^2, x^4, \ldots\}$ is an eigenbasis under the operator $h = -\frac{1}{2}(xD + Dx)$, with eigenvalues $-\frac{1}{2}, -\frac{5}{2}, -\frac{9}{2}, \ldots$. Moreover, it is clearly generated by the highest weight vector 1. So this is a highest weight representation of highest weight $-\frac{1}{2}$.

(d) We know that all finite-dimensional irreducible representations of $\mathfrak{sl}_2$ are highest weight representations, with nonnegative highest weights. Part (b) above shows that $(U\mathfrak{sl}_2)^\eta$ does not admit such highest weight representations, hence it admits no finite-dimensional irreducible representations. By the isomorphism of (a), neither does $\mathrm{Weyl}(V)^G$. For the final statement, note that any finite-dimensional representation has an irreducible quotient, so the existence of a finite-dimensional representation implies the existence of a finite-dimensional irreducible representation. □

SOLUTION TO EXERCISE 1.10.5. The first statement follows because a surjection of finite-dimensional spaces is an isomorphism if and only if the dimensions are equal. The second statement follows for the same reason, applied to each weight space individually. □

SOLUTION TO EXERCISE 1.10.6. (a) Notice that the defining relations form a Gröbner basis: $xy - yx - z$, $yz - zy - x$, and $zx - xz - y$. Therefore, using the lexicographical ordering on monomials with $x < y < z$, we can uniquely reduce any element to a linear combination of monomials of the form $x^a y^b z^c$, just as for $\Bbbk[x, y, z]$. (That is, since we have a Gröbner basis for a filtered algebra A whose associated graded set is a Gröbner basis for the associated graded algebra $B = \mathrm{gr}(A)$, we conclude that the canonical surjection $B \to \mathrm{gr}(A)$ of Exercise 1.3.4 is an isomorphism.)

It is clear that this algebra is the enveloping algebra of the three-dimensional Lie algebra with basis x, y, z satisfying $[x, y] = z$, $[y, z] = x$, and $[z, x] = y$. This must be isomorphic to $\mathfrak{sl}_2$ since the latter is the unique three-dimensional Lie algebra $\mathfrak{g}$

which is semisimple, i.e., satisfying $[\mathfrak{g}, \mathfrak{g}] = \mathfrak{g}$. Alternatively, we can explicitly write an isomorphism with $\mathfrak{sl}_2$, by the assignment $x = \frac{1}{2}(e - f)$, $y = \frac{1}{2}(e + f)$, and $z = \frac{1}{2}h$.

(b) For example, one can get a deformation which is not flat by setting $[x, y] = x$, $[y, z] = 0$, and $[z, x] = y$. This does not satisfy the Jacobi identity: $[x, [y, z]] + [y, [z, x]] + [z, [x, y]] = 0 + 0 + [z, x] = y$. Thus it will not be flat: the element y is generated by these relations, and hence so is x since $[x, y] = x$. Thus the quotient by these relations is $\Bbbk[z]$.

(c) A Gröbner basis is again given by the defining relations, whose associated graded relations are the defining relations for the skew product $B := \mathrm{Sym}(\Bbbk^2) \rtimes \mathbb{Z}/2 = \Bbbk[x, y] \rtimes \mathbb{Z}/2$, independently of λ. Thus, letting A denote the Cherednik algebra we define here, we get an isomorphism $B \to \mathrm{gr}(A)$, which says (by our definition) that A is a flat deformation of B for all λ.

(d) For the associated graded relations, we still get a semidirect product $\Bbbk[x, y] \rtimes \mathbb{Z}/2$, where now the action of $\mathbb{Z}/2$ is by $\pm \mathrm{Id}$ on x, but is trivial on y. But for the Cherednik algebra, we will get the relation

$$0 = [z, 1] = [z, [xy - yx]] = [z, x]y + x[z, y] - [z, y]x - y[z, x] = 2z(xy - yx),$$

and together with the relation $xy - yx = 1$ we get $z = 0$. Since $z^2 = 1$ we get $1 = 0$, so that the algebra defined by these relations is the zero algebra. □

SOLUTION TO EXERCISE 1.10.7. We identify $\mathrm{Weyl}(V)$ with Weyl_n for dim $V = 2n$. To see that $\mathrm{Weyl}_n \rtimes \mathbb{Z}/2$ is simple, first note that if you have an ideal generated by a nonzero element of the form $f \in \mathrm{Weyl}_n$, then it is the unit ideal, because Weyl_n itself is simple (this is easy to verify: take commutators of f sequentially with generators x_i or y_i until we get a nonzero element of $\Bbbk$). Writing $\mathbb{Z}/2 = \{1, \sigma\}$, we also have $(f\sigma) = (f\sigma)\sigma = (f) = (1)$ for $\sigma \in \mathbb{Z}/2$. Now take an ideal of the form $(a + b\sigma)$ for nonzero $a, b \in \mathrm{Weyl}_n$. Again by taking commutators with generators, we obtain that there is an element of the form $(1 + c\sigma)$ for some $c \in \mathrm{Weyl}_n$. If $c = 0$, we are done. Otherwise, $[x, 1 + c\sigma] = (xc + cx)\sigma$, which cannot be zero since $\mathrm{gr}(xc + cx) = \mathrm{gr}(xc) = \mathrm{gr}(x)\mathrm{gr}(c) \neq 0$. We then get from above that $((xc + cx)\sigma) = (1)$ and hence our ideal is the unit ideal.

For $\mathrm{Weyl}_n^{\mathbb{Z}/2}$, we can also do an explicit computation; alternatively, we can follow the hint. Namely, for $f \in \mathrm{Weyl}_n^{\mathbb{Z}/2}$, note that $f = efe$. Thus,

$$\mathrm{Weyl}_n^{\mathbb{Z}/2} \cdot f \cdot \mathrm{Weyl}_n^{\mathbb{Z}/2} = e(\mathrm{Weyl}_n \rtimes \mathbb{Z}/2)(efe)(\mathrm{Weyl}_n \rtimes \mathbb{Z}/2)e$$
$$= e(\mathrm{Weyl}_n \rtimes \mathbb{Z}/2)(f)(\mathrm{Weyl}_n \rtimes \mathbb{Z}/2)e$$
$$= e(\mathrm{Weyl}_n \rtimes \mathbb{Z}/2)e = \mathrm{Weyl}_n^{\mathbb{Z}/2}. \qquad \square$$

SOLUTION TO EXERCISE 1.10.8. Following the hint, we see that if $\mathrm{gr}(C_\bullet)$ is exact, then so must be $C_\bullet$. Thus, in the situation at hand, since $\mathrm{gr}(Q_\bullet) \to \mathrm{gr}(M)$ is exact, so is $Q_\bullet \to M$. □

SOLUTION TO EXERCISE 1.10.11. (a) Given a codimension-one ideal, we have the algebra morphism $B \to B/B_+ = \Bbbk$, and given the algebra morphism, we can take its kernel, which has codimension one.

(b) It is clear that $TV/(TV)_+ = \Bbbk$, so by (a) this is an augmentation.

(c) Let $V \subseteq B_+$ be any generating subspace (e.g., simply $V = B_+$); we then have a canonical surjection of augmented algebras, $TV \twoheadrightarrow B$. Since this is compatible with the augmentation, the kernel must be in the augmentation ideal, $(TV)_+$, so we can let R be this kernel (or any generating subspace of the kernel).

(d) We follow the hint. The last two differentials become zero by construction, since they involve multiplying V on $\Bbbk$. Then the first homology is V.

(e) The second assertion (about the case that R is spanned by homogeneous elements) follows because we can assume that R has minimal Hilbert series among subspaces spanned by homogeneous elements which generate the same ideal, call it J. This is equivalent to the given condition $R \cap (RV + VR) = \{0\}$, for R spanned by homogeneous elements (hence, R a graded subspace of TV), which we can see because, for all $m \geq 2$, R_m is a complementary subspace in J_m to $J_m \cap (R_{\leq (m-1)})$.

We now consider the assertion $\mathrm{Tor}_A(\Bbbk, \Bbbk) \cong R$. For the complex in the hint to be well-defined, we need to show that the multiplication map $TV \otimes R \to TV \cdot R$ is injective (hence an isomorphism). To see this, for a contradiction, suppose that the kernel, call it K, is nonzero, and that $m \geq 0$ is the least nonnegative integer such that the projection of the kernel to $T^m V \otimes R$ is nonzero. By the assumption $(RV + VR) \cap R = \{0\}$ (in fact we only need $(TV)_+ R \cap R = \{0\}$), we must have $m \neq 0$. Take an element $f \in K$ projecting to a nonzero element of $T^m V \otimes R$. Let (v_i) be a basis of V and write $f = \sum_i v_i f_i$ for $f_i \in TV \otimes R$. Then we clearly have $f_i \in K$ as well for all i; but some f_i must have a nonzero projection to $T^{m-1} V \otimes R$, a contradiction.

Let K be the kernel of the map $B \otimes R \to B \otimes V$, and consider its inverse image $\tilde{K} \subseteq TV \otimes R \subseteq TV$. Then it is clear that $\tilde{K} = (TV \cdot R) \cap (TV \cdot R \cdot TV_+)$. Thus we can let $S = \tilde{K}$ (or any left TV-module generating subspace thereof), and we obtain the extension of the resolution given in the hint. Applying $\otimes_B \Bbbk$, and using the assumption $R \cap (RV + VR) = \{0\}$, we get a complex $S \to R \to V \twoheadrightarrow \Bbbk$. We claim that the maps are zero. The map $R \to V$ is zero by the assumption $R \subseteq ((TV)_+)^2$. The final step, showing that $S \to R$ is zero, uses the full strength of the assumption $R \cap (RV + VR) = \{0\}$. Namely, we can assume $S = \tilde{K} = (TV \cdot R) \cap (TV \cdot R \cdot TV_+)$. Then the kernel of $S \to R$ is $(TV_+ \cdot R) \cap (TV \cdot R \cdot TV_+)$. This is all of S, though, because if $\sum_i f_i r_i = \sum_i g_i r'_i h'_i$ for $f_i, g_i \in TV$, $h_i \in TV_+$, and $r_i, r'_i \in R$, then we can write $f_i = \lambda_i + f_i^+$ for $f_i^+ \in TV_+$, and we then have $\sum_i \lambda_i r_i = -\sum_j f_j^+ r_j + \sum_i (g_i r'_i h'_i) \in R \cap (RV + VR)$, which must be zero. Hence $f_i = f_i^+$, and $(TV_+ \cdot R) \cap (TV \cdot R \cdot TV_+) = (TV \cdot R) \cap (TV \cdot R \cdot TV_+)$, so the map $S \to R$ is zero, as desired. $\qquad\square$

SOLUTION TO EXERCISE 1.10.13. This follows from the previous exercise; namely, in the graded case, we can certainly assume $B = TV/(R)$ where R is spanned by homogeneous elements in degrees ≥ 2. It follows from part (d) that $V \cong \mathrm{Tor}_1(\Bbbk, \Bbbk)$. If R is minimal, then it follows from part (e) that $R \cong \mathrm{Tor}_2(\Bbbk, \Bbbk)$. Hence, if $\mathrm{Tor}_2(\Bbbk, \Bbbk)$

is concentrated in degree two, B is quadratic by taking minimal R. The converse is immediate from part (e) above. $\qquad\square$

SOLUTION TO EXERCISE 1.10.15. If we apply $\otimes_B \Bbbk$ to (1.O), we immediately get that (2) implies (1). We show that (1) implies (2). Take a minimal graded projective resolution as in (1.O), where by minimal we mean that the Hilbert series of the V_i are minimal (which uniquely determines the resolution, inductively constructing $B \otimes V_i \to B \otimes V_{i-1}$ so that $h(V_i; t)$ is minimal). Then V_i is in degrees $\geq i$ for all i. We show by induction that V_i is in degree exactly i for all i. Indeed, if not, and m is minimal such that V_m is not only in degree m, then applying $\otimes_B \Bbbk$ to the minimal resolution, we obtain that $\mathrm{Tor}_m(\Bbbk, \Bbbk) \cong V_m$ is not concentrated in degree m, a contradiction. $\qquad\square$

SOLUTION TO EXERCISE 1.10.16. (a) Taking associated graded, we see that $\mathrm{gr}_3(E \otimes V \cap (V + \Bbbk) \otimes E) \subseteq \mathrm{gr}_3(E \otimes V) \cap \mathrm{gr}_3((V + \Bbbk) \otimes E) = R \otimes V \cap V \otimes R$.

(b) The fact that the sequences are complexes is straightforward, and the exactness at $B \otimes V$, B, $A \otimes V$, and A was already observed in Exercise 1.10.11.(d). Now, restrict the first complex (which is a complex of graded modules) to degrees ≤ 3. Then, its exactness follows from Exercise 1.10.11.(e), since in this case the degree ≤ 3 part of $(R)V \cap (TV)_+ \cdot R$ is just $RV \cap VR$, which is S by definition.

(c) Note that the maps $A \otimes T \to A \otimes E$ and $B \otimes S \to B \otimes R$ are injective in degrees ≤ 3. Restricting to degrees ≤ 3, the second complex deforms the first except possibly at $A \otimes T$. Moreover, in this situation, the associated graded of the kernel of $A \otimes E \to A \otimes V$ must be S in degree three; since E deforms R, this implies that the kernel lies in $(V \oplus \Bbbk) \otimes E$, and hence must be $(V + k)E \cap EV$ (here the parentheses do *not* denote an ideal).

(d) The solution is outlined in the problem sheet. $\qquad\square$

SOLUTION TO EXERCISE 2.3.2. Here we note that, if B is finite-dimensional, then $B[[\hbar]] = B \otimes_{\Bbbk} \Bbbk[[\hbar]]$, the ordinary tensor product (allowing only finite linear combinations). Explicitly, if $b_1, \ldots, b_n$ is a basis of B, then $B[[\hbar]]$ is a free $\Bbbk[[\hbar]]$-module with basis $b_1, \ldots, b_n$. Then, (2.E) follows from the fact that $bf \star b'g = (b \star b')fg$ for $b, b' \in B$ and $f, g \in \Bbbk[[\hbar]]$.

On the other hand, if B is infinite-dimensional, then this argument does not apply: knowing $b \star b'$ only will determine $(b_1 f_1 + \cdots + b_m f_m) \star (c_1 g_1 + \cdots + c_n g_n)$ by $\Bbbk[[\hbar]]$-linearity, for $b_1, \ldots, b_m, c_1, \ldots, c_n \in B$ and $f_1, \ldots, f_m, g_1, \ldots, g_n \in \Bbbk[[\hbar]]$. So multiplying two series with coefficients linearly independent in B will not be determined by $\Bbbk[[\hbar]]$-linearity from the star product on B, i.e., (2.E) need not hold. (We will not attempt to construct a counterexample, however.) $\qquad\square$

SOLUTION TO EXERCISE 2.3.10. (a) This follows immediately from the formula: in particular, $x_i \star y_j = x_i y_j + \frac{1}{2}\hbar\delta_{ij}$, $y_j \star x_i = x_i y_j - \frac{1}{2}\hbar\delta_{ij}$, $x_i \star x_j = x_i x_j$, and $y_i \star y_j = y_i y_j$. Also, $f \star z_i = f z_i = z_i \star f$ for all f.

(b) This follows because $e^{\frac{1}{2}\hbar\pi} = \sum_{n \geq 0} \frac{1}{n!}\frac{1}{2^n}\hbar^n\pi^n$, and π^n decreases degree by $2n$, so the sum evaluated on any element $f \otimes g$ is finite.

(c) Since we have a star product defined over $\Bbbk[\hbar]$, i.e., $(\mathcal{O}(X)[\hbar], \star)$, we can quotient by the ideal $(\hbar - 1)$ (which is not the unit ideal) and get back a star product which is obtained from the above by setting $\hbar = 1$; in particular it is clearly a filtered deformation of the undeformed product.

(d) By part (a), the given map is a homomorphism, and it is clearly surjective. To test injectivity, it suffices to show that the associated graded morphism is injective, but the latter morphism is the identity. Uniqueness follows because the x_i, y_i, and z_i generate the source (they also generate the target, of course).

(e) If we show that the map, call it Φ, is an algebra morphism, then it will automatically invert the morphism of (d), since it sends the generators x_i, y_i, and z_i to the generators x_i, y_i, and z_i. We have to show that, for generators $f_i, g_j \in \{x_1, \ldots, x_m, y_1, \ldots, y_m, z_1, \ldots, z_{n-2m}\}$, that, with multiplication taken in the Weyl algebra,

$$\Phi(f_1 \cdots f_n)\Phi(g_1 \cdots g_p) = \Phi(f_1 \cdots f_n \star g_1 \cdots g_p). \tag{II.A}$$

The LHS can be expanded as

$$\frac{1}{n!p!} \sum_{\sigma \in S_n, \tau \in S_p} f_{\sigma(1)} \cdots f_{\sigma(n)} g_{\tau(1)} \cdots g_{\tau(p)}. \tag{II.B}$$

Now we can attempt to symmetrize this by applying the relations of the Weyl algebra. More precisely, recall that an n, p-shuffle is a permutation $\theta \in S_{n+p}$ such that $\theta(1) < \cdots < \theta(n)$ and $\theta(n+1) < \cdots < \theta(n+p)$. Let $\mathrm{Sh}_{n,p} \subseteq S_{n+p}$ be the set of all such shuffles, which has size $(n+p)!/(n!p!)$. For each $\theta \in \mathrm{Sh}_{n,p}$, and each summand $f_{\sigma(1)} \cdots f_{\sigma(n)} g_{\tau(1)} \cdots g_{\tau(p)}$ of the above, we can rearrange the terms according to the shuffle θ, by moving first $f_{\sigma(n)}$ to its proper place, then $f_{\sigma(n-1)}$, etc.; each time we move an f_i past a g_j, we apply the relation $f_i g_j = g_j f_i + [g_j, f_i]$. Since the g_j and f_i are generators of the Weyl algebra, $[g_j, f_i] \in \Bbbk[[\hbar]]$. Doing this, we obtain the following identity. Define the following index set we will use for the summation that results:

$$\mathrm{Pairs}_{n,p} := \{(I, J, \iota) \mid I \subseteq \{1, \ldots, n\}, J \subseteq \{1, \ldots, p\}, \iota : I \to J \text{ bijective}\},$$

which is the set of partial pairings between $\{1, \ldots, n\}$ and $\{1, \ldots, p\}$. Define the subset

$$\mathrm{Pairs}_\theta = \{(I, J, \iota) \in \mathrm{Pairs}_{n,p} \mid \iota(i) < \theta(i), \forall \in i\},$$

obeying the condition $\iota(i) < \theta(i)$. For each $1 \leq i \leq n$, let $[f_i]$ denote f_i if $i \notin I$, and otherwise let it denote $[f_i, g_{\iota(i)}] = \{f_i, g_{\iota(i)}\}$. Let $\{g_j\}$ denote g_j if $j \notin J$ and 1 otherwise. We then get

$$f_1 \cdots f_n g_1 \cdots g_p = \sum_{(I,J,\iota)\in\mathrm{Pairs}_\theta} \theta\left(\prod_{i=1}^{n} [f_i] \prod_{j=1}^{n} \{g_j\}\right),$$

where the θ rearranges the following $n+p$ terms according to the permutation. Call the RHS R_θ. We can then write the LHS as $(n!p!/(n+p)!) \sum_{\theta \in \mathrm{Sh}_{n,p}} R_\theta$, a symmetrization. Applying this to (II.B), one can check that we get the RHS of (II.A) identically. Namely, both yield the following expansion:

$$\sum_{(I,J,\iota)\in\mathrm{Pairs}_{n,p}} 2^{-|I|}\,\mathrm{symm}\left(\prod_{i=1}^{n}[f_i]\prod_{j=1}^{n}\{g_j\}\right),$$

where now we take the product in the symmetric algebra and apply $\mathrm{symm}: \mathrm{Sym}\,V \to \mathrm{Weyl}(V)$. This is so because, by symmetry, both sides have to be a sum of the above form but possibly with coefficients $c_{(I,J,\iota)}$ replacing $2^{-|I|}$; the LHS of (II.A) has these coefficients being the probability that, in a random ordering of $f_1, \ldots, f_n, g_1, \ldots, g_p$, we have f_i occurring before $g_{\iota(i)}$ (which is $2^{-|I|}$), and the RHS is $2^{-|I|}$ by definition. See also [115, §4.3] for a similar proof in the context of a Moyal product on algebras defined from quivers. □

SOLUTION TO EXERCISE 2.3.11. Since π^2 annihilates $v \otimes w$ when v, w are generators, it is immediate that any quantization will satisfy the relations of part (a). Therefore the map given is an isomorphism. The fact that it is the identity modulo $\hbar$ is equivalent to the statement that the associated graded morphism is the identity endomorphism of $\mathcal{O}(\mathbb{A}^n)[\![\hbar]\!]$, which is immediate from the definition. □

SOLUTION TO EXERCISE 2.7.2. As above, if A is finite-dimensional, then $A[\![\hbar]\!] = A \otimes_{\Bbbk} \Bbbk[\![\hbar]\!]$, so the space of $\Bbbk[\![\hbar]\!]$-linear endomorphisms of A is identified with $\mathrm{End}_{\Bbbk}(A) \otimes \Bbbk[\![\hbar]\!]$, and every element therein is continuous. Note that, if A is infinite-dimensional, then a $\Bbbk[\![\hbar]\!]$-linear automorphism need not be continuous. □

SOLUTION TO EXERCISE 2.8.1. (a) Here is a bimodule resolution of $\mathrm{Sym}\,V$:[1]

$$0 \to \mathrm{Sym}\,V \otimes \textstyle\bigwedge^{\dim V} V \otimes \mathrm{Sym}\,V \to \mathrm{Sym}\,V \otimes \textstyle\bigwedge^{\dim V-1} V \otimes \mathrm{Sym}\,V \to \cdots$$
$$\to \mathrm{Sym}\,V \otimes V \otimes \mathrm{Sym}\,V \to \mathrm{Sym}\,V \otimes \mathrm{Sym}\,V \twoheadrightarrow \mathrm{Sym}\,V,$$

$$f \otimes (v_1 \wedge \cdots \wedge v_i) \otimes g \mapsto \sum_{j=1}^{i}(-1)^{j-1}(fv_j) \otimes (v_1 \wedge \cdots \hat{v}_j \cdots \wedge v_i) \otimes g$$
$$+ (-1)^j f \otimes (v_1 \wedge \cdots \hat{v}_j \cdots \wedge v_i) \otimes (v_j g).$$

Cutting off the $\mathrm{Sym}\,V$ term and applying $\mathrm{Hom}_{(\mathrm{Sym}\,V)^e}(-, \mathrm{Sym}\,V \otimes \mathrm{Sym}\,V)$, we get the same complex, up to the isomorphism $\bigwedge^{\dim V-i} V^* \cong \bigwedge^i V$, which comes from contracting with a nonzero element of $\bigwedge^{\dim V} V$. Thus, we compute that $\mathrm{HH}^\bullet(\mathrm{Sym}\,V, \mathrm{Sym}\,V \otimes \mathrm{Sym}\,V) = \mathrm{Sym}\,V[-\dim V]$, as desired.

(b) The latter complex deforms to give a resolution of $\mathrm{Weyl}(V)$, given by the same formula as above:

$$0 \to \mathrm{Weyl}(V) \otimes \textstyle\bigwedge^{\dim V} V \to \mathrm{Weyl}(V) \otimes \textstyle\bigwedge^{\dim V-1} V \to \cdots$$
$$\to \mathrm{Weyl}(V) \otimes \mathrm{Weyl}(V) \twoheadrightarrow \mathrm{Weyl}(V),$$

---

[1]It actually can be directly obtained from the given resolution of $\Bbbk$, if one notices that $\mathrm{Sym}\,V$ is a Hopf algebra: to go from the above resolution to the bimodule one, one applies the functor of induction from $\mathrm{Sym}\,V$ to $(\mathrm{Sym}\,V)^{\otimes 2}$ (using the Hopf algebra structure); for the opposite direction, one applies the functor $\otimes_{\mathrm{Sym}\,V}\Bbbk$ (which exists for arbitrary augmented algebras).

$$f \otimes (v_1 \wedge \cdots \wedge v_i) \otimes g \mapsto \sum_{j=1}^{i} (-1)^{j-1} (fv_j) \otimes (v_1 \wedge \cdots \hat{v}_j \cdots \wedge v_i) \otimes g$$
$$+ (-1)^j f \otimes (v_1 \wedge \cdots \hat{v}_j \cdots \wedge v_i) \otimes (v_j g).$$

As pointed out, to see it is a resolution, we only need to verify it is a complex, which is easy. Now, if we cut off the last term $\text{Weyl}(V)$ and apply $\text{Hom}_{\text{Weyl}(V)^e}(-, \text{Weyl}(V))$, just as before we get the same complex as before applying this functor, so the homology will again be concentrated in degree d: $HH^{\bullet}(\text{Weyl}(V), \text{Weyl}(V) \otimes \text{Weyl}(V)) = \text{Weyl}(V)[-\dim V]$, as desired.

(c) The fact that this forms a complex, call it $P_{\bullet}$ (cutting off the $U\mathfrak{g}$, setting $P_0 = U\mathfrak{g} \otimes U\mathfrak{g}$), is a straightforward verification. Then, setting $Q^{\bullet} := \text{Hom}_{U\mathfrak{g}^e}(P_{\bullet}, U\mathfrak{g} \otimes U\mathfrak{g})$, we may not any longer get an isomorphic complex (this is related to the fact that $U\mathfrak{g}$ is not Calabi–Yau in general, as we will see in the next exercise sheet), but the associated graded complex is still the resolution of $\text{Sym}\, V$, and hence it is still a deformation of the resolution of $\text{Sym}\, V$. Thus, by the same argument as before, $Q^{\bullet}$ is a resolution of its $(\dim V)$-th cohomology, $\text{HH}^{\dim V}(U\mathfrak{g}, U\mathfrak{g} \otimes U\mathfrak{g})$, which is a filtered deformation of the bimodule $\text{Sym}\, V$ (i.e., it is a filtered $U\mathfrak{g}$-bimodule whose associated graded $\text{Sym}\, V$-bimodule is $\text{Sym}\, V$).

(d) The first paragraph is straightforward to verify explicitly following the details given.

For the final assertion, the first isomorphism is immediate from the first paragraph. For the second, we can write $U\mathfrak{g}^{ad} = Z(U\mathfrak{g}) \oplus W$ where the center, $Z(U\mathfrak{g})$, is the sum of all trivial subrepresentations, and W is the sum of all nontrivial irreducible subrepresentations. Thus, $H^{\bullet}_{CE}(\mathfrak{g}W) = 0 = H_{\bullet}^{CE}(\mathfrak{g}W)$, and we obtain the final isomorphism since $Z(U\mathfrak{g}) = Z(U\mathfrak{g}) \otimes_{\Bbbk} \Bbbk$ (regarding the latter as the vector space $Z(U\mathfrak{g})$ tensored with the trivial representation $\Bbbk$). $\qquad\square$

SOLUTION TO EXERCISE 2.8.3. Since the deformation of the quadratic relation is by a scalar term only, denoting by $E \subseteq V^{\otimes 2} + \Bbbk$ the deformed relation, we have (referring to Exercise 1.10.16) that $T = E \otimes V \cap (V + \Bbbk) \otimes E$ must equal $E \otimes V \cap V \otimes E$. Next note that $S = R \otimes V \cap V \otimes R$ is spanned by half of the undeformed Jacobi identity, $[x, y] \otimes z + [y, z] \otimes x + [z, x] \otimes y$. Therefore, the map $\text{gr}(T) \to S$ is an isomorphism if and only if the given (deformed) Jacobi identity is satisfied. $\qquad\square$

SOLUTION TO EXERCISE 2.8.4. (a) Given only A together with $\phi : A/\hbar A \xrightarrow{\sim} B$, A is generated by $\phi^{-1}(V)$ as a topological $\Bbbk[[\hbar]]$-module. This follows because it is so generated modulo $\hbar$, and A is $\hbar$-adically complete as it is a topologically free $\Bbbk[[\hbar]]$-module. So we have a continuous surjection $q : TV[[\hbar]] \to A$. Now, for every element $r \in R$, we have $q(r) \in \hbar A$, so there exists a power series $r' \in \hbar TV[[\hbar]]$ such that $q(r + \hbar r') = 0$. Let $\{r_i\}$ be a basis of R and let $\{r'_i\}$ be as before. Let E be the span of the r'_i. Then $E \to TV[[\hbar]] \to TV$ is an isomorphism onto R. We also have an obvious surjection $\psi : TV[[\hbar]]/(E) \to A$. Moreover, the composition of this with the quotient to $A/\hbar \cong B$ clearly is the identity on V. We have only to show that ψ is an isomorphism. To see this, consider the $\hbar$-adic filtration. We get the surjection

$\mathrm{gr}(\psi) : T V[\![\hbar]\!]/(R) \to \mathrm{gr}_\hbar(T V[\![\hbar]\!]/(E)) = \mathrm{gr}_\hbar(A)$. It suffices to show that this is an isomorphism. Composing with the isomorphism $\mathrm{gr}_\hbar(A) \cong B[\![\hbar]\!]$, we obtain a surjection $T V[\![\hbar]\!](R) \to B[\![\hbar]\!]$, which is nothing but the canonical isomorphism. So $\mathrm{gr}(\psi)$ is an isomorphism, and hence so is ψ.

(b) Most of the details here are provided; we leave to the reader to fill in the missing ones. $\qquad\square$

SOLUTION TO EXERCISE 2.8.6. Suppose A is a graded formal deformation over $\Bbbk[\![\hbar]\!]$ of a Koszul algebra B. Since B is Koszul, it admits a graded free bimodule resolution $P_\bullet \to B$, with each P_i generated in degree i. The same argument as in Exercise 1.10.15.(d) (taking now the $\hbar$-adic filtration, instead of the weight filtration) shows that $P_\bullet \to B$ deforms to a graded bimodule resolution $P_\bullet^\hbar \to A$, with each $P_i^\hbar$ free over A of the same rank as P_i over B. Moreover, $P_i^\hbar$ is also generated in degree i, so A is Koszul over $\Bbbk[\![\hbar]\!]$ in the sense given. $\qquad\square$

SOLUTION TO EXERCISE 3.7.4. (a) We follow the hint. To see σ is a $\Bbbk$-algebra homomorphism, note that $\sigma(ab) \cdot 1 = 1 \cdot ab = (1 \cdot a) \cdot b = \sigma(a) \cdot 1 \cdot b = \sigma(a)\sigma(b) \cdot 1$. Then $\sigma(ab) = \sigma(a)\sigma(b)$ because M is a free left module. It is clear that $\sigma(1) = 1$, and easy to check that σ is $\Bbbk$-linear. Then, let τ be defined by $h \cdot 1 = 1 \cdot \tau(h)$. Then it is clear that $\sigma \circ \tau = \tau \circ \sigma = \mathrm{Id}$.

(b) This follows immediately from (a), noting that, as a graded module on either side, $M \cong A$ (since M is generated in degree zero), and σ must be a graded automorphism. $\qquad\square$

SOLUTION TO EXERCISE 3.7.5. This follows from the hint: we only need to observe that the given action on (3.E) is indeed compatible with the differential and outer bimodule structure. More generally, if M is a $A^e \otimes B$-module, then $\mathrm{HH}^\bullet(A, M)$ always has a canonical B-module structure; in this case we can view $M = A \otimes A$ as an $A^e \otimes A^e$-module, where the first A^e acts via the outer action and the second A^e via the inner action. $\qquad\square$

SOLUTION TO EXERCISE 3.7.11. (a) We computed $\mathrm{HH}^\bullet(\mathrm{Sym}\, V, \mathrm{Sym}\, V \otimes \mathrm{Sym}\, V)$ and $\mathrm{HH}^\bullet(\mathrm{Weyl}(V), \mathrm{Weyl}(V) \otimes \mathrm{Weyl}(V))$ in the last exercise sheet, as vector spaces. If we take care of the bimodule action, we see that $\mathrm{HH}^{\dim V}(\mathrm{Sym}\, V, \mathrm{Sym}\, V \otimes \mathrm{Sym}\, V)$ really is $\mathrm{Sym}\, V$ as a bimodule, since the bimodule complex computing this is isomorphic to the original one computing $\mathrm{Sym}\, V$ as a bimodule. The same is true for $\mathrm{Weyl}(V)$.

(b) The resulting complex is still exact because $\Bbbk[\Gamma]$ is flat over $\Bbbk$ (in fact, free), and we haven't changed the differential by applying $\otimes_\Bbbk \Bbbk[\Gamma]$. We only need to check we get a complex of $A \rtimes \Gamma$-bimodules, and this follows directly from the definition of the bimodule structure. For the final assertion, note that, for general $\Gamma < \mathrm{GL}(V)$, we can still consider $\mathrm{Sym}\, V \rtimes \Gamma$, and the above yields a bimodule resolution. Applying $\mathrm{Hom}_{(\mathrm{Sym}\, V)^e}(-, \mathrm{Sym}\, V)$, we get a complex computing $\mathrm{HH}^\bullet(\mathrm{Sym}\, V, \mathrm{Sym}\, V \otimes \mathrm{Sym}\, V)$, and we get that this is $(\mathrm{Sym}\, V)^\sigma[-d]$, where σ is the automorphism $\sigma(v \otimes \gamma) = \det(\gamma)(v \otimes \gamma)$. So this is trivial if $\Gamma < \mathrm{SL}(V)$, which yields that in this case $\mathrm{Sym}\, V \rtimes \Gamma$ is trivial.

REMARK. For a general algebra B and automorphism σ, we have $B \cong B^\sigma$ as bimodules if and only if σ is inner, i.e., of the form $\sigma(a) = x^{-1}ax$ for some invertible $x \in B$. In the above situation of $B = \operatorname{Sym} V \rtimes \Gamma$, and $\sigma(v \otimes \gamma) = \det(\gamma)(v \otimes \gamma)$, one can check that σ is not inner when Γ is not in $\operatorname{SL}(V)$, so that actually $\operatorname{Sym} V \rtimes \Gamma$ is Calabi–Yau (for Γ finite) if and only if $\Gamma < \operatorname{SL}(V)$. $\qquad\square$

SOLUTION TO EXERCISE 3.8.2. Let $\Gamma_n := \sum_{i=1}^n \varepsilon^i \gamma_i$. Working in $\Bbbk[\varepsilon]/(\varepsilon^{n+2})$, we have $\frac{1}{2}[\mu + \Gamma_n, \mu + \Gamma_n] = \delta \varepsilon^{n+1}$, for some $\delta \in C^2(A, A)$. Moreover, we claim that δ is a cocycle. Using $[\mu, -] = d(-)$, we have

$$d(\tfrac{1}{2}[\mu + \Gamma_n, \mu + \Gamma_n]) = [d\Gamma_n, \mu + \Gamma_n] = [d\Gamma_n, \Gamma_n] = [[\mu, \Gamma_n], \Gamma_n].$$

Then, since $[\Gamma_n]$ is a multiple of ε and $[\mu, \Gamma_n] + \frac{1}{2}[\Gamma_n, \Gamma_n]$ is a multiple of ε^{n+1}, the RHS equals (modulo ε^{n+2}):

$$-\left[\tfrac{1}{2}[\Gamma_n, \Gamma_n], \Gamma_n\right] = 0, .$$

by the Jacobi identity (which holds on cochains identically). Thus δ is indeed a three-cycle, and defines a cohomology class $[\delta] \in \operatorname{HH}^3(A)$.

Now, to extend Γ_n to an $(n+1)$-st order deformation $\Gamma_{n+1} = \Gamma_n + \varepsilon^{n+1}\gamma_{n+1}$, we need to solve the equation $\frac{1}{2}[\mu + \Gamma_{n+1}, \mu + \Gamma_{n+1}] = 0$ (modulo ε^{n+2}), which simplifies to

$$d(\gamma_{n+1}) + \delta = 0.$$

Hence, the condition to extend Γ_n to an $(n+1)$-st order deformation is the condition that the cohomology class $[\delta] \in \operatorname{HH}^3(A)$ is zero.

Finally, the space of choices of γ_{n+1} is equal to $d^{-1}(\delta)$, which is an affine space on the space of Hochschild two-cocycles. If we apply an automorphism of the form $\Phi = \operatorname{Id} + \varepsilon^{n+1}(f)$, we get that $\Phi \circ (\mu + \Gamma_n) \circ (\Phi^{-1} \otimes \Phi^{-1})$ is nothing but $\mu + \Gamma_n + [\varepsilon^{n+1}f, \Gamma_n] = \mu + \Gamma_n - \varepsilon^{n+1}(df)$. So, up to these gauge equivalences, the space of extensions to an $n+1$-st order deformation Γ_{n+1} is an affine space on $\operatorname{HH}^2(A)$ (provided it is nonempty, i.e., δ is a coboundary). $\qquad\square$

SOLUTION TO EXERCISE 3.10.1. We computed $\operatorname{HH}^\bullet(U\mathfrak{g}, U\mathfrak{g} \otimes U\mathfrak{g})$ as a vector space in the last exercise sheet. A little more work computes the bimodule structure. We find that $\operatorname{HH}^{\dim V}(U\mathfrak{g}, U\mathfrak{g} \otimes U\mathfrak{g}) \cong U\mathfrak{g}^\sigma$ for σ the automorphism $\sigma(x) = x - \operatorname{tr}(\operatorname{ad}(x))$ because $\operatorname{tr}(ad(x))$ coincides with the action of $\operatorname{ad}(x)$ on $\bigwedge^{\dim V} V$. Therefore $U\mathfrak{g}$ is twisted Calabi–Yau, and actually Calabi–Yau when $\operatorname{tr}(ad(x)) = 0$ for all x (i.e., $\mathfrak{g}$ is unimodular). For the final statement, note that, if $\mathfrak{g} = [\mathfrak{g}, \mathfrak{g}]$, then for some $x_1^{(i)}, x_2^{(i)} \in \mathfrak{g}$, we have $x = \sum_i [x_1^{(i)}, x_2^{(i)}]$, and hence $\operatorname{tr}(\operatorname{ad}(x)) = \sum_i \operatorname{tr}([\operatorname{ad}(x_1^{(i)}), \operatorname{ad}(x_2^{(i)})]) = 0$. It is clear that the adjoint action, hence also its trace, is zero on an abelian Lie algebra, and we conclude that all reductive Lie algebras are unimodular. Finally, since the trace of a nilpotent linear map is zero, nilpotent Lie algebras are unimodular. $\qquad\square$

REMARK. If $\operatorname{tr}(\operatorname{ad}(x)) \neq 0$ for some x, then $U\mathfrak{g}^\sigma$ is not isomorphic to $U\mathfrak{g}$, since the automorphism σ is not inner. In fact, the only invertible elements of $U\mathfrak{g}$ are scalars, since $fg = 1$ implies $\operatorname{gr}(f)\operatorname{gr}(g) = 1$ as $\operatorname{Sym}\mathfrak{g}$ has no zero divisors. (We remark that we could alleviate this by completing $U\mathfrak{g}$ at the augmentation ideal $(\mathfrak{g})$, in which case all elements not in the ideal are invertible; however, σ does not preserve this ideal so it still cannot be inner.) So $U\mathfrak{g}$ is Calabi–Yau if and only if $\mathfrak{g}$ is unimodular.

SOLUTION TO EXERCISE 3.10.3. (a) The details for the first isomorphism (for Hochschild cohomology) are already given in the hint. We omit the similar computation for Hochschild homology.

(b) Write $A \rtimes \Gamma = \bigoplus_{g \in \Gamma} A \cdot g$ as an A-bimodule. For every pair of elements $g, h \in \Gamma$, we have an isomorphism by conjugation, $\mathrm{Ad}(g) : \mathrm{HH}^\bullet(A, A \cdot h) \xrightarrow{\sim} \mathrm{HH}^\bullet(A, A \cdot (ghg^{-1}))$. Therefore, letting $\mathcal{C}_h := \mathrm{Ad}(\Gamma) \cdot h$ denote the conjugacy class, we obtain, as a Γ-representation,

$$\mathrm{HH}^\bullet(A, A \cdot \mathcal{C}_h) \cong \mathrm{Ind}^\Gamma_{Z_h(\Gamma)} \mathrm{HH}^\bullet(A, A \cdot h),$$

and taking Γ-invariants, we get $\mathrm{HH}^\bullet(A, A \cdot h)^{Z_h(\Gamma)}$. Summing over all conjugacy classes yields the desired formula. Now, for an arbitrary Γ-representation V, we have

$$V \otimes (\Bbbk \cdot \mathcal{C}_g) \cong \mathrm{Ind}^\Gamma_{Z_h(\Gamma)} V,$$

and taking Γ-invariants, we get $(V \otimes (\Bbbk \cdot \mathcal{C}_g))^\Gamma \cong V^{Z_h(\Gamma)}$. Applying this to the above, and summing over all conjugacy classes, yields the desired formula.

(c) The necessary details are in the hint (we omit the corresponding ones for Hochschild homology).

(d) The necessary details here are also given; see [1] for the full details (it is only a few pages). $\qquad\square$

SOLUTION TO EXERCISE 4.1.5. First, we verify the final assertion, that the skew-symmetrization of a dg (right) pre-Lie product is a Lie bracket. This is because the pre-Lie identity, upon skew-symmetrization, becomes a multiple of the Jacobi identity, and the compatibility with the differential carries over to the skew-symmetrization.

Going back to the situation of $C^\bullet(A)[1]$ for A an associative algebra, we have to check that $\circ$ is a derivation and that it satisfies the right pre-Lie identity. The derivation property follows from the fact that $d(\gamma \circ \eta)$ and $d(\gamma) \circ \eta + (-1)^{|\gamma|}\gamma \circ (d\eta)$, applied to $a_1 \otimes \cdots \otimes a_{m+n}$, are both

$$\sum_{i=1}^{m} (-1)^{(i-1)(n+1)} \left(a_1 \gamma \left(a_2 \otimes \cdots \otimes a_i \otimes \eta(a_{i+1} \otimes \cdots \otimes a_{i+n}) \otimes a_{i+n+1} \otimes \cdots \otimes a_{m+n} \right) \right.$$

$$+ \sum_{k=1}^{i-1} (-1)^k \gamma \left(a_1 \otimes \cdots \otimes a_k a_{k+1} \cdots \otimes a_i \otimes \eta(a_{i+1} \otimes \cdots \otimes a_{i+n}) \otimes a_{i+n+1} \otimes \cdots \otimes a_{m+n} \right)$$

$$+ \sum_{k=i}^{i+n-1} (-1)^k \gamma \left(a_1 \otimes \cdots \otimes a_i \otimes \eta(a_{i+1} \otimes \cdots \otimes a_k a_{k+1} \cdots \otimes a_{i+n}) \otimes a_{i+n+1} \otimes \cdots \otimes a_{m+n} \right)$$

$$+ \sum_{k=i+n}^{m+n-1} (-1)^k \gamma \left(a_1 \otimes \cdots \otimes a_i \otimes \eta(a_{i+1} \otimes \cdots \otimes a_{i+n}) \otimes a_{i+n+1} \otimes \cdots \otimes a_k a_{k+1} \cdots \otimes a_{m+n} \right)$$

$$\left. + (-1)^{m+n} \gamma \left(a_1 \otimes \cdots \otimes a_i \otimes \eta(a_{i+1} \otimes \cdots \otimes a_{i+n}) \otimes a_{i+n+1} \otimes \cdots \otimes a_{m+n-1} \right) a_{m+n} \right).$$

The basic point is that the terms on the LHS are all printed above: they involve multiplying adjacent components either inside γ or inside η with a sign. The corresponding terms on the RHS all occur with the same sign, and the RHS also has some terms which cancel pairwise, of the form $\pm\gamma(a_1 \otimes \cdots \otimes a_i \eta(a_{i+1} \otimes \cdots \otimes a_{i+n}) \otimes \cdots \otimes a_{m+n})$ and similarly replacing the middle by $\eta(a_1 \otimes \cdots \otimes a_{i+n-1})a_{i+n}$. One could (and should) think of this diagrammatically, where η takes n successive inputs to one output, γ takes m successive inputs to one output, and the multiplication map takes two successive inputs to one output, and then both the LHS and RHS express as the same signed combination of diagrams.

A similar, but more involved, direct computation shows that the pre-Lie identity is satisfied (see [109]). The basic idea is that, diagrammatically, both sides are the ways of applying η to some successive inputs and θ simultaneously to other successive inputs, and finally applying γ to the result. $\qquad\square$

SOLUTION TO EXERCISE 4.1.6. (a) The identity is again a similar explicit computation; we refer to [109] for details. Both sides diagrammatically give the result of applying γ_1 and γ_2 each to different blocks of successive inputs (so γ_1 to the leftmost or rightmost $|\gamma_1|$ inputs, and γ_2 to the other inputs) and then multiplying the result, with an appropriate sign. On the RHS, all other ways of applying γ_1, γ_2, and the multiplication cancel pairwise.

One can similarly verify the identity that the cup product is compatible with the differential, so it descends to a binary operation on cohomology. Using the identity, we see that the LHS is a coboundary, hence the cup product is symmetric on cohomology, as desired.

Note that, since the cup product is associative on cochains, it is also on cohomology, so we get a commutative graded algebra structure on Hochschild cohomology.

(b) Note that the Gerstenhaber bracket is the skew-symmetrization of the circle product, and so $d[\gamma_1, \gamma_2] - [d\gamma_1, \gamma_2] - (-1)^{|\gamma_1|}[\gamma_1, d\gamma_2]$ is the skew-symmetrization of the LHS, which is zero. So the Gerstenhaber bracket is indeed compatible with the differential, which implies we obtain a Gerstenhaber bracket on Hochschild cohomology.

We remark that one can similarly directly verify the Leibniz identity for the Gerstenhaber bracket on Hochschild cohomology (see [109]). $\qquad\square$

SOLUTION TO EXERCISE 4.4.1. Since we have assumed $G < \mathsf{GL}_n$ and $\mathfrak{g} < \mathfrak{gl}_n$ and $\Bbbk = \mathbb{R}$ or $\mathbb{C}$, we can write

$$\gamma \cdot d(\iota \circ \gamma) = \exp(\beta) \cdot d(\exp^{-\beta}) = \mathrm{Ad}(\exp^{\beta})(d) = \exp(\mathrm{ad}(\beta))(d)$$
$$= \frac{\exp(\mathrm{ad}(\beta)) - 1}{\mathrm{ad}(\beta)}(-d\beta),$$

where $\mathrm{Ad}(\exp^{\beta})(d)$ and $\exp(\mathrm{ad}(\beta))(d)$ are just two formal expressions (standing for certain infinite linear combinations of terms of the form $\beta^a d\beta^b$, which are equal because $\mathrm{Ad}(\exp^{-}) = \exp(\mathrm{ad}(-))$ as formal series). This implies (4.E). $\qquad\square$

SOLUTION TO EXERCISE 4.5.5. From $\pi \in \bigwedge^2 \text{Vect}(X)$ we define the bracket $\{f, g\}$ as $\pi(f \otimes g)$. The skew-symmetry of the bracket is immediate from the skew-symmetry of π, and the Leibniz rule follows from the fact that $\text{Vect}(X)$ acts by derivations on $\mathcal{O}(X)$.

We need to check that $[\pi, \pi] = 0$ if and only if the Jacobi identity is satisfied. Write $\pi = \sum_i \xi_i^{(1)} \otimes \xi_i^{(2)}$, a skew-symmetric element of $\text{Vect}(X)^{\otimes 2}$ (which is identified with the image of $\sum_i \xi_i^{(1)} \wedge \xi_i^{(2)}$ under the skew-symmetrization map $\bigwedge^2 \text{Vect}(X) \to \text{Vect}(X)^{\otimes 2}$). Then we have

$$[\pi, \pi] = 4 \sum_{i,j} [\xi_i^{(1)}, \xi_j^{(1)}] \wedge \xi_j^{(2)} \wedge \xi_i^{(2)}.$$

Let $\Phi = \sum_{i,j} [\xi_i^{(1)}, \xi_j^{(1)}] \otimes \xi_j^{(2)} \otimes \xi_i^{(2)}$. Then

$$\Phi(f \otimes g \otimes h) = \{\{f, g\}, h\} - \{\{f, h\}, g\}.$$

Skew-symmetrizing and multiplying by 4, we get

$$[\pi, \pi](f \otimes g \otimes h) = -\tfrac{8}{3}(\{f, \{g, h\}\} + \{g, \{h, f\}\} + \{h, \{f, g\}\}),$$

which implies that $[\pi, \pi] = 0$ if and only if the Jacobi identity is satisfied. $\square$

SOLUTION TO EXERCISE 4.8.9. Given a dgla morphism $\phi : \mathfrak{g} \to \mathfrak{h}$, we need to show that the induced map $\phi^* : \hat{S}(\mathfrak{h}[1])^* \to \hat{S}(\mathfrak{g}[1])^*$ is a dg algebra morphism. It is clear that it is an algebra morphism, since this only requires ϕ to be linear. We need to show that ϕ^* is compatible with the differential. This follows from the fact that $\mathfrak{g} \to \mathfrak{h}$ is both compatible with the differential and with the Lie bracket, since the differential d_{CE} is the sum of two terms, one corresponding to each. Namely, for $\xi \in \mathfrak{h}[1]^*$ and $a, b \in \mathfrak{g}$, we have

$$\phi^* \circ d_{CE}(\xi)(a \otimes b) = \xi([\phi(a), \phi(b)]) = \xi(\phi([a, b])) = d_{CE} \circ \phi^*(\xi)(a \otimes b),$$

$$\phi^* \circ d_{CE}(\xi)(a) = \xi(d\phi(a)) = \xi(\phi(da)) = d_{CE} \circ \phi^*(\xi)(a). \qquad \square$$

SOLUTION TO EXERCISE 4.14.1. The Poisson bivector has degree -1, or equivalently, $|\pi(f \otimes g)| = |f| + |g| - 1$ when f and g are homogeneous. Hence, the only possible graphs that can be nonzero applied to $v \otimes w$ for $v, w \in \mathfrak{g}$ are those corresponding to the product, vw, the Poisson bracket, $\{v, w\} = \pi(v \otimes w)$, and finally $\pi^2(v \otimes w)$. But, π^2 is symmetric (since π is skew-symmetric), and hence $\pi^2(v \otimes w - w \otimes v) = 0$; similarly $vw - wv = 0$. Therefore the only graph that can contribute to $v \star w - w \star v$ is the one corresponding to the Poisson bracket, $\{v, w\} = [v, w]$. We conclude that $v \star w - w \star v = c\hbar[v, w]$ for some $c \in \Bbbk$. Then, the fact that we get a deformation quantization implies that $c = 1$. $\square$

SOLUTION TO EXERCISE 4.14.2. The necessary details are given already in the exercise sheet. $\square$

SOLUTION TO EXERCISE 4.14.4. The details below (and more) are all taken from [156].

(i) To see that $B_{W_m}(\pi, \ldots, \pi)$ has order m, note that by construction it has m arrows pointed at the vertex labeled f and each arrow corresponds to differentiation. Moreover, since π has degree -1, this operator has degree $-m$; an operator of degree $-m$ and order m must be a constant-coefficient operator.

(ii) The formula is equivalent to the statement that, if we apply the wheel to the function x^m placed at the vertex f (for $x \in \mathfrak{g}$), we obtain $\text{tr}((\text{ad}\,x)^m)$. Note that $\pi(x \otimes -) = \text{ad}(x)(-)$. Let x_i be a basis of $\mathfrak{g}$, and write $\text{ad}(x) = \sum_{i,j} c_{ij} x_i \partial_j$ for some $c_{ij} \in \mathbb{k}$. Applying the wheel to x^m yields

$$\sum_{i_1, i_2, \ldots, i_m} c_{i_1 i_2} c_{i_2 i_3} \cdots c_{i_m i_1},$$

but this is nothing but $\text{tr}((\text{ad}\,x)^m)$, as desired.

(iii) This follows immediately from the single wheel ($k = 1$) case, by the definition of B_Γ.

(iv) Kontsevich's isomorphism must be expressed as a formal sum of graphs, and the graphs must have exactly one vertex labeled f (since the underlying linear map of our isomorphism is the restriction of a linear map from a single copy of $\text{Sym}\,\mathfrak{g}$ to $\text{Sym}\,\mathfrak{g}[[\hbar]]$). In the graphs that appear, each vertex labeled π cannot have both of its arrows pointing to the same vertex, since π is skew-symmetric. Also, each vertex labeled π can be the target of at most one arrow, since π is linear: in more detail, if we write $\pi = \sum_{j,k} f_{jk} \partial_j \wedge \partial_k$ for $f_{jk} \in \mathfrak{g}$, applying more than one partial derivative to f_{jk} would yield zero, hence $B_\Gamma = 0$ if Γ is a graph with a vertex labeled π which is the target of multiple arrows. Since every vertex labeled π is the source of two arrows, the only possibility (to have $B_\Gamma \neq 0$) is to have every vertex labeled π pointing to both the vertex labeled f and one other vertex, such that each vertex labeled π is the target of exactly one arrow. Such graphs are the union of wheels, so the result follows from (iii).

(v) The stated results imply that the isomorphism is a sum of the form

$$x \mapsto \sum_{m_1, \ldots, m_k \text{ even}} \frac{1}{k!} c_{m_1} \cdots c_{m_k} \text{tr}((\text{ad}\,x)^{m_1}) \cdots \text{tr}((\text{ad}\,x)^{m_k}),$$

which coincides with the given formula (since the $m_1, \ldots, m_k$ can occur in all possible orderings, e.g., 2, 4 and 4, 2 both occur). $\qquad\square$

SOLUTION TO EXERCISE 5.1.1. More generally, if $f_1, \ldots, f_k \in \mathcal{O}(\mathbb{A}^n)$ cut out a variety X of dimension $n - k$, then we can form

$$\Xi_X := i_{\partial_1 \wedge \cdots \wedge \partial_n}(df_1 \wedge \cdots \wedge df_k) \in \wedge^{n-k}_{\mathcal{O}(\mathbb{A}^n)} \text{Vect}(\mathbb{A}^n),$$

a $(n - k)$-polyvector field on $\mathbb{A}^n$. Note that, by construction, $i_{\Xi_X}(df_i) = 0$ for all i.

We claim that Ξ_X is unimodular in the sense that each vector field of the form $\xi = i_{\Xi_X}(dh_1 \otimes \cdots \otimes dh_{n-k-1}) \in \text{Vect}(\mathbb{A}^n)$ is divergence-free, i.e., $L_\xi(\text{vol}) = 0$ with $\text{vol} = dx_1 \wedge \cdots \wedge dx_n$ the standard volume form on $\mathbb{A}^n$. Equivalently, $i_{\partial_1 \wedge \cdots \wedge \partial_n}(\alpha)$ is unimodular whenever α is an exact $(n - 1)$-form. This is a standard local computation, and it only uses that α is closed: writing $\alpha = \sum_i f_i dx_1 \wedge \cdots \wedge \widehat{dx_i} \wedge \cdots \wedge dx_n$, we have

$i_{\partial_1 \wedge \cdots \wedge \partial_n}(d\alpha) = \sum_i df_i/dx_i$, which is the same as the divergence of $i_{\partial_1 \wedge \cdots \wedge \partial_n}(\alpha) = \sum_i f_i \partial_i$.

By construction, Ξ_X is parallel to X (since $f_1, \ldots, f_k$ vanish there); algebraically this is saying that we have a well-defined map,

$$\Xi_X|_X : \mathcal{O}(X)^{\otimes(n-k)} \to \mathcal{O}(X), \quad (g_1 \otimes \cdots \otimes g_{n-k}) \mapsto i_{\Xi_X}(dg_1 \otimes \cdots \otimes dg_{n-k}),$$

which is skew-symmetric and a derivation in each component. As a result, so is $[\Xi_X, \Xi_X]$, which on X is obviously zero (in the case $\dim X \geq 2$ this is because its degree is greater than $\dim X$; for $\dim X = 1$ this is clear). Thus $[\Xi_X, \Xi_X] = 0$. In the case $\dim X = 2$, this implies that Ξ_X is a Poisson bivector on $\mathbb{A}^n$ (by Proposition 4.5.4). Finally, to see that f_i are central in this bracket, we recall the identity $i_{\Xi_X}(df_i) = 0$ above, but $i_{\Xi_X}(df_i)$ is the Hamiltonian vector field $\xi_f := \{f, -\}$. □

III. Symplectic reflection algebras

SOLUTION TO EXERCISE 1.9.1. Let $z = \sum_{g \in G} f_g \cdot g$ be an element in the centre of $C[V] \rtimes G$. Choose some $g \neq 1$. Since $G \subset GL(V)$, there exists $h \in C[V]$ such that $g(h) \neq h$. Then

$$[h, z] = \sum_{g \in G} f_g(h - g(h)) \cdot g = 0$$

implies that $f_g = 0$ for all $g \neq 1$. Therefore $Z(C[V] \rtimes G)$ is contained in $C[V]$. But it is clear that if z belongs to both $Z(C[V] \rtimes G)$ and $C[V]$, then it is contained in $C[V]^G$. On the other hand one can easily see that $C[V]^G$ is contained in $Z(C[V] \rtimes G)$. □

SOLUTION TO EXERCISE 1.9.2. Taking away the relation $[y, x] = 1$ from $(y-s)x = xy + s$ gives $sx = -s$ and hence $x = -1$. Then the relation $[y, x] = 1$ implies that $1 = 0$. □

SOLUTION TO EXERCISE 1.9.4. At $t = 0$, the algebra $\mathcal{D}_t(\mathfrak{h}_{\text{reg}}) \rtimes W$ is equal to $C[\mathfrak{h}_{\text{reg}} \times \mathfrak{h}^*] \rtimes W$. Therefore the image of $eH_{0,c}(W)e$ is contained in

$$e(C[\mathfrak{h}_{\text{reg}} \times \mathfrak{h}^*] \rtimes W)e \simeq C[\mathfrak{h}_{\text{reg}} \times \mathfrak{h}^*]^W.$$

This is a commutative ring. □

SOLUTION TO EXERCISE 1.9.5. Part (1): Since the isomorphism $\mathfrak{h}^* \xrightarrow{\sim} \mathfrak{h}$ given by $x \mapsto \tilde{x} = (x, -)$ is W-equivariant, the only thing to check is that the commutation relation

$$[y, x] = tx(y) - \sum_{s \in S} c(s)\alpha_s(y)x(\alpha_s^\vee)s, \quad \forall\, y \in \mathfrak{h}, \ x \in \mathfrak{h}^*$$

still holds after applying $\widetilde{(-)}$ everywhere. This follows from two trivial observations. First,

$$x(y) = \tilde{y}(\tilde{x}) = (x, \tilde{y}) = (y, \tilde{x}),$$

and secondly, possibly after some rescaling, we have $(\alpha_s, \alpha_s) = (\alpha_s^\vee, \alpha_s^\vee) = 2$ and $\tilde{\alpha}_s = \alpha_s^\vee$, $\tilde{\alpha}_s^\vee = \alpha_s$. For part (2), just follow the hint. □

SOLUTION TO EXERCISE 2.11.3. Part (1): Choose some $0 \neq x \in \mathfrak{h}^* \subset C[\mathfrak{h}]$ and take $M = H_c(W)/H_c(W) \cdot (\mathbf{eu} - x)$. This module cannot be a direct sum of its generalized eigenspaces.

Part (2): Let L be a finite-dimensional $H_c(W)$-module. Then, it is a direct sum of its generalized $\mathbf{eu}$-eigenspaces because $\mathbf{eu} \in \mathrm{End}_C(L)$ and a finite-dimensional complex vector space decomposes as a direct sum of generalized eigenspaces under the action of any linear operator. If $l \in L_a$ for some $a \in C$, then the relation $[\mathbf{eu}, y] = -y$ implies that $y \cdot l \in L_{a-1}$. Hence $y_1 \cdots y_k \cdot l \in L_{a-k}$. But L is finite-dimensional which implies that $L_{a-k} = 0$ for $k \gg 0$. Hence $\mathfrak{h}$ acts locally nilpotently on L. □

SOLUTION TO EXERCISE 2.11.4. Since M is finitely generate as a $C[\mathfrak{h}]$-module, we may choose a finite-dimensional, $\mathbf{eu}$- and $\mathfrak{h}$-stable subspace M_0 of M such that M_0 generates M as a $C[\mathfrak{h}]$-module. Therefore, there is a surjective map of $C[\mathfrak{h}]$-modules $C[\mathfrak{h}] \otimes_C M_0 \to M$, $a \otimes m \mapsto am$. This can be made into a morphism of $C[\mathbf{eu}]$-modules by defining $\mathbf{eu} \cdot (a \otimes m) = [\mathbf{eu}, a] \otimes m + a \otimes \mathbf{eu} \cdot m$. If $f_0(t) \in \mathbb{Z}[x^a \mid a \in C]$ is the character of M_0 then the character of $C[\mathfrak{h}] \otimes M_0$ is given by

$$\frac{1}{(1-t)^n} f_0(t) \in \bigoplus_{a \in C} t^a \mathbb{Z}[[t]]$$

and hence $\mathrm{ch}(M) \in \bigoplus_{a \in C} t^a \mathbb{Z}[[t]]$ too. □

SOLUTION TO EXERCISE 2.11.5. Let $M \in \mathcal{O}$. We can decompose M as a $C[\mathbf{eu}]$-module as

$$M = \bigoplus_{\bar{a} \in C/\mathbb{Z}} M^{\bar{a}}$$

where $M^{\bar{a}} = \bigoplus_{b \in \bar{a}} M_b$. It suffices to show that each $M^{\bar{a}}$ is a $H_c(W)$-submodule of M. But if $x \in \mathfrak{h}^*$ and $m \in M_b$ then $x \cdot m \in M_{b+1}$ and $b \in \bar{a}$ iff $b + 1 \in \bar{a}$. A similar argument applies to $y \in \mathfrak{h}^*$ and $w \in W$. Thus $M^{\bar{a}}$ is a $H_c(W)$-submodule of M. □

SOLUTION TO EXERCISE 2.11.6. Exercise 2.11.5 implies that

$$\mathcal{O} = \bigoplus_{\lambda \in \mathrm{Irr}(W)} \mathcal{O}^{\bar{c}_\lambda}$$

with $\Delta(\lambda) \in \mathcal{O}^{\bar{c}_\lambda}$. In this situation Claim 2.7.3 is applicable. It implies that $\Delta(\lambda) = L(\lambda)$ and hence category $\mathcal{O}$ is semi-simple. □

SOLUTION TO EXERCISE 2.11.9. (1) If $c \notin \frac{1}{2} + \mathbb{Z}$ then category $\mathcal{O}$ is semi-simple. The commutation relation of x and y implies that

$$[y, x^n] = \begin{cases} nx^{n-1} - 2cx^{n-1}s & n \text{ odd} \\ nx^{n-1} & n \text{ even} \end{cases}$$

This implies that

$$L(\rho_0) = \frac{C[x] \otimes \rho_0}{x^{2m+1} C[x] \otimes \rho_0}, \quad L(\rho_1) = \Delta(\rho_1)$$

when $c = \frac{1}{2} + m$ for some $m \in \mathbb{Z}_{\geq 0}$. Similarly,

$$L(\rho_0) = \Delta(\rho_0), \quad L(\rho_1) = \frac{\mathbb{C}[x] \otimes \rho_1}{x^{2m+1}\,\mathbb{C}[x] \otimes \rho_1}$$

when $c = \frac{-1}{2} - m$ for some $m \in \mathbb{Z}_{\geq 0}$.

For part (2), we have $\mathbf{eu} = xy - cs$. Therefore $c_0 = -c$ and $c_1 = c$. Thus, $\rho_0 \leq_c \rho_1$ if and only if $2c \in \mathbb{Z}_{\geq 0}$. Similarly, $\rho_1 \leq_c \rho_0$ if and only if $2c \in \mathbb{Z}_{\leq 0}$. For all other c, ρ_0 and ρ_1 are incomparable.   □

SOLUTION TO EXERCISE 2.11.10. Both these steps are direct calculations. You should get $\alpha = \frac{c}{2}(2 - c)$.   □

SOLUTION TO EXERCISE 2.11.7. By the proof of Corollary 1.6.3, we know that c is aspherical if and only if

$$I := \mathsf{H}_c(W) \cdot e \cdot \mathsf{H}_c(W)$$

is a proper two-sided ideal of $\mathsf{H}_c(W)$. If this is the case then there exists some primitive ideal J such that $I \subset J$. Hence Ginzburg's result implies that there is a simple module $L(\lambda)$ in category $\mathcal{O}$ such that $I \cdot L(\lambda) = 0$. But this happens if and only if $e \cdot L(\lambda) = 0$.   □

SOLUTION TO EXERCISE 2.11.9(3). The only aspherical value for $\mathbb{Z}_2$ is $c = -\frac{1}{2}$.   □

SOLUTION TO EXERCISE 3.14.1. The Young diagram with residues is

0					
1					
2	3	0			
3	0	1	2	3	0
0	1	2	3	0	1

Then $F_2|\lambda\rangle = q\,|(6, 6, 3, 2, 1)\rangle$, $K_1|\lambda\rangle = |\lambda\rangle$ and

$$E_4|\lambda\rangle = q^{-2}|(6, 5, 3, 1, 1)\rangle + q^{-1}|(6, 6, 2, 1, 1)\rangle + |(6, 6, 3, 1)\rangle.   \square$$

SOLUTION TO EXERCISE 3.14.3. By adding an infinite number of zeros to the end of $\lambda \in \mathcal{P}$, we may consider it as an infinite sequence $(\lambda_1, \lambda_2, \dots)$ with $\lambda_i \geq \lambda_{i+1}$ and $\lambda_N = 0$ for all i and all $N \gg 0$. Define $I(\lambda)$ by $i_k = \lambda_k + 1 - k$. It is clear that this rule defines a bijection with the required property.   □

SOLUTION TO EXERCISE 3.14.4. We have

$$\mathcal{G}([4]) = [4] + q[3,1] + q[2,1,1] + q^2[1,1,1,1], \quad \mathcal{G}([3,1]) = [3,1] + q[2,2] + q^2[2,1,1],$$

$$\mathcal{G}([2,2]) = [2,2] + q[2,1,1], \qquad\qquad \mathcal{G}([2,1,1]) = [2,1,1] + q[1,1,1,1],$$

$$\mathcal{G}([1,1,1,1]) = [1,1,1,1]$$

and

$$\mathcal{G}([5]) = [5] + q[3,1,1] + q^2[1,1,1,1,1], \qquad \mathcal{G}([4,1]) = [4,1] + q[2,1,1,1],$$

$$\mathcal{G}([3,2]) = [3,2] + q[3,1,1] + q^2[2,2,1], \qquad \mathcal{G}([2,2,1]) = [2,2,1],$$

$$\mathcal{G}([3,1,1]) = [3,1,1] + q[2,2,1] + q[1,1,1,1,1], \quad \mathcal{G}([2,1,1,1]) = [2,1,1,1],$$

$$\mathcal{G}([1,1,1,1,1]) = [1,1,1,1,1]. \qquad\qquad \square$$

SOLUTION TO EXERCISE 3.14.6.　　(1) In this case, the numbers $e_{\lambda,\mu}(1)$, c_λ and dim λ are:

$\mu\backslash\lambda$	(5)	(4,1)	(3,2)	(3,1,1)	(2,2,1)	(2,1,1,1)	(1,1,1,1,1)
(5)	1	0	0	0	0	0	0
(4,1)	0	1	0	0	0	0	0
(3,2)	0	0	1	0	0	0	0
(3,1,1)	-1	0	-1	1	0	0	0
(2,2,1)	1	0	0	-1	1	0	0
(2,1,1,1)	0	-1	0	0	0	1	0
(1,1,1,1,1)	0	0	1	-1	0	0	1
c_λ	$-\frac{65}{2}$	-15	$-\frac{9}{2}$	0	$\frac{19}{2}$	20	$\frac{75}{2}$
dim λ	1	4	5	6	5	4	1

If we define $\mathrm{ch}_\lambda(t) := (1-t)^5 \cdot \mathrm{ch}(L(\lambda))$, then

$$\mathrm{ch}_{(5)}(t) = t^{-\frac{65}{2}} - 6t + 5t^{\frac{19}{2}}, \quad \mathrm{ch}_{(4,1)}(t) = 4t^{-15} - 4t^{20},$$

$$\mathrm{ch}_{(3,2)}(t) = 5t^{-\frac{9}{2}} - 6t + t^{\frac{75}{2}}, \quad \mathrm{ch}_{(3,1,1)}(t) = 6t - 5t^{\frac{19}{2}} - t^{\frac{75}{2}},$$

$$\mathrm{ch}_{(2,2,1)}(t) = 5t^{\frac{19}{2}}, \quad \mathrm{ch}_{(2,1,1,1)}(t) = 4t^{20}, \quad \mathrm{ch}_{(1,1,1,1,1)}(t) = t^{\frac{75}{2}}.$$

(2) It suffices to calculate the 2-, 3- and 5-cores of the partitions of 5. For $r = 2$ we get

$$\{[5], [3,2], [2,2,1], [1,1,1,1,1]\}, \quad \{[4,1], [2,1,1,1]\},$$

where the 2-cores are [1] and [2,1] respectively. For $r = 3$ we get

$$\{[5], [2,2,1], [2,1,1,1]\}, \quad \{[4,1], [3,2], [1,1,1,1,1]\},$$

where the 3-cores are [2] and [1,1] respectively. For $r = 5$ we get

$$\{[5], [4,1], [2,1,1,1], [1,1,1,1,1]\}, \quad \{[3,2]\}, \quad \{[2,2,1]\},$$

where the 5-cores are $\varnothing$, [3,2] and [2,2,1] respectively (for an explanation of what is going on in this final example read [32]). $\qquad \square$

SOLUTION TO EXERCISE 3.14.7. The argument is essentially identical to that for $\mathrm{ch}(L(\lambda))$. We use the polynomials $e_{\lambda,\mu}(q)$ to express the character $\mathrm{ch}_W(L(\lambda))$ in terms of the character of the standard modules $\Delta(\mu)$. Then, we just need to calculate $\mathrm{ch}_W(\Delta(\lambda))$. Note that 1) $\mathbb{C}[\mathfrak{h}] \otimes \lambda$ is a graded W-module 2) the **eu**-character of $\Delta(\lambda)$ is simply the graded character of $C[\mathfrak{h}] \otimes \lambda$ multiplied by t^{c_λ}. Hence, it suffices to describe $C[\mathfrak{h}] \otimes \lambda$ as a graded W-module. This factors as

$$C[\mathfrak{h}] \otimes \lambda = C[\mathfrak{h}]^W \otimes C[\mathfrak{h}]^{coW} \otimes \lambda$$

and hence

$$\mathrm{ch}_W(C[\mathfrak{h}] \otimes \lambda) = \frac{1}{\prod_{i=1}^n (1 - t^i)} \mathrm{ch}_W(C[\mathfrak{h}]^{coW} \otimes \lambda).$$

We have

$$\mathrm{ch}_W(C[\mathfrak{h}]^{coW} \otimes \lambda) = \sum_{\mu \in \mathrm{Irr}(W)} \left(\sum_{i \in \mathbb{Z}} [C[\mathfrak{h}]_i^{coW} \otimes \lambda : \mu] t^i \right) [\mu] = \sum_{\mu \in \mathrm{Irr}(W)} f_{\lambda, \mu}(t)[\mu].$$

In fact, for $W = \mathfrak{S}_4$, one can explicitly calculate the generalized fake polynomials:

$\mu \backslash \lambda$	(4)	(3, 1)	(2, 2)
(4)	1	$t + t^2 + t^3$	$t^2 + t^4$
(3, 1)	$t + t^2 + t^3$	$1 + t + 2t^2 + 2t^3 + 2t^4 + t^5$	$t + t^2 + 2t^3 + t^4 + t^5$
(2, 2)	$t^2 + t^4$	$t + t^2 + 2t^3 + t^4 + t^5$	$t + 2t^3 + t^5$
(2, 1, 1)	$t^3 + t^4 + t^5$	$t + 2t^2 + 2t^3 + 2t^4 + t^5 + t^6$	$1 + 2t^2 + t^3 + t^4 + t^5$
(1, 1, 1, 1, 1)	t^6	$t^3 + t^4 + t^5$	$t + t^3$

$\mu \backslash \lambda$	(2, 1, 1)	(1, 1, 1, 1)
(4)	$t^3 + t^4 + t^5$	t^6
(3, 1)	$t + 2t^2 + 3t^3 + 2t^4 + t^6$	$t^3 + t^4 + t^5$
(2, 2)	$t + t^2 + 2t^3 + t^4 + t^5$	$t^2 + t^4$
(2, 1, 1)	$1 + t + 2t^2 + 2t^3 + 2t^4 + t^5$	$t + t^2 + t^3$
(1, 1, 1, 1, 1)	$t + t^2 + t^3$	1

$\square$

SOLUTION TO EXERCISE 4.10.2. Let $z = x^2$ so that $C[x]^{\mathbb{Z}_2} = C[x^2] = C[z]$. The ring $\mathcal{D}(\mathfrak{h})^W$ is generated by $x^2, x\partial_x$ and ∂_x^2 and $\mathcal{D}(\mathfrak{h}/W) = C\langle z, \partial_z \rangle$. Since

$$x\partial_x(z^n) = x\partial_x(x^{2n}) = 2nx^{2n} = 2nz^n$$

and

$$\partial_x^2(z^n) = 2n(2n - 1)z^{n-1}$$

we see that $\phi : \mathcal{D}(\mathfrak{h})^W \to \mathcal{D}(\mathfrak{h}/W)$ sends $x\partial_x$ to $2z\partial_z$ and ∂_x^2 is sent to $\partial_z(4z\partial_z - 2)$. Then ∂_z is not in the image of ϕ so it is not surjective. A rigorous way to show this is as follows: the morphism ϕ is filtered. Therefore, it induces a morphism on associated graded; this is the map

$$C[A, B, C]/(AC - B^2) \to C[D, E], \quad A \mapsto D, \quad B \mapsto 2DE, \quad C \mapsto 4DE^2.$$

This is a proper embedding. $\square$

SOLUTION TO EXERCISE 4.10.3. Part (1). Under the Dunkl embedding,

$$y = \partial_x - \frac{c}{x}(1 - s),$$

which implies that $\partial_x \cdot \rho_0 = 0$ and $\partial_x \cdot \rho_1 = (2c/x)e_1$. So the $\mathbb{Z}_2$-equivariant local systems on $\mathbf{C}^\times$ corresponding to $\Delta(\rho_0)[\delta^{-1}]$ and $\Delta(\rho_1)[\delta^{-1}]$ are given by the differential equations $\partial_x = 0$ and $\partial_x - 2c/x = 0$ respectively. Now we need to construct the corresponding local systems on $\mathbf{C}^\times/\mathbb{Z}_2$. If $z := x^2$ then $\partial_z = (1/2x)\partial_x$ and $\Delta(\rho_0)[\delta^{-1}]^{\mathbb{Z}_2}$, resp. $\Delta(\rho_1)[\delta^{-1}]^{\mathbb{Z}_2}$ has basis $a_0 = 1 \otimes \rho_0$, resp. $a_1 = x \otimes \rho_1$, as a free $\mathbb{C}[z^{\pm 1}]$-module. We see that

$$\partial_z \cdot a_0 = 0, \quad \partial_z \cdot a_1 = \frac{1+2c}{2z}a_1.$$

Since the solutions of these equations are 1 and $z^{\frac{1+2c}{2}}$, the monodromy of these equations is given by $t \mapsto 1$ and $t \mapsto -\exp(2\pi\sqrt{-1}ct)$ respectively. Therefore $\mathsf{KZ}(\Delta(\rho_0))$ is the one-dimensional representation $\mathbb{C}\,b_0$ of $\pi_1(\mathbb{C}^\times/\mathbb{Z}_2)$ defined by $T \cdot b_0 = b_0$ and $\mathsf{KZ}(\Delta(\rho_1))$ is the one-dimensional representation $\mathbb{C}\,b_1$ of $\pi_1(\mathbb{C}^\times/\mathbb{Z}_2)$ defined by $T \cdot b_1 = -\exp(2\pi\sqrt{-1}c)b_1$ respectively.

Part (2). If $c = \frac{1}{2} + m$ for some $m \in \mathbb{Z}_{\geq 0}$ then $P_{KZ} = P(\rho_1)$. If $c = -\frac{1}{2} - m$ for some $m \in \mathbb{Z}_{\geq 0}$ then $P_{KZ} = P(\rho_0)$. Otherwise, $P_{KZ} = P(\rho_0) \oplus P(\rho_1)$. $\qquad\square$

SOLUTION TO EXERCISE 5.9.2. The space $X_c(G_2)$ is never smooth. The Poincaré polynomial of $\mathbb{C}[\mathfrak{h}]^{\mathrm{co}G_2}$ is

$$(1 - t^2)(1 - t^6)/(1 - t)^2 = 1 + 2t + 2t^2 + \cdots + 2t^5 + t^6.$$

The polynomials $t^2 + t^4$ and $t + t^5$ do not divide this polynomial in $\mathbb{Z}[t, t^{-1}]$. Therefore

$$\dim L(\mathfrak{h}_1), \dim L(\mathfrak{h}_2) < 12$$

for any parameter c, which implies the claim. $\qquad\square$

IV. Noncommutative resolutions

SOLUTION TO EXERCISE 1.5.2. (2) S_0 is generated by x^r, y^r and xy, so it is isomorphic as a ring to $\mathbb{C}[a, b, c]/(ab - c^r)$. As an S_0-module, for $i > 0$ we have S_i is generated by x^i and y^{r-i} (depending on conventions). This leads to the quiver

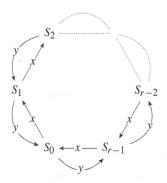

(3) For $\frac{1}{3}(1, 1)$ S_0 is generated as a ring by x^3, x^2y, xy^2, y^3, whereas S_1 is generated as an S_0-module by x and y, and S_2 is generated as an S_0-module by x^2, xy, y^2. The

quivers are

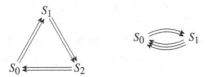

For $\frac{1}{5}(1,2)$ the generators of the ring (again, up to conventions) are x^5, x^3y, xy^2, y^5. As S_0-modules, S_1 is generated by x and y^3, S_2 by x^2 and y, S_3 by x^3, xy, y^4, and S_4 by x^4, x^2y, y^2. The quivers are

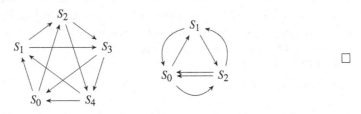

SOLUTION TO EXERCISE 1.5.3. (1), (2) and (4) are CM. (3) is not.   □

SOLUTION TO EXERCISE 1.5.4. (1) (a), (b) are not CM, whereas (c) and (d) are. (2) (a), (b), (c) are not CM, whereas (d) is CM, as are (u, x) and (u, y). The module (u^2, ux, x^2) is not. Part (3) is similar.   □

SOLUTION TO EXERCISE 1.5.5. All hypersurfaces are Gorenstein, so this implies that (1), (2) and (3) in Exercise 1.5.4 are Gorenstein, as are (1), (4) in 1.5.3. The ring in (2) in 1.5.3 is not Gorenstein.   □

SOLUTION TO EXERCISE 2.5.1. Label the algebras A, B, C, D from left to right.

Question	A	B	C	D
(1)	$\mathbb{C}[x, y]$	$\mathbb{C}[x, y]^{\frac{1}{2}(1,1)}$	$\mathbb{C}[x, y, z]^{\frac{1}{3}(1,2)}$	$\mathbb{C}[x, y, z]^{\frac{1}{3}(1,1)}$
(2)a	$M = R \oplus (x, y)$	$M = S_0 \oplus S_1$	$M = S_0 \oplus S_1$	$M = S_0 \oplus S_1$
(2)b	Not CM	Yes CM	Yes CM	Yes CM
(3)	No	Yes	Yes	Yes
(4)	2, 1	2, 2	∞, ∞	2, 3
(5)			resolutions are periodic	
(6)	2	2	∞	3
(7)		NCCR		
(8)	$\mathcal{O}_{\mathbb{P}^1}(-1)$	$\mathcal{O}_{\mathbb{P}^1}(-2)$	X_1	$\mathcal{O}_{\mathbb{P}^1}(-3)$

where for (2)a we use the notation from Exercise 1.5.2, and in (8) X_1 is one of the partial resolutions of $\mathbb{C}[x, y]^{\frac{1}{3}(1,2)}$ containing only one curve. The fact that X_1 has only hypersurface singularities is the phenomenon that explains the periodicity in (5).   □

SOLUTION TO EXERCISE 2.5.2. (1) Set $M_1 := (u, x-1)$, $M_2 := (u, x(x-1))$ and $M_3 := (u, x^2(x-1))$. The main calculation is

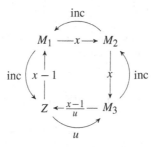

from which, after some work, the presentation follows.

(2) We can check whether $\Lambda \cong \operatorname{End}_Z(Z \oplus M_1 \oplus M_2 \oplus M_3)$ is a NCCR by localizing to the maximal ideals, and there (up to Morita equivalence) we find the NCCR from 2.5.1.

(3) There are many. Set $M := Z \oplus (u, x-1) \oplus (u, x(x-1)) \oplus (u, x^2(x-1))$, the module from (1), then for example taking $N := Z \oplus (u, x) \oplus (u, x-1)$ we see that $\operatorname{End}_Z(N)$ is morita equivalent to $\operatorname{End}_Z(M)$, since add $M =$ add N. □

SOLUTION TO EXERCISE 2.5.3. (2) As right modules, the projective resolutions of the simples S_1, S_2, S_3 are

$$0 \to e_3 A \to (e_1 A)^2 \to e_3 A \oplus e_2 A \to e_1 A \to S_1 \to 0,$$

$$0 \to e_2 A \to e_1 A \oplus e_3 A \to e_2 A \to S_2 \to 0,$$

$$0 \to (e_3 A)^{\oplus 2} \to e_2 A \oplus (e_1 A)^{\oplus 2} \to e_3 A \to S_3 \to 0.$$

See [240, 6.9] for the general form of the projective resolutions. Part (1) is similar. For (3), the key point is that $Z(\Lambda)$ is not Gorenstein, and in general we can't apply Auslander–Buchsbaum unless Λ is a Gorenstein $Z(\Lambda)$-order, which it is not. □

SOLUTION TO EXERCISE 2.5.4. Label the algebras A, B, C, D from left to right.

Question	A	B	C	D
(2)	$\dfrac{\mathbb{C}[u,v,x,y]}{(uv-xy)}$	$\dfrac{\mathbb{C}[u,v,x,y]}{(uv-x^2)}$	$\dfrac{\mathbb{C}[u,v,x,y]}{(uv-(x-y^2)(x+y^2))}$	$\mathbb{C}[x,y,z]^{\frac{1}{3}(1,1,1)}$
(3)	e.g. consider $R \oplus (u,x)$	e.g. consider $R \oplus (u,x)$	e.g. consider $R \oplus (u,x+y^2)$	Consider $S_0 \oplus S_1 \oplus S_2$
(4)	$\mathcal{O}_{\mathbb{P}^1}(-1)^{\oplus 2}$	$\mathcal{O}_{\mathbb{P}^1}(-2) \oplus \mathcal{O}_{\mathbb{P}^1}$	Y_1	$\mathcal{O}_{\mathbb{P}^2}(-3)$

where for (3)D we use the notation from Exercise 1.5.2, and in (4) Y_1 is the blowup of the ideal $(u, x+y^2)$, which forms one half of the Pagoda flop. The first three examples capture the phenomenon of Type A contractions in 3-folds; in example A the curve has width 1, in Example C the curve has width 2 (changing the 2 to n in the relations gives the example with width n), and in example B the curve has width ∞. □

SOLUTION TO EXERCISE 2.5.5. (3) With R a complete local CM ring of dimension three, the general result is that $\mathrm{Hom}_R(M, N) \in \mathrm{CM}\, R$ if and only if $\mathrm{depth}_R \mathrm{Ext}^1_R(M, N)$ is positive. See for example [141, 2.7]. □

SOLUTION TO EXERCISE 2.5.6. (3) The general theorem due to Watanabe is that if G has no complex reflections, then the invariant ring is always CM, and it is Gorenstein if and only if $G \leq \mathrm{SL}(n, \mathbb{C})$. □

SOLUTION TO EXERCISE 3.2.1. Label the algebras A, B, C, D from left to right.

	A	B	C	D
$\theta = (-1, 1)$	$\mathcal{O}_{\mathbb{P}^1}(-1)$	$\mathcal{O}_{\mathbb{P}^1}(-2)$	Y_1	$\mathcal{O}_{\mathbb{P}^1}(-3)$
$\theta = (1, -1)$	$\mathbb{C}^2$	$\mathcal{O}_{\mathbb{P}^1}(-2)$	Y_2	Z_1

where Y_1 is one of the partial resolutions of the $\frac{1}{3}(1, 2)$ singularity containing only one curve, Y_2 is the other, and Z_1 is a scheme with two components, one of which is $\mathcal{O}_{\mathbb{P}^1}(-3)$. □

SOLUTION TO EXERCISE 3.2.2. Label the algebras A, B, C from left to right. Then

	A	B	C
$\theta = (-1, 1)$	$\mathcal{O}_{\mathbb{P}^1}(-1)^{\oplus 2}$	$\mathcal{O}_{\mathbb{P}^1}(-2) \oplus \mathcal{O}_{\mathbb{P}^1}$	Y_1
$\theta = (1, -1)$	$\mathcal{O}_{\mathbb{P}^1}(-1)^{\oplus 2}$	$\mathcal{O}_{\mathbb{P}^1}(-2) \oplus \mathcal{O}_{\mathbb{P}^1}$	Y_2

where Y_1 is blowup of $\mathrm{Spec}\,\mathbb{C}[u, v, x, y]/(uv - (x - y^2)(x + y^2))$ at the ideal $(u, x + y^2)$ and Y_2 is blowup at the ideal $(u, x - y^2)$. In examples A and C, the two spaces are abstractly isomorphic, but not isomorphic in a compatible way over the base; they are examples of flops. □

SOLUTION TO EXERCISE 3.2.3. Since $\theta_0 = -\theta_1 - \theta_2$, the stability condition is determined by the pair (θ_1, θ_2). The chamber structure is

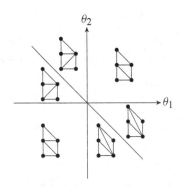

where in each chamber is the toric fan corresponding to the quiver GIT for that chamber. There are thus three crepant resolutions. □

SOLUTION TO EXERCISE 3.2.4. The invariants are generated by $R_1 := as$, $R_2 := at = bs$, $R_3 := bt$, v and w, and abstractly the invariant ring is isomorphic to

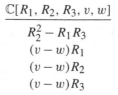

$$\frac{\mathbb{C}[R_1, R_2, R_3, v, w]}{\begin{array}{c} R_2^2 - R_1 R_3 \\ (v - w) R_1 \\ (v - w) R_2 \\ (v - w) R_3 \end{array}}$$

☐

SOLUTION TO EXERCISE 4.9.1. (1) No. Take for example $R = \mathbb{C}[x, y]$ and $M = R \oplus (x, y)$. (2) See [21, 4.1].   ☐

SOLUTION TO EXERCISE 4.9.2. (1) Note that $\operatorname{End}_R(R \oplus M) \in \operatorname{CM} R$ implies that $M \in \operatorname{CM} R$. Because of this, $\Lambda_{\mathfrak{p}} \cong \operatorname{End}_{R_{\mathfrak{p}}}(R_{\mathfrak{p}} \oplus M_{\mathfrak{p}})$ implies that for any prime $\mathfrak{p}$ not in the singular locus, $\Lambda_{\mathfrak{p}} \cong M_n(R_{\mathfrak{p}})$ for some n. On the other hand, finite global dimension is preserved under localization, so if $\mathfrak{p}$ is in the singular locus then $\Lambda_{\mathfrak{p}}$ cannot be a matrix algebra over $R_{\mathfrak{p}}$ (since they have infinite global dimension). Thus the Azumaya locus equals the nonsingular locus. (2) Note in (1) that both loci equal the locus on which M is not free. This can be described as the support of the module $\operatorname{Ext}_R^1(M, \Omega M)$, and hence is closed. (3) The K_0 group of $\operatorname{End}_R(R \oplus M)$, which is finitely generated since the global dimension is finite (and R is now local), surjects onto the class group of R.   ☐

SOLUTION TO EXERCISE 4.9.3. $\mathbb{C} \oplus \mathbb{C}[x]$.   ☐

SOLUTION TO EXERCISE 4.9.4. See [141, 5.4].   ☐

SOLUTION TO EXERCISE 4.9.5. The most direct way is to establish (using for example the snake lemma) that $\operatorname{End}_Y(\mathcal{O}_Y \oplus \mathcal{L}) \cong \operatorname{End}_R(R \oplus (x, y))$ via the global sections functor, where $R = \mathbb{C}[x, y]$. From there, in the presentation the arrow a corresponds to multiplication by x taking an element from R to (x, y), similarly the arrow b to multiplication by y, and the arrow t corresponds to the inclusion of the ideal (x, y) into R.   ☐

SOLUTION TO EXERCISE 5.5.1. (2) is the singularity $uv = x^2 y$ from Example 3.1.5 and Exercise 3.2.3. For the remainder, see [247, §16].   ☐

SOLUTION TO EXERCISE 5.5.3. By Artin–Verdier theory, taking the torsion-free lift $\mathcal{M}$ of M to the minimal resolution, there are short exact sequences $0 \to \mathcal{O}^{\oplus r} \to \mathcal{M} \to \mathcal{O}_D \to 0$ and $0 \to \mathcal{M}^* \to \mathcal{O}^{\oplus r} \to \mathcal{O}_D \to 0$ where D is a divisor transversal to the curve corresponding to $\mathcal{M}$. Taking the pullback of these sequences, the middle exact sequence splits giving an exact sequence $0 \to \mathcal{M}^* \to \mathcal{O}^{\oplus 2r} \to \mathcal{M} \to 0$, proving the statement.   ☐

SOLUTION TO EXERCISE 5.5.5. The Ext groups are isomorphic, since the Knörrer functor gives an equivalence of categories $\underline{\operatorname{CM}}R \simeq \underline{\operatorname{CM}}R'$, known as Knörrer periodicity. See for example [247, §12].   ☐

SOLUTION TO EXERCISE 6.1.1. (1) 8, (2) 11, (3) ∞. The general result is that the path algebra is finite-dimensional if and only if there is no oriented cycle. $\square$

SOLUTION TO EXERCISE 6.1.3. (1) $k[x]/(x^3-1)$. When $k = \mathbb{C}$ this is isomorphic to $k \oplus k \oplus k$ which can be viewed as just a quiver with three dots. (2) $k[x, y]$. (3) $k \oplus k[x]$. $\square$

SOLUTION TO EXERCISE 6.1.4. (1) One vertex, n loops, no relations. (2) One vertex, n loops, the commutativity relations. (3) Draw the quivers side by side, and take the union of the relations. (4) Since $\mathbb{C}G$ is semisimple, it is a direct product of matrix rings. Combine answers for (3) above and 6.1.5(1) below. Alternatively, work up to morita equivalence, where $\mathbb{C}G$ is then just a finite number of dots, with no relations. (5) One vertex, number of loops=number of generators, then the finite number of relations. $\square$

SOLUTION TO EXERCISE 6.1.5. (1) is an easy extension of 6.0.13(2). (2) then follows as in 6.0.25(2) and 6.0.23. (3) Consider the functor $\mod k \to \mod M_n(k)$

$$V \mapsto \quad V \underset{1}{\overset{1}{\rightleftarrows}} V \underset{1}{\overset{1}{\rightleftarrows}} V \cdots V \underset{1}{\overset{1}{\rightleftarrows}} V$$

This is clearly fully faithful, and is essentially surjective by (2). $\square$

SOLUTION TO EXERCISE 6.1.6. Take an arbitrary simple module, which is necessarily finite-dimensional. View as a quiver representation, with vector space V and loop corresponding to a linear map $f: V \to V$ such that $f^n = 0$. Consider the kernel, then this gives a submodule, so necessarily the kernel must be everything (since f cannot be injective). Thus f must be the zero map, so V must be one-dimensional else the representation decomposes. $\square$

SOLUTION TO EXERCISE 6.1.7. View any such simple as a finite-dimensional representation of the quiver. Thus X and Y are linear maps from a finite-dimensional vector space to itself, and they must satisfy the relation $XY - YX = 1$. Taking the trace of this equation gives $0 = n$, which is a contradiction. $\square$

Bibliography

[1] J. Alev, M. A. Farinati, T. Lambre, and A. L. Solotar, *Homologie des invariants d'une algèbre de Weyl sous l'action d'un groupe fini*, J. Algebra **232** (2000), 564–577.

[2] S. Ariki, *Finite dimensional Hecke algebras*, pages 1–48 in *Trends in representation theory of algebras and related topics*, EMS Ser. Congr. Rep., Eur. Math. Soc., Zürich, 2008.

[3] D. Arnal, *Le produit star de Kontsevich sur le dual d'une algèbre de Lie nilpotente*, C. R. Acad. Sci. Paris Sér. I Math. **327** (1998), no. 9, 823–826.

[4] M. Artin, *Some problems on three-dimensional graded domains*, Representation theory and algebraic geometry (Waltham, MA, 1995), Cambridge Univ. Press, Cambridge, 1997, pp. 1–19.

[5] M. Artin and W. F. Schelter, *Graded algebras of global dimension 3*, Adv. in Math. **66** (1987), no. 2, 171–216.

[6] M. Artin, L. W. Small and J. J. Zhang, *Generic flatness for strongly noetherian algebras*, J. Algebra, **221** (1999), 579–610.

[7] M. Artin and J. T. Stafford, *Noncommutative graded domains with quadratic growth*, Invent. Math. **122** (1995), no. 2, 231–276.

[8] M. Artin and J. T. Stafford, *Semiprime graded algebras of dimension two*, J. Algebra **227** (2000), no. 1, 68–123.

[9] M. Artin, J. Tate, and M. Van den Bergh, *Some algebras associated to automorphisms of elliptic curves*, The Grothendieck Festschrift, Vol. I, Birkhäuser, Boston, 1990, pp. 33–85.

[10] M. Artin, J. Tate, and M. Van den Bergh, *Modules over regular algebras of dimension 3*, Invent. Math. **106** (1991), no. 2, 335–388.

[11] M. Artin and M. Van den Bergh, *Twisted homogeneous coordinate rings*, J. Algebra **133** (1990), no. 2, 249–271.

[12] M. Artin and J.-L. Verdier, *Reflexive modules over rational double points*, Math. Ann. **270** (1985), no. 1, 79–82.

[13] M. Artin and J. J. Zhang, *Noncommutative projective schemes*, Adv. Math. **109** (1994), no. 2, 228–287.

[14] M. Artin and J. J. Zhang, *Abstract Hilbert schemes*, Algebr. Represent. Theory **4** (2001), no. 4, 305–394.

[15] M. Auslander, *On the dimension of modules and algebras, III: Global dimension*, Nagoya Math. J. **9** (1955), 67–77.

[16] M. Auslander, *Representation dimension of Artin algebras*, lecture notes, Queen Mary College, London, 1971.

[17] M. Auslander, *Functors and morphisms determined by objects*, pages 1–244 in *Representation theory of algebras* (Philadelphia, 1976) volume 37 in *Lecture Notes in Pure Appl. Math.*, Dekker, New York, 1978.

[18] M. Auslander, *Isolated singularities and existence of almost split sequences*. pages 194–242 in *Representation theory, II* (Ottawa, 1984), volume 1178 of *Lecture Notes in Math.*, Springer, Berlin, 1986.

[19] M. Auslander, *Rational singularities and almost split sequences*. Trans. Amer. Math. Soc. **293** (1986), no. 2, 511–531.

[20] M. Auslander and D. A. Buchsbaum, *Homological dimension in local rings*, Trans. Amer. Math. Soc. **85** (1957), 390–405.

[21] M. Auslander and O. Goldman, *Maximal orders*, Trans. Amer. Math. Soc. **97** (1960), 1–24.

[22] M. Auslander, I. Reiten, and S. O. Smalo, *Representation theory of Artin algebras*, volume 36 of Cambridge Studies in Advanced Mathematics, *Cambridge University Press*, Cambridge, 1997.

[23] H. Bass, *On the ubiquity of Gorenstein rings*, Math. Z. **82** (1963), 8–28.

[24] A. A. Beilinson, *The derived category of coherent sheaves on* $\mathbb{P}^n$, Selecta Math. Soviet. **3** (1983/84), no. 3, 233–237.

[25] I. N. Bernstein, I. M. Gelfand, and S. I. Gelfand, *A certain category of* $\mathfrak{g}$-*modules*, Funkcional. Anal. i Priložen. **10** (1976), no. 2, 1–8.

[26] A. Beilinson, V. Ginzburg, and W. Soergel, *Koszul duality patterns in representation theory*, J. Amer. Math. Soc. **9** (1996), no. 2, 473–527.

[27] G. Bellamy, *On singular Calogero–Moser spaces*, Bull. Lond. Math. Soc., **41** (2):315–326, 2009.

[28] G. Bellamy and M Martino, *On the smoothness of centres of rational Cherednik algebras in positive characteristic*, Glasgow Math. J. **55** (2013), no. A, 27–54.

[29] G. Bellamy and T. Schedler, *A new linear quotient of* $\mathbf{C}^4$ *admitting a symplectic resolution*, Math. Z., **273** (2013), 753–769.

[30] Y. Berest and O. Chalykh, *Quasi-invariants of complex reflection groups*, Compos. Math. **147** (2011), no. 3, 965–1002.

[31] Y. Berest and O. Chalykh, *Recollement of deformed preprojective algebras and the Calogero-Moser correspondence*, Mosc. Math. J. **8** (2008), no. 1, 21–37.

[32] Y. Berest, P. Etingof, and V. Ginzburg, *Finite-dimensional representations of rational Cherednik algebras*, Int. Math. Res. Not. (2003), **(19)**:1053–1088.

[33] R. Berger and R. Taillefer, *Poincaré-Birkhoff-Witt deformations of Calabi-Yau algebras*, J. Noncommut. Geom. **1** (2007), 241–270.

[34] G. M. Bergman, *The diamond lemma for ring theory*, Adv. in Math. **29** (1978), no. 2, 178–218.

[35] R. Bezrukavnikov and M. Finkelberg, *Wreath Macdonald polynomials and categorical McKay correspondence*, Camb. J. Math. **2** (2014), no. 2, 163–190.

[36] R. Bezrukavnikov and I. Losev, *Etingof conjecture for quantized quiver varieties*, arXiv:1309.1716, 2013.

[37] R. Bezrukavnikov and D. Kaledin, *Fedosov quantization in algebraic context*, Mosc. Math. J. **4** (2004) no. 3, 559–592.

[38] R. Bocklandt, *Graded Calabi-Yau algebras of dimension 3*, J. Pure Appl. Algebra **212** (2008), no. 1, 14–32.

[39] R. Bocklandt, T. Schedler and M. Wemyss, *Superpotentials and Higher Order Derivations*, J. Pure Appl. Algebra **214** (2010), no. 9, 1501–1522.

[40] A. Braverman and D. Gaitsgory, *Poincaré–Birkhoff–Witt theorem for quadratic algebras of Koszul type*, J. Algebra, **181**(2) (1996), 315–328.

[41] T. Bridgeland, *Equivalences of triangulated categories and Fourier-Mukai transforms*, Bull. London Math. Soc. **31** (1999), no. 1, 25–34.

[42] T. Bridgeland, *Flops and derived categories*. Invent. Math. **147** (2002), no. 3, 613–632.

[43] T. Bridgeland and S. Iyengar, *A criterion for regularity of local rings*, C. R. Math. Acad. Sci. Paris **342** (2006), no. 10, 723–726.

[44] T. Bridgeland, A. King and M. Reid, *The McKay correspondence as an equivalence of derived categories*, J. Amer. Math. Soc. **14** (2001), no. 3, 535–554.

[45] T. Bridgeland and A. Maciocia, *Fourier-Mukai transforms for K3 and elliptic fibrations*, J. Algebraic Geom. **11** (2002), no. 4, 629–657.

[46] E. Brieskorn, *Singular elements of semi-simple algebraic groups*, Actes du Congrès International des Mathématiciens (Nice, 1970), Tome 2, Gauthier-Villars, Paris, 1971, pp. 279–284.

[47] M. Broué, G. Malle, and R. Rouquier, *Complex reflection groups, braid groups, Hecke algebras*, J. Reine Angew. Math., **500** (1998), 127–190.

[48] K. A. Brown, *Symplectic reflection algebras*, pages 27–49 in *Proceedings of the All Ireland Algebra Days, 2001* (Belfast), number 50, 2003.

[49] K. A. Brown and K. R. Goodearl, *Homological aspects of Noetherian PI Hopf algebras of irreducible modules and maximal dimension*, J. Algebra, **198** (1997) 240–265.

[50] K. A. Brown and K. R. Goodearl, *Lectures on algebraic quantum groups, Advanced Courses in Mathematics, CRM Barcelona*, Birkhäuser, Basel, 2002.

[51] K. A. Brown and I. Gordon, *Poisson orders, symplectic reflection algebras and representation theory*, J. Reine Angew. Math., **559** (2003), 193–216.

[52] K. A. Brown and I. G. Gordon, *The ramification of centres: Lie algebras in positive characteristic and quantised enveloping algebras*, Math. Z., **238(4)** (2001), 733–779.

[53] K. A. Brown and C. R. Hajarnavis, *Homologically homogeneous rings*, Trans. Amer. Math. Soc. **281** (1984), no. 1, 197–208.

[54] W. Bruns and J. Herzog, *Cohen-Macaulay rings*, volume 39 in *Cambridge Studies in Advanced Mathematics*, Cambridge University Press, Cambridge, 1998.

[55] J.-L. Brylinski, *A differential complex for Poisson manifolds*, J. Differential Geom. **28** (1988), no. 1, 93–114.

[56] R. O. Buchweitz, *Maximal Cohen–Macaulay modules and Tate–cohomology over Gorenstein rings*, preprint, 1986. https://tspace.library.utoronto.ca/handle/1807/16682

[57] D. Calaque, C. A. Rossi, and M. Van den Bergh, *Căldăraru's conjecture and Tsygan's formality*, Ann. of Math. (2) **176** (2012), no. 2, 865–923.

[58] D. Calaque and M. Van den Bergh, *Global formality at the G_∞-level*, Mosc. Math. J. **10** (2010), no. 1, 31–64, 271.

[59] D. Calaque and M. Van den Bergh, *Hochschild cohomology and Atiyah classes*, Adv. Math. **224** (2010), no. 5, 1839–1889.

[60] F. Calogero, *Solution of the one-dimensional N-body problems with quadratic and/or inversely quadratic pair potentials*, J. Math. Phys., **12** (1971), 419–436.

[61] H. Cassens and P. Slodowy, *On Kleinian singularities and quivers*, Progr. Math. **162** (1998), 263–288.

[62] T. Cassidy and M. Vancliff, *Generalizations of graded Clifford algebras and of complete intersections*, J. Lond. Math. Soc. (2) **81** (2010), no. 1, 91–112.

[63] D. Chan, *Lectures on orders*, available at web.maths.unsw.edu.au/ danielch.

[64] D. Chan and C. Ingalls, *The minimal model program for orders over surfaces*, Invent. Math. **161** (2005), no. 2, 427–452.

[65] K. Chan, E. Kirkman, C. Walton, and J. J. Zhang, *Quantum binary polyhedral groups and their actions on quantum planes*, arXiv:1303.7203, 2013.

[66] K. Chan, C. Walton, Y. H. Wang, and J. J. Zhang, *Hopf actions on filtered regular algebras*, J. Algebra **397** (2014), 68–90.

[67] C. Chevalley, *An algebraic proof of a property of Lie groups*, Amer. J. Math. **63** (1941), 785–793.

[68] C. Chevalley, *Invariants of finite groups generated by reflections*, Amer. J. Math., **77** (1955), 778–782.

[69] M. Chlouveraki, I. Gordon, and S. Griffeth, *Cell modules and canonical basic sets for Hecke algebras from Cherednik algebras*, pages 77–89 in *New trends in noncommutative algebra*, volume 562 of *Contemp. Math.*, Amer. Math. Soc., Providence, RI, 2012.

[70] E. Cline, B. Parshall, and L. Scott, *Finite-dimensional algebras and highest weight categories*, J. Reine Angew. Math., **391** (1988), 85–99.

[71] A. M. Cohen, *Finite quaternionic reflection groups*, J. Algebra, **64** (1980), 293–324.

[72] A. Craw, *Quiver representations in toric geometry*, arXiv:0807.2191, 2008.

[73] A. Craw and A. Ishii, *Flops of G-Hilb and equivalences of derived categories by variation of GIT quotient*. Duke Math. J. **124** (2004), no. 2, 259–307.

[74] W. Crawley-Boevey, P. Etingof, and V. Ginzburg, *Noncommutative geometry and quiver algebras*, Adv. Math. **209** (2007), no. 1, 274–336.

[75] W. Crawley-Boevey and M. P. Holland, *Noncommutative deformations of Kleinian singularities*, Duke Math. J., **92** (1998), 605–635.

[76] H. Dao and C. Huneke, *Vanishing of Ext, cluster tilting modules and finite global dimension of endomorphism rings*, Amer. J. Math. **135** (2013), no. 2, 561–578.

[77] B. Davison, *Superpotential algebras and manifolds*, Adv. Math. **231** (2012), no. 2, 879–912.

[78] M. De Wilde and P. B. A. Lecomte, *Existence of star-products and of formal deformations of the Poisson Lie algebra of arbitrary symplectic manifolds*, Lett. Math. Phys. **7** (1983), no. 6, 487–496.

[79] C. Dean and L. W. Small, *Ring theoretic aspects of the Virasoro algebra*, Comm. Algebra **18** (1990), no. 5, 1425–1431.

[80] P. Deligne, *Équations différentielles à points singuliers réguliers*, volume 163 of *Lecture Notes in Mathematics*, Springer, Berlin, 1970.

[81] J. Diller and C. Favre, *Dynamics of bimeromorphic maps of surfaces*, Amer. J. Math. **123** (2001), no. 6, 1135–1169.

[82] G. Dito, *Kontsevich star product on the dual of a Lie algebra*, Lett. Math. Phys. **48** (1999), no. 4, 307–322.

[83] J. Dixmier, *Enveloping algebras*, volume 14 of *North-Holland Mathematical Library*, North-Holland, Amsterdam, 1977.

[84] V. Dlab and C. M. Ringel, *Representations of graphs and algebras*, Department of Mathematics, Carleton University, Ottawa, Ont., 1974, Carleton Mathematical Lecture Notes, No. 8.

[85] V. Dolgushev, *All coefficients entering kontsevich's formality quasi-isomorphism can be replaced by rational numbers*, arXiv:1306.6733, 2013.

[86] V. Dolgushev, D. Tamarkin, and B. Tsygan, *The homotopy Gerstenhaber algebra of Hochschild cochains of a regular algebra is formal*, J. Noncommut. Geom. **1** (2007), no. 1, 1–25.

[87] P. Donovan and M. R. Freislich, *The representation theory of finite graphs and associated algebras*, Carleton University, Ottawa, Ont., 1973, Carleton Mathematical Lecture Notes, No. 5.

[88] V. G. Drinfeld, *Degenerate affine Hecke algebras and Yangians*, Funktsional. Anal. i Prilozhen., **20** (1986), 69–70.

[89] V. G. Drinfeld, *On quadratic commutation relations in the quasiclassical case*, Selecta Math. Soviet. **11** (1992), no. 4, 317–326.

[90] J. Du, B. Parshall, and L. Scott, *Quantum Weyl reciprocity and tilting modules*, Comm. Math. Phys., **195** (1998), 321–352.

[91] C. F. Dunkl and E. M. Opdam, *Dunkl operators for complex reflection groups*, Proc. London Math. Soc. (3), **86** (2003), 70–108.

[92] D. Eisenbud, *Commutative algebra, with a view toward algebraic geometry*, Springer, New York, 1995, Graduate texts in mathematics, No. 150.

[93] D. Eisenbud and J. Harris, *The geometry of schemes*, volume 197 of *Graduate Texts in Mathematics*, Springer, New York, 2000.

[94] H. Esnault, *Reflexive modules on quotient surface singularities*, J. Reine Angrew. Math. **362** (1985), 63–71.

[95] P. Etingof, *Calogero-Moser Systems and Representation Theory*, Zurich Lectures in Advanced Mathematics, European Mathematical Society (EMS), Zürich, 2007.

[96] P. Etingof, *Lecture notes on Cherednik algebras*, lecture notes, 2009. http://www-math.mit.edu/ etingof/ 18.735notes.pdf

[97] P. Etingof, *Cherednik and Hecke algebras of varieties with a finite group action*, arXiv:math/0406499, 2004.

[98] P. Etingof, *Exploring noncommutative algebras via representation theory*, arXiv:math/0506144, 2005.

[99] P. Etingof and V. Ginzburg, *Symplectic reflection algebras, Calogero-Moser space, and deformed Harish-Chandra homomorphism*, Invent. Math. **147** (2002), no. 2, 243–348.

[100] ———, *Noncommutative complete intersections and matrix integrals*, Pure Appl. Math. Q. **3** (2007), no. 1, 107–151.

[101] P. Etingof and V. Ginzburg, *Noncommutative del Pezzo surfaces and Calabi-Yau algebras*, J. Eur. Math. Soc. (JEMS) **12** (2010), no. 6, 1371–1416.

[102] P. Etingof and D. Kazhdan, *Quantization of Lie bialgebras I*, Selecta Math (N.S.) **2** (1996), no. 1, 1–41.

[103] P. Etingof and S. Montarani, *Finite dimensional representations of symplectic reflection algebras associated to wreath products*, Represent. Theory **9** (2005), 457–467.

[104] P. Etingof and C. Walton, *Semisimple Hopf actions on commutative domains*, Adv. Math. **251** (2014), 47–61.

[105] B. V. Fedosov, *A simple geometrical construction of deformation quantization*, J. Differential Geom. **40** (1994), no. 2, 213–238.

[106] W. Fultonj *Young tableaux*, volume 35 of *London Mathematical Society Student Texts*, Cambridge University Press, Cambridge, 1997.

[107] P. Gabriel, *Unzerlegbare Darstellungen, I*, Manuscripta Math. **6** (1972), 71–103; correction, ibid. **6** (1972), 309.

[108] I. M. Gelfand and V. A. Ponomarev, *Model algebras and representations of graphs*, Func. Anal. and Applic. **13** (1979), no. 3, 1–12.

[109] M. Gerstenhaber, *The cohomology structure of an associative ring*, Ann. of Math. (2) **78** (1963), 267–288.

[110] E. Getzler, *Lie theory for nilpotent L_∞-algebras*, Ann. of Math. (2) **170** (2009), no. 1, 271–301.

[111] V. Ginzburg, *On primitive ideals*, Selecta Math. (N.S.), **9**(3) (2003), 379–407.

[112] V. Ginzburg, *Calabi-Yau algebras*, arXiv:math/0612139, 2006.

[113] V. Ginzburg, N. Guay, E. Opdam, and R. Rouquier, *On the category $\mathcal{O}$ for rational Cherednik algebras*, Invent. Math., **154** (2003), 617–651.

[114] V. Ginzburg and D. Kaledin, *Poisson deformations of symplectic quotient singularities*, Adv. Math., **186** (2004), 1–57.

[115] V. Ginzburg and T. Schedler, *Moyal quantization and stable homology of necklace Lie algebras*, Mosc. Math. J. **6** (2006), no. 3, 431–459.

[116] G. Gonzalez-Sprinberg and J-L. Verdier, *Construction géométrique de la correspondance de McKay*, Ann. Sci. École Norm. Sup. (4) **16**, (1983) no. 3, 409–449.

[117] K. R. Goodearl and R. B. Warfield, Jr., *An introduction to noncommutative Noetherian rings*, second ed., volume 61 of *London Mathematical Society Student Texts*, Cambridge University Press, Cambridge, 2004.

[118] I. G. Gordon, *Baby Verma modules for rational Cherednik algebras*, Bull. London Math. Soc., **35** (2003), 321–336.

[119] I. G. Gordon, *On the quotient ring by diagonal invariants*, Invent. Math., **153** (2003), 503–518.

[120] I. G. Gordon, *Symplectic reflection algebras*, pages 285–347 in *Trends in representation theory of algebras and related topics*, EMS Ser. Congr. Rep., Eur. Math. Soc., Zürich, 2008.

[121] I. G. Gordon, *Rational Cherednik algebras*, pages 1209–1225 in *Proceedings of the International Congress of Mathematicians. Volume III*, New Delhi, 2010, Hindustan Book Agency.

[122] I. G. Gordon and M. Martino, *Calogero-Moser space, restricted rational Cherednik algebras and two-sided cells*, Math. Res. Lett., **16** (2009), 255–262.

[123] I.G. Gordon and J.T. Stafford, *Rational Cherednik algebras and Hilbert schemes*, Adv. Math. **198**, (2005), no. 1, 222–274.

[124] I.G. Gordon and J.T. Stafford, *Rational Cherednik algebras and Hilbert schemes. II. Representations and sheaves*, Duke Math. J. **132**, (2006) no. 1, 73–135.

[125] N. Guay, *Projective modules in the category $\mathcal{O}$ for the Cherednik algebra*, J. Pure Appl. Algebra, **182** (2003), 209–221.

[126] S. Gutt, *An explicit $*$-product on the cotangent bundle of a Lie group*, Lett. Math. Phys. **7** (1983), 249–258.

[127] Harish-Chandra, *On some applications of the universal enveloping algebra of a semisimple Lie algebra*, Trans. Amer. Math. Soc. **70** (1951), 28–96.

[128] R. Hartshorne, *Algebraic geometry*, volume 52 of *Graduate Texts in Mathematics*, Springer, New York, 1977.

[129] T. Hayashi, *Sugawara operators and Kac-Kazhdan conjecture*, Invent. Math., **94** (1988), 13–52.

[130] J.-W. He, F. Van Oystaeyen, and Y. Zhang, *Cocommutative Calabi-Yau Hopf algebras and deformations*, J. Algebra **324** (2010), no. 8, 1921–1939.

[131] L. Hille and M. Van den Bergh, *Fourier-Mukai transforms*, pages 147–177 in *Handbook of tilting theory*, volume 332 of *London Math. Soc. Lecture Note Ser.*, 2007.

[132] V. Hinich, *Homological algebra of homotopy algebras*, Comm. Algebra **25** (1997), no. 10, 3291–3323.

[133] R. R. Holmes and D. K. Nakano, *Brauer-type reciprocity for a class of graded associative algebras*, J. Algebra, **144** (1991), 17–126.

[134] R. Hotta, K. Takeuchi, and T. Tanisaki, *D-modules, perverse sheaves, and representation theory*, volume 236 of *Progress in Mathematics*, Birkhäuser, Boston, 2008.

[135] J. E. Humphreys, *Introduction to Lie algebras and representation theory*, volume 9 of *Graduate Texts in Mathematics*, Springer, New York, 1978.

[136] J. E. Humphreys, *Reflection groups and Coxeter groups*, volume 29 of *Cambridge Studies in Advanced Mathematics*, Cambridge University Press, Cambridge, 1990.

[137] D. Huybrechts, *Fourier-Mukai transforms in algebraic geometry*, Oxford Mathematical Monographs, Oxford University Press, Oxford, 2006.

[138] O. Iyama, *Higher-dimensional Auslander–Reiten theory on maximal orthogonal subcategories*, Adv. Math. **210** (2007), no. 1, 22–50.

[139] O. Iyama and I. Reiten, *Fomin-Zelevinsky mutation and tilting modules over Calabi-Yau algebras*, Amer. J. Math. **130** (2008), no. 4, 1087–1149.

[140] O. Iyama and R. Takahashi, *Tilting and cluster tilting for quotient singularities*, Math. Ann. **356** (2013), no. 3, 1065–1105.

[141] O. Iyama and M. Wemyss, *Maximal Modifications and Auslander-Reiten Duality for Non-isolated Singularities*, Invent. Math. **197** (2014), no. 3, 521–586.

[142] O. Iyama and M. Wemyss, *On the Noncommutative Bondal–Orlov Conjecture*, J. Reine Angew. Math. **683** (2013), 119–128.

[143] O. Iyama and M. Wemyss, *Singular derived categories of $\mathbb{Q}$-factorial terminalizations and maximal modification algebras*, Adv. Math. **261** (2014), 85–121.

[144] O. Iyama and M. Wemyss, *Reduction of triangulated categories and Maximal Modification Algebras for cA_n singularities*, arXiv:1304.5259, 2013.

[145] David A. Jordan, *The graded algebra generated by two Eulerian derivatives*, Algebr. Represent. Theory **4** (2001), no. 3, 249–275.

[146] M. Kapranov and E. Vasserot, *Kleinian singularities, derived categories and hall algebras*, Math. Ann. **316**, (2000) no. 3, 565–576.

[147] M. Kashiwara, T. Miwa, and E. Stern, *Decomposition of q-deformed Fock spaces*, Selecta Math. (N.S.), **1** (1995), 787–805.

[148] M. Kashiwara and R. Rouquier, *Microlocalization of rational Cherednik algebras*, Duke Math. J. **144** (2008) no. 3, 525–573.

[149] D. Kazhdan, B. Kostant, and S. Sternberg, *Hamiltonian group actions and dynamical systems of Calogero type*, Comm. Pure Appl. Math., **31** (1978), 481–507.

[150] D. S. Keeler, *Criteria for σ-ampleness*, J. Amer. Math. Soc. **13** (2000), no. 3, 517–532.

[151] D. S. Keeler, D. Rogalski, and J. T. Stafford, *Naïve noncommutative blowing up*, Duke Math. J. **126** (2005), no. 3, 491–546.

[152] B. Keller, *Derived categories and their uses*, pages 671–701 in *Handbook of algebra*, vol. 1, North-Holland, Amsterdam, 1996. http://www.math.jussieu.fr/ keller/publ/dcu.pdf

[153] A. D. King, *Moduli of representations of finite-dimensional algebras*, Quart. J. Math. Oxford Ser. (2) **45** (1994), no. 180, 515–530.

[154] E. Kirkman, J. Kuzmanovich, and J. J. Zhang, *Gorenstein subrings of invariants under Hopf algebra actions*, J. Algebra **322** (2009), no. 10, 3640–3669.

[155] M. Kontsevich, *Deformation quantization of algebraic varieties*, Lett. Math. Phys. **56** (2001), no. 3, 271–294, EuroConférence Moshé Flato 2000, Part III (Dijon).

[156] M. Kontsevich, *Deformation quantization of Poisson manifolds*, Lett. Math. Phys. **66** (2003), no. 3, 157–216.

[157] G. R. Krause and T. H. Lenagan, *Growth of algebras and Gelfand-Kirillov dimension*, revised ed., American Mathematical Society, Providence, RI, 2000.

[158] P. Kronheimer, *ALE gravitational instantons*, D. Phil. thesis, University of Oxford, (1986).

[159] A. Lascoux, B. Leclerc, and J. Thibon, *Hecke algebras at roots of unity and crystal bases of quantum affine algebras*, Comm. Math. Phys., **181** (1996), 205–226.

[160] L. Le Bruyn, *Quotient singularities and the conifold algebra*, lecture notes, 2004, http://win.ua.ac.be/~lebruyn/b2hd-LeBruyn2004e.html.

[161] B. Leclerc and J.-Y. Thibon, *Canonical bases of q-deformed Fock spaces*, Internat. Math. Res. Notices (1996), no. 9, 447–456.

[162] H. Lenzing, *Weighted projective lines and applications*, pages 153–187 in *Representations of algebras and related topics*, EMS Ser. Congr. Rep., Eur. Math. Soc., Zürich, 2011.

[163] G. Leuschke, *Endomorphism rings of finite global dimension*, Canad. J. Math. **59** (2007), no. 2, 332–342.

[164] T. Levasseur, *Some properties of noncommutative regular graded rings*, Glasgow Math. J. **34** (1992), no. 3, 277–300.

[165] J.-L. Loday, *Cyclic homology*, volume 301 of *Grundlehren der Mathematischen Wissenschaften*, Springer, Berlin, 1998.

[166] J.-L. Loday and B. Vallette, *Algebraic operads*, volume 346 of *Grundlehren der Mathematischen Wissenschaften*, Springer, Heidelberg, 2012.

[167] D.-M. Lu, J. H. Palmieri, Q.-S. Wu, and J. J. Zhang, *Regular algebras of dimension 4 and their A_∞-Ext-algebras*, Duke Math. J. **137** (2007), no. 3, 537–584.

[168] G. Lusztig, *Canonical bases arising from quantized enveloping algebras*, J. Amer. Math. Soc., **3** (1990), 447–498.

[169] M. Markl, S. Shnider, and J. Stasheff, *Operads in algebra, topology and physics*, volume 96 of *Mathematical surveys and monographs*, American Mathematical Society, 2000.

[170] R. Martin, *Skew group rings and maximal orders*, Glasgow Math. J. **37** (1995), no. 2, 249–263.

[171] O. Mathieu, *Homologies associated with Poisson structures*, pp. 177–199 in *Deformation theory and symplectic geometry* (Ascona, 1996), volume 20 in *Math. Phys. Stud.*, Kluwer, Dordrecht, 1997.

[172] J. C. McConnell and J. C. Robson, *Noncommutative Noetherian Rings*, volume 30 of *Graduate Studies in Mathematics*, American Mathematical Society, Providence, RI, revised edition, 2001.

[173] J. McKay, *Graphs, singularities, and finite groups*, Proc. Sympos. Pure Math. **37** (1980), 183–186.

[174] D. Milicic, *Lectures on derived categories*, http://www.math.utah.edu/~milicic/Eprints/dercat.pdf.

[175] K. Nagao, *Derived categories of small toric Calabi-Yau 3-folds and curve counting invariants*, Q. J. Math. **63** (2012), no. 4, 965–1007.

[176] Y. Namikawa, *Poisson deformations of affine symplectic varieties*, Duke Math. J., **156** (2011), 51–85.

[177] L. A. Nazarova, *Representations of quivers of infinite type*, Izv. Akad. Nauk SSSR Ser. Mat. **37** (1973), 752–791.

[178] A. Neeman, *The Grothendieck duality theorem via Bousfield's techniques and Brown representability*, J. Amer. Math. Soc. **9** (1996), no. 1, 205–236.

[179] D. Orlov, *Triangulated categories of singularities and D-branes in Landau-Ginzburg models*, Proc. Steklov Inst. Math. **246** (2004), no. 3, 227–248.

[180] D. Orlov, *Triangulated categories of singularities and equivalences between Landau-Ginzburg Models*, Sb. Math. **197** (2006), no. 11–12, 1827–1840.

[181] A. Polishchuk and L. Positselski, *Quadratic algebras*, volume 37 in *University Lecture Series*, American Mathematical Society, Providence, RI, 2005.

[182] I. Reiten and M. Van den Bergh, *Two-dimensional tame and maximal orders of finite representation type.*, Mem. Amer. Math. Soc. **408** (1989).

[183] I. Reiten and M. Van den Bergh, *Noetherian hereditary abelian categories satisfying Serre duality*, J. Amer. Math. Soc. **15** (2002), no. 2, 295–366.

[184] M. Reyes, D. Rogalski, and J. J. Zhang, *Skew Calabi-Yau algebras and homological identities*, Adv. Math. **264** (2014), 308–354.

[185] J. Rickard, *Morita theory for derived categories*, J. London Math. Soc. (2) **39** (1989), 436–456.

[186] D. Rogalski, *Generic noncommutative surfaces*, Adv. Math. **184** (2004), no. 2, 289–341.

[187] D. Rogalski and Susan J. Sierra, *Some projective surfaces of GK-dimension 4*, Compos. Math. **148** (2012), no. 4, 1195–1237.

[188] D. Rogalski, S. J. Sierra, and J. T. Stafford, *Classifying orders in the Sklyanin algebra*, Algebra and Number Theory **9** (2015), 2056–2119.

[189] D. Rogalski and J. T. Stafford, *Naïve noncommutative blowups at zero-dimensional schemes*, J. Algebra **318** (2007), no. 2, 794–833.

[190] D. Rogalski and J. T. Stafford, *A class of noncommutative projective surfaces*, Proc. Lond. Math. Soc. (3) **99** (2009), no. 1, 100–144.

[191] D. Rogalski and J. J. Zhang, *Canonical maps to twisted rings*, Math. Z. **259** (2008), no. 2, 433–455.

[192] D. Rogalski and J. J. Zhang, *Regular algebras of dimension 4 with 3 generators*, pp. 221–241 in *New trends in noncommutative algebra*, volume 562 of *Contemp. Math.*, Amer. Math. Soc., Providence, RI, 2012.

[193] J. J. Rotman, *An introduction to homological algebra*, Universitext, Springer, New York, second edition, 2009.

[194] R. Rouquier, *Representations of rational Cherednik algebras*, pages 103–131 in *Infinite-dimensional aspects of representation theory and applications*, volume 392 of *Contemp. Math.*, Amer. Math. Soc., Providence, RI, 2005.

[195] R. Rouquier, *Dimensions of triangulated categories*, J. K-Theory **1** (2008), no. 2, 193–256.

[196] R. Rouquier, *q-Schur algebras and complex reflection groups*, Mosc. Math. J., **8** (2008), 119–158, 184.

[197] T. Schedler, *Hochschild homology of preprojective algebras over the integers*, arXiv:0704.3278v1, 2007. Accepted to *Adv. Math*.

[198] J. P. Serre, *Faisceaux algébriques cohérents*, Ann. of Math. (2) **61** (1955), 197–278.

[199] P. Shan, *Crystals of Fock spaces and cyclotomic rational double affine Hecke algebras*, Ann. Sci. Éc. Norm. Supér. (4), **44** (2011), 147–182.

[200] B. Shelton and M. Vancliff, *Schemes of line modules. I*, J. London Math. Soc. (2) **65** (2002), no. 3, 575–590.

[201] G. C. Shephard and J. A. Todd, *Finite unitary reflection groups*, Canadian J. Math., **6** (1954), 274–304.

[202] S. J. Sierra, *Rings graded equivalent to the Weyl algebra*, J. Algebra **321** (2009), no. 2, 495–531.

[203] S. J. Sierra, *Classifying birationally commutative projective surfaces*, Proc. Lond. Math. Soc. (3) **103** (2011), no. 1, 139–196.

[204] S. J. Sierra and C. Walton, *The universal enveloping algebra of the Witt algebra is not noetherian*, Adv. Math. **262** (2014), 239–260.

[205] P. Slodowy, *Four lectures on simple groups and singularities*, volume 11 of *Communications of the Mathematical Institute*, Rijksuniversiteit Utrecht, 1980.

[206] P. Slodowy, *Simple singularities and simple algebraic groups*, volume 815 of *Lecture Notes in Mathematics*, Springer, Berlin, 1980.

[207] S. P. Smith, *Subspaces of non-commutative spaces*, Trans. Amer. Math. Soc. **354** (2002), no. 6, 2131–2171.

[208] S. P. Smith, *Maps between non-commutative spaces*, Trans. Amer. Math. Soc. **356** (2004), no. 7, 2927–2944.

[209] S. P. Smith, *A quotient stack related to the Weyl algebra*, J. Algebra **345** (2011), 1–48.

[210] S. P. Smith and J. T. Stafford, *Regularity of the four-dimensional Sklyanin algebra*, Compositio Math. **83** (1992), no. 3, 259–289.

[211] S. P. Smith and Michel Van den Bergh, *Noncommutative quadric surfaces*, J. Noncommut. Geom. **7** (2013), no. 3, 817–856.

[212] T. A. Springer, *Linear algebraic groups*, second ed., *Modern Birkhäuser Classics*, Birkhäuser, Boston, 2009.

[213] J. T. Stafford and J. J. Zhang, *Examples in non-commutative projective geometry*, Math. Proc. Cambridge Philos. Soc. **116** (1994), no. 3, 415–433.

[214] J. T. Stafford and M. Van den Bergh, *Noncommutative curves and noncommutative surfaces*, Bull. Amer. Math. Soc. (N.S.) **38** (2001), no. 2, 171–216.

[215] D. R. Stephenson, *Noncommutative projective geometry*, unpublished lecture notes.

[216] D. R. Stephenson, *Artin-Schelter regular algebras of global dimension three*, J. Algebra **183** (1996), no. 1, 55–73.

[217] D. R. Stephenson, *Algebras associated to elliptic curves*, Trans. Amer. Math. Soc. **349** (1997), no. 6, 2317–2340.

[218] D. R. Stephenson and J. J. Zhang, *Growth of graded Noetherian rings*, Proc. Amer. Math. Soc. **125** (1997), no. 6, 1593–1605.

[219] D. R. Stephenson and J. J. Zhang, *Noetherian connected graded algebras of global dimension 3*, J. Algebra **230** (2000), no. 2, 474–495.

[220] R. Steinberg, *Invariants of finite reflection groups*, Canad. J. Math., **1** (1960), 616–618.

[221] R. Steinberg, *Differential equations invariant under finite reflection groups*, Trans. Amer. Math. Soc., **11** (1964), 392–400.

[222] D. Tamarkin, *Quantization of Lie bilagebras via the formality of the operad of little disks*, pp. 203–236 in *Deformation quantization* (Strasbourg, 2001), volume 1 of *IRMA Lect. Math. Theor. Phys.*, de Gruyter, Berlin, 2002.

[223] D. E. Tamarkin, *Formality of chain operad of little discs*, Lett. Math. Phys. **66** (2003), no. 1-2, 65–72.

[224] J. Tate and M. Van den Bergh, *Homological properties of Sklyanin algebras*, Invent. Math. **124** (1996), no. 1-3, 619–647.

[225] R. W. Thomason and T. Trobaugh, *Higher algebraic K-theory of schemes and of derived categories*, pages 247–435 in *The Grothendieck Festschrift*, vol. III, volume 88 in *Progr. Math.*, Birkhäuser, Boston, 1990.

[226] D. Uglov, *Canonical bases of higher-level q-deformed Fock spaces and Kazhdan-Lusztig polynomials*, pages 249–299 in *Physical combinatorics* (Kyoto, 1999), volume 191 of *Progr. Math.*, Birkhäuser, MA, 2000.

[227] R. Vale, *On category $\mathcal{O}$ for the rational Cherednik algebra of the complex reflection group $(\mathbb{Z}/\ell\mathbb{Z}) \wr S_n$*, Thesis, 2006.

[228] M. Van den Bergh, *A relation between Hochschild homology and cohomology for Gorenstein rings*, Proc. Amer. Math. Soc. **126** (1998), no. 5, 1345–1348.

[229] M. Van den Bergh, *Blowing up of non-commutative smooth surfaces*, Mem. Amer. Math. Soc. **154** (2001), no. 734, x+140.

[230] M. Van den Bergh, *Three-dimensional flops and noncommutative rings*, Duke Math. J. **122** (2004), no. 3, 423–455.

[231] M. Van den Bergh, *Non-commutative crepant resolutions*, pages 749–770 in *The legacy of Niels Henrik Abel*, Springer, Berlin, 2004.

[232] M. Van den Bergh, *Noncommutative quadrics*, Int. Math. Res. Not. IMRN (2011), **no. 17**, 3983–4026.

[233] M. Van den Bergh, *On global deformation quantization in the algebraic case*, arXiv:math/0603200, 2006.

[234] M. Van den Bergh, *Calabi-Yau algebras and superpotentials*, Selecta Math. (N.S.) **21** (2015), no. 2, 555–603.

[235] M. Van den Bergh and L. de Thanhoffer de Völcsey, *Calabi-Yau deformations and negative cyclic homology*, arXiv:1201.1520, 2012.

[236] F. Van Oystaeyen and L. Willaert, *Grothendieck topology, coherent sheaves and Serre's theorem for schematic algebras*, J. Pure Appl. Algebra **104** (1995), no. 1, 109–122.

[237] M. Varagnolo and E. Vasserot, *On the decomposition matrices of the quantized Schur algebra*, Duke Math. J., **100** (1999), 267–297.

[238] M. Verbitsky, *Holomorphic symplectic geometry and orbifold singularities*, Asian J. Math., **4** (2000), 553–563.

[239] C. A. Weibel, *An introduction to homological algebra*, volume 38 of *Cambridge Studies in Advanced Mathematics*. Cambridge University Press, Cambridge, 1994.

[240] M. Wemyss, *Reconstruction Algebras of Type A*, Trans. Amer. Math. Soc. **363** (2011), 3101–3132.

[241] S. Wilcox, *Supports of representations of the rational Cherednik algebra of type A*, 2010.

[242] T. Willwacher, *A note on Br-infinity and KS-infinity formality*, arXiv:1109.3520, 2011.

[243] G. Wilson, *Collisions of Calogero-Moser particles and an adelic Grassmannian*, Invent. Math., **133** (1998), 1–41.

[244] J. Wunram, *Reflexive modules on quotient surface singularities*, Math. Ann. **279** (1988), no. 4, 583–598.

[245] A. Yekutieli, *Deformation quantization in algebraic geometry*, Adv. Math. **198** (2005), no. 1, 383–432.

[246] A. Yekutieli and J. J. Zhang, *Serre duality for noncommutative projective schemes*, Proc. Amer. Math. Soc. **125** (1997), no. 3, 697–707.

[247] Y. Yoshino, *Cohen-Macaulay modules over Cohen-Macaulay rings*, volume 146 of *London Mathematical Society Lecture Note Series*, Cambridge University Press, Cambridge, 1990.

[248] X. Yvonne, *A conjecture for q-decomposition matrices of cyclotomic v-Schur algebras*, J. Algebra, **304** (2006), 419–456.

[249] J. J. Zhang, *Twisted graded algebras and equivalences of graded categories*, Proc. London Math. Soc. (3) **72** (1996), no. 2, 281–311.

[250] J. J. Zhang, *Non-Noetherian regular rings of dimension 2*, Proc. Amer. Math. Soc. **126** (1998), no. 6, 1645–1653.

[251] J. J. Zhang and J. Zhang, *Double Ore extensions*, J. Pure Appl. Algebra **212** (2008), no. 12, 2668–2690.

[252] J. J. Zhang and J. Zhang, *Double extension regular algebras of type (14641)*, J. Algebra **322** (2009), no. 2, 373–409.

Index

Printed in the United States
by Baker & Taylor Publisher Services

Printed in the United States
By Bookmasters